ECOLOGY OF TELEOST FISHES

Fish and Fisheries Series

VOLUME 24

Amongst the fishes, a remarkably wide range of fascinating biological adaptations to diverse habitats has evolved. Moreover, fisheries are of considerable importance in providing human food and economic benefits. Rational exploitation and management of our global stocks of fishes must rely upon a detailed and precise insight of the interaction of fish biology with human activities.

The *Fish and Fisheries Series* aims to present authoritative and timely reviews which focus on important and specific aspects of the biology, ecology, taxonomy, physiology, behaviour, management and conservation of fish and fisheries. Each volume will cover a wide but unified field with themes in both pure and applied fish biology. Although volumes will outline and put in perspective current research frontiers, the intention is to provide a synthesis accessible and useful to both experts and non-specialists alike. Consequently, most volumes will be of interest to a broad spectrum of research workers in biology, zoology, ecology and physiology, with an additional aim of the books encompassing themes accessible to non-specialist readers, ranging from undergraduates and postgraduates to those with an interest in industrial and commercial aspects of fish and fisheries.

Applied topics will embrace synopses of fishery issues which will appeal to a wide audience of fishery scientists, aquaculturists, economists, geographers and managers in the fishing industry. The series will also contain practical guides to fishery and analysis methods and global reviews of particular types of fisheries.

Books already published are listed at the end of this book. The Publisher and Series Editor would be glad to discuss ideas for new volumes in the series.

ECOLOGY OF TELEOST FISHES

SECOND EDITION

ROBERT J. WOOTTON

Institute of Biological Sciences
The University of Wales
Aberystwyth, UK

KLUWER ACADEMIC PUBLISHERS

DORDRECHT / BOSTON / LONDON

Library of Congress Cataloging in Publication Card Number: 98-70538

ISBN 0 412 84590 3(HB)   0 412 64200 X (PB)

Published by Kluwer Academic Publishers,
P.O. Box 17, 3300 AA Dordrecht, The Netherlands.

Sold and distributed in North, Central and South America
by Kluwer Academic Publishers,
101 Philip Drive, Norwell, MA 02061, U.S.A..

In all other countries, sold and distributed
by Kluwer Academic Publishers Group,
P.O. Box 322, 3300 AH Dordrecht, The Netherlands.

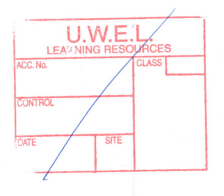

05-0104-500 ts

Printed in The Netherlands

CONTENTS

PREFACE TO THE SECOND EDITION

This second edition of the *Ecology of Teleost Fishes* retains the aims and structure of the first edition

It is intended for final year undergraduates and postgraduates, but may be useful for research workers, who wish to see their research in a wider context. As before, it assumes a knowledge of contemporary ecological ideas as described in textbooks like those by Krebs (1994) and Begon *et al.* (1996), and a knowledge of fish biology such as that provided in Bone *et al.* (1995) and Moyle and Cech (1996). A pleasing feature of writing a second edition has been to see the way in which themes emphasized in the first edition have developed and have contributed to our contemporary understanding of fish ecology. The recognition of the importance of understanding the responses of individual fish is illustrated by the development of individual-based models that simulate patterns of growth, mortality and reproduction in populations. Bioenergetics studies of fish growth and food consumption have advanced to the point that they are being used as management tools. The adaptations of teleost fishes to their environment continue to be a source of wonder.

As I hope the end of the last chapter makes clear, the future for the subject looks particularly promising. Unfortunately, in the increasingly bizarre world of British universities, early retirement may be thought to be a suitable reward for activities such as writing a textbook. It would be a great pity not to have the opportunity to review the progress that fish ecology makes over the next few years. It has been a privilege to read so much good research, although only a small proportion is referred to explicitly in the text.

I would like to thank again all those who contributed advice on the first edition. For this second edition, John Gee again provided valuable guidance on aspects of community ecology and support in other ways. The constructive comments by an anonymous reviewer of the first edition were extremely valuable when planning the second edition. Chuck Hollingworth was again a thorough, constructive and sympathetic editor. I am grateful to Nigel Balmforth and Martin Tribe of Chapman & Hall for their patience in seeing this edition through to completion. As in the first edition, all the errors are my responsibility.

Maureen Wootton has been a continual source of support and encouragement.

R.J. Wootton
Aberystwyth, 1998

PREFACE TO THE FIRST EDITION

This book grew out of a series of lectures on the ecology of teleosts which I give to final year honours students in zoology and aquatic biology. It has been written because there is no recent text on the ecology of teleosts that reflects the advances made in the past two decades in areas such as feeding ecology, bio-energetics or life-history strategies. The intended audience includes advanced undergraduates, post-graduates and research workers. It assumes a knowledge of general ecology such as that given in the recent text-books by Krebs (1985) and Begon, Harper and Townsend (1986) and of fish biology such as that pro-vided in Bond (1979) or Moyle and Cech (1982). The text, though referring where relevant to applied problems in fish ecology, is intended to complement rather than compete with volumes exemplified by Pitcher and Hart (1981), which have as their central theme the biological problems posed by fisheries or aquaculture.

The present volume is intended not as a self-contained and self-sufficient account of teleost ecology but as an instrument to provoke its readers into their own explorations and interpretations of the literature and of contemporary research programmes. It emphasizes my belief that the key to understanding the ecology of any group of animals is an understanding of how individual animals react to environmental conditions by altering their allocations of time and resources to their different activities, and the consequences of these altera-tions for the reproductive success of individuals. It also reflects my belief that natural selection operates predominantly at the level of individuals and is a pervasive and powerful process. Thirdly, it emphasizes that teleosts are not 'primitive' or 'lower' vertebrates, a fallacy commonly held by undergraduates. The text, perhaps inevitably, reflects my research experiences with freshwater, north temperates fishes. It is also heavily biased toward studies published in North America and the U.K., a restriction imposed by ease of access to journals and the author's lack of knowledge of languages other than English. Perhaps the best fate for this volume is that it provokes another and more comprehen-sive account of the ecology of this wonderfully diverse group of animals – the teleost fishes.

R. J. Wootton
1989

1

INTRODUCTION

1.1 THE DIVERSITY OF TELEOST FISHES

The Mesozoic era lasted about 155 million years and ended some 70 million years before the present. During this era three important vertebrate lineages evolved. The land saw the appearance of the endothermic birds and mammals, the waters saw the rise of the teleost fishes. Teleost fishes now dominate both marine and fresh waters. These fishes (and hereafter fishes will refer only to teleosts unless otherwise qualified) are represented by approximately 24 000 living species (Table 1.1) and account for about half of all vertebrate species (Nelson, 1994). Of the living species, about 58% are marine, 41% live in fresh waters and 1% migrate between fresh and salt waters (Moyle and Cech, 1996).

The teleosts have mastered life in water, colonizing virtually every aquatic habitat. Liquid water occurs from the altitude of the permanent snow-line on land to the abyssal depths of the oceans. Fishes are found at altitudes in excess of 4000 m on the Tibetan Plateau (Cao *et al.*, 1981) and in the Andes (Payne, 1986). At these heights the partial pressure of oxygen is only about 60% of that at sea level and there are extreme variations in temperature and light intensity (Price, 1981). Fast-flowing upland streams, with their characteristic fishes adapted to turbulent waters, are transformed as the gradient slackens into sediment-laden, meandering lowland rivers. This hydrological change is accompanied by an increase in the number of fish species. The motion of riverine water helps to keep it well oxygenated. In still fresh waters, deoxygenation can develop and

this does restrict the distribution of fishes. The deep African Great Lakes, Malawi and Tanganyika, contain several hundred species of fishes, but these are confined to the upper 200 m because the deeper hypolimnion is devoid of oxygen (Lowe-McConnell, 1975, 1987). Although the seas have a relatively constant salinity of about 35% (conductivity of 46.0 mS cm^{-1}), the salinities of waters of continental areas vary widely (Heisler, 1984). Fishes are found in the blackwater rivers of the Amazon Basin, which are acidic (pH 3.8–4.9) and close to being distilled water (conductivity 20–30 µS cm^{-1}) (Lowe-McConnell, 1987). At the other extreme, a cichlid species, *Oreochromis grahami*, has been introduced into Lake Nakuru, Kenya, which is highly saline (conductivity 162.5 mS cm^{-1}) (Payne, 1986). In the littoral and estuarine waters at the margin of sea and land, salinities can change rapidly with changes in tide-level or precipitation, yet these transitional habitats are well populated by fishes. All depths of the seas have been colonized. In the abyssal depths of the oceans, fluctuations in the temperature or oxygen content of the water are negligible, but fish are living at pressures of several hundred atmospheres (Hochachka and Somero, 1984). There are fishes in water that reaches nearly 40 °C in lakes in East Africa (Lowe-McConnell, 1987) while species in the Antarctic Ocean live their whole lives at temperatures below 0 °C (DeVries, 1980).

Associated with this wide range of habitats, fishes show a profusion of body shapes (Fig. 1.1) and life-history patterns (Breder and Rosen, 1966; Nelson, 1994). The smallest vertebrate is a teleost, a goby that reaches

Table 1.1 Orders of living teleost fishes with numbers of families, genera and species*

Order	Families	Genera	Species
Osteoglossiformes	6	29	217
Elopiformes	2	2	8
Albuliformes	3	8	29
Anguilliformes	15	141	738
Saccopharyngiformes	4	5	26
Clupeiformes	5	83	357
Gonorhynchiformes	4	7	35
Cypriniformes	5	279	2 662
Characiformes	10	237	1 343
Siluriformes	34	412	2 405
Gymnotiformes	6	23	62
Esociformes	2	4	10
Osmeriformes	13	74	236
Salmoniformes	1	11	66
Stomiiformes	4	51	321
Ateleopodiformes	1	4	12
Aulopiformes	13	42	219
Myctophiformes	2	35	241
Lampridiformes	7	12	19
Polymixiiformes	1	1	5
Percopsiformes	3	6	9
Ophidiiformes	5	92	355
Gadiformes	12	85	482
Batrachoidiformes	1	19	69
Lophiiformes	16	65	297
Mugiliformes	1	17	66
Atheriniformes	8	47	285
Beloniformes	5	38	191
Cypriniodontiformes	8	88	807
Stephanoberyciformes	9	28	86
Beryciformes	7	28	123
Zeiformes	6	20	39
Gasterosteiformes	11	71	257
Synbranchiformes	3	12	87
Scorpaeniformes	25	266	1 271
Perciformes	148	1 496	9 293
Pleuronectiformes	11	123	570
Tetraodontiformes	9	100	339
Totals	426	4 097	23 637

*Source: Nelson, J.S.(1994). Fishes of the World, 3rd ed. © John Wiley & Sons Inc. Reprinted with permission of John Wiley & Sons Inc.

sexual maturity at a length of 8 mm (Miller, 1984), while in the Amazon River system the arapaima, *Arapaima gigas*, may reach lengths of 2.5 m (Moyle and Cech, 1996).

1.2 DEFINING THE PROBLEM

In view of this extraordinary diversity, how can an account of the ecology of fishes be

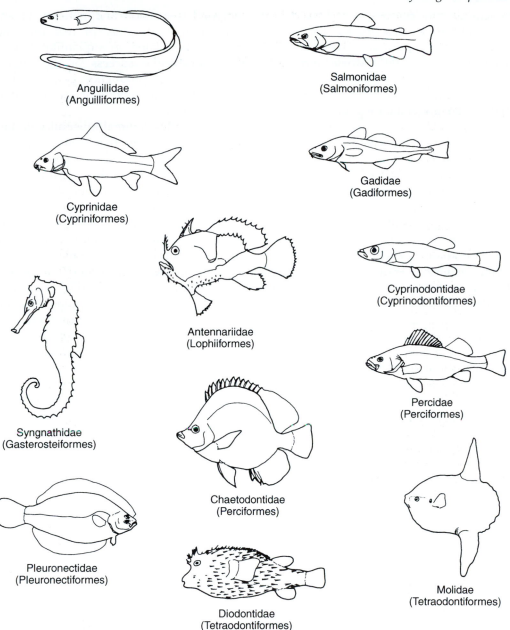

Fig. 1.1 Some examples of the diversity of body form in teleost fishes. Redrawn from Nelson (1994).

given? One approach is to describe the fishes that are found in important habitats such as the open sea, the littoral zone of seas, coral reefs, lakes and rivers. These descriptions then lead on to questions of the relationship between the characteristics of the habitats and the fish species found in them (Wootton, 1992a). This text does not follow this approach for two reasons. The first is that several good books describe the fish faunas associated with important aquatic environments (Bone *et al.*, 1995; Moyle and Cech, 1996). The second is the belief that a key to understanding the ecology of any group of animals is an understanding of the ecology of the individuals that make up the group (Lomnicki, 1988). Two central questions in ecology are: what factors cause fluctuations in the abundance of a population of animals, and what factors determine how many different species occur in the assemblage of species found in a habitat? This text is based on the assumption that an understanding of why a population changes in abundance over time requires a knowledge of the effects of the environment on the individuals in that population. An understanding of the species composition of an assemblage of fishes in an environment depends on knowing how individual fish in the populations of the species that form that assemblage are affected by their environment, including the presence of individual fish of the other species. This focus on the individual has stimulated the deployment of computer models, called 'individual-based' simulations, to gain insights into the population dynamics of fish populations (Van Winkle *et al.*, 1993).

Ecology is often defined as the study of organisms in relation to their environment. Such a definition begs the questions of which relationships between organisms and their environment are of biological importance and how this importance is to be judged. The biological success of an individual is measured by its success in being represented genetically in the next generation. Clearly the crucial problem of ecology is to determine the environmental factors that influence, in a predictable way, the number of offspring that an individual produces in its lifetime. Those subjects which form the substance of modern ecological studies including – predator–prey relationships, inter- and intraspecific competition, patterns of spatial and

temporal dispersion and patterns of energy acquisition – are only of real relevance if they can be related, in a quantitative way, to the reproductive success of the organisms under study. Reproduction is truly "the axis about which the biology of species revolves" (Meien, 1939).

An individual animal must allocate limited resources in a way that ensures its genetic representation in the next generation (Sibly and Calow, 1986). The critical limiting resources are food and time. Food provides energy and nutrients. A simple way of emphasizing the central importance of reproduction is to model the individual animal as an input–output system which is mapping an input in the form of food into an output in the form of progeny (Fig. 1.2) (Pianka, 1976). To achieve this mapping, the animal has to develop, maintain and protect a framework, the somatic body, which is responsible for the mechanics of transforming food into offspring. This imposes a cost in terms of the energy and nutrients that are allotted to the somatic body rather than to progeny. Time is required for foraging, but there are other demands on the time available to an individual (Fig. 1.3). It may be spent on reproductive activities and on sheltering from predators or unfavourable abiotic conditions. The allocation of time, energy and nutrients takes place in a changing environment. These environmental changes take place over time or space, and often both.

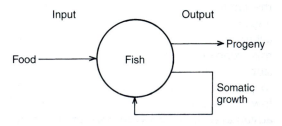

Fig. 1.2 Individual fish as an input–output system mapping food into progeny. Redrawn from Wootton (1986).

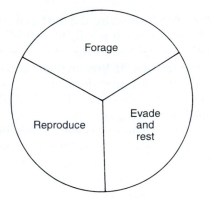

Fig. 1.3 Schematic representation of time budgeting by a fish. Area of circle represents 24 h. A change in the time allocated to one activity is at the expense of time allocated to other activities.

They may be changes in abiotic factors such as temperature, or they may be changes in biotic factors such as the density of predators. The problem is to determine the responses that individuals make in terms of their allocation of time and energy (nutrients) in the face of environmental variability. The allocation patterns, through their effects on survival and reproduction, will determine the lifetime reproductive success of individuals, that is their fitness.

1.3 ORGANIZATION OF THE TEXT

There are constraints on the possible range of responses to a changing environment, which relate partly to the physical characteristics of the environment and partly to the morphological, physiological and biochemical characteristics of vertebrates. Chapter 2 illustrates this concept of constraints through discussions of the body shape of fishes in relation to propulsion, the problems imposed by the low solubility of oxygen in water and the advantages and disadvantages of water as a medium for the transmission of information to the sense organs. The subsequent five chapters form the core of the text. Chapter 3 describes feeding ecology in detail because the rate of feeding determines what energy and nutrients are available for allocation. Chapter 4 introduces the concept of an energy budget as a quantitative technique for describing the allocation of energy. It also examines a method of classifying the effects that abiotic factors can have on metabolism. Some energy is used in locomotion and so Chapter 5 considers the patterns of movement that fishes show both temporally and spatially. The allocation of resources to somatic growth is discussed in Chapter 6 and that to reproduction in Chapter 7. Interactions with other fish may affect the reproductive success of an individual. Between individuals in a population, such interactions may include cannibalism or intraspecific competition, while an individual may encounter individuals of other species in the form of predators, pathogens or interspecific competitors (Chapters 8 and 9). The pattern of allocation of resources and time determines the survival and reproductive success of the individual, and so the contribution of that individual to changes in abundance of the population of which it is a unit (Chapter 10). Chapter 11 discusses the evolution of the pattern of allocation for a population in relation to its abiotic and biotic environment. The reproductive success of the individual fish in populations of different species living in the same area will determine how many of these species coexist in an assemblage over an extended period of time (Chapter 12). This order of chapters corresponds approximately to the sequence in which the ecology of a species can be analysed.

In places in the text there is an emphasis on the development of quantitative models to describe aspects of the ecology of fishes. The advantage of such models is that they make specific predictions that can be tested against observations on real fishes. The successes and failures of such predictions indicate the extent to which an understanding of the phenomena under study has been achieved. In many cases the failures of a model are as informative as its successes. In developing models, it is necessary to recognize the capacity of fishes to respond adaptively to environmental changes, and this response forms the subject of the rest of this Introduction.

1.4 ADAPTIVE RESPONSE TO ENVIRONMENTAL CHANGE

A characteristic of vertebrates is that they tend to buffer the effects of environmental changes by a variety of homeostatic mechanisms. A familiar example is the ability of birds and mammals to maintain a constant core body temperature in the face of fluctuations in the environmental temperature. A common but biologically naive error is to assume that teleost fishes are in some way biologically more primitive than the endothermic birds and mammals because, with a few exceptions, they do not regulate body temperature. In evolutionary terms, fishes are as recent as birds and mammals, and compared with these endothermic vertebrates many fishes show great flexibility in their response to environmental change in traits such as growth, age at first reproduction and maximum life span. Irrespective of their habitats and lifestyles, fishes will be confronted with changes in their environment. What suites of responses do teleosts show in the face of a changing environment? To what extent do homeostatic mechanisms buffer the effects of a changing environment in teleosts?

The nature of the responses of organisms to changing environmental circumstances is a central problem in biology. Slobodkin and Rapoport (1974) provide an enlightening general discussion of this problem. Organisms can be viewed as playing a game against Nature, in which success is judged not by how large the winnings are, but rather by how long the player can stay in the game. Individual organisms stay in the game if they leave offspring and so contribute to the gene pool formed by the next generation. An individual loses if it fails to reproduce. A population (gene pool) loses the game when it becomes extinct. If this analogy is correct, then the optimal response of an organism to a change in its environment is to minimize the cost to it, in terms of a reduction in fitness, of a recurrence of such a change. To do this, the organism must make some sort of adaptive response to the change. Each organism can be thought of as having a genotype for which there corresponds an optimal set of environmental conditions and succession of conditions that result in maximal survivorship and fecundity. Any deviation from these conditions will cause a reduction in survivorship and fecundity, but that reduction will be lessened the more effectively the organism makes an adaptive response to the change (Slobodkin and Rapoport, 1974). In most populations, individuals will vary in their capacity to mount an adaptive response, because they vary genetically.

The nature of the adaptive response will depend on the time scale of the environmental change in relation to the generation length of the organism. Over shorter time scales, the adaptive response will be made by the individual organism, that is at the level of the phenotype. A short-term change may be evaded or its effects ameliorated by a behavioural response. If the change persists, then biochemical and physiological changes resulting in the acclimatization of the organism will become important. Over a longer time scale, changes in growth rate, age at maturity and other life-history traits will become apparent. Fishes can show great plasticity in such traits. This is in contrast to groups such as insects, birds and mammals in which growth is determinate and there is a characteristic adult size. Over even longer time scales, the responses made by individuals will be detected in the demographic characteristics of the population, with changes in birth and death rates. Eventually, the population may adapt to a long-term

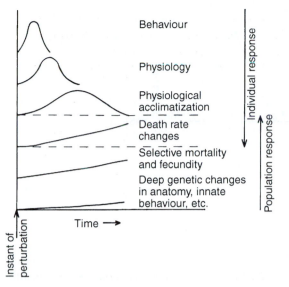

Behaviour

Physiology

Physiological acclimatization

Death rate changes

Selective mortality and fecundity

Deep genetic changes in anatomy, innate behaviour, etc.

Individual response

Population response

Time →

Instant of perturbation

Fig. 1.4 Diagrammatic and simplified representation of the events following an environmental perturbation. Redrawn from Slobodkin and Rapoport (1974).

knowledge is required for the prediction of the effects of both natural changes and those changes caused by human activity.

Two possible responses to environmental change result in changes in the distribution of traits in a population. Individuals in a population may show phenotypic plasticity. This means the expression of a trait, for example size at first reproduction, changes with the environment in which an individual is raised (Roff, 1992). Although this change may simply reflect physiological effects of no adaptive significance, the more interesting case is where the change represents an adaptive response to the environment (Roff, 1992). The reaction norm is the set of phenotypes produced by a given genotype across a range of environmental conditions (Chapter 11). The second possibility is that some genotypes in the population are better adapted to the changed environmental conditions than others, so that the representation of genotypes in the population changes through the process of natural selection.

change through alterations in gene and genotype frequencies. The environmental change has acted as an evolutionary factor causing changes in the gene pool of the population through the process of natural selection (Fig. 1.4).

All these responses are probably initiated as soon as individuals in the population detect an environmental change, either directly through their neurosensory system, or indirectly though effects on biochemical and physiological processes. If the period of the environmental change is sufficiently short, the homeostatic responses of the individuals may be so effective that no demographic or evolutionary changes can be detected. However, it is important to know the capacity of populations to adjust to environmental change though the responses of individuals within their lifetimes and through changes in the genetic composition of the population caused by the differential reproductive success of individuals. Such

Examples of homeostatic responses in fishes

The relevance of these general ideas to fishes is illustrated by a brief discussion of the consequences of a change in water temperature on individual fish and on populations. (The effects of temperature and other abiotic factors are described in more detail in Chapter 4.) Hochachka and Somero (1984) describe the biochemical strategies by which organisms can adapt to extreme or changing environmental conditions. Their book provides an exciting picture of the powerful metabolic systems that organisms can call upon in an indifferent but capricious and potentially hazardous environment.

Temperature, because of its effect on the rate of chemical reactions, is perhaps the most pervasive of abiotic environmental factors. Endotherms evade the effects of changes in temperature by maintaining a constant core

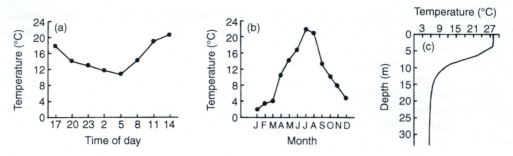

Fig. 1.5 Examples of temperature changes in water bodies. (a) Diel changes in salt-marsh pools, St Lawrence Estuary in June. Redrawn from Reebs *et al.* (1984). (b) Seasonal changes in surface water of Llyn Frongoch, Mid-Wales. (c) Vertical changes in Crooked Lake, Indiana in July. Redrawn from Wetzel (1983).

temperature. But this thermal homeostasis exerts a significant cost through a higher rate of metabolism. Teleosts are typically ectothermic. The rare exceptions include the tunas and their relatives. These large-bodied, pelagic, marine species can maintain elevated core body temperatures because of morphological adaptations (Stevens and Neill, 1978; Moyle and Cech, 1996).

The ectothermy of teleosts is understandable. Water is an exacting medium in which to maintain a constant, elevated body temperature. Because water has a high specific heat and is dense (compared with air), it acts as a heat sink (Graham, 1983; Denny, 1993). Water effectively absorbs heat generated during the course of normal metabolism of fish. The gills of fish present a large surface area to the water. Only a short distance separates the blood circulating through the gills and the water passing over them (Chapter 2). This arrangement allows an efficient exchange of gases between blood and water, but it also provides a path for the dissipation of body heat, in addition to loss over the general body surface.

Changes in temperature occur less abruptly in water than in air, but fish can be exposed to large temperature variations. In shallow water, temperatures can fluctuate by several degrees during the diel cycle (Fig. 1.5(a)).

There are major seasonal changes, especially at moderate and high latitudes (Fig. 1.5(b)). Fish may move through a thermocline between layers of water that differ considerably in temperature (Fig. 1.5(c)). The temperature of the fish's body will closely track that of the water. What mechanisms do fish use to minimize the cost of such changes in temperature?

Fish can show behavioural thermoregulation (Murray, 1971; Beitinger and FitzPatrick, 1979). When introduced into a thermal gradient, fish will move through the gradient, but will spend most of their time within a relatively narrow range of temperatures (pages 78–80). The consequences of such temperature selection for survival, growth and reproduction still have to be determined. Fish have thermosensory-receptors to detect a gradient in temperature. This sensory system is probably a component of the general innervation of the skin (Murray, 1971). Behavioural studies have shown that a temperature difference as small as 0.03 °C is detected by some species (Murray, 1971). Such sensitivity potentially allows a fish to make fine discriminations within a temperature gradient and adjust its position accordingly. The fish could detect a change in temperature at an early stage and then move in a way that minimizes the cost of that change.

If a fish cannot avoid the change in temperature by a behavioural response, then biochemical mechanisms that minimize the cost of the change become important (Hochachka and Somero, 1984). These mechanisms include changes within cells in the concentrations of micromolecules that alter the environment in which enzymes are functioning. There may be changes in the concentrations of enzymes that serve to minimize the effect of temperature change on the overall rate of metabolism. For some reactions in a metabolic pathway, there may be multiple enzyme systems, with one form of the enzyme working better over one part of the temperature range while another form works better over a different range. These multiple enzymes are coded by different genetic loci and may have evolved by gene duplication. Hochachka and Somero (1984) give examples of each of these mechanisms in some detail. These biochemical details are not of direct interest ecologically, but the important outcome is that the effect of temperature on metabolic rate is at least partially buffered.

As the temperature change persists, the fish acclimates as its physiology becomes adjusted to the new temperature regime. This process (or acclimatization if more than one abiotic environmental factor is changing) involves changes in the biochemical mechanisms such as those outlined above and may also involve changes in structural components such as lipid membranes.

When changes in temperature are predictable, for example those associated with the change in seasons at high latitudes, the required biochemical and physiological acclimation of the fish may come under the control of anticipatory, physiological mechanisms. Some fish that live at high latitudes produce antifreeze molecules, which inhibit ice formation in the body fluids (DeVries, 1980). In the winter flounder, *Pseudopleuronectes americanus*, the degradation of the antifreeze is under the control of the pituitary organ (Hew and Fletcher, 1979).

Both temperature and photoperiod are implicated in the control of the disappearance of antifreeze in some northern fishes (Duman and DeVries, 1974). The photoperiodic control helps to prevent a premature loss of antifreeze being stimulated by an unseasonal period of warming. The change in photoperiod is a more reliable cue for seasonal changes than are changes in temperature.

All the mechanisms described so far allow individual fish to avoid or ameliorate the effects of a temperature change on their metabolic processes. However, within a natural population there is considerable genetic variation (Lewontin, 1974). At many genetic loci, there may be two or more alleles present in the population: these are polymorphic loci. An individual may be homozygous or heterozygous at a polymorphic locus. Such genetic variation may have implications for the effect of a temperature change on the population as a whole. Some genotypes may be able to respond to a particular temperature change more effectively than others, that is the cost of the change is lower than for other genotypes.

A study of the locus that codes for the enzyme lactate dehydrogenase B (LDH-B) in the killifish, *Fundulus heteroclitus*, strongly suggested that different alleles at a given locus may not be neutral to the effect of a temperature change (Place and Powers, 1979; Powers *et al.*, 1983). This cyprinodont species is found along the eastern seaboard of North America from Newfoundland to Florida. Along this coastline, there is a steep gradient in temperature, a 1 °C change in annual mean temperature for every degree of latitude. At the LDH-B locus, there are two alleles, B^a and B^b, so that the possible genotypes are B^aB^a (homozygous for B^a), B^bB^b (homozygous for B^b) and B^aB^b (heterozygous). In northern populations, there is a high frequency of the B^b allele whereas in the south the frequency of the B^a allele is high (Fig. 1.6). Biochemical, physiological and developmental studies suggested that

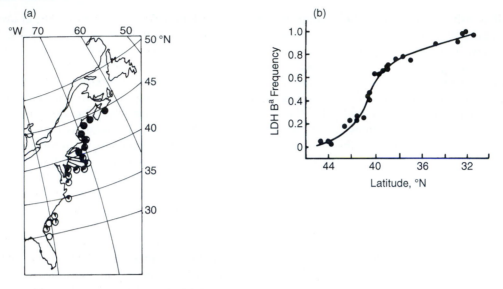

Fig. 1.6   Cline in LDH-B allelic frequencies in *Fundulus heteroclitus* along Atlantic seaboard of North America. (a) Geographic variation in allelic frequencies. Open areas of circles, B^a allele frequencies; closed areas, B^b allele frequencies. Redrawn from Place and Powers (1979). (b) Latitudinal variation in frequency of B^a allele. Redrawn from Powers *et al.* (1983).

this cline in the frequency of the alleles of LDH-B is adaptive, with the genotype B^bB^b having an advantage at lower temperatures. In this example, the temperature change is along a geographical axis, so the adaptation to minimize the cost of the change is seen at the level of populations. This geographical pattern of the frequency of the two alleles has presumably come about at least partly because the different genotypes have different patterns of survivorship and fecundity in the different temperature regimes (Powers *et al.*, 1986, 1993). The study of *F. heteroclitus* provides a powerful example of the application of the techniques of molecular biology to ecological and evolutionary problems.

A population may respond to a long-term change in temperature through a change in its allelic frequencies. The homeostatic character of such evolutionary changes was illustrated by a study of the enzyme lactate dehydrogenase (LDH) in four species of bar-

racuda, *Sphyraena*, which live along the west coast of the Americas (Graves and Somero, 1982). Although similar both in body form and general ecology, the species occupy different temperature regimes because of their geographical distribution (Fig. 1.7). The most northerly species, *S. argentea*, and the most southerly species, *S. idiastes*, experience similar regimes. The LDH from *S. argentea* and *S. idiastes* are indistinguishable both electrophoretically and in their enzyme kinetics. The enzymes are most similar from the species that geographically are most distant, but which experience similar temperature regimes. An important characteristic of an enzyme, which helps to define its kinetic properties, the K_m value, is almost identical in the LDH from the four species when the value is estimated at the relevant mid-range temperature experienced by each species. The evolutionary changes in LDH of these four closely related species have maintained the

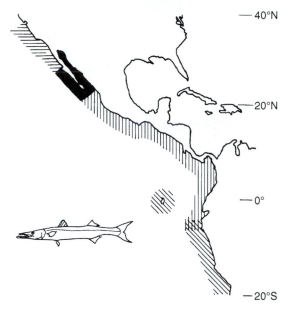

Fig. 1.7 Distribution of four eastern Pacific barracudas. Horizontal hatching, *Sphyraena argentea*; solid, *S. lucasana*; vertical hatching, *S. ensis*; diagonal hatching, *S. idiastes*. Redrawn from Graves and Somero (1982).

kinetic characteristics of the enzyme even though the species experience different temperature regimes.

1.5 SUMMARY AND CONCLUSIONS

1. Teleost fishes account for nearly half of all known vertebrate species. They have representatives in virtually all aquatic systems and an enormous variety in the ways they make a living.

2. A key to understanding the ecology of fishes is a knowledge of how individual fish allocate resources and time among activities related to maintenance, growth and reproduction in the face of a changing environment.

3. Fishes are highly evolved vertebrates with well-developed homeostatic capacities that tend to buffer the effects of environmental change. In addition to homeostatic responses that occur within the lifetime of individuals, populations react to environmental change with changes in the composition of the gene pool under the action of natural selection. Examples are given of the possible responses to a change in temperature.

4. Both for ease of modelling and perhaps because there is still a tendency to think of fishes as lower vertebrates, the assumption is usually made – either explicitly or more often implicitly – that the fishes can be regarded as responding passively to environmental changes whether these are abiotic or biotic. An important aspect of future studies must be to define the extent to which fishes buffer the effects of environmental changes on lifetime reproductive success and the point at which such buffering capacity collapses.

2

ENVIRONMENTAL AND ORGANISMIC CONSTRAINTS

2.1 INTRODUCTION

An individual fish confronted with a changing environment has available biochemical, physiological, behavioural and morphological mechanisms that will, to a greater or lesser extent, buffer any adverse effects of change. Each individual inherits from its parents a genetic constitution that defines the total range of its capacity to respond to change. The nature and the range of the responses have evolved through the process of natural selection during the history of the gene pool of which the individual fish is a member. This evolution has involved an interaction between challenges presented by the characteristics of the environment and the biological characteristics of the individuals in the lineage. These environmental and biological characteristics impose constraints on what responses evolve and so constrain the ecological roles that individuals within a lineage can play. This concept of constraints is illustrated by considering the problems of locomotion, respiration and obtaining sensory information in water.

2.2 BODY FORM AND LOCOMOTION

Modes of locomotion

The mode of swimming of a fish will have consequences for its foraging, its ability to escape from predators or unfavourable environmental conditions and its reproductive behaviour. The physical properties of water are an influential determinant of the mode of life of fishes because of the relationship between body form and effective propulsion in water. As a medium for locomotion, water has both advantages and disadvantages (Lindsey, 1978). Compared with air, water is about 800 times denser and 60 times more viscous (Denny, 1993). The density of water means that it is a buoyant medium. Fishes do not require strong internal skeletal structures to support their weight against gravitational forces and need to do relatively little work to keep from sinking. In addition, most teleosts have a swim bladder that renders them neutrally buoyant so that they can hold a position at any level in the water column with only a small expenditure of energy. Consequently, fishes can readily exploit the three available spatial dimensions.

A fish moves by exerting a thrust against the water with movements of its body or fins (Videler, 1993). Its motion is opposed by drag generated by frictional and inertial (pressure) forces. Frictional drag arises because of the viscosity of water. A layer of water is dragged along as the fish moves. Inertial drag is generated because although dense, water is also a yielding medium and there is a distortion of the flow around the body of the fish as it moves (Webb, 1975, 1988; Videler, 1993). A Reynolds number describes the relative importance of inertial and viscous forces acting on a body such as a fish moving through a fluid. The Reynolds number is calculated as $\rho L U \mu$, where ρ is the density of the liquid, L is length of the moving body, U is the velocity of the moving body and μ is the viscosity of the medium. Viscous forces are of major importance up to a Reynolds number of 30, inertial forces prevail when the number exceeds 200, and values between 30 and 200 define an intermediate zone (Fuiman and Webb, 1988).

These values are important because teleosts typically produce large numbers of small eggs (Chapter 7). Larvae are small when they hatch, only a few mm long, and consequently experience the intermediate regime. Growth and morphological development rapidly take them into the regime dominated by inertial forces (Webb and Weihs, 1986; Fuiman and Webb, 1988). With high viscous forces, motion ceases when swimming ceases. The mortality rate of these small larvae is close to that of immobile eggs (Chapter 10). High inertial forces oppose sudden acceleration and turning, but characteristics of body shape such as streamlining reduce inertial drag and the fish can glide forward at the end of a swimming bout (Webb, 1975, 1978a; Videler, 1993). The rapid growth in length of larvae may be an adaptation that takes them out of a regime dominated by viscous forces into one dominated by inertial forces (Müller and Videler, 1996).

The two principal modes of locomotion in fishes are undulatory and oscillatory motion. In the former, the thrust that drives the fish forward is generated by undulatory waves passing down either the body of the fish or a long median fin. In oscillatory motion, the paired fins are used to generate the thrust by acting either as oars or as wings. The types of locomotion seen in fishes are named after the types of fish that show that mode of locomotion (Fig. 2.1) (Lindsey, 1978). This can be illustrated with the classification for undulatory propulsion.

Anguilliform locomotion is the mode of swimming characteristic of eels such as *Anguilla*. The thrust-generating waves pass down most of the body, maintaining a relatively large amplitude which tends to

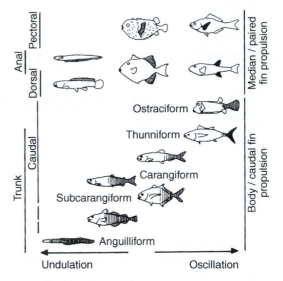

Fig. 2.1 Modes of forward swimming in fishes. Density of shading indicates propulsive contributions of body and fins. Redrawn and modified from Lindsey (1978).

increase towards the tail. The body is usually long and thin. This mode of swimming is not associated with high speeds, and fishes using it are frequently found near the bottom. This mode of locomotion is also often typical of larvae swimming at low Reynolds numbers, with, as in herring larvae, extended pectoral fins (Videler, 1993). Subcarangiform locomotion is similar to anguilliform locomotion except that the amplitude of the undulations is small at the anterior of the fish but increases greatly in the posterior half or third of the body. Salmonids and many cyprinid fishes show this mode of locomotion. Carangiform locomotion is seen in fish such as the jack, *Caranx*, herring, *Clupea*, and mackerel, *Scomber*. The undulations are largely confined to the posterior third of the body. The caudal peduncle is narrow and the caudal fin is stiff and sharply notched. Carangiform locomotion is seen in fishes that cruise continuously in a relatively unbounded environment such as the open ocean.

Thunniform locomotion, named after the tuna, *Thunnus*, is a more specialized form of carangiform locomotion in which the caudal peduncle is thin and the caudal fin high and stiff with a characteristic semilunate shape. Only the peduncle and fin show significant undulatory motion: the rest of the body is relatively stiff. Indeed, the thunniform mode can be analysed as an oscillatory form of locomotion, in which the caudal fin oscillates (Webb, 1988). Ostraciform locomotion is shown by fishes that have a rigid body. The caudal fin, pivoting on the caudal peduncle, develops the thrust. This type of locomotion is typically found in fishes like the boxfish, *Ostracion*, that live in structurally complex habitats such as coral reefs. Modes of locomotion in which the propulsive force is generated by either median or paired fins are shown in Fig. 2.1.

The use of high-speed film or video is allowing more precise, quantitative descriptions of the swimming of fish (Videler, 1993). Such analyses should, over time, replace this descriptive classification.

An ecomorphological analysis of locomotion

Ecomorphology is the study of the relationships between morphological features of an organism and its ecology. At its most ambitious, it seeks to predict ecology from the morphology. Relationships between body shape, mode of locomotion and the ecology of fish were defined by Webb (1984a,b). He recognized three functional locomotory mechanisms. These are body, caudal fin periodic propulsion (BCF periodic), body, caudal fin transient propulsion (BCF transient) and median and paired fin propulsion (MPF). In BCF periodic, there are cyclically repeated patterns of locomotory wave generation which provide thrust over periods from about a second to several weeks, that is for sprinting or for cruising. In BCF transient, the wave generation is brief and non-cyclical,

providing thrust for fast starts or powered turns. In MPF, the paired fins generate the thrust for locomotion, typically providing low speed but high manoeuvrability.

Webb (1984a,b) has described the body shapes that are probably optimal for BCF periodic and BCF transient propulsion. In BCF periodic propulsion, a lunate tail maximizes thrust. This tail is high, but has a low surface area (i.e. a high aspect ratio). There is a narrow caudal peduncle that minimizes any sideways thrust and a large, anterior body depth and mass which minimizes recoil of the head end. The body is relatively rigid and streamlined. The characteristics that maximize thrust are compatible with the features that minimize drag. The thunniform body shape approaches this optimal morphology (Fig. 2.2), while the carangiform body shape has many of these optimal features.

For BCF transient propulsion, thrust is maximized by a large body depth throughout the length of the body, especially caudally. There is a flexible body that allows large-amplitude propulsive movements. Drag is minimized by a small dead weight, that is a small non-muscle mass to be accelerated. For BCF transient propulsion, the morphological characteristics that maximize thrust are not the same as those that minimize drag. Consequently, there is a tendency to evolve either as a thrust maximizer, such as a sculpin, *Cottus*, or as a drag minimizer such as the pike, *Esox lucius*. *Cottus* has a deep body accentuated by the dorsal and anal fins but a relatively large dead weight. *Esox* has a low dead weight but a shallow head silhouette and a body only relatively deep at the caudal end compared with a sculpin (Fig. 2.2).

The optimal features for BCF periodic propulsion are not compatible with those for BCF transient propulsion. Fish that are adapted for cruising have a relatively poor performance in sudden acceleration or powered turns. Fish adapted for BCF transient propulsion have a poor performance in sustained swimming. Morphological features that are adaptive for one mode of propulsion constrain performance in other modes.

Specializations for BCF periodic propulsion are also incompatible with the median and paired fin (MPF) mode of propulsion, because the fins in the specialist cruiser are usually reduced to relatively stiff hydroplanes which provide hydrodynamic lift (there is a tendency for BCF periodic propulsion specialists to lack a swim bladder). The optimal design for MPF propulsion, which frequently depends on the oscillation of the paired pectoral fins, has not been defined as clearly as the designs for the BCF modes. MPF propulsion is probably an adaptation for efficient low-speed swimming and high manoeuvrability in structurally complex environments such as rocky shores or beds of vegetation. Many species that usually show MPF propulsion can switch temporarily to BCF periodic propulsion to move away from danger or to maintain station in a current.

Fish specialized for each of these three mechanisms of propulsion can be pictured at the apices of a triangle that crudely represents the swimming modes of fishes (Fig. 2.2). This triangle, originally described by Webb (1984b), shows that there is a mutual exclusion of the optimal designs for BCF periodic, BCF transient and MPF propulsion. These optimal designs reflect the constraints that the physical properties of water impose on locomotory performance in relation to body shape.

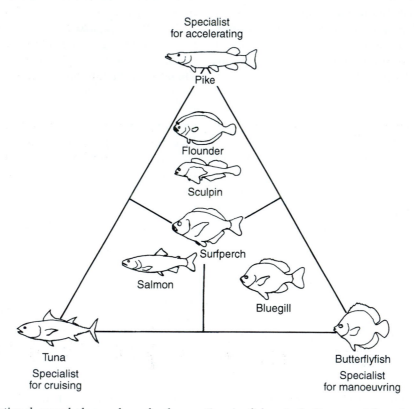

Fig. 2.2 Functional morphology plane for locomotion in fishes including specialists (at apices) and generalists (towards centre). Redrawn and modified from Webb (1984b).

An important feature of Webb's analysis is the link between locomotion and foraging (Webb, 1984b). If food is widely dispersed, the fish must move at speeds that sample the greatest volume of water for the lowest expenditure of energy. This would favour a BCF periodic specialist cruising at relatively low speeds. When food is locally abundant, the predator must cope with any attempts of the prey to escape. This would favour a BCF transient specialist, which by sudden rapid acceleration could minimize the chances of escape. Ambush predators such as the pike are typically more successful at capturing their prey than are fishes that attempt to chase down their prey (Webb, 1984a). However, an ambush is often made easier if the habitat is structurally complex, affording cover. Such a habitat would favour specialists in MPF propulsion. Rocky shores or coral reefs usually have a fish fauna with several representatives that have MPF propulsion as their usual method of locomotion. Such a specification may also be favoured where the prey does not attempt to escape, but occupies a habitat in which the exploitation of prey requires accurate and complex manoeuvring by the predator.

Many species cannot be regarded as specialists in locomotion. These fishes have body forms that represent a compromise between more than one mode of propulsion. Typical locomotory generalists are species with sub-carangiform locomotion, such as salmonids, many cyprinids and some percids. Webb (1984a) has suggested that such generalists are numerous because the specialists exploit only a small proportion of the food resources. The BCF specialists tend to rely on larger food items, but these are rare in the environment, leaving the smaller items to be exploited by the locomotor generalists (Fig. 2.3). MPF specialists, with their restricted ability to flee rapidly from predators, are largely confined to foraging in or near cover, again leaving a portion of the food resource underexploited. The success of locomotor generalists feeding on particulate prey can be enhanced by morphological adaptations unrelated to locomotion, such as protrusible

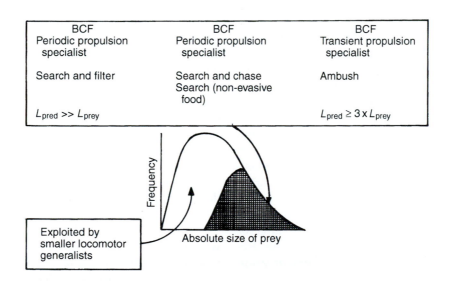

Fig. 2.3 Diagrammatic illustration of the distribution of food items in a habitat, showing probable resource exploitation by body and caudal fin (BCF) propulsion specialists and generalists. Redrawn from Webb (1984b).

jaws. This and other relationships between morphology and feeding are described in Chapter 3.

Videler (1993) has extended Webb's analysis by recognizing other categories including economic swimmers such as the pelagic sunfish, *Mola mola* (Fig. 1.1), and the laterally compressed flatfishes (Pleuronectiformes) that live associated with the bottom. Again, there is a relationship between body morphology and locomotion.

Even in locomotor generalists, body form reflects environmental differences. A comparison of juvenile Atlantic salmon, *Salmo salar*, living in two tributaries of the Miramichi River in eastern Canada found differences in morphology. The fish from the tributary that had the higher flow velocities tended to have a more spindle-shaped body and larger paired fins. This difference may be related to the problem of maintaining station in fast water. Juvenile salmon extend their paired fins as hydrofoils to maintain a position on the substratum and larger fins would be more effective in generating the negative lift required. The differences in body morphology between the two populations have a genetic basis and are not merely a phenotypic effect of the different flow regimes. Selection has favoured the evolution of slightly different morphologies within the same species within the same watershed (Ridell and Leggett, 1981; Ridell *et al.*, 1981).

This study of the salmon also illustrates that fishes not only have to move through water but may also have to hold station in flowing water. In many cases, fish avoid fast-moving water by sheltering in slack areas behind or under shelter. In species whose typical habitat is fast-flowing or turbulent water, the species of torrential streams or exposed rocky shores, the paired fins are frequently modified to act as suckers or grapples (Webb, 1988).

The relationship between body form, locomotion and reproductive success is complex. Body form largely determines swimming performance. This in turn will influence the ability of the fish to escape predators and to forage. Body form will also determine the energy spent on locomotion and so not available to be spent on reproduction. A large mass of locomotory muscle means that a high priority has to be given to its growth and maintenance and so a lower priority may have to be given to gonadal growth. The implications of locomotor specialization for the life-history patterns of the specialists have yet to be systematically explored.

2.3 RESPIRATION AND GILL STRUCTURE

Gill structure and size

Fishes are dependent on aerobic respiration, except for short periods of anaerobic respiration during sprinting or, for a few species, longer periods at extremely low oxygen levels (Hochachka, 1980; Hochachka and Somero, 1984). For animals with a high oxygen demand, water has several disadvantages. Oxygen has a low solubility in water: a given volume of water contains only about one-thirtieth of the oxygen in an equivalent volume of air. This solubility decreases with an increase in temperature, yet the rate of metabolism of fishes and so their demand for oxygen tends to increase with temperature (Chapter 4). The high density and viscosity of water means that fishes have to work harder to move it over the gills than do terrestrial vertebrates to move air into lungs (Lindsey, 1978).

The rate of diffusion of oxygen from the water into the circulatory system can be described.by the relationship:

$$R = DA \, \Delta p \, / \, d \qquad (2.1)$$

where R is the rate of diffusion, D is the diffusion constant, the value of which depends on the material through which the oxygen is diffusing, A is the area across which the diffusion takes place, Δp is the difference in the partial pressures of oxygen in the water and in the blood, and *d* is the distance over which the oxygen has to diffuse (Alexander, 1974). The morphology of fish gills ensures a large surface area. The gill arches carry numerous gill filaments, the surface area of which is increased by secondary lamellae (Fig. 2.4). The difference in partial pressure is maintained by a countercurrent system. This means that fully oxygenated water is adjacent to partially oxygenated blood in the efferent blood vessels, while partially deoxygenated water is adjacent to the least-oxygenated blood in the afferent vessels. The distance over which the oxygen has to diffuse is small, sometimes less than 1 μm (Hughes, 1984). This morphological arrangement has the disadvantage that it provides a large surface area and a short pathway for the exchange of ions with the external environment. The fish, unless the external environment is isotonic with its body fluids, has to do work to control the osmotic and ionic

properties of its body fluids (Chapter 4). The area of the gill surface must reflect a compromise between the requirements for respiration and the costs of osmoregulation. There is also the potential incompatibility between presenting a large respiratory surface to the water and maintaining streamlining of the head end of the fish. The bony opercula that cover the gills in teleost fishes maintain the streamlining.

There is a general correlation between the mode of life and gill structure (Hughes, 1984). Fast-swimming, oceanic species such as tunas have gills with a large surface area resulting from a high total filament length and a high frequency of lamellae. Such fish frequently ventilate the gills passively by ram ventilation, keeping the mouth open as they swim so that water enters the mouth, then passes over the gills and leaves by the opercular slits. In fishes that show moderate activity, the gill arches are usually well developed, with filaments of average length. In such species, ventilation of the gills usually involves rhythmic respiratory movements in which buccal and opercular suction pumps are equally important in maintaining a flow of water over the gills (Hughes, 1974, 1984). In sluggish fishes, the gills are more poorly developed, with short filaments. In such fishes, the opercular suction pump is often more important than the buccal pump.

The gill area in a given species is a function of body size. An allometric relationship of the form:

$$A = aW^b \qquad (2.2)$$

where *A* is gill area, *W* is body weight and a and b are parameters, can be used to describe the relationship (Hughes, 1984). Although there is no value for b that is common to all fishes, a typical value is about 0.8 (Hughes, 1984). Bigger fish of a given species have a smaller relative gill area. The metabolic rates of fish have a similar allometric relationship to body size (see Chapter

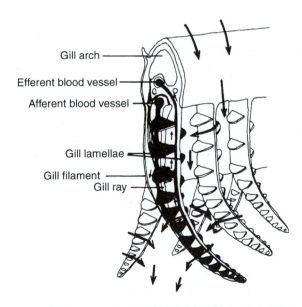

Fig. 2.4 Diagram of portion of gill arch of a fish. Thick arrows show direction of water flow, thin arrows direction of blood flow (note countercurrent arrangement). Redrawn from Hughes and Morgan (1973).

Gill arch
Efferent blood vessel
Afferent blood vessel
Gill lamellae
Gill filament
Gill ray

4). Pauly (1981, 1994) suggested that the allometric relationships between gill area, metabolic rate and body weight determine the typical pattern of growth for a species because growth can occur only if there is a sufficient supply of oxygen (Chapter 6). The value of the other parameter, a, is an index of gill area of a fish of a given size. Thus, fast-swimming, oceanic fishes tend to have high values for this parameter (Fig. 2.5). Fishes that can also breathe air, typically have low values for the same parameter.

For larval fish, experiencing water as a viscous medium (page 14), the respiratory surface is the skin. The boundary layer around the gill filaments would be too thick to allow effective gas exchange (Müller and Videler, 1996). As larvae grow, the surface-to-volume ratio declines, so cutaneous respiration becomes less effective. However, the gill area relative to body volume increases and the gills become the functional respiratory organ.

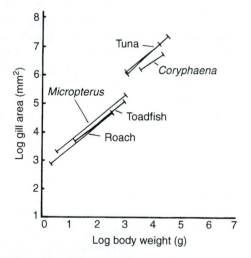

Fig. 2.5 Relationship between gill area and body weight (both area and weight shown in $\log_{10}$ units). Redrawn from Hughes (1984); two species of tuna are shown.

Adaptations to low oxygen concentration

There are environments in which the oxygen concentration in the water drops even below the low value that represents saturation. In deep tropical lakes, the hypolimnion is often permanently anoxic because the climatic conditions are so stable that thermal stratification of the lake does not break down during the year. Consequently, the oxygen content of deeper waters is never renewed. Fishes are essentially excluded from this zone except for short excursions (Payne, 1986). In shallow, still, warm waters that contain decaying organic material, the oxygen demand exerted by decomposition results in deoxygenation. In some waters, the deoxygenation is greatest during the night, because during the day photosynthesis by aquatic plants helps to oxygenate the water while at night plant respiration depletes the oxygen. In a shallow Amazonian floodplain lake, over a day, oxygen concentrations in the surface waters reached 2.4–4.5 mg l^{-1} (32.4–60.7% saturation at 30 °C) at 1500 h, but dropped to less than 0.1 mg l^{-1} (1.3% saturation) between 0300 h and 0600 h (Saint-Paul and Soares, 1987). At higher latitudes, small lakes may become deoxygenated during the winter if a thick covering of ice and snow prevents diffusion of oxygen into the water. In one small lake in northern Wisconsin, the median oxygen level in the winter was 0.3 mgl^{-1} (Klinger *et al.*, 1982). An added complication is that in habitats where the oxygen concentration is extremely low, toxic reducing substances such as hydrogen sulphide tend to build up, making the environment even more hostile.

Fishes show a variety of physiological, behavioural and morphological adaptations for coping with deoxygenated water. (The effect of oxygen concentration on metabolism is discussed in Chapter 4.) Some fishes survive low oxygen levels by using the surface film of water, which is relatively rich in oxygen because of diffusion from the atmosphere. Species that use this strategy for coping with a hypoxic environment usually outnumber species that can breathe air (Kramer *et al.*, 1978). Species that use the surface film do not always show significant morphological adaptations, although the exploitation of the oxygen-rich surface film is probably improved by features such as a flattened head, upturned fleshy mouth and small body size (Kramer, 1983). As the oxygen concentration in the water declines, the fish begin to spend more time at the surface, utilizing the surface film while still maintaining some activities away from the surface. When the concentration of oxygen in the main water body approaches zero, the fish spend virtually all their time at the surface (Kramer, 1983, 1987).

In waters that are low in oxygen for long periods, the fish fauna may include several species that can breathe air. Fish species that only use air breathing are rare; most air breathers can also use aquatic respiration and possess respiratory organs appropriate to both aerial and aquatic respiration. The capacity to use atmospheric oxygen has evolved independently many times in the teleosts and a variety of adaptations for air breathing are found. In some species, normal gills are used, for example *Hypostomus* in swamps in South America. In other species, parts of the gills are modified for air breathing. In the catfish, *Clarias*, the dorsal parts of the gills have squat secondary lamellae carried on structures that do not collapse in air. In several species, the swim bladder has become modified as a lung. The South American characin, *Erythrinus*, has a posterior chamber of the swim bladder modified for the rapid absorption of oxygen. Even the alimentary canal may have an area specialized for respiration: a bubble of air is swallowed and passes through the canal until it comes to rest in this area. Catfishes such as *Plecostomus* show this adaptation (Alexander, 1974; Hughes, 1984).

Air breathing is obligatory in a few fishes including some swamp eels (Synbranchiformes). The fish must have access to the air–water interface even when the water is fully oxygenated. Other species are facultative air breathers, using aquatic respiration when the water has a high oxygen content but aerial respiration as it declines. In some species the switch to air breathing occurs when some threshold level of oxygen is reached, whereas other species use some air breathing over a wide range of aquatic oxygen tensions (Kramer, 1983, 1987).

Any form of respiration has some costs associated with it. As the oxygen content of the water changes, a fish should adopt the mode of respiration that minimizes the cost of that change (Kramer, 1987). Even in water saturated with oxygen, the fish must do work to move water over the gill surfaces. For a fish at rest, up to 10% of metabolic expenditure may go to support gill ventilation. This energy cost will tend to rise as the oxygen content of the water declines because the fish must drive more water over the gills to compensate for the lower oxygen content of the water. Other costs may be incurred if the fish has to use the surface film for respiration. There is the energy cost of swimming to the surface from deeper water. The time spent at the surface is not available for other activities such as feeding or courtship that take place in deeper water. The risk of predation from fish-eating birds and other terrestrial predators may also be increased. The costs of air breathing can include the additional cost of developing the specialized morphological features necessary for this mode of respiration. Kramer (1983, 1987) discussed the costs associated with air and aquatic breathing. For species that can use both modes of respiration, Kramer and his associates predicted and have partly confirmed that the proportional uptake of oxygen from the air should increase with a decline in the oxygen content of the water, but decrease with the depth at which the fish normally lives. Increased risk of predation by birds should increase the proportional uptake of oxygen by aquatic respiration. A serious energy limitation would favour increased uptake from the air because less work is required to move air across a respiratory surface.

2.4 SENSORY CAPACITIES

The physical characteristics of water also impose constraints on the sensory organs of fishes, although in comparison with air, water may also be an advantageous medium for some sensory modalities (Denny, 1993). The properties of water largely determine the distance over which a sensory system can effectively obtain information. Fishes have an array of sensory systems, which provide them with information about their environments. A highly developed central nervous system processes the incoming information and allows the fish to make appropriate homeostatic responses to changing environmental conditions (Bone *et al.*, 1995).

The sensory modalities familiar from studies of terrestrial vertebrates are also important to fishes and include vision, olfaction, gustation, hearing, touch and temperature sensitivity. In addition, fishes have a sensory organ related to hearing, the lateral line (acoustico-lateralis) system, which enables them to detect disturbances in the water. Some fishes can also detect changes in the electric field around their body and some can both generate and detect such fields. Four of these senses can each provide a fish with information about the spatial disposition of its environment. Under conditions in which visibility is poor such as in caves or in deep water, hearing, the lateral line system or, if present, the electrical sense may substitute for vision (Lythgoe, 1979).

In relation to its sensory capacities, a fish can be thought of as lying at the focal point of a multi-layered volume. The outermost boundary of this volume represents the greatest distance over which the fish can receive information about its environment. In many fish, this will be information in the form of sound. A boundary closer to the fish represents the distance over which the fish can receive visual information. Other boundaries represent the limits for the lateral line system or the electroreceptors. The positions of these boundaries may change as the fish grows or as the environment changes. For objects close to the fish, several sensory modalities may provide information, but objects distant from the fish may be perceived only through a single modality. The ability of the fish to respond to the object depends on the interrelationship between the sensory systems and the environment and on the capacity of the fish to analyse information that may be received through one or more sensory systems.

Vision

As a medium for the transmission of light, water compares unfavourably with air (Denny, 1993). Yet most species of teleosts have well-developed eyes. The presence of cones in the retina of many species suggests colour vision, and behavioural tests have confirmed that fish are capable of discrimination on the basis of hue (Guthrie and Muntz, 1993). In this respect, teleosts are more similar to birds than to mammals, which often depend heavily on hearing and olfaction. Even under the most favourable conditions, sufficient light for vision penetrates water only to a depth of about 1000 m (Lythgoe, 1979). Water exerts a strong filtering effect on the wavelengths of light, an effect that depends on the nature of the water (Loew and McFarland, 1990). In open oceanic conditions, where the water contains little sediment or phytoplankton, short-wavelength blue light penetrates to the greatest depth. In fertile waters that have high densities of phytoplankton containing chlorophyll or in waters containing sediments, light of longer wavelengths penetrates deepest and such waters appear greenish or brownish (Lythgoe, 1979). Fresh waters are often strongly coloured so that even habitats separated by a few metres may differ greatly in the spectral quality of the light (Levine *et al.*, 1980). These differences in the quality of light must be appreciated when interpreting the ecological significance of colours for fishes.

There are broad correlations between the visual pigments present in a species and the optical properties of the water in which that species is typically found. Deep-sea fishes have retinas dominated by rod cells, sensitive to low light intensities (scotopic). Usually a single rod pigment is present, which absorbs maximally at short wavelengths, 470 to 490 nm. Species from shallower waters have two or three cone cells, sensitive to higher light intensities (photopic), in addition to rods. Fishes from the blue waters characteristic of oceanic and deep coastal waters and coral reefs have two cone cell visual pigments with wavelengths of maximum absorption in the range 450–550 nm. The fish are most sensitive to blues and greens. Fish from shallow coastal waters or fresh water can have three cone cell pigments with absorption maxima in the range 450 nm (blues) to nearly 650 nm (orange-red) (Lythgoe, 1979; Guthrie and Muntz, 1993). Fish that live near the water surface may even have pigments sensitive in the ultraviolet (355–360 nm), although UV light is strongly absorbed by water (Guthrie and Muntz, 1993).

A sample of tropical freshwater fishes illustrated a broad correlation between their visual pigments and mode of life (Fig. 2.6) (Levine *et al.*, 1980). Species grouped as diurnal surface dwellers have shortwave-shifted sets of pigments with absorption maxima in the range 415–574 nm. Midwater species have pigments with maxima over a spectral range of 450–620 nm. A group of species, characterized as crepuscular or predaceous, have longwave-sensitive pigments

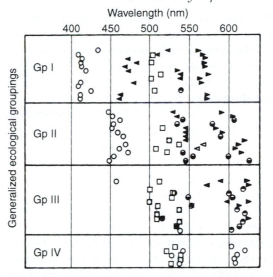

Fig. 2.6 Distribution of cone types in four groups of tropical freshwater fishes. Group I, diurnal surface dwellers; group II, midwater species; group III, crepuscular and predaceous species; group IV, bottom dwellers. Symbols: □, rods; ○, single cones; ◀, short-wave member of paired cones; ▶, long-wave member of paired cones; ◖, twin cones when no other types of paired cones are present; ◗, twin cones when other types of paired cones are also present. Redrawn from Levine *et al.* (1980).

in the range 500–620 nm. Benthic species lack shortwave- and often middlewave-sensitive cones.

Visual acuity is a measure of the ability to resolve two objects that lie close together. Guthrie and Muntz (1993) have suggested a correlation between the mode of life of the fish and its acuity because acuity is a factor determining the ability to detect small objects. A predaceous sunfish, *Lepomis*, has an acuity of 4′ (minutes of arc) subtended at the retina, whereas the goldfish, *Carassius*, which is partly herbivorous, has an acuity of 15′ (Schwassman, 1974).

As fish grow larger, their acuity improves (Li *et al.*, 1985). Larval fish have small eyes, with a small number of retinal receptor cells

(Kotrschal *et al.*, 1990). In the larvae of shallow-water species, these are cone cells. As the young fish grows, the number of cones increases and there is an increasing number of rods. As a result, acuity improves and there is an increase in photopic and scotopic sensity. Thus, visual competence is size dependent, with that competence improving most rapidly during the rapid, early growth. Note that this improvement is also accompanied by an improvement in locomotory performance (page 14).

Visual contrast decreases rapidly in water (Guthrie and Muntz, 1993). In the clearest water, objects further than 40 m from the fish are unlikely to be seen even if they subtend an angle at the retina that should make them visible (Lythgoe, 1979). Vision gives fishes information only about close objects, though for many species that information is rich in detail.

Hearing

Sound is potentially an important source of information about the environment for fishes because, in contrast to its relatively poor transmission of light, water transmits sound well (Denny, 1993). In comparison with air, water is denser and less elastic and transmits sound at about 4.8 times its velocity through air (Popper and Coombs, 1980). This transmission has a low rate of attenuation. Hawkins (1993) noted that an underwater explosion could be detected halfway round the world.

With an appropriate detecting and analysing system, both the direction and distance of a sound source from the receiver can be identified. Experiments with cod, *Gadus morhua*, in open-water cages showed that the fish could locate the position of a sound source. The fish discriminated between loudspeakers separated both horizontally and vertically, and they orientated towards a source (Hawkins, 1993).

The sensitivity of fishes is restricted to

sound of low frequencies, typically below 2–3 kHz (Hawkins, 1993). Within this restricted range, sensitivity can be high. Fishes that have a close connection between the swim bladder and the inner ear have high sensitivities and are able to respond to the sound pressure signal. The ostariophysan fishes, a group that includes the Cypriniformes and Siluriformes (catfishes), provide an example. In the ostariophysans, a chain of small bones, the Weberian apparatus, links the swim bladder to the ear. Fishes lacking a close connection between ear and swim bladder detect the particle-velocity component of the sound signal and are less sensitive (Fig. 2.7) (Tavolga, 1971; Popper and Coombs, 1980; Hawkins, 1993).

In fish that feed nocturnally or live in turbid water, morphological adaptations that enhance the detection of sound may compensate for the poor quality of visual information. The squirrelfish, *Myripristis*, is a nocturnal feeder in which there is a link between the swim bladder and the inner ear. This link consists of anterior projections of

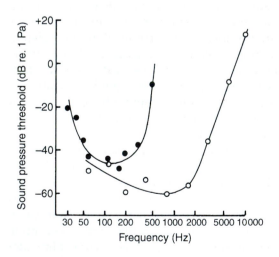

Fig. 2.7 Audiograms for a non-ostariophysan fish (●), the cod, *Gadus morhua*, and an ostariophysan fish (○), the catfish, *Ictalurus nebulosus*. Modified from Hawkins (1993).

the bladder that abut against a window in a portion of the skull that encloses the inner ear. Those fishes that live in murky waters and have evolved electroreceptors have also evolved some type of swim bladder–inner ear connection, for example the Mormyridae, Gymnarchidae and Gymnotidae (Popper and Coombs, 1980).

Lateral line system

The physical properties of water are also important for the lateral line sensory system. This detects low-frequency water displacements, with maximum sensitivity in the range 30–150 Hz, or net current flow as low as 0.025 mm s^{-1} (Bleckmann, 1993). The lateral line system allows the fish to detect disturbances created by other animals including potential prey, predators or other members of a shoal. The wake left behind by a swimming fish may be detectable several seconds later and when the fish is some metres away. Some surface-feeding fish have well-developed lateral line systems on the top of the head. Potential prey trapped in the surface film at the air–water interface generate water disturbances by their struggles, and the fish detect and locate the prey by these signals (Bleckmann, 1993). Such signals attenuate rapidly with distance, so information is transmitted over only a short range. Fish may also locate and identify stationary objects using their lateral line sense, because of the effect of the object on the water movements caused by the swimming of the fish. A blind minnow, *Phoxinus phoxinus*, can detect a glass filament with a diameter of 0.25 mm at a distance of 10 mm. The blind cave fish, *Anoptichthys jordani*, differentiated between pairs of grids of vertical bars when the difference between bar intervals was 1.25 mm. The ability of the fish to make these discriminations seems to depend on its swimming around the object while keeping a narrow gap between itself and the object (Bleckmann, 1993).

Electroreception

This sensory system is probably derived from the lateral line system. Electroreceptors are found in, for example, the Mormyridae (Osteoglossiformes), the Gymnotiformes and the Siluriformes. These are fish of turbid fresh waters or they are nocturnal feeders operating in conditions in which vision provides little information. There are two types of electroreceptor, passive and active. Passive receptors detect electric fields generated by objects in the environment. For example, the catfish, *Ictalurus*, can detect the bioelectric stimuli produced by a live prey fish hidden in the substratum.

Active receptors detect distortions in the electric field that is actively generated by specialized electric organs of the fish itself. Objects that differ in electrical conductivity from water cause distortions in the electric field generated. The fish can use these distortions to detect and locate the object. As with the lateral line system, the electroreceptors can also be important in communication with conspecifics (Bleckmann, 1993; Moller, 1995).

Electroreception has the disadvantage that an electric field attenuates rapidly with distance (Denny, 1993) and so electroreception is useful only over short distances.

Olfaction and gustation

Both olfactory and gustatory stimuli may also provide important information about the environment, but chemical signals are easily diffused by water, and in the absence of a current may propagate only slowly and provide little directional information (Popper and Coombs, 1980). However, chemosensory information can be important in feeding (Chapter 3), in orientation during migration (Chapter 5), in reproduction (Chapter 7) and in response to predators (Chapter 8) (Hara, 1993).

2.5 SUMMARY AND CONCLUSIONS

1. The constraints imposed by environmental and biological characteristics define the adaptations that are possible. Adaptation and constraint form the framework within which each individual fish allocates its time and resources.

2. Body shape, mode of locomotion and lifestyle are closely interrelated. Optimum morphologies for sprinting and cruising (e.g. tuna), acceleration and powered turns (e.g. pike) and manoeuvrability (e.g. butterflyfish) are being defined. The optimum designs, defined by the hydrodynamic characteristics of water, are mutually incompatible. Many species are generalists, compromising aspects of their locomotory performance.

3. The low solubility of oxygen in water imposes the need for a large respiratory surface, which is provided by the gills. This surface area also, necessarily, acts as a site of heat and ion exchange. Some aquatic environments become hypoxic. Fish may respond to hypoxia behaviourally (aquatic surface respiration), biochemically (anaerobic respiration), or with morphological adaptations for aerial respiration. Both aquatic and aerial respiration have costs and benefits associated with them, which may change with oxygen concentration, depth of water, presence of predators and food supply.

4. Although water is often a poor transmitter of light, many fishes have colour vision. The visual pigments present in the retina relate to the light properties of the waters in which the fish usually live. Hearing and the lateral line receptors that detect disturbances in the water are also important for fishes. Some species can also detect changes in the electric field around their body caused by other objects. The visual, lateral line and electroreceptive systems are effective only over short distances because of the transmission properties of water.

5. Improvements in locomotion, respiratory organs and sensory capacities take place rapidly during the early free-living stages of fish, when fish are growing rapidly. These changes mean that the performance of fish is size dependent.

6. The interaction of constraint and adaptation makes it possible to predict in a general way the mode of life of individuals from a knowledge of their morphology and physiology.

3

FEEDING

3.1 INTRODUCTION

Chapter 1 described an individual fish as a system that converts food into progeny. The survival, growth and reproduction of a fish depend on the income of energy and nutrients generated by its feeding activities. Although fishes have a considerable capacity to resist starvation (Love, 1980) and many species normally cease to feed at times during their life cycle (Wootton, 1979), the capacity of individuals to survive such periods depends on their ability to lay down reserves which can be mobilized when feeding ceases. The size of such reserves will reflect feeding success. An ecological analysis of feeding must answer basic questions including: what is eaten, when is it eaten, where is it eaten and how much is eaten?

3.2 TROPHIC CATEGORIES IN FISHES

Fishes occupy virtually every possible trophic role, from herbivorous species such as the menhaden, *Brevoortia tyrannus*, feeding on unicellular algae, to secondary and tertiary carnivores, for example the pike which eats other fish, amphibians, birds and mammals (Keenleyside, 1979; Gerking, 1994). Some species form part of the decomposer food chain, utilizing detritus or scavenging carcasses. There is even a cichlid that feeds on decomposing hippopotamus faeces! Other unusual methods of obtaining food have also evolved. The archer fish (Toxotidae) can spit out a jet of water to knock terrestrial insects into the water. Angler fish (Lophiiformes) attract prey close with a moveable lure

formed by a modification of a ray of the dorsal fin. A general classification of trophic categories is shown in Table 3.1.

Although this Table provides a useful summary of the range of feeding habits, it should not be taken as a rigid classification. Many species show great flexibility in their trophic ecology (Keenleyside, 1979; Dill, 1983; Gerking, 1994). Some fish assemblages contain significant proportions of omnivores (Winemiller, 1990). Examples of this trophic flexibility occur throughout this chapter and the concept is discussed later (pages 60–62). The diversity and flexibility of fish diets can generate complex food webs. This was illustrated in a study of the feeding habits of fish in swamp and stream habitats in Venezuela and Costa Rica (Winemiller, 1990). For one locality, over 1200 trophic links were identified.

3.3 MORPHOLOGICAL ADAPTATIONS FOR FEEDING

Correlations between morphology and diet

Morphological traits can be guides to the trophic ecology of a species because the traits determine how a fish can feed and so what it can eat. Fish that are not phylogenetically closely related may show convergent evolution in their morphologies because they feed on similar types of food.

Studies of two assemblages of lake-dwelling fishes help to illustrate the correlation between morphology and diet. The feeding ecology of species living in Lake Opinicon, Ontario, has been extensively described

Table 3.1 Major trophic categories in teleost fishes*

---

1. Detritivores, e.g. *Tilapia* spp. (Cichlidae), *Puntius* spp. (Cyprinidae)
2. Scavengers, e.g. *Anguilla* (Anguillidae) (opportunistically)
3. Herbivores
   3.1 Grazers, e.g. *Hypostomus* (Loricariidae)
   3.2 Browsers, e.g. *Ctenopharyngodon* (Cyprinidae)
   3.3 Phytoplanktivores, e.g. *Tilapia* spp. (Cichlidae)
4. Omnivores, e.g. *Rutilus* (Cyprinidae)
5. Carnivores
   5.1 Benthivores
       a. Picking at relatively small prey, e.g. *Gasterosteus* (Gasterosteidae)
       b. Disturbing, then picking at prey, e.g. *Sufflamen* (Balistidae)
       c. Picking up substrate and sorting prey, e.g. *Lethrinops* (Cichlidae)
       d. Grasping relatively large prey, e.g. *Balistes* (Balistidae)
   5.2 Zooplanktivores
       a. Filter feeders, e.g. *Engraulis* (Engraulidae) feeding on nauplii
       b. Particulate feeders, e.g. *Engraulis* feeding on adult zooplankters
   5.3 Aerial feeders, e.g. *Toxotes* (Toxotidae)
   5.4 Piscivores
       a. Ambush hunters, e.g. *Cottus* (Cottidae)
       b. Lurers, e.g. *Lophius* (Lophiidae)
       c. Stalkers, e.g. *Esox* (Esocidae)
       d. Chasers, e.g. *Salmo* (Salmonidae)
       e. Ectoparasites, including scale eaters, e.g. *Exodon* (Characidae) and fin
          eaters, e.g. *Belonophago* (Citharinidae)

---

*Modified after Keenleyside (1979).

(Keast and Webb, 1966; Keast and Welsh, 1968; Keast, 1970, 1978). The fishes of Lake Opinicon belong to several teleostean orders (Fig. 3.1). Lakes Victoria, Malawi and Tanganyika form the Great Lakes of Africa. They contain an extraordinary profusion of cichlids. These fish belong to a single Perciform family, the Cichlidae. They have shown such a wide adaptive radiation in the African Great Lakes that within this single family "all available feeding niches. including some unexpected ones, appear to have been exploited by one or other of these fishes" (Fryer and Iles, 1972).

Body form

Body form in both faunas ranges from that characteristic of fishes specialized for rapid acceleration (Chapter 2) such as the piscivorous pike (Fig. 3.2) to forms characteristic of highly manoeuvrable fishes like the deep-bodied bluegill sunfish, *Lepomis macrochirus* (Centrarchidae), a generalist feeder of Lake Opinicon (Fig. 3.3).

Mouth shape and position

The position, shape and size of the mouth are related to diet. In fishes three major categories of feeding can be recognized (Liem, 1980, 1991). All fish employ inertial sucking at some stage in their lives. Water containing food is sucked into the mouth by a rapid increase in the volume of the buccal cavity as the mouth opens. Some species also use ram feeding: the fish swims with its mouth open and overtakes the food. A third feeding

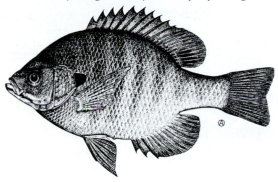

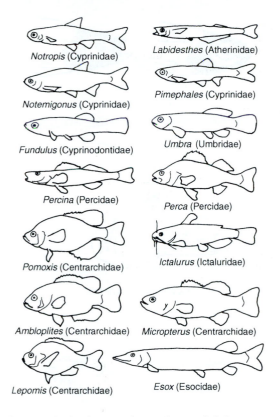

Fig. 3.1 Body forms of members of fish assemblage in Lake Opinicon, Ontario (Canada). Redrawn from Keast and Webb (1966).

Fig. 3.3 Body form of a generalist feeder, the bluegill sunfish, *Lepomis macrochirus* (adult length about 180 mm). Note body form relatively specialized for manoeuvrability (see Chapter 2). Reproduced with permission of authors from Scott and Crossman (1979).

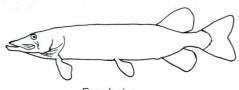

Esox lucius

Rhamphochromis longiceps

Fig. 3.2 Body form in two unrelated piscivores, pike, *Esox lucius*, from L. Opinicon and a cichlid, *Rhamphochromis longiceps*, from L. Malawi. Redrawn from Keast and Webb (1966) and from Fryer and Iles (1972).

mode is manipulation, which includes such techniques as biting, scraping, rasping, gripping and clipping. The ability to protrude the jaw is common in the more evolutionarily advanced fishes (Fig. 3.4). The advantages of this in feeding have yet to be fully defined (Motta, 1984). Protrusion may momentarily but crucially increase the rate of approach of the predator to its prey. Protrusion may also increase the distance from which a prey can be sucked. It may decrease the rotation of the lower jaw required to close the mouth once the prey is captured (Motta, 1984; Osse, 1985). The ability to protrude the jaw may also confer an advantage in specific circumstances such as obtaining benthic prey or food from otherwise inaccessible places (Alexander, 1967; Osse, 1985).

Fishes feeding at the surface or in the middle of the water column frequently have a dorso-terminal or terminal mouth. Two examples from Lake Opinicon are the golden shiner, *Notemigonus crysoleucas*, a cyprinid with a protrusible jaw (Fig. 3.5), and the brook silverside, *Labidesthes sicculus*, an atherinid. Pelagic, zooplanktivorous cichlids of Lake Malawi have a terminal mouth,

Closed mouth

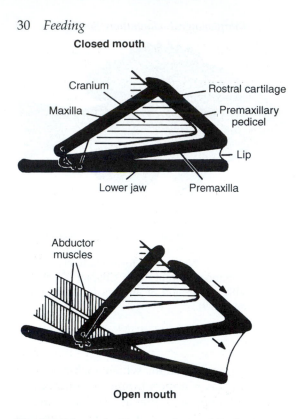

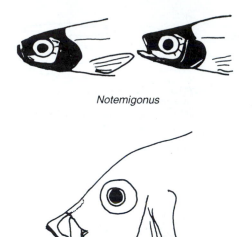

Notemigonus

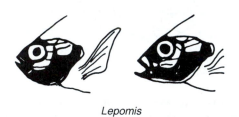

Haplochromis cyaneus

Open mouth

Lepomis

Fig. 3.4 Schematic diagram of a model of jaw protrusion in a cichlid. Arrows show direction of movement of various bones; position of ligaments shown by lines. Redrawn from Fryer and Iles (1972).

Fig 3.5 Some examples of jaw protrusion from L. Opinicon (*Notemigonus* and *Lepomis*) and L. Victoria (*Haplochromis cyaneus*). Redrawn from Keast and Webb (1966) and from Fryer and Iles (1972).

which forms a protrusible tube (Fig. 3.5). In Lake Opinicon the bluegill sunfish is a generalized feeder eating zooplankton, insect larvae, crustaceans, molluscs and fish fry. It has a narrow terminal mouth, which is tubular when protruded (Fig. 3.5). The major piscivores of Lake Opinicon are the pike and the largemouth bass, *Micropterus salmoides*, a centrarchid (Fig. 3.1). These have a wide mouth gape and a strong jaw. In pike there is also a noticeable flattening of the head (Fig. 3.2). Piscivorous cichlids of the African Great Lakes also have mouths with a large gape and their head shape gives some species a pike-like appearance (Fig. 3.2). The bluntnose minnow, *Pimephales notatus*, of

Lake Opinicon feeds on organic detritus and invertebrates from the substrate. Its mouth is ventroterminal. A bottom-feeder from Lake Malawi, *Lethrinops furcifer*, has a highly protractile mouth, which it uses to vacuum-clean through the substrate to collect chironomid larvae (Fig. 3.6).

Marginal and pharyngeal teeth

Fish may carry teeth on the tongue, the marginal bones of the jaw, the palatal bones and the pharyngeal bones. The marginal teeth of

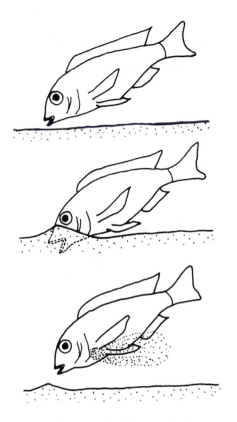

Fig. 3.6 Feeding of *Lethrinops furcifer*, a sand-digging cichlid from L. Malawi, showing 'vacuum cleaning' of substrate. Redrawn from Fryer and Iles (1972).

the fishes of Lake Opinicon, with the exception of the cyprinids, which lack them, are relatively unspecialized cones, straight or recurved. In the pike, the marginal and palatal bones are angled so prey can only move down the throat. In contrast, the African Great Lakes cichlids show a diversity of marginal teeth, the shape of which can be related to diet (Fig. 3.7).

The pharyngeal bones in the throat are well developed in the African Great Lakes cichlids (Fryer and Iies, 1972; Liem, 1973; Greenwood, 1984). There are two sets of bones, one in the roof of the throat and the other on the floor. Any food passing through to the oesophagus must pass between the

upper and lower sets as though passing between a pair of millstones (Fig. 3.8). The surfaces of the bones carry teeth, the shape of which correlates with diet. Examples of the pharyngeal dentition of some cichlids are shown in Fig. 3.7. The convergent evolution in tooth shape in, for example, those cichlids that feed by crushing molluscs is striking. In Lake Opinicon, the pumpkinseed sunfish, *Lepomis gibbosus*, feeds on molluscs and isopods. Its stout, flattened pharyngeal teeth act as a grinding mechanism. The bluegill sunfish, which feeds on a wide range of small invertebrates, has the surface of its pharyngeal bones covered with fine, needle-like teeth (Keast, 1978).

Gill rakers

Rakers are forward-directed projections from the inner margins of the gill arches. Their shape and abundance are related to diet. Fish that feed on small food particles usually have numerous long, fine rakers (Fig. 3.9) whereas fish feeding on large particles have fewer, shorter, blunter rakers. The brook silverside of Lake Opinicon has abundant long rakers, whereas the more generalist feeders, the sunfish, have rakers that are short and few in number (Keast, 1978). However, Gerking (1994) noted that fish with short rakers do feed on zooplankton and fish with long rakers do feed on benthic prey.

In fishes like the menhaden, that filter feed on phytoplankton or small zooplankton, the gill rakers probably sieve the food particles from the respiratory current (Friedland, 1985). The detailed mechanism of the sieving action and the collection of food is not fully understood (Lauder, 1983). In filter-feeding cichlids, *Oreochromis* spp., the gill rakers and filaments entrap food particles bound in mucus. When this material is passed back, it is sorted and raked by pharyngeal teeth (Dempster *et al.*, 1995). In some filter-feeding fishes, the rakers may not play any direct role. The food is trapped in mucus clumps,

Fig. 3.7 Relationship between shape of marginal and pharyngeal teeth and diet for cichlids from L. Malawi (trophic status of 'eye biter' uncertain). Redrawn and simplified from Fryer and Iles (1972).

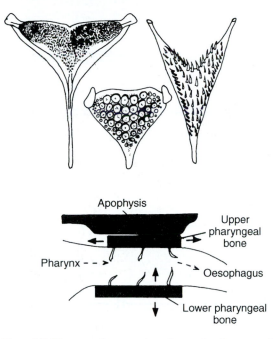

Fig. 3.8 Upper: three examples of pharyngeal bones from African Great Lakes cichlids showing interspecific variation in form of pharyngeal teeth. Lower: the pharyngeal mill of the cichlids. Solid arrows show direction of movement of pharyngeal bones; broken arrows show food path. Redrawn from Fryer and Iles (1972).

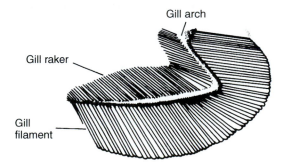

Fig. 3.9 Gill arch showing arrangement of gill rakers in planktivorous fish. Redrawn and modified from Zaret (1980).

which pass over the pharyngeal teeth to the oesophagus (Gerking, 1994).

In those zooplanktivores that feed by snapping up their prey, such as the lake

whitefish, *Coregonus clupeaformis*, the size of prey ingested is not a simple function of the sieve mesh of the rakers (Seghers, 1974a; Zaret, 1980; Langeland and Nost, 1995). The precise role during feeding of the rakers in these and other particulate-feeding fishes remains unclear (O'Brien, 1987).

Alimentary canal

There is a correlation between the diet and the gut length relative to body length (Kapoor *et al.*, 1975). This is illustrated by the relative gut lengths of carnivorous, omnivorous and herbivorous cichlids from Lake Tanganyika (Fig. 3.10). Fish consuming high-quality food can process it with a gut that is shorter than their total length. Fish having diets including a high proportion of material that resists digestion, such as cellulose or lignin, have guts that are several times longer than their body length. The Indian cyprinid, *Labeo horie*, feeds on detritus and has a gut length that is 15 to 21 times its body length (Bond, 1979).

The relationship between gut length (*GL*) and body length (*L*) can be described by the allometric relationship: $GL = aL^b$. This relationship held for fish from a freshwater creek in South Carolina (Ribble and Smith, 1983) and a forest stream in Panama (Kramer and Bryant, 1995a,b). In most species, the exponent b was significantly greater than unity (positive allometry), indicating that the relative gut length increased with an increase in fish length. This increase in relative gut length may maintain the relative absorptive area as the fish grows. By comparing fish of similar length but different diets, Kramer and Bryant (1995b) found that relative gut length tended to decrease from herbivores to omnivores to carnivores. But, among the omnivorous species, there was no correlation between the length of the gut and the proportion of plant material in the diet.

Carnivorous fishes usually have a large stomach, which in some species has pyloric

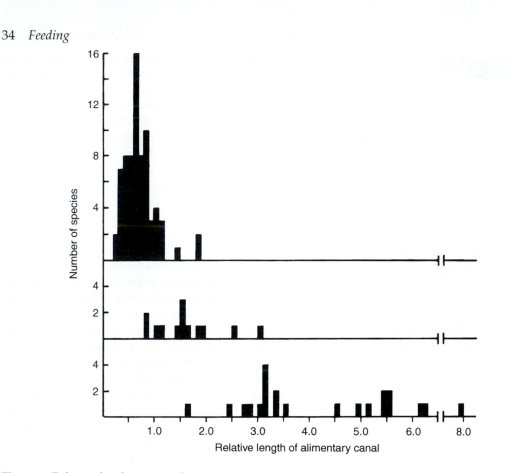

Fig. 3.10 Relationship between relative gut length (see text) and diet in a sample of African Great Lake cichlids. Top, carnivores; middle, omnivores; bottom, herbivores. Redrawn from Fryer and Iles (1972).

caeca associated with it. These are blind-ending tubes that arise as outgrowths of the pyloric region of the gut.

Plant material, because of the presence of cellulose and lignins, is harder to digest than animal flesh. Nevetheless, herbivory is common in some habitats, for example coral reefs and some tropical fresh waters. There are a variety of morphological features associated with herbivory in fishes (Horn, 1989; Choat, 1991). Some species have a thin-walled stomach, which can generate a low pH, together with a long intestine. In other species, including mullet (Mugilidae), there is a muscular, grinding stomach. Some herbivorous species, including representatives of the Cyprinidae, lack stomachs. They have well-developed pharyngeal bones which can physically break up plant material. In the Kyphosidae, a southern hemisphere marine family, there is a caecum in the hind gut. This caecum contains microflora, which help to break down plant material.

Ecomorphological hypothesis

Although correlations between morphology and diet provide some insight into the feeding ecology of fishes, correlation does not necessarily identify causation. Two further lines of evidence provide evidence for a causal relationship between morphology and diet. Biomechanical studies analyse the mechanisms by which morphological traits lead to particular methods of feeding. The principles of physics and engineering are applied to biological structures (Wainwright and Richard, 1995). Secondly, phylogenetic information on a group of species can be used to identify the pattern of evolution of morphological traits and diet. In a study of coral reef wrasse (Labridae), species were characterized as feeding on hard prey or evasive prey (Westneat, 1995). Species were also characterized by the transmission of force and motion during feeding, which relates to morphological features of the jaw. Diet and mechanical characteristics were then mapped onto a phylogenetic tree of the wrasse. The mapping suggested that a combination of feeding on evasive prey and possession of particular jaw dynamics had evolved independently within the wrasse, perhaps four times. This provides good evidence that diet and morphology are causally rather than coincidentally linked.

A descriptive approach to relating morphology and diet is also being replaced by the use of multivariate statistical techniques. These allow the positions of species to be plotted in a morphospace, defined on the basis of several, quantitatively measured morphological traits. Species that fall close together in this morphospace can be compared to see if their diets are also similar. For an assemblage of fish in Sri Lankan streams, a multivariate analysis correlated species with short guts, dorsoterminal or terminal mouths and lacking barbels with a diet of invertebrates. Species with long guts, ventral or subterminal mouths and barbels correlated with a herbivorous diet (Wikramanayake, 1990). Species that fall close to each other in a morphospace, but which are not closely related phylogenetically, illustrate convergence of morphology. An example is the convergence in body form of stalking predators from habitats in Alaska, Central America, South America and Africa (Wine-

miller, 1991). Such convergence at least partly reflects the constraints on body form that are imposed by the hydrodynamic properties of water (Chapter 2).

3.4 DIET COMPOSITION

Although morphology can provide circumstantial evidence of the diet of a fish, the inferences must be confirmed by more direct evidence of what is eaten.

Describing dietary composition

A description of the composition of the diet should indicate the relative importance of the items eaten. The flexibility of feeding habits can make even this starting point difficult to achieve – for example, the threespine stickleback, *Gasterosteus aculeatus*, in a small Welsh reservoir eats about 20 different categories of food (Allen and Wootton, 1984). Diet can rarely be studied by directly observing feeding behaviour and identifying what is eaten. Usually the diet is sampled by extracting the gut contents, either by killing the fish and dissecting out the gut or by flushing out the contents of the gut. Several methods are used to provide a quantitative description of such samples (Hynes, 1950; Windell, 1971; Hyslop, 1980); probably no one method is entirely satisfactory.

The simplest method estimates the frequency of occurrence in stomachs. After identification of the food categories present, the number of stomachs in which a given category occurs is expressed as a percentage of the total number of stomachs sampled. This method only provides information on the presence or absence of categories and not on their relative numbers or bulk. In the numerical method, the number of items in each food category is counted in all stomachs in the sample. The importance of a category is then usually estimated by expressing the number of items in that category as a percentage of the total number of items counted in all the stomachs. This method emphasizes the importance of small and numerous items such as zooplankton. It cannot be used when the diet contains significant proportions of plant material or detritus, categories that do not include discrete, individual prey. The volumetric and gravimetric techniques emphasize the bulk of the food categories. In the former, the volume of each category in each stomach is estimated. In the latter, the weight of each category is measured. In both methods, the relative importance of a food category can be expressed as a percentage of the total volume or weight of all the categories present in the samples. The points method is essentially a modification of the volumetric method that is quicker to use. The investigator allocates points to each food category in proportion to its contribution to the total volume of the stomach contents. The allocation of points is subjective for it is based on a visual assessment of contribution. This subjectivity of the method can be partially reduced by correcting the points score for the degree of fullness of the stomach (Hyslop, 1980; Allen and Wootton, 1984).

Any of these methods provides a general picture of the composition of the diet, and frequently one that is similar to that provided by the other methods (Hynes, 1950; Pollard, 1973). A fuller picture can be obtained by using the numerical method together with one of the three methods that emphasize the contribution to the volume or weight of the food consumed (Hyslop, 1980). A simple graphical plot of frequency of occurrence against number (or bulk, whichever is relevant) can be used to identify specialized and generalist feeding patterns (Costello, 1990). The statistical analysis of such data presents major problems. There may be many food categories in the diet, some of which are present only in small amounts. Fish sampled at the same time and place may have stomach contents that are very different. Some food categories are quickly digested and so difficult to detect.

Other categories, for example insect larvae or crustaceans with chitinous exoskeletons, remain identifiable over longer periods of time. The difficulties of rigorously analysing quantitative data obtained by the analysis of stomach contents have yet to be fully resolved (Crow, 1982). However, a judicious choice of sampling and statistical methods should provide an interpretable picture of diet.

Measurement of diet selectivity in natural populations

The diet will reflect what food is available in the environment. A fish is a sampling device with the contents of its gut representing a sample of what is available. Does a fish take food items strictly in the proportions in which they occur in the environment or does it show some selectivity? Several methods for measuring such selectivity have been developed. An early and commonly used example is the index of electivity (Ivlev, 1961) defined as:

$$E = (r_i - p_i)/(r_i + p_i) \qquad (3.1)$$

where E is the index of electivity, r is the relative abundance of prey category i in the gut, and p_i is the relative abundance of prey category i in the environment. The index can take values between -1 and $+1$, with negative values indicating avoidance or inaccessibility of prey category i, and positive values active selection. A value of zero indicates random selection. Although commonly used, the index of electivity has serious faults (Strauss, 1979; Lechowicz, 1982; Chesson, 1983). Its value depends not only on the behaviour of the forager but on the numbers of each food type present. This usually precludes comparing values for the index obtained at different sites or at different times of the year.

Chesson (1983) suggests a measure of preference for a food item i, a_i, which is defined by the relationship:

$$P_i = a_i n_i \bigg/ \sum_{j=1}^{m} a_j n_j \qquad (3.2)$$

where P_i is the probability that the next food item consumed is of type i, n_i is the number of type i available and m is the total number of types. Each a_i can be interpreted as the proportion of the diet that could consist of type i if all food types were present in equal numbers in the environment. Table 3.2 gives the formulae for calculating a, under different conditions.

An alternative approach is to use statistical techniques to compare the composition of the diet with what is available in the environment. In a study of predation of the alewife, *Alosa pseudoharengus*, on zooplankton, Kohler and Ney (1982) used a nonparametric statis-

tical test to compare the rank order of prey in the diet with their rank order in the environment. MacDonald and Green (1986) used a multivariate parametric analysis to identify prey selection by an assemblage of benthic-feeding fishes including cod, winter flounder and American plaice, *Hippoglossoides platessoides*, off the coast of New Brunswick (Canada). Their technique, canonical analysis, allowed a statistical comparison of the faunal composition of benthic samples with the faunal composition of the diet of each species.

Both indices and the statistical techniques assume that the gut samples and habitat samples accurately reflect the relative abundance of prey consumed and in the environment respectively (Kohler and Ney, 1982). A

Table 3.2 Formulae for calculating preference for prey item i, a_i in Chesson's (1983) index of prey preference

1. No food depletion, n_i assumed constant:

$$\hat{a}_i = \frac{r_i / n_i}{\sum\limits_{j=1}^{m} (r_j / n_j)}, \quad i = 1, \dots, m$$

2. Food depletion, n_i not assumed constant:

$$\hat{a}_i = \frac{\log_e \{[n_i(0) - r_i] / n_i(0)\}}{\sum\limits_{j=1}^{m} (\log_e \{[n_j(0) - r_j] / n_j(0)\})}, \quad i = 1, \dots, m$$

3. Order of selection of items by consumer known, but only first prey item taken is recorded, observation is repeated for k consumers:

$$\hat{a}_i = \frac{k_i / n_i}{\sum\limits_{j=1}^{m} (k_j / n_j)}, \quad i = 1, \dots, m$$

Where:
a_i = preference for prey type i
n_i = number of items of type i present in the environment
$n_i(0)$ = number of items of type i present in the environment at the beginning of foraging bout
r_i = number of items of type i in consumer's diet
k_i = number of consumers whose first food item was of type i
K = number of consumers, i.e.

$$K = \sum_{i=1}^{m} k_i$$

m = number of types in the environment

weakness of indices or statistical analyses of selection is that they give no indication of the mechanisms responsible for any selection identified. They are guides for identifying situations in which diet selection may be taking place and so for stimulating studies which uncover the behavioural processes that result in the selection.

Factors that determine prey selection

There are two complementary approaches to predicting prey selection by fish. The first studies the mechanisms by which prey are detected and ingested. This approach develops mechanistic models of prey selection. The second approach, usually called optimal foraging theory, predicts prey selection in terms of the outcome of that selection for the fish. This outcome is usually assessed as the rate of energy intake, but other currencies can be used. Models that predict outcome are called functional models. The mechanistic approach to prey selection will be considered first.

For an item to appear in the diet of a fish it has to be detected (encountered), approached, selected, manipulated and ingested (Fig. 3.11) (Lima and Dill, 1990; Hart, 1993). Most studies on prey detection and acceptance by fishes use species that detect their prey visually and feed on living prey such as zooplankton (Lazzaro, 1987; O'Brien, 1987), benthic invertebrates or fish. The factors that determine the diet composition of species that rely on non-visual cues to find food, or of detritivores and herbivores generally receive less attention. Given the importance of herbivorous and detritivorous cyprinids and cichlids to aquaculture in developing countries such as India and China, this will change.

Prey availability

For prey to appear in the diet of a fish they must be available and accessible given the constraints of the morphology and sensory capacities of the fish. This was illustrated by an experiment on rainbow trout, *Oncorhynchus mykiss*. The trout hunted for amphipods placed at different densities on the bottom of an aquarium. Different sorts of litter placed on the bottom changed the quality of the substratum. This provided greater or lesser concealment for the amphipods. Although the number of prey attacked increased linearly with an increase in their density, the proportion of prey attacked at a given density decreased with an increase in the concealment provided by the litter (Fig. 3.12) (Ware, 1972).

A field study complementing this experiment demonstrated this importance of refuges from predation for the prey. In Marion Lake, British Columbia, even under favourable conditions, less than 30% of the amphipod population was at the substratum–water interface and thus exposed to predation by rainbow trout. At low temperatures in winter, the vulnerable portion of the amphipod population fell as low as 2%, the remaining 98% being concealed within the substratum (Ware, 1973). Ecological studies that seek to analyse the availability of prey to a predatory fish have to estimate not only the absolute abundance of the prey, but also that proportion of the prey population that is at risk of predation.

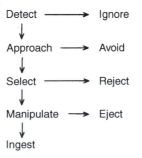

Fig. 3.11 Events in a feeding sequence.

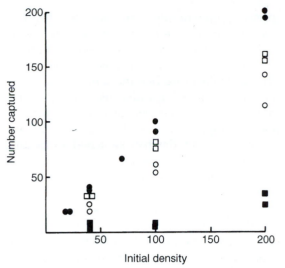

Fig. 3.12 Effect of litter on capture success of rainbow trout, *Oncorhynchus mykiss*, feeding on amphipods (*Hyallela*). Symbols: ●, no litter; □, large litter covering ≈. 6% of bottom; ○, large litter covering ≈. 15% of bottom; ■, smaller litter covering 100% of bottom. Redrawn from Ware (1972).

Prey and predator characteristics in prey selection

Selection may occur because of differences in the detectability of potential prey. This form of selection is related to the sensory capacities of the predator and the ability of the prey to avoid detection. The predator may detect the prey then fail to capture it because of morphological or behavioural constraints. The predator may detect the prey and be able to ingest it, but reject it in favour of another prey.

For fish predators hunting by sight, the important visual characteristics of the prey are size, contrast with the background and movement. Other relevant characteristics may be shape, colour and oddity (Wootton, 1984a; Gerking, 1994).

The maximum distance at which a predatory fish can see a potential prey, the reac-

tion distance, increases with the prey size. Rainbow trout can detect an amphipod 4 mm in length at a mean distance of 180 mm but a 9 mm amphipod at 350 mm (Ware, 1972). Larger prey are at more risk because the area or volume within which they can be seen is greater. This reactive field volume model of prey detection is illustrated in Fig. 3.13 (Eggers, 1977, 1982).

A complication to this story occurs when two or more prey are simultaneously visible to the predator. A prey may be apparently larger than another that is larger in absolute terms. If the smaller prey is closer to the fish, its image on the retina can be larger than that of the more distant prey. Experimental studies on sunfish and on the threespine stickleback showed that fish may select the apparently larger of two prey even when it is absolutely smaller (O'Brien *et al.*, 1976; Gibson, 1980). When the prey were dense, selection was on the basis of absolute rather than apparent size (Gibson, 1980). Selection was also based on absolute size when the prey were close to the fish, or when the apparently larger prey were not directly in front of the fish but off to one side (O'Brien *et al.*, 1985; O'Brien, 1987). For the apparent size model to operate, prey must be sufficiently abundant for more than one to be

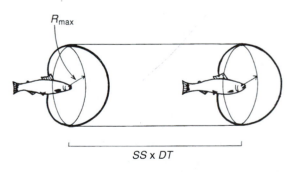

Fig. 3.13 Reactive field model showing volume of detection for a visually hunting planktivorous fish. R_{max}, maximum lateral distance of perception; SS, swimming speed; DT, time engaged in search. Redrawn from Eggers (1977).

visible simultaneously. At low prey densities, the simple reactive field volume model is appropriate.

The size of predator is also relevant because there are ontogentic changes in the retina, which result in an improvement in visual acuity as the fish increases in size (Hairston *et al.*, 1982) (Chapter 2). The improvement is greatest for small fish; for example in bluegill sunfish, the improvement is greatest in fish up to 50 mm in length. This change means that larger fish can discriminate between the sizes of prey better than small fish. However, sunfish less than 20 mm in length were more selective when feeding on *Daphnia* than predicted by the apparent size model (Walton *et al.*, 1992). The small sunfish had poor success when attacking large *Daphnia*.

The consequences for prey populations of selection by size are shown by the effects of the introduction of planktivorous fishes into waters that previously lacked them (Zaret, 1980; O'Brien, 1987). Sporley Lake in Michigan was cleared of fish by poisoning, allowed to recover and stocked with rainbow trout; fathead minnows, *Pimephales promelas*, and smelt were also introduced (Galbraith, 1967). As the numbers of fish increased, the composition of the zooplankton assemblage changed. There was reduction in the mean size of the mature *Daphnia*, a replacement of larger by smaller daphnid species, and a subsequent replacement of daphnids by smaller cladocerans such as *Bosmina*. Such changes in the size spectrum of prey species can have effects on the fish species that can coexist in the same habitat (discussions in Chapters 9 and 12).

Reaction distance increases with an increase in visual contrast between the prey and the background (Ware, 1971). An elegant demonstration of the importance of contrast was provided by Zaret (1972, 1980). The freshwater planktivore, *Melaniris chagresi*, an atherinid from Central America, eats the cladoceran *Ceriodaphnia cornuta*. The fish preys preferentially on one of the two morphs of the cladoceran present in Gatún Lake, Panama. There is no difference in body size between the two morphs, but the morph that is eaten more has the larger black compound eye. Feeding India ink particles to the morph with the less conspicuous eye made it more conspicuous than the morph with the larger eye, and the fish switched to feeding on the morph containing ink.

As light levels decline, the contrast between prey and background usually declines. Visually hunting fish show a sigmoidal relationship between light intensity and feeding intensity. As light levels increase from total darkness, the feeding intensity shows little change until a threshold level of light intensity is reached. Feeding intensity then increases rapidly to a maximum with any further increase in light intensity (Dabrowski, 1982a). Larval roach, *Rutilus rutilus*, continue to feed on zooplankton at extremely low light intensities, but this ability may depend on the lateral line system rather than sight (Dabrowski, 1982b).

Reaction distance is reduced in turbid water (Moore and Moore, 1976). This effect of water turbidity on the visibility of prey may be reflected in the species composition in a habitat. The turbidity of the water of floodplain lakes of the Orinoco River system in Venezuela was a good predictor of the species composition of the lakes (Rodriguez and Lewis, 1997). There was a decline in the representation of visually orientated fish with a decline in water transparency.

Movement is the third important prey characteristic. The reaction distance of rainbow trout feeding on pieces of blanched liver was significantly greater when the liver was kept in motion (Ware, 1973). Mysids are important in the diet of the littoral fifteen-spine stickleback, *Spinachia spinachia*. The stickleback prefers moving mysids to stationary ones unless the prey are moving slowly. The frequency of attempted or completed feeding responses by the stickleback increased with the speed of movement of the mysid up to a speed of about 30 mm s^{-1}, but then declined with a further increase (Kislalioglu and Gibson, 1976a).

Size, contrast and movement are in most situations the most important prey characteristics for a predator hunting by sight. But other factors can become relevant.

When feeding on chironomid larvae, threespine sticklebacks preferred red to pale larvae. The sticklebacks showed a preference for prey that were red, long, thin and moving (Ibrahim and Huntingford, 1989). When attacking a swarm of *Daphnia*, threespine sticklebacks turned their attention to *Daphnia* that differed in colour from others (Ohguchi, 1981). Prey that are behaving oddly, perhaps because of injury or deformity, may also be more at risk (Curio, 1976).

Having detected a potential prey, the predator has to pursue, capture and ingest it. Some fish predators chase their prey, whereas others ambush them. Capture success can depend on the body form and mode of locomotion of the predator (Chapter 2). Tunas, which are specialized for cruising, may catch only 10–15% of the fish they strike at, whereas the pike, a specialist ambush piscivore, can catch 70–80% of their intended prey. Generalist predators such as trout and perch (*Perca*) have success rates of about 40–50% (Webb, 1984b). The ability of prey to avoid the tactics of the predator can also be important in determining diet composition. Zooplanktivorous fish often take cladocerans in preference to copepods, perhaps because copepods have a more erratic mode of swimming and are less easy to capture (Zaret, 1980; O'Brien, 1987).

After capture, prey palatability can be a factor. Threespine sticklebacks used to feeding on *Tubifex* worms nevertheless showed a strong preference for enchytraeid worms when these were also made available. The presence of the enchytraeid worms significantly reduced the risk of predation to the *Tubifex* (Beukema, 1968).

Diet will also reflect morphological constraints including mouth size and the spacing of the gill rakers. Coho salmon fry, *Oncorhynchus kisutch*, 40–80 mm in length, ingested a smaller proportion of prey as prey width increased from 2 mm to 6 mm (Fig. 3.14; no prey larger than 5 mm were caught). For a coho of a given length, the proportion of prey ingested decreased rapidly with an

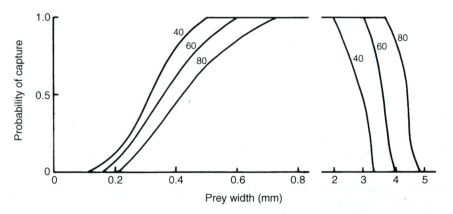

Fig. 3.14 Model of effect of prey width on probability of successful capture by fish of three lengths, 40, 60 and 80 mm. Model is based on data from juvenile coho salmon, *Oncorhynchus kisutch*. Redrawn from Dunbrack and Dill (1983).

increase in prey width above a critical width. This critical width is greater for the bigger fry because their gape increases (Dunbrack and Dill, 1983). Fry of the Atlantic salmon showed a similar effect of prey size on the chances of rejection (Wankowski, 1979). For both species of salmon fry, the probability of ingestion also decreased when prey were below a critical width. The spacing of the gill rakers or the difficulty of seeing small prey may set this lower limit. A mechanistic model of prey selection by coho fry feeding on drifting invertebrates, based on the relationship between prey size and the probability of ingestion (Fig. 3.14), gave a good prediction of the diet of coho in a stream (Dunbrack and Dill, 1983).

As prey size increases, the predator has to pay an increasing cost in the time taken to handle the prey. With an increase in the size of prey relative to the predator's mouth size, handling time increases, slowly at first, but then sharply as the prey size approaches the gape size (Fig. 3.15). Experiments on species of sunfish and the fifteenspine stickleback demonstrated this relationship (Werner, 1974; Kislalioglu and Gibson, 1976b). Handling time may also increase as the fish becomes satiated (Ware, 1972; Werner, 1974; Kislalioglu and Gibson, 1976b). The increase in handling time with prey size is important for two reasons. Firstly, there is not an infinite supply of time for foraging, and time spent handling a prey cannot be spent seeking and pursuing other prey. Secondly, handling time is a factor that determines the profitability of a prey to a predator, a point taken up again on page 44.

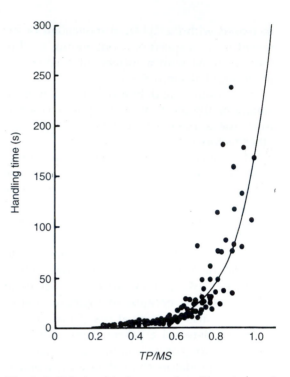

Fig. 3.15 Relationship between handling time and prey thickness relative to mouth diameter (*TP / MS*) for fifteenspine sticklebacks, *Spinachia spinachia*, feeding on mysids. Redrawn from Kislalioglu and Gibson (1976b).

Prey digestibility

Even after ingestion, some prey may not contribute to the diet because they are not successfully digested. Ostracods are microcrustaceans that are protected by a carapace that encloses the whole body. Some ostracods are able to pass through the alimentary canal of a fish and emerge alive (Victor *et al.*, 1979). Herbivorous fishes usually do not have specialized zones of the gut in which cellulose and lignin are digested by a microbial flora. Instead they rely on breaking open plant cell walls mechanically or chemically (Horn, 1989). Herbivorous cyprinids rely on the mechanical action of their pharyngeal teeth (Sillah, 1981), but herbivorous cichlids can achieve highly acidic conditions in their stomach (Moriarty and Moriarty, 1973). However, some intact plant cells are defecated.

Effects of experience of the predator

Fish may have to learn that an object is edible and this may cause a time lag between the appearance of a prey item and its effective exploitation by the fish (Werner *et al.*, 1981). Rainbow trout exposed to novel prey, for example blanched pieces of liver, took several days to accept the food (Ware, 1971). Most trout did not approach it until the fourth day of presentation and took two additional days to complete the development of the feeding response. Even then, the feeding performance of the trout improved further. The distance from which the trout approached the food increased with experience to about twice that at which naive fish reacted. The trout showed a reduction in their reaction distance back to the level of naive fish when the liver was dyed black. This distance improved with experience, suggesting that the fish had to learn the prey characteristics (Ware, 1971). With experience of the prey, fifteenspine sticklebacks improved their success when attacking crustaceans such as *Gammarus* (Croy and Hughes, 1991).

The pattern of movement of a foraging fish is influenced by experience. Threespine sticklebacks hunting for *Tubifex* in a maze consisting of linked hexagonal cells progressively modified their search paths with experience of the maze (Beukema, 1968). This change in behaviour increased the effectiveness with which they searched the maze and encountered prey. An encounter with a *Tubifex* worm also influences a stickleback's behaviour. If the worm is eaten, the stickleback tends to stay in the area of discovery, increasing its intensity of searching and its frequency of approaching the substrate. This change in behaviour leads to area-restricted searching. If the worm is rejected, the fish tends to move away and initially decrease the intensity of its searching, leading to area-avoided searching (Thomas, 1974, 1977). Rainbow trout switched from searching behaviour to undirected swimming when their rate of prey capture fell below about 0.06 captures s^{-1} (Ware, 1972).

The problems of observing fish foraging in the field make it difficult to assess the importance of these experimental effects on diet composition. Fish from the same population sampled at the same time can have significant differences in their diet (Bryan and Larkin, 1972). These may reflect differences in the experiences of the fish as they forage. Rainbow trout can be trained to show a preference for a familiar food, but this bias is weak and may not be an important factor in determining food selection under natural conditions (Bryan, 1973). However, the effects of prior experience may cause difficulties when hatchery-reared fish are stocked into natural waters, because the fish have to change from feeding on artificial, usually pelleted food to feeding on natural prey.

Diet selection in herbivorous fishes

Herbivorous fishes range from species that filter feed on phytoplankton, species that graze on algal mats that encrust substrates such as coral, to browsers that bite or tear at vegetation (Horn, 1989; Choat, 1991; Gerking, 1994). Some freshwater, tropical species take advantage of periods when rivers flood the adjacent floodplain to feed on fruits and flowers falling from trees into the flood waters (Goulding, 1980).

Fish feeding on phytoplankton have two possible methods of selecting their diet. The first is behavioural. The fish may initiate or cease filtering depending on the species composition of the phytoplankton. The second mechanism is mechanical. The filtering system traps a restricted size range of algal cells (Gerking, 1994).

Selection by browsers and grazers may relate to several qualities of plants (Horn, 1989; Gerking, 1994). These include digestibility, the degree of calcification of the algae, the toughness of the plants and the production by plants of secondary metabolites that deter attack by herbivores. On coral reefs in Hawaii, the damselfish, *Stegastes fasciolus*, defends a territory in which it grazes a mat dominated by filamentous algae (Hixon and Brostoff, 1996). The presence of the fish affects the species composition of the mat. Statistically, the fish showed some preference for algal species over others. However, the fish grazed relatively unselectively on the algal species in a well-established mat within a territory. Outside the territories of the damselfish, the algal mat is exposed to grazing by schooling parrotfishes (Scaridae) and surgeonfishes (Acanthuridae). Here, the plant assemblage is reduced to crusts and mats of bluegreen algae because of the removal of all erect algae.

Optimal foraging

Models that seek to predict dietary composition from a knowledge of the sensory, morphological and behavioural characteristics of the forager are called mechanistic models, because the mechanisms that lead to the selection of prey are defined. The reactive field model and apparent size model used to predict the dietary composition of planktivores (pages 39–40) provide examples. A complementary set of functional models of foraging behaviour seek to predict the diet composition by assuming that the action of natural selection on the physiology, morphology and behaviour involved in feeding maximizes the Darwinian fitness of the forager. Because of the difficulties of directly measuring the effect of foraging behaviour on fitness, it is usually assumed that the forager attempts to maximize the rate of food consumption per unit time. Most models assume that the net rate of food consumption is maximized, where net food consumption is measured as the gross energy content of the food less the energy cost of acquiring it. These optimal foraging models do not seek to describe the mechanisms by which the forager achieves this maximization, but only to predict outcomes such as what is eaten and where is searched for prey (Stephens and Krebs, 1986). Hart (1993) and Gerking (1994) review the application of optimal foraging theory to fishes.

Prey selection

If the net or gross energy intake per unit time is to be maximized, then the most profitable prey will be those for which the cost of prey capture is minimized. One measure of this cost is the time taken to handle the prey (h) divided by the weight (or better the energy content) of the prey (r), that is h/r. This measure of cost can be further modified to incorporate the time costs of searching and capture. The relationship between cost, prey size and predator size has been estimated for bluegill sunfish (Werner, 1974) and the fifteenspine stickleback (Kislalioglu and Gibson, 1976b). In both species, the cost first decreases with an increase in prey size because the handling time increases only slowly with an increase in prey size (Fig. 3.15). Then cost increases as the handling time increases sharply with a further increase in prey size until the prey become too large to be handled by the fish. This U-shaped curve means that for a predator of a given size, there is a prey size that minimizes cost (Fig. 3.16). Larger predators have a bigger optimal prey size and a wider range of prey sizes that impose costs close to the minimum (Fig. 3.16). Such estimates of the costs (or their inverse, profitabilities) of the prey can then be used to predict the optimal composition of the diet. For the fifteenspine stickleback, there was a high correlation between the predicted optimal prey size and the mean prey size taken by fish in the wild (Table 3.3). There was also evidence that the range of prey sizes taken increased with an increase in the size of the fish (Kislalioglu and Gibson, 1976b).

Classical optimal foraging theory predicts that from an array of prey items, the optimal forager should always select the most-profitable prey when it is encountered. As the rate of encounter with that prey declines, the next-most-profitable prey should be included, and so on. The diet composition should be expanded to include less-profitable prey

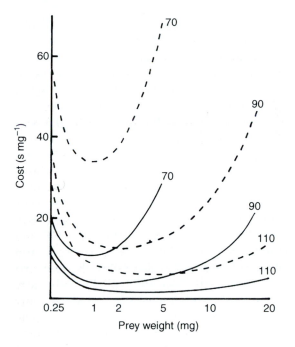

Fig. 3.16 Relationship between time cost of prey capture and prey size for fifteenspine sticklebacks feeding on mysids. Cost curves are estimated for fish 70, 90 and 110 mm in length. Continuous line, first prey taken; broken line, sixth prey taken. Redrawn from Kislalioglu and Gibson (1976b).

Table 3.3 Predicted optimum dimensions for mysid prey of *Spinachia spinachia* compared with mean prey size in the field*

	Fish length (mm)					
	70	80	90	100	110	120
Mouth size (mm)	2.14	2.50	2.86	3.22	3.58	3.94
Optimum prey size (mm)	7.5	9.0	10.5	12.0	13.5	15.0
Mean prey size in the field (mm)	7.4	8.9	10.4	12.0	13.5	15.0

*Source: Kislalioglu and Gibson (1976a).

as the density of the more-profitable prey declines, but the optimal diet is not influenced by changes in the density of less-profitable prey. A test of this prediction using bluegill sunfish feeding on *Daphnia* showed that at low prey densities, *Daphnia* of different sizes were taken in the same proportions as they were encountered. At higher prey densities, the bluegill preyed more selectively on the bigger *Daphnia* available. Although this result is predicted by optimal foraging theory, the bluegill did not completely stop feeding on the smaller *Daphnia* (Werner and Hall, 1974). Brown trout, *Salmo trutta*, feeding in an experimental stream had a diet much closer to the predicted optimal diet than to a random selection of the prey supplied, but it did not reach the predicted composition (Ringler, 1979). The trout improved their feeding performance over a period of days, showing an effect of experience. In contrast, perch, *Perca fluviatilis*, preying on *Daphnia* or *Chaoborus* larvae either separately or together, fed about equally on both types of prey when these were together. The profitabilities of the two prey types when available separately suggested that an optimally foraging perch should select only *Chaoborus* from the mixture (Persson, 1985).

The apparent failure of fish to select a predicted optimal diet may have several causes (Stephens and Krebs, 1986; Hart, 1993). One common problem is that the experimental design violates the basic assumptions of the simple model of optimum prey selection. A failure may indicate that the variable assumed to be maximized has been incorrectly identified: the fish is maximizing something else. Fish may lack the information required to make an optimal choice. Another possible explanation is that there is a lack of suitable genetic variation in the population so an optimal solution cannot evolve through natural selection. One of the great advantages of a good model is that deviations from its predicted outcomes can help to identify important components of the process being modelled.

Optimal foraging theory has been used with conspicuous success by Werner and his co-workers as a guide to organizing empirical evidence and suggesting future research in a study of the interrelationships between sunfishes in North American lakes (Werner and Mittelbach, 1981). This programme has used a powerful combination of theoretical, experimental and field studies in an exemplary way. The results are treated again in Chapters 8 and 9, which discuss interspecific interactions, but an example of the approach applied to a single species, the bluegill sunfish, is relevant here (Mittelbach, 1981a,b; Werner and Mittelbach, 1981). An expression for the net profitability of prey to the sunfish

was derived in terms of net energy intake. An important variable is the rate of encounter with prey of a given size in a given habitat. Laboratory experiments in which bluegill were allowed to hunt for prey in three different types of habitat provided data from which encounter rates were predicted from a knowledge of prey size, prey density and fish size (Fig. 3.17). The abundance and size spectrum of prey available in a small lake were estimated and this information was used to predict the optimal diet for bluegill of different lengths. These predicted optimal diets were then compared with the diets observed in the lake. There was a striking correspondence between the predicted and observed diets, although the correspondence between the observed diet and the size

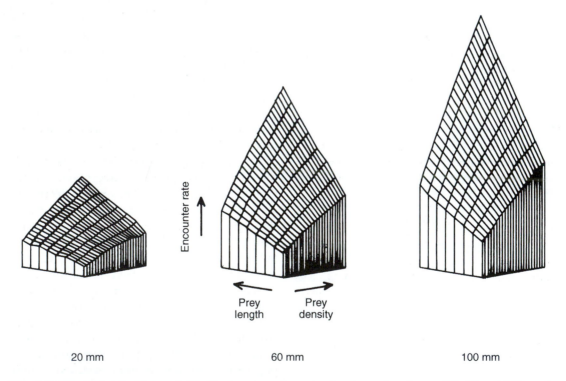

Fig. 3.17 Effect of prey size and density on encounter rates for three size classes (20, 60 and 100 mm) of bluegill sunfish, *Lepomis macrochirus*, foraging on damselfly naiads. Redrawn from Werner and Mittelbach (1981).

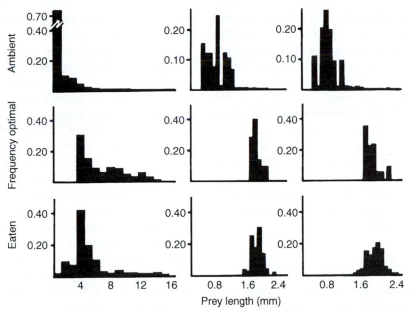

Fig. 3.18 Size-frequency distribution of prey in vegetation and open water in Lawrence Lake, Michigan, compared with predicted optimal and observed diets of bluegill sunfish, *Lepomis macrochirus*. Top row, prey distribution in: (left) vegetation, May; (centre) plankton, July; (right) plankton, August. Middle row, predicted optimal diet of bluegill sunfish in May, July and August respectively. Bottom row, observed diets of 101–150 mm bluegill sunfish in May, July and August respectively. Redrawn from Werner and Mittelbach (1981).

spectrum of prey available in the habitats sampled was weak (Fig. 3.18). Optimal foraging theory had been used successfully to predict the pattern of selective predation by bluegills.

Classical optimal foraging theory assumes that the forager has perfect knowledge of the profitabilities of different prey but does not specify how such information is achieved. The rules which foragers use to approach an optimal diet may depend on a learning process or on relatively simple 'rules of thumb' such as always select the largest prey visible (Stephens and Krebs, 1986). The importance and definition of such rules is now an active field of research.

A second weakness of the classical theory is that it does not predict the effect on diet of changes in the feeding motivation of the fish.

Changes in motivation will reflect the physiological state of the fish such as the degree of fullness of the stomach or the level of fat reserves (Hart, 1993). Fifteenspine stickleback approaching satiation showed an increase in prey handling time. This caused a shift in the cost curve because prey of all sizes became less profitable. Experimental observations suggested that the range of prey sizes taken decreased with satiation as the fish became more selective (Kislalioglu and Gibson, 1976b). These changes with satiation may reflect a change in the benefit that accrues to the fish by continuing to forage compared with the benefits of doing something else, for example being more vigilant (Heller and Milinski, 1979).

If each prey item is sufficiently large to cause a significant reduction in the feeding motivation of the fish, the simple optimal foraging theory may not be adequate. The model did not accurately predict prey selection by threespine sticklebacks feeding on different sizes of the isopod, *Asellus*. The pattern of selection was influenced by the degree of stomach fullness. A model using stochastic dynamic programming was developed to predict the prey selection by the stickleback (Hart and Gill, 1993). The important characteristic of this class of model is that it is dynamic, allowing for changes in the state of the animal as a consequence of the behaviour performed.

Choice of foraging location

Optimal foraging theory also addresses the problem of where to feed if the food is distributed in patches. Most theoretical studies of this problem have considered an individual forager (Charnov, 1976). A solution to the problem, the marginal value theorem, is that the animal should leave a patch when its rate of capture of prey immediately before leaving equals the average rate of capture for the food patches in the habitat (Charnov, 1976; Hart, 1993). The solution assumes the forager knows the distribution of food patches and the average rate of capture. Again the theory predicts the optimal behaviour of the animal, but does not specify how the animal achieves the optimal solution. The relevance to the theory in its simple form for studies of fish foraging is open to question. In many species, fish forage in shoals and the behaviour of an individual in a shoal is influenced by that of the other fish (page 94) (Pitcher and Parrish, 1993). Even when fish are foraging for food that has a patchy distribution, such as zooplankton, the position of the patches and the density of prey in them vary, making impossible the instantaneous assessments of patch quality required by the marginal value theorem. The theory is more relevant when the position of patches is predictable. The aggregation of mullet around an outflow of raw sewage provides a striking example of the ability of fish to detect rich patches of food.

However, there is evidence both from experimental and from field studies that fishes do respond to differences in their rate of return as they forage in different regions of their environment. The ideal free model developed by Fretwell and Lucas (1970) predicts the distribution of animals between habitats that differ in resource levels when individuals are competing for those resources (Chapter 5). The model predicts that in an environment in which food is distributed in patches of different densities, fish should distribute themselves so that no fish is able to increase its feeding rate by switching to another patch (Fretwell and Lucas, 1970; Milinski, 1986b; Sutherland, 1996). In its simplest form, the model predicts that all individuals should achieve the same rate of return. The model assumes that the fish have perfect information about the profitabilities of each patch and that competition with conspecifics places no constraints on the movement of the fish.

The densities of an armoured catfish, *Ancistrus spinosus*, in pools in a small Panamanian stream were inversely correlated with the density of the forest canopy over the pools (Power, 1984). Shading by trees restricts the growth of the algae, which the catfish scrape from rocks. The growth rates and survivorship of pre-reproductive catfish did not vary with the degree of shading of the pools, suggesting that the fish distributed themselves in relation to the growth rates of the algae so that individuals had similar rates of food intake.

Threespine sticklebacks presented with two patches of *Daphnia* that differed in profitability distributed themselves between the patches in a ratio that closely approached that predicted by the ideal free distribution, although it took the fish a few minutes to achieve the ratio (Milinski, 1979). A detailed analysis of feeding rates showed that the fish did not obtain equal rewards. Some were competitively superior to others, so the distribution was not following the simplest form of the ideal free distribution (Milinski, 1984b). When two sticklebacks were presented simultaneously with one large and one small *Daphnia*, over a series of presentations a good competitor took more of the large daphnids (Milinski, 1982). Mechanical constraints or simple optimal foraging rules do not solely determine prey selection by sticklebacks. The presence of a conspecific also exerts an effect, at least on the poorer competitor.

In stream fishes, the effects of current velocity may influence the distribution of fishes between food patches (Tyler and Gilliam, 1995). As current speed increases, the energy required to maintain a position in a stream increases (Chapter 4). A faster current may bring more drifting food past the fish per unit time, but the prey may be more difficult to capture at high current velocities. A study in an experimental stream with blacknose dace, *Rhinichthys atratulus*, found that the distribution of fish between food patches was predicted better by a model modified to include the costs imposed by current speed than by the simple ideal free distribution model (Tyler and Gilliam, 1995).

Theory based on the principles of optimal foraging predicted with some accuracy the movement of bluegill sunfish between habitats that contain different types of prey (Werner and Mittelbach, 1981; Werner *et al.*, 1983a). The changing profitabilities of the habitats to the bluegills were predicted from observations of seasonal changes in the densities and sizes of prey. As the predicted profitabilities of the habitats changed, the observed diet of the bluegills changed, indicating they moved into the more profitable habitat (Fig. 3.19). The bluegills did not move into the more profitable habitat if piscivores were present. Chapter 8 discusses this and other examples of the effect of the presence of piscivores on the distribution of their prey. The success of these bluegill

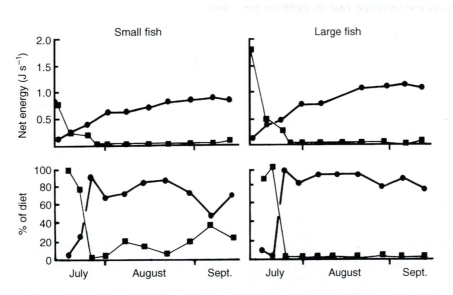

Fig. 3.19 Seasonal pattern of predicted habitat profitabilities (above) and observed habitat use by bluegill sunfish, *Lepomis macrochirus*, in experimental ponds (below). Habitat use is expressed as percentage of diet obtained from habitat. Small size class (left), 34–50 mm; large size class (right), 71–97 mm; ●, sediments; ■, open water. Redrawn from Werner and Mittelbach (1981).

studies depended on detailed quantitative information on individual foraging behaviour, gathered in the context of a clear theoretical framework which indicates what sort of data are likely to be valuable.

3.5 TEMPORAL CHANGES IN DIET COMPOSITION

The diet of fish usually changes as they grow because of the morphological changes that accompany growth. Superimposed on these ontogenetic changes in diet, there may be changes during the diel cycle, or seasonal changes.

Ontogenetic changes

As fish grow, and in most species they continue to grow throughout life (Chapter 6), they show changes in their diet and their susceptibility to predation. These ontogenetic changes are probably central to an understanding of the ecology of fishes (Chapter 10) (Werner and Gilliam, 1984; Werner, 1986).

The first year of life is a time of rapid growth for fish (Chapter 6) and as they grow, they can handle bigger prey. This is often a period when diet changes rapidly. The diets of larval and juvenile cyprinids in Lake Constance, Switzerland, clearly showed these changes (Fig. 3.20). Thereafter, in some species ontogenetic changes in diet occur gradually, whereas in others, changes occur abruptly. The diets of centrarchid species in Lake Opinicon, Ontario, illustrated this (Keast, 1978, 1980, 1985).

In Lake Opinicon, the first prey taken by the sunfish larvae were copepod nauplii and small cyclopoid copepods. As the young fish grew in length from 4.5 mm to 20 mm, the diversity of the diet and the size of prey taken increased, with the larval fish taking larger cladocerans. Largemouth bass larvae as small as 7 mm ate chironomid larvae, a feat made possible by the relatively large mouth of this species. As the sunfish grew

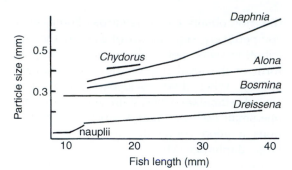

Fig. 3.20 Size of food particles in guts of young cyprinids of Lake Constance. Redrawn from Hartmann (1983).

over the first summer of life, the diet continued to consist mainly of cladocerans, but the larger juveniles also took chironomid larvae. By autumn, bluegill sunfish had an essentially adult diet. Over the next years of life, the bluegills' diet shifted gradually from being dominated by cladocerans to one that included trichopteran larvae, small anisopteran nymphs and even some vegetable material. In contrast, after their first year of life the rock bass, *Ambloplites rupestris*, exhibited three distinct age-specific diets. In their second year, the rock bass consumed mainly chironomid larvae and other small prey. By the fourth year, anisopteran nymphs and other arthropods of a similar size dominated the diet. The diet of older fish consisted of decapods and small fish.

Most of these ontogenetic changes probably reflect morphological and maturational changes, particularly the increase in mouth size and the improvements in locomotory (Chapters 2, 4) and sensory abilities (Chapter 2). With these improvements, the size of the reactive field increases and the size of the prey that can be ingested increases (Gerking, 1994).

Other factors can also be important, including age-specific changes in the use of habitats, as observed in surf perch, *Embiotoca*, on the coast of California (Schmitt and Holbrook, 1984). In Lake Opinicon, several species including rock bass and sunfish occupy inshore open water as larvae, but as juveniles move into weed beds. Such movements may relate to changes in the level of danger from invertebrate predators and vertebrate piscivores (Chapters 8, 9).

Diel changes in diet composition

A fish of a given size may still show detectable changes in diet composition over time, including changes over the diel cycle. For example, in Lake Opinicon the pumpkinseed sunfish is primarily a diurnal feeder, but shows some feeding activity at night. A study of its stomach contents over a 24 h period in summer (Keast and Welsh, 1968) illustrated how diet can change over a day: in the afternoon the fish fed on chironomid larvae, anisopteran nymphs and bivalves; in the early morning they ate chironomid pupae, amphipods and trichopteran larvae; and in the late morning they were eating chironomid larvae, bivalves, zygopteran nymphs and isopods. The changes probably reflect changes in the activity, and hence the vulnerability, of the prey. They also illustrate that the diet of the pumpkinseed sunfish can be catholic, although in other lakes it is primarily a molluscivore (Mittelbach, 1984).

Seasonal changes in diet composition

Seasonal changes in food availability may be caused by changes in the habitats available for foraging, changes resulting from the life-history patterns of food organisms and changes caused by the feeding activities of the fish themselves. An example of the first is provided by the seasonal pattern of inundation of the forest floor in the Amazonian river systems, which provides access to feeding areas for fish moving out of the main river channels (Goulding, 1980). Many insect species with aquatic larval or nymphal stages have an annual life cycle. The juvenile stages

metamorphose into adults and leave the water, depriving predators of that food supply until the next generation becomes vulnerable. Intensive predation by fish may cause a reduction in prey density, making them less vulnerable to further predation.

In Lake Opinicon, two patterns of seasonal change in diet were identified (Keast, 1978). Species that had a catholic diet showed changes in its taxonomic composition. Species that had a more specialized diet showed changes in the proportions rather than the composition. The diet of the pumpkinseed sunfish included a significant proportion of isopods in May when isopods were most frequent in the benthos. Amphipods were well represented in the diet from August onwards when they were abundant. In autumn, isopods were again important in the diet although this was not correlated with their high numerical abundance, but rather with a predominance of large-bodied individuals in the isopod populations. The seasonal representation of some prey in fish diets changes little. Several of the Lake Opinicon fishes, for example the sunfish, ate chironomid larvae throughout the main feeding period between May and November.

Most studies of seasonal (and diel) changes in diet composition have been purely descriptive. Good, reliable descriptions of diet often require large sampling programmes and many hours of analysis of diet composition. The next difficult challenge for fish ecologists is to develop mechanistic and functional models of foraging that predict such changes and provide an explanation for them.

3.6 FACTORS THAT DETERMINE THE RATE OF FOOD CONSUMPTION

Effects of prey, conspecifics and predators

Density of prey

The relationship between the number of prey attacked per predator and the density of the

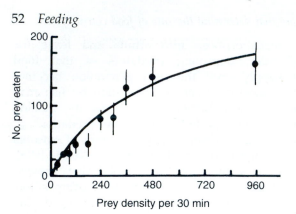

Fig. 3.21 Mean functional response curve of ten brown trout, *Salmo trutta*, feeding on drifting prey (brine shrimp) showing relationship between number of prey eaten (mean and SE) and prey density. Line fitted using Holling's (1959) disc equation. Redrawn from Ringler and Brodowski (1983).

prey is called the functional response curve. As the density of prey increases, the number attacked increases (Fig. 3.21), but the functional response curve is asymptotic (Chapter 8) (Ringler and Brodowski, 1983). Two factors may produce the asymptote. Firstly, the fish may become satiated. Secondly, the time available for foraging is restricted. The predatory fish takes up some of that available time capturing and handling each prey. At a sufficiently high prey density, all the time available for foraging will be taken up (Holling, 1959).

Although the number attacked increases with density, prey at high densities are not always preferentially attacked. A threespine stickleback confronted with a swarm of *Daphnia* attacks the densest part of the swarm when it is hungry. As it becomes satiated, it switches to attacking less dense regions of the swarm, although its rate of reward is lower (Milinski, 1977a,b; Heller and Milinski, 1979). A plausible explanation for the observation is that a high prey density imposes a high confusion cost. The attacker has more difficulty in focusing on

any one prey because there are numerous similar prey moving irregularly in its field of view and so has to attend closely to the task (Ohguchi, 1981). But this is at the cost of paying less attention to other events in the environment such as the approach of a predator (Heller and Milinski, 1979).

Effect of conspecifics

The presence of conspecifics can have both beneficial and adverse effects on food consumption. Many species forage in shoals. Experimental studies suggested that when food is distributed in patches, a fish can increase its rate of feeding by joining a shoal (Pitcher and Parrish, 1993). The median time taken by individual European minnows, and goldfish, *Carassius auratus*, to find a randomly positioned food item decreased with shoal size (Fig. 3.22) (Pitcher *et al.*, 1982). Fish foraging in a shoal recognize when another member of the shoal finds food. Threespine sticklebacks that see a conspecific assuming the head-down posture characteristic of a fish taking food off the bottom rush into the same area (Wootton, 1976). An individual minnow in a shoal devoted more time to foraging than to predator-avoidance behaviour such as hiding in weed beds than when it was on its own (Fig. 3.23) (Magurran and Pitcher, 1983). As the number of fish in a shoal increases, the potential for competition for the available food also increases.

This effect of competition is seen most clearly when dominance hierarchies develop in groups, such that dominant individuals have better access to food than subordinate fish. In groups of rainbow trout and Arctic charr, *Salvelinus alpinus*, the formation of hierarchies increased individual variation in the rate of food consumption (Chapter 6) (McCarthy *et al.*, 1992; Jobling and Baardvik, 1994). In charr, the provision of particular water currents, which probably reduced the

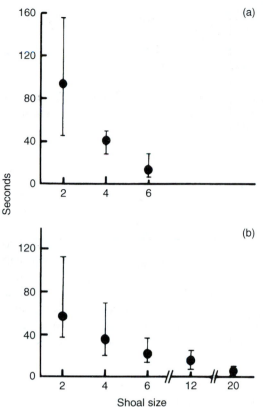

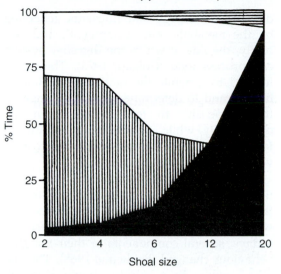

Fig. 3.22 Relationship between shoal size and median time spent before a randomly located food item was found by an individual fish: (a) goldfish, *Carassius auratus*; (b) minnows, *Phoxinus phoxinus*. Medians and interquartiles for 16 replicates shown. Redrawn from Pitcher (1986).

Fig. 3.23 Effect of shoal size in minnow, *Phoxinus phoxinus*, on time spent foraging (black tone), in weeds (vertical shading), swimming in mid-water (horizontal shading) and swimming at bottom (open). Redrawn from Pitcher (1986). Note horizontal scale.

intensity of social interactions, ameliorated the effect on individual variations in food consumption.

Effect of predators

The presence of piscivores can reduce the food consumption of a fish because the fish is less likely to forage. In the presence of a potential predator, juvenile coho salmon reduced the distance they swam to catch a fly (Dill, 1983). A similar reduction in attack distance occurred in juvenile Atlantic salmon

in the presence of a piscivore (Metcalfe *et al.*, 1987) (Chapter 8).

The increased attentiveness required to overcome any confusion effect to exploit the high prey density detracts from the ability to be vigilant to other things such as the approach of a piscivorous fish or bird (Milinski, 1984a, 1986a). Food-deprived threespine sticklebacks exposed to a model of an avian piscivore switched from directing most attacks at the dense part of a swarm of *Daphnia* to attacking a less dense region (Milinski and Heller, 1978).

Hunger and appetite

If food is abundant and there are no other confounding factors such as the presence of predators, the rate of food consumption will be determined by the feeding motivation of the fish. Two systems control the rate of consumption: systemic demand, that is the demand for energy and nutrients generated by the metabolic rate (Chapter 4), and secondly, the rate at which the digestive system can process food (Colgan, 1973). These two interact to generate the motivational state of hunger and to determine appetite. Hunger is the propensity to feed when given the opportunity, while appetite is the quantity of food consumed before the fish ceases to feed voluntarily.

The distinction between hunger and appetite is illustrated by experiments with the threespine stickleback. As the period for which sticklebacks were starved increased to 4 days, several components of their foraging behaviour changed (Beukema, 1968). The distance swum, the number of bouts of search swimming, the proportion of discovered prey that were grasped and the proportion of grasped prey that were eaten all increased. Fish also directed more feeding attempts at inedible objects. These changes indicated increasing hunger. The amount of food eaten during the first 8 h after food was supplied in excess did not increase as the preceding period of starvation increased from 0 to 4 days. However, the proportion of the total that was eaten in the first hour did increase. In brown trout, *Salmo trutta*, appetite (measured as the amount of food consumed in 15 minutes) remained close to zero as the number of hours since the last meal increased until a threshold period of deprivation was reached (Elliott, 1975a). Beyond that threshold, the amount eaten increased rapidly to an asymptote (Fig. 3.24).

Rate of gastric evacuation

The amount of food required to satiate a fish is related to the state of distension of the stomach (or foregut in fish that lack stomachs). The rate of food consumption is dependent on the rate at which the stomach contents are evacuated. Studies on the brown trout (Elliott, 1972) and fingerling sockeye salmon, *Oncorhynchus nerka* (Brett and Higgs,

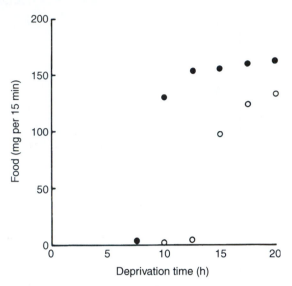

Fig. 3.24 Effect of deprivation time and temperature on appetite (food consumed per 15 min) of brown trout, *Salmo trutta*, at two temperatures: ○, 6.8 °C; ●, 15.1 °C. Redrawn from Elliott (1975b).

1970) suggested that the rate of evacuation could be described with a simple exponential model:

$$dS/dt = -kS \qquad (3.3)$$

where dS/dt is the instantaneous rate of evacuation, S is the weight of stomach contents and k is the rate constant. This model states that the rate of evacuation is proportional to the weight of remaining contents, which means that the absolute rate of evacuation is highest immediately after the meal (Fig. 3.25). An exponential model describes gastric evacuation in many other species, although in some there is a time lag between the ingestion of the meal and the exponential phase of evacuation. For fish that take large meals, at long intervals, the rate of evacuation may be linear rather than exponential. Other models of evacuation may be appropriate in some circumstances (Jobling, 1986; Bromley, 1994).

The rate of evacuation increases with tem-

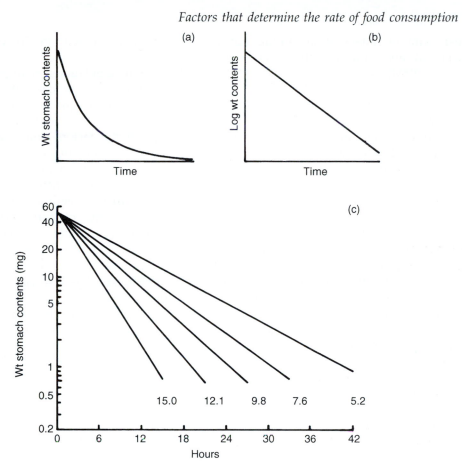

Fig. 3.25 Exponential gastric evaculation in brown trout, *Salmo trutta*. (a) Arithmetic plot of decline in stomach contents; (b) semi-log plot of decline; (c) effect of temperature over range 5.2–15.0 °C on rate of evacuation. Redrawn from Elliott (1972).

perature (Fig. 3.25). In brown trout, the time taken to evacuate 90% of a meal of amphipods (*Gammarus*) decreased from 24 h at 5.2 °C to 8 h at 15.0 °C (Elliott, 1972). The quality of the food also influences the rate of evacuation: food with a low energy content is evacuated faster than food with a high energy content (Jobling, 1980). However, if a meal consists of a mixture of different prey, the rate of evacuation of each type of prey is not independent of the other items (Persson, 1984).

In brown trout and sockeye salmon, appetite increased rapidly when about 75–90% of the previous meal had been evacuated (Brett, 1971a; Elliott, 1975b). These species eat medium-sized discrete food items such as invertebrates or small fish. In contrast, some filter-feeding fishes such as the menhaden may never become satiated, but ingest food continuously. Similarly detritivores such as the striped mullet, *Mugil cephalus*, which are exploiting a resource with a low energy content, may feed almost continuously, relying on a rapid turnover of the gut contents (Pandian and Vivekanandan, 1985). Herbivorous, young grass carp, *Ctenophar-yngodon idella*, supplied with duckweed, fed almost without interruption throughout a 24 h period, even during a 10 h dark period (Cui *et al.*, 1993).

Effect of temperature and other abiotic factors

Temperature affects the maximum rate of consumption through its effects on the rate of gastric evacuation (Fig. 3.25) and its effects on systemic demand (Chapter 4). At low temperatures fish may cease to feed, but as the temperature increases, the rate of consumption also increases up to a maximum. Any further increase in temperature is marked by a rapid decrease in consumption (Fig. 3.26) (Elliott, 1975a,b, 1981). The optimum temperature for feeding is the temperature at which the highest rate of consumption occurs. In brown trout, the increase in consumption with temperature reflected an increase both in the amount eaten in one meal and in the number of meals eaten in a day (Elliott, 1975a,b). The decline in consumption at high temperatures was a systemic effect because the rate of gastric

evacuation continued to increase (Elliott, 1972).

In natural populations, changes in the rate of consumption can be correlated with water temperature, although the causal effects of temperature may be confounded with those of other abiotic factors and changes in the availability of food. In Lake Opinicon, bluegills do not eat in winter, but commence feeding in spring when the temperature reaches 8–10 °C. Feeding ceases in late autumn. The rate of consumption by the winter flounder in a New England salt pond also increased with temperature, ranging from a mean of 1.27% of dry body weight in April (6.5 °C) to 3.31% in September (22 °C) (Worobec, 1984).

Abiotic factors other than temperature also influence the rate of consumption. In the threespine stickleback, there was a significant reduction in consumption as the pH of the water declined from 5.5 to 4.5 (Faris, 1986). Light may have a direct effect: as the ambient light levels fall, the fish is unable to perceive food and feeding ceases (page 40). Photoperiod may also influence the endocrine system and so indirectly cause changes in hunger or appetite (Peter, 1979; Matty and Lone, 1985).

Effect of body weight

The relationship between the maximum weight of food consumed over a period of time, typically 24 h, and the weight of the fish is described by:

$$C = aW^b \qquad (3.4)$$

or in the linear form:

$$\log C = \log a + b \log W \qquad (3.5)$$

where C is weight of food consumed, W is weight of fish and a and b are parameters that can be estimated by regression analysis. For the brown trout, Elliott (1975a,b) showed that this relationship describes both the weight of food consumed during a meal and

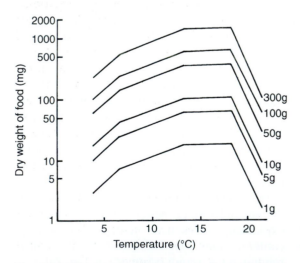

Fig. 3.26 Effect of temperature on dry weight of food eaten in a meal by brown trout, *Salmo trutta*, of live weights over range 1–300 g. Redrawn from Elliott (1975a).

Table 3.4 Values* for weight exponent (b) in relationship between food consumption (C) and body weight (W): $C = aW^b$

Species	Temp. (°C)	Food	b
Ctenopharyngdon idella	23	Lettuce	0.80
		Tubifex + lettuce	0.49
Esox spp.	5–30	Minnows	0.82
Gasterosteus aculeatus	3–19	*Enchytraeus* spp.	0.93
Phoxinus phoxinus	7–19	*Tubifex* spp.	0.56
	5–15	*Enchytraeus* spp.	0.81
Oncorhynchus mykiss	8–16	Fish	0.84
	12	Pelleted food	1.1
	5–26	Formulated diet	0.70–0.77
Salmo trutta	4–22	*Gammarus*	0.73–0.77
Solea solea	10–26	Mussel meat	0.40–0.94

*Source: Cui (1987).

the total weight consumed in a day. The value of the exponent, b, is usually less than one (Table 3.4). As fish grow, the weight of food they consume relative to their body weight decreases, although the absolute weight consumed increases.

The effects that temperature and body weight have on voluntary food consumption mean that if fish are being fed artificially, as in a hatchery or fish farm, the ration provided has to be adjusted as the fish grow and as the water temperature changes.

Effect of physiological state

Major changes in consumption can occur as the physiological state changes. Juvenile Atlantic salmon provide striking examples of changes in appetite in relation to physiological state which have consequences for the pattern of growth and timing of migration from fresh water to the sea (page 130). In overwintering juvenile salmon, appetite increased when lipid reserves fell, but then declined once the reserves had been replenished (Metcalfe and Thorpe, 1992). In fish that have been deprived of food for several days, the rate of consumption when the

deprived fish are supplied with excess of food is significantly higher than the rate of fish continually exposed to excess food (Jobling, 1994). European minnows showed this hyperphagic response after a deprivation period of 16 days (Russell and Wootton, 1992).

In some species, the rate of feeding decreases or ceases as the fish become reproductively active. Well-known examples are the anadromous salmonids, which cease to feed during their upstream spawning migration – although they continue to strike at prey, making them vulnerable to capture by anglers. (This suggests the paradox that the salmon are hungry but have no appetite!) During the upstream migration, the gut undergoes degenerative changes which prevent any normal food processing (Brett, 1983).

Some marine fishes also cease to feed during their spawning migration or on the spawning grounds, for example the herring, *Clupea harengus* (Iles, 1974, 1984). Iles' analysis of the feeding and reproductive cycle of the North Sea herring is important because it illustrated the potential that fish have to control their rate of food consumption in the face of changes in the external environment. The period of intensive feeding begins when temperatures are still low, but ceases when temperatures are relatively high and while food is still plentiful. Iles (1984) argued that the lack of a close relationship between the feeding activity of the herring and either temperature or food availability indicated that to a surprising extent the herring is independent of its thermal environment.

Methods of estimating food consumption in natural populations

Estimates of the rate of consumption by fish in natural populations are essential for three reasons: to assess the demands that fish make on their food resources; to assess the extent to which survival, growth and reproduction are limited by food availability; and to estimate the energy and nutrients available for allocation between maintenance, growth and reproduction. Although in clear waters, divers can record rates of biting by individual fish (Polunin, 1988), rates of consumption can rarely be estimated directly. Indirect methods must be adopted. There are several methods. The first method relies on estimating the rate of passage of food through the gut and so gives an instantaneous estimate of consumption (Elliott and Persson, 1978; Eggers, 1979). The second method integrates consumption over a relatively long period and estimates it by calculating the rate that would give the growth observed over that period (Wootton, 1986). A third method uses the flux in a radioactive isotope with a relatively long half-life, such as caesium 137, to estimate rates of food consumption (Forseth *et al.*, 1992).

A popular method for estimating food consumption assumes that if the consumption rate over a fixed period of time is constant, the rate at which the weight of the stomach contents changes can be written as:

$$dS/dt = C - kS \qquad (3.6)$$

where S denotes the weight of contents, dS/dt the rate of change of stomach contents, C = rate of food consumption and k is a rate constant assuming an exponential pattern of gastric evacuation (Elliott and Persson, 1978; Eggers, 1979). By integration, the following expression is derived:

$$C_t = [(S_t - S_0 e^{-kt})kt] / (1 - e^{-kt}) \qquad (3.7)$$

where C is the weight consumed in the time interval 0 to t, S_0 and S_t are the weights of stomach contents at times 0 and t respectively, and k is the evacuation rate (Elliott and Persson, 1978). Worobec (1984) gave formulae for putting confidence intervals on the estimate of C. This method can be used to estimate the daily consumption by taking samples of fish at intervals of 2–3 h throughout a 24 h period. The daily consumption, FD, is given as:

$$FD = \Sigma \, C_t. \qquad (3.8)$$

When the weight of stomach contents at the end of the 24 h period is the same as at the start, a simple estimate of FD can be obtained as:

$$FD = 24 S_m k \qquad (3.9)$$

where S_m is the mean weight of stomach contents and k is the rate of evacuation with units h^{-1} (Eggers, 1979). Field observations (Allen and Wootton, 1984) and a comparative study (Boisclair and Leggett, 1988) suggested that this simpler method can give estimates of daily consumption similar to those obtained with the Elliott and Persson method. From their study of pumpkinseed sunfish, Boisclair and Marchand (1993) suggested that the use of Eggers' model applied to the complete gut contents gave reasonably stable and precise estimates. The simpler method has the advantage of requiring a reduced sampling effort (Bromley, 1994).

Fish must be killed or stomach-pumped to obtain samples of stomach contents. A method of estimating food consumption that avoids these traumatic procedures depends on the quantitative collection of the faeces produced by the fish over a known time period (Allen and Wootton, 1983). Laboratory experiments provide information on the relationship between the weight of faeces produced and the weight of food consumed, which is then used to predict consumption from faecal production. For this method, the food used in the laboratory study must resemble the food the fish are consuming naturally.

Although not strictly applicable to natural populations, a method useful for estimating food consumption by individual fish living in groups in tanks or cages relies on the presence of material opaque to X-rays in the food. This material, for example tiny glass balls, is added to the food in known quantities. After feeding, fish are captured, then X-rayed. The number of X-ray-opaque balls in the gut of the fish is counted on the resulting radiograph. The use of this technique has allowed the detection of individual variation in rates of consumption in groups of fish (McCarthy *et al.*, 1992).

A problem with methods that estimate the rate of consumption over a period of about a day is that there may be considerable day-to-day variation in the rate. In field samples, fish with empty guts are regularly recorded (Gerking, 1994). Groups of largemouth bass kept at a constant temperature with food constantly present showed highly significant changes in their daily consumption rates (Smagula and Adelman, 1982). Such variations may be more typical of piscivores, eating large prey. Fish eating smaller prey may show less irregular patterns of daily consumption (Boisclair and Leggett, 1989a). Problems caused by day-to-day variations in consumption may be avoided if food consumption is estimated from growth over relatively long periods.

Methods that estimate consumption from growth fall into two categories. In the first, empirical relationships between growth and food consumption determined in laboratory studies are used to predict food consumption from growth rates observed in a natural population. In the second category are those methods that assume that a balanced budget, usually for energy or nitrogen, can be calculated for individual fish. The input from the food must equal losses through defecation, excretion and metabolism plus the gain in the form of an increase in the energy (or nitrogen) content of the fish (Chapter 4). This gain is measured as growth (Chapter 6).

Winberg (1956) developed a simple formulation of an energy budget, which suggested that food consumption could be estimated as:

$$0.8C = P + R \qquad (3.10)$$

where C is food consumed over the time interval, P is the growth and R is total metabolism over the interval, all measured in energy units (Chapter 4). Frequently R is estimated as twice the metabolic rate of a resting fish measured under laboratory conditions (Chapter 4). Modifications and refinements of Winberg's original equation have been developed (Majkowski and Waiwood, 1981), but the original formulation is still used. Bioenergetics models developed to predict growth, described in Chapter 6, can also be adapted to predict rates of consumption when growth rates are known.

Gerking (1962, 1972) estimated food consumption in a population of bluegill sunfish from a nitrogen budget by combining experimental studies on the relationship between nitrogen retention (growth) and consumption with observations on the growth rate in the natural population.

Each method makes assumptions that may not be valid for the population under study. Most methods also require that information obtained under laboratory conditions be extrapolated to a natural population. How reliable are estimates provided by these methods? Studies comparing estimates of natural rates of food consumption obtained by more than one method give an ambiguous answer.

Estimates of food consumption by perch in

Windermere comparing results from Winberg's energy budget equation with estimates using the formula, $FD = 24S_m k$, were in reasonable agreement (Craig, 1978). Both methods described a similar pattern of seasonal change in consumption. In contrast, Minton and McLean (1982) compared estimates of consumption based on the Winberg method and from an analysis of stomach contents for a population of the piscivorous sauger, *Stizostedion canadense*, in a reservoir in Tennessee. For the period March to October, the estimates agreed well, but over the winter months, the rate of consumption estimated from the stomach contents was much higher than that estimated from growth. A study of pike in Lac Ste Anne, Manitoba, also found poor agreement between estimates based on stomach contents and on a detailed energy budget: only four of 18 estimates were within 20% of each other (Diana, 1983a). A study estimated the rate of food consumption by a population of threespine sticklebacks in a small Welsh lake by using various methods – those of Elliott and Persson (1978), the rate of faecal production, growth rate and by calculating a balanced energy budget (Allen and Wootton, 1984; Wootton, 1984a, 1986). With one exception, the methods gave estimates for the annual consumption per fish of between 2.3 and 4.2 g wet weight. Although this similarity suggests that independent methods can give comparable results, it is partly coincidental because the methods gave different pictures of seasonal changes (Fig. 3.27) (Wootton. 1986). Nevertheless, the study suggested that as methods are refined, reasonably accurate estimates of consumption can be achieved, although the use of more than one method is advisable.

3.7 FLEXIBILITY IN THE FEEDING ECOLOGY OF FISHES

Any attempts at generalizations on the feeding ecology of fishes confront the pro-

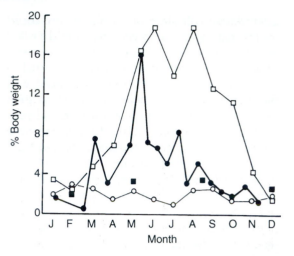

Fig. 3.27 Estimates of seasonal changes in daily rate of food consumption (% body fresh weight) by an average-size threespine stickleback, *Gasterosteus aculeatus*, in Llyn Frongoch, Mid-Wales. Estimates from: □, rate of faecal production; ●, growth rate; ■, Elliott and Persson's (1978) method; ○, weight of stomach contents (assuming two meals consumed per day). Redrawn from Wootton (1986).

blems posed by the trophic flexibility that many species show. The ontogenetic, seasonal and diel changes in diet already described provide examples of this flexibility. Further exampes are provided by comparisons between individuals. Fish from the same population sampled at approximately the same time and place can have significantly different diets (Bryan and Larkin, 1972; Ringler, 1983). Some of these differences may reflect individual variation in physiological and morphological characteristics (Ringler, 1983).

Even for the bluegill sunfish, a species commonly described as a generalist feeder, there is some evidence of specialization by individual fish. In a pond experiment, the bluegills initially foraged in vegetation, but when this habitat became depleted of prey they switched to more profitable habitats. Some individuals specialized as foragers in open water and others as foragers on benthic invertebrates living in sediments (Werner *et al.*, 1981). A comparison of the feeding behaviour and morphology of bluegills from a small lake in Michigan identified individual fish that tended to be specialized for foraging in either a littoral or an open-water habitat (Ehlinger and Wilson, 1988). An experimental study showed that the bluegills selected foraging habitats and allocated searching effort in relation to differences in return rates between habitats (Ehlinger, 1990). The bluegill modified their foraging according to the nature of the habitat. In open water, when feeding on *Daphnia*, the bluegill hovered for short periods of time and moved quickly between hovers. In vegetation, where prey were more difficult to detect, the hovers lasted longer and the fish manoeuvred precisely between hovers. However, individuals varied in their ability to adjust their behaviour, and these differences were correlated with differences in the length of the pectoral fins, a correlation between behaviour and morphology. Thus, even a generalist population was made up of partial specialists (Chapter 9).

Another source of variation is the behavioural flexibility of individual fish that move to habitats of greater profitability or change their preferences (Werner *et al.*, 1981). Individual brown trout fed meal-worm larvae and tent caterpillars in different proportions showed changes in preference both within a 30 min period and over a period of days (Ringler, 1985). Sometimes individual preferences result in fish with bizarre diets. A brown trout caught in an upland Welsh lake had about 150 000 ostracods in its gut (Whatley, 1983), although these microcrustaceans normally form only a minor component of the diet of trout in that lake (Al-Shamma, 1986).

Fish may switch from one type of prey to another as the relative abundance of the prey types changes (Murdoch and Oaten, 1975). Changes in diet are well known in generalist feeders such as sunfish, perch and trout (Keast, 1978), but they also occur in species that seem to be specialized for exploiting particular foods. In Lake Victoria, cichlids, which are typically zooplanktivores, insectivores or detritivores, switched to feeding on a diatom, *Melosira*, when it became superabundant (Witte, 1984).

Ontogenetic and seasonal changes in diet frequently involve a change in its taxonomic composition. A rock bass shifts from feeding on zooplankton as a larva and juvenile, to feeding on the larvae and nymphs of invertebrates and finally ends up as a piscivore. Some species that are herbivorous as adults include animal food in their diet when they are young. The proportion of plant and animal food in the diet can change seasonally. *Cichlasoma panamense* is a stream-living Central American cichlid. In a population living in a small Panamanian stream, the proportion of algae in the diet greatly increased during the dry season (Townshend, 1984). Such changes make it difficult to categorize many species of fish in terms of the well-established ecological concept of trophic levels, i.e. herbivore, primary carnivore, secondary carnivore or detritivore (cf. Table 3.1). An alternative method of analysing the trophic relationships of many assemblages of fishes may be a description of the rate of exploitation of food items of different sizes by fishes of different sizes (Parsons and Le Brasseur, 1970; Werner and Gilliam, 1984) (Chapter 12).

Within a species, differences in the diet composition at different stages in the life history or between populations can be reflected in morphological differences (Kapoor *et al.*, 1975; Weatherley and Gill, 1987). The striped mullet feeds on microalgae and decaying plant detritus; its gut often contains quantities of inorganic sediment. A comparison of two populations showed that the increase in gut length as fish length increased was significantly greater in the population with a diet including a high proportion of plant detritus than in the population having a diet that included a high proportion of benthic and epibenthic diatoms (Odum, 1970). Sibly (1981) used a cost–benefit model to predict that within a population, relative gut length would change with a change in the quality of the diet. A possible example of this is shown by *C. panamense*. The change to herbivory in the dry season was associated with an increase in relative gut length in the population (Townshend, 1984).

Morphological differences between individuals can reflect phenotypic plasticity, genetic differences or a combination of both (Chapter 1). In *C. managuense*, fish raised on different diets in the laboratory developed differences in their oral jaws (Meyer, 1987). The development of crushing pharyngeal teeth of the L. Victoria cichlid, *Astatoreochromis alluaudi*, varied depending on whether the fish were fed snail or snail-less diets (Barel *et al.*, 1991). Pumpkinseed sunfish from lakes in Michigan and Wisconsin showed a relationship between the development of the pharyngeal bones and associated muscle used to crush prey and the proportion of molluscs in their diet (Mittelbach *et al.*, 1992).

Some morphological differences between populations reflect an evolutionary process. The Cowichan drainage system in British Columbia includes several lakes which range in surface area from 1.4 to 6180 ha. Three-spine sticklebacks from these lakes show significant interpopulation variation in morphological features related to their trophic ecology, including the length of the upper jaw and the number and length of the gill rakers (Lavin and McPhail, 1985). Sticklebacks from the smallest lakes have a morphology and behavioural repertoire that enables them to feed on benthic prey more effectively than fish from the larger lakes, whereas the latter are more effective at feeding in the water column (Lavin and McPhail, 1986). These interpopulation differences have probably evolved since the last

Pleistocene glaciation, that is in the last 10 000 years. This gives an estimate of the rate of evolution of ecologically significant morphological and behavioural characteristics (Chapter 9).

A spectacular example of evolutionary plasticity in relation to feeding is the extraordinary adaptive radiation of cichlids in the African Great Lakes (Fig. 3.7). This single family within restricted geographical and physical localities has evolved to exploit virtually every type of resource that is exploited by the teleosts as a group (Fryer and Iles, 1972). The processes that have led to this evolutionary flowering are still under active discussion (Echelle and Kornfield, 1984), but the African Great Lakes cichlids provide a unique testbed for ecological and evolutionary theories on the biology of feeding. Sadly, the cichlid fauna of Lake Victoria is threatened by the combined consequences of overfishing and the piscivorous Nile perch, *Lates niloticus*, introduced into the lake in about 1960 (Barel *et al.*, 1985; Pitcher and Hart, 1995).

3.8 SUMMARY AND CONCLUSIONS

1. Fishes show a wide adaptive radiation in their feeding habits. They occupy many trophic roles from detritivores to carnivores. There is often a correlation between morphological traits and trophic role because morphology determines how a fish can feed. Body shape, mouth morphology, teeth, gill rakers and the structure of the alimentary canal can all be related, in a general way, to diet.

2. Several methods are available for describing diet composition on the basis of either the number or the bulk of food items present in a diet. No single method is entirely satisfactory, and the best methods for the statistical analysis of data on diets are still under debate.

3. The presence of a food item in a diet

depends on its availability, its detection by the fish and its selection as food. For many carnivorous fishes, including zoo-planktivores, vision is the most important sense for prey detection. Prey selection depends also on the mechanical ability of the fish to manipulate the prey successfully. Selection may depend also on the profitability of the prey. Optimal foraging theory has been used with some success to predict the diet composition of some species on the basis of prey profitability.

4. Factors involved in the selection of food by herbivores or detritivores are less well understood, but may include nutritional value, ease of ingestion and, in plants, the presence of secondary metabolites.

5. Diet composition often changes as a fish grows (ontogenetic diet changes), so a fish may adopt several trophic roles in its lifetime. Diel and seasonal changes in diet may also occur.

6. The rate of food consumption is determined by the availability of food and the motivational state of the fish. The rate is affected by temperature and other abiotic factors, by body size and by the physiological condition of the fish.

7. Estimates of the rate of consumption by fish in natural populations can be obtained from calculations based on the rate at which the gut is evacuated or from estimates of the amount of food required to provide the energy and nutrient requirements of the fish.

8. Many fishes are flexible in their choice of foods, responding to changes in the availability or profitability of potential prey. Even species with specialized morphological adaptations related to feeding may show surprising flexibility in their choice of food.

9. There is evidence that morphological and behavioural traits related to feeding may show rapid evolutionary change. This reflects the central importance of feeding as the activity that provides the resources required for maintenance, growth and reproduction.

10. The goal of studies of the ecology of feeding must be to develop models that predict the composition and quantity of the food consumed. The quality and amount consumed determine the energy available to a fish. The patterns of allocation of this energy form the subject of the next four chapters.

4

BIOENERGETICS

4.1 INTRODUCTION

The energy in ingested food has one of two fates. Some is dissipated in waste products or as heat and some is incorporated as chemical energy in new tissue. Metabolic processes generate the heat losses. These processes release the energy in the food to do useful work, such as that done in tissue function and repair, synthesizing new tissue and swimming. The processes that result in the dissipation of energy can be grouped together as maintenance. New tissue may take three forms: structural tissue such as muscle, reserves such as visceral fat, and thirdly, gametes. The income of energy (and nutrients) will be limited by time, availability of food and the capacity of the gut to process food (Chapter 3). How should this limited income be allocated among maintenance, growth and reproduction? What pattern of allocation will maximize the lifetime production of offspring? This chapter starts the discussion of these questions by first introducing the concept of an energy budget and then describing the effects of abiotic environmental factors on the maintenance item in the energy budget using a classification developed by Fry (1971). The following three chapters examine the allocation of time and energy in relation to patterns of movement (Chapter 5), growth (Chapter 6) and reproduction (Chapter 7).

Allocation of energy or nutrients can be described by developing budgets for those components of the income for which a mass-balance equation can be written of the form:

$$Income = Expenditure + Storage. \quad (4.1)$$

The laws of conservation of energy and matter apply to living organisms, so mass-balance equations can be written for energy or for elements such as nitrogen, calcium or phosphorus. Most attention has been paid to the development of energy budgets, but other budgets will have to be calculated in situations where a nutrient such as phosphorus rather than energy may be the limiting resource.

4.2 STRUCTURE OF AN ENERGY BUDGET

For an individual fish, an energy budget for a defined period of time takes the basic form:

$$C = P + R + F + U \quad (4.2)$$

where C is the energy content of the food consumed over the time period, P is the energy in growth and gametes which together form the production, R is the energy lost in the form of the heat produced during metabolism, F is the energy lost in faeces and U is the energy lost in excretory products, particularly nitrogenous products such as ammonia and urea (Brafield, 1985; Jobling, 1994; Lucas, 1996). It is now conventional to use joules (J) to measure energy, but in many bioenergetics studies energy was measured in calories (4.184 J = 1 cal). The rate of energy flow per unit time is power and can be measured as $J s^{-1}$, which is equal to watts (W).

R, F and U are the maintenance components in the budget. P is the gain component and can be subdivided:

$$P = P_s + P_r \qquad (4.3)$$

where P_s is the energy gain in the form of growth (structural and reserves) and P_r is the energy gain in the form of gametes (Kitchell, 1983). If the energy in the food is less than the energy expended in maintenance, P will be negative, that is degrowth occurs (Chapter 6).

Winberg (1956) developed an early form of an energy budget for fishes (page 59), but a more comprehensive model developed by Warren and Davis (1967) is the basis of most subsequent studies and forms the framework for the present chapter (Fig. 4.1). It has the disadvantage that it does not represent pathways by which information on rates of energy expenditure on maintenance or production can be fed back in a way that alters the rate of consumption. The model seems to suggest that a fish maintains no regulatory control over its pattern of energy allocation. Studies, particularly on the life-history patterns of salmonids, provide evidence of regulation (Metcalfe and Thorpe, 1992).

Nevertheless, the model has formed the basis for bioenergetics models of growth and consumption in natural populations which can, with care, be used in the development of management policies for such populations (Hansen *et al.*, 1993)

The magnitudes of the maintenance components of energy budgets in fishes and the effects of environmental factors on these components are explored, using the framework provided by the Warren and Davis model. Kooijman (1993) has developed an alternative theory of dynamic energy budgets, but this has yet to be applied systematically to fishes.

4.3 COMPONENTS OF THE ENERGY BUDGET

Measurement of the components

A variety of direct and indirect calorimetric methods can be used to measure the components of the energy budget (Brafield, 1985; Jobling, 1994; Lucas, 1996). Aerobic respiration is a controlled burning of organic mate-

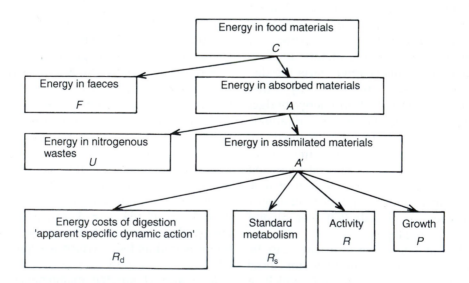

Fig. 4.1 Energy partitioning by an individual fish based on Warren and Davis' (1967) model. Redrawn from Wootton (1984a).

rial using oxygen as the oxidizing agent. The energy content of samples of food, solid faeces, flesh and gametes is measured most directly by burning dried material in an oxygen atmosphere in a bomb calorimeter and measuring the heat evolved. Fresh or dried material can also be oxidized using a dichromate solution. An alternative approach is to calculate the energy content from the proportions of protein (23.6 kJ g^{-1}), carbohydrate (17.2 kJ g^{-1}) and lipid (39.5 kJ g^{-1}) in the dried material. Bomb calorimetry, wet oxidation and the chemical composition gave similar estimates for the energy content of dried, homogenized bodies of perch (Craig *et al.*, 1978).

Because of technical difficulties, the metabolic heat losses, which represent the cost of maintenance, have rarely been measured by direct calorimetry. They are usually estimated from the rate of oxygen consumption by fish in a respirometer. The rate of oxygen consumption is converted to a rate of energy expenditure using an oxycalorific equivalent that depends on the food composition. For fish, which have a metabolism based on protein and lipid rather than on carbohydrate, a suitable value is about 13.6 J mg^{-1} oxygen. An improved estimate of metabolic expenditure is obtained if the production of carbon dioxide and ammonia are measured at the same time as the rate of oxygen consumption (Brafield, 1985). The energy losses by nitrogenous excretion can be estimated by measuring the amounts of ammonia and urea produced, because these account for almost all the nitrogenous waste products of a fish (Elliott, 1976a; Brafield, 1985). The energy losses are about 24.85 J mg^{-1} ammonia-nitrogen and 23.05 J mg^{-1} urea-nitrogen excreted (Elliott, 1976a).

It is not yet technically possible to measure all the components of the energy budget simultaneously, because the energy content of a fish's body can be measured only after the fish has been killed. Growth, which is the change in the total energy content of the body over the time interval for which the budget is being developed, can only be estimated. Studies that have measured the accuracy of energy budgets of individual fish usually find some imbalance between the input and the sum of the outputs. A study of the cichlid, *Oreochromis mossambicus*, found that the fraction of the energy of the food consumed over a period of 30 days accounted for in the budget ranged from 72.8% to 105% (Musisi, cited in Brafield, 1985). With adequate techniques, the components of an energy budget can be measured with sufficient accuracy to present a clear picture of the pattern of energy allocation.

Faecal losses – the absorption efficiency

Faeces consist of unabsorbed food with traces of mucus, sloughed gut cells, catabolized enzymes and bacteria. The energy content of the faeces is lost to the fish, although it may be important to organisms that form part of the decomposer food web. Absorption efficiency (sometimes called digestibility) measures the proportion of the food energy that is retained by the fish, that is:

$$AE = 100 \, (C - F) / C \qquad (4.4)$$

where AE is absorption efficiency, C is the energy content of the food consumed and F is the energy content of the faeces. Similar indices of digestibility may be calculated for dry weights, protein (or nitrogen), lipid or carbohydrate. It is usually simple to collect solid faeces, but soluble material may leach from them. Several techniques are used to obtain estimates of absorption efficiency, including the use of inert, indigestible markers such as chromic oxide which are included in the ingested food in a known proportion (Talbot, 1985). The increase in the proportion of the marker in the faeces relative to its proportion in the food gives a measure of the proportion of food absorbed.

The most important factor affecting

absorption efficiency is the quality of the food ingested. Fish feeding on animal food usually have higher absorption efficiencies than herbivores. Brown trout feeding on gammarids had an absorption efficiency that ranged from 70% to 90% (Elliott, 1976a). Fish feeding on prey that lack chitinous exoskeletons or bony endoskeletons have even higher efficiencies. Threespine sticklebacks feeding on enchytraeid worms (oligochaetes) absorbed over 90% of the energy consumed (Allen and Wootton, 1983). Such high values are typical of carnivores (Brett and Groves, 1979). Absorption efficiencies of herbivorous fish tend to be lower. *Sarpa salpa*, found on rocky coasts of southern Europe and much of Africa, had an energy assimilation efficiency of 65% when fed a green alga, *Ulva lactuca* (Gerking, 1984). In Lake George, East Africa, the cichlid, *Oreochromis niloticus*, feeds largely on phytoplankton (Moriarty and Moriarty, 1973). Assimilation of carbon rose during the day, from a low of about 30–40% when feeding started in the morning to a high of 70–80% at the end of the feeding cycle, with a daily average of about 43%. Absorption became more efficient as the stomach contents became more acid during the feeding cycle. In Lake Sibayi, Republic of South Africa, *Oreochromis mossambicus* feeds on benthic detrital aggregate which has an average organic content of 43% and an energy content of 17.0 J mg^{-1} organic matter (Bowen, 1979). The mean absorption efficiencies for juveniles were: energy 42%, protein 46%, total carbohydrate 35%.

Both animal and plant proteins are absorbed at high efficiencies – over 90% for animal proteins and over 80% for plant proteins (Brett and Groves, 1979; Gerking, 1984; Horn and Neighbor, 1984). Lipids also tend to have high digestibilities, but only 30–40% of the carbohydrate content of food may be absorbed (Brett and Groves, 1979).

Temperature and the amount of food eaten can also affect absorption efficiency. Brown trout feeding on *Gammarus* showed an increase in efficiency with an increase in temperature, but the efficiency decreased with an increase in ration (Elliott, 1976a).

Excretory losses – assimilation efficiency

Energy losses as nitrogenous excretory products, mostly ammonia and urea, come from two sources. Endogenous nitrogen excretion is the result of catabolism of protein in energy-yielding respiration. In fish, protein is a usual substrate for respiration. Exogenous nitrogen excretion is related to feeding. It results from the elimination of excess protein or during the adjustment of the balance of amino acids. The distinction is schematized (Brett and Groves, 1979):

$$\text{Consumed N} = \text{Absorbed N} + \text{Faecal N} \qquad (4.5)$$

$$\text{Absorbed N} = \text{Metabolizable N} + \text{Excreted N (mostly exogenous)} \qquad (4.6)$$

$$\text{Metabolizable N} = \text{Retained N} + \text{Excreted N (mostly endogenous)} \qquad (4.7)$$

Endogenous nitrogen excretion is measured approximately as the rate of excretion by a fish deprived of food. The rate is a function of body weight (W) of the form:

$$dN/dt = a\ W^b \qquad (4.8)$$

where dN/dt is the rate of nitrogen excretion, a and b are constants. The exponent b is typically significantly below 1.0, indicating that the excretion rate is lower per unit of body weight in larger fish (Gerking, 1955).

Exogenous nitrogen excretion is measured as the increase in ammonia and urea excretion after a meal. When sockeye salmon fry were fed a meal, the rate of excretion of ammonia increased rapidly from a basal level, reaching a maximum 4–4.5 h after the meal, then declined. The exogenous nitrogen excretion represented about 27% of the nitrogen ingested (Brett and Zala, 1975). In brown trout, the proportion of food energy ingested lost as ammonia and urea excretion decreased as the ration increased, but increased with temperature (Elliott, 1976a). The average losses ranged from 4–6% at 4 °C to 11–15% at 15 °C. The effects of temperature and ration size on the proportion lost in nitrogenous excretion were the opposite to those on the proportion of energy lost as faeces.

Because of methodolgical differences in separating faecal and nitrogenous losses, it is frequently expedient to combine the losses through faeces and nitrogenous excretion. An assimilation efficiency can be calculated as:

$$\text{Assimilation efficiency} = 100[C - (F + U)] / C \qquad (4.9)$$

where C, F and U are defined above (page 65).

In the brown trout, the assimilation efficiency changed little over a wide range of temperatures and ration sizes because these two factors have opposite effects on faecal and nitrogenous excretion (Elliott, 1976a). The value for assimilation efficiency in brown trout feeding on *Gammarus*, 70–75%, was lower than the 80% suggested by Winberg's (1956) early bioenergetics model. The effects of food quality and quantity and temperature on absorption and assimilation efficiencies show that they cannot always be regarded as constant, but may vary with changes in environmental conditions.

Metabolic losses

The energy in the faeces and excretory products is lost to the individual without being used to do useful work. Metabolism is the sum of the reactions that yield the energy the organism utilizes for activities (Fry, 1971). Activities are what the fish does with the useful energy released in metabolism before it is dissipated as heat. If the rate of metabolism falls below a minimum, the fish dies. There is also a maximum sustainable rate of metabolism, which the fish can exceed only by accumulating an oxygen debt. As in other vertebrates, sustained activity depends on aerobic respiration with its requirement for oxygen. In a fish swimming slowly and continuously, the muscles used, mainly red muscle, are operating aerobically. An oxygen debt is incurred during short bursts of fast swimming such as might be used to chase a prey or escape a predator. Burst swimming is powered by the white body muscle in which anaerobic metabolism predominates (Hochachka, 1980; Wardle, 1993; Jobling, 1995). White muscle fatigues rapidly as lactate, an end product of anaerobic respiration, accumulates. The fish must then break down the lactate via the aerobic respiratory pathway. Fish may also resort to anaerobic respiration in hypoxic water and so develop an oxygen debt to be repaid when the fish returns to water of a higher oxygen concentration. An exception may be provided by some cyprinids, which use anaerobic respiration in hypoxic conditions yet do not accumulate an oxygen debt (Hochachka, 1980; Smit, 1980). Such fish may excrete ethanol as one of the end products.

Three levels of metabolism are defined by fish physiologists (Fry, 1971; Brett and Groves, 1979). Standard metabolism (R_s) is the rate of energy expenditure by a resting, unfed fish. It is approximately equivalent to the minimum rate required to keep the fish alive. This rate is low in fishes in comparison with basal metabolic rates of the endothermal birds and mammals, some one-tenth to one-thirtieth that of mammals and one-hundredth that of small birds (Brett and Groves, 1979). Maximum active metabolism (R_{max}) is the maximum sustainable aerobic rate. The difference between maximum active and standard metabolism, $R_{max} - R_s$, defines the fish's scope for activity. Lying between the standard rate and maximum active rate is the routine rate of metabolism (R_{rout}), which is the metabolic rate of an unfed fish showing spontaneous swimming.

Standard metabolism

This level is difficult to measure because,

within the confines of a respirometer or calorimeter, fish usually show spontaneous swimming (Weatherley and Gill, 1987). An estimate can be obtained by measuring metabolic rate over a range of swimming speeds, then extrapolating to the metabolic rate expected when the fish is stationary (Fry, 1971). The relationship between standard metabolism, R_s, and fish weight (W) is described by:

$$R_s = aW^b \qquad (4.10)$$

where a and b are constants and the value of the exponent, b, is usually less than 1.0, indicating that the relative rate of standard metabolism, R_s / W, declines as the fish increases in size. There is interspecific variation in the value of b, but a mean value of 0.86 has been suggested for fish (Glass, 1969; Brett and Groves, 1979).

Routine and active metabolism

A similar function can also be used to describe the relationship between routine metabolic rate and body weight, with the weight exponent b taking a value similar to that for standard metabolism (Evans, 1984). Because fish differ in the amount of spontaneous activity that they show, direct comparisons between different studies of routine respiration are difficult unless the amount of swimming is recorded accurately and in a way that makes comparison possible. Respirometers have been designed that force the fish to swim, at a speed determined by the experimenter, while the active rate of metabolism is recorded (Beamish, 1978). Such respirometers can be used to measure the rate of energy expenditure of fish swimming at known speeds, including the maximum speed sustainable by a fish without incurring an oxygen debt. The increase in metabolic rate with swimming speed has been described by the function:

$$\log R = a + bV \qquad (4.11)$$

where R is the rate of energy expenditure (or oxygen consumption) per unit weight and V is the swimming speed expressed as fish lengths (L) per unit time. The intercept, a, in this expression is the logarithm of standard metabolic rate, that is of the rate of energy expenditure when $V = 0$ (Beamish, 1978; Priede, 1985). The metabolic cost of swimming at a given speed is estimated as the difference between the metabolic rate at that speed and standard metabolic rate. The relationship can be used to estimate energy expenditure during behaviour such as migration or territorial defence (Chapter 5).

The optimum swimming speed that minimizes the work done per metre covered can also be calculated from the relationship between metabolic rate and speed (Videler, 1993). In terms of lengths per second, the optimum speed decreases with an increase in fish size. A compilation of values ranged from 5.8 $L\ s^{-1}$ for a larval cyprinid to 0.8 $L\ s^{-1}$ for an adult whitefish, *Coregonus artidii* (Videler, 1993). Optimum swimming speed in metres per second is positively correlated with body mass. The cost of transport at the optimum speed (expressed as the energy required to move a unit of mass through 1 metre) decreases with an increase in body size, a relationship relevant to the effect of body size on the propensity of fish to migrate (Chapter 5).

Maximum active metabolism

The metabolic rate at the maximum sustainable swimming speed of an unfed fish, its active metabolic rate R_{max}, provides one estimate of the maximum power that the fish can generate to do useful work without incurring an oxygen debt. The relationship between active metabolic rate and body weight may be close to direct proportionality, that is in the relationship: $R_{max} = a\ W^b$, b takes a value close to 1.0 (Brett and Groves, 1979). However, the aerobic red muscle usually accounts for only a small proportion of the aerobic capacity in fish (Goolish, 1991). Consequently, in some conditions, a fish may have a higher maximum aerobic metabolic rate than that measured during sustained swimming, for example during food processing in small fast-growing fish.

Swimming speeds during sustained swimming are less than 6–7 L s^{-1}, and typically 3–4 L s^{-1} (Videler, 1993). In burst swimming, powered by anaerobic metabolism, speeds of 20 L s^{-1} can be reached. These speeds can only be maintained for seconds before fatigue and represent rates of energy expenditure as much as 40 times more costly than swimming at sustained speeds. The ability to escape from the attack of a predator is a matter of life or death. Maximizing speed of movement will take precedence over efficiency of movement.

Apparent specific dynamic action

When a fish is fed, its metabolic rate increases because the events and processes asso-

ciated with feeding involve the fish in energy expenditure. This increased metabolism is signalled by a rapid increase in the rate of oxygen consumption when a meal is eaten. Respiration reaches a peak within a few hours and then drops back slowly to a level similar to that before the meal was eaten (Fig. 4.2). This increased expenditure is often called apparent specific dynamic action (SDA) (Beamish, 1974). The importance of apparent SDA is that ingestion, digestion and post-digestive processes impose a significant metabolic expenditure that has to be met before other useful work can be done.

Apparent SDA has two components. The first is the work done during the process of digestion, including increased muscular activity and the secretion of digestive enzymes. The second, sometimes called the

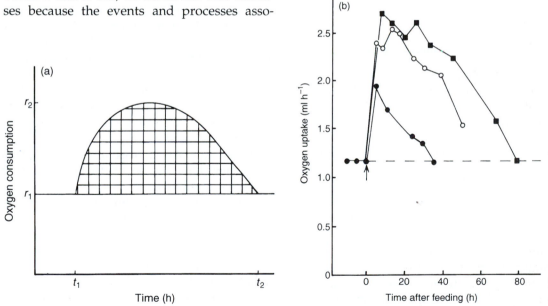

Fig. 4.2 Apparent specific dynamic action (heat increment of feeding). (a) Schematic diagram showing the rise in oxygen consumption after a meal. Apparent SDA is measured by shaded area. r_1, routine rate; r_2, peak rate; t_1, time of feeding; t_2, termination of SDA. Redrawn from Jobling (1981). (b) Effect of meal size on mean apparent SDA in plaice, *Pleuronectes platessa*, fed white fish paste. ●, 0.25 ml paste ($N = 1$); ○, 0.5 ml paste ($N = 7$); ■ 1.0 ml paste ($N = 5$); arrow, time of meal. Broken line indicates the routine rate. Redrawn from Jobling and Spencer Davies (1980).

heat increment, probably represents the work that is done in processing and transforming the products of digestion. These processes include the deamination of proteins (Brett and Groves, 1979) and include the metabolic costs of the synthesis of new tissue (Jobling, 1981, 1983a, 1985a).

Apparent SDA is related to the size of the ration consumed and the composition of the food. For a given type of food, the apparent SDA often increases linearly with ration. A linear relationship was found for the herbivorous cichlid, *Tilapia rendalli*, feeding on a macrophyte, *Ceratophyllum*, and for the largemouth bass feeding on fish (Beamish, 1974; Caulton, 1978). High-protein diets cause a higher apparent SDA than diets high in lipid (Medland and Beamish, 1985). About 15% of the energy in the food ingested is expended on apparent SDA: a range of mean values from 9.5% to 19% was reported by Jobling (1981) in a list that includes both herbivores and carnivores.

Ontogenetic changes in metabolic rates

Most experimental studies of the relationship between metabolic rate and body weight have used juvenile and adult fish, after the period of rapid development and growth that characterizes early life history. There is evidence that during ontogeny, the weight exponent, b, in the relation, $R = aW^b$, changes (Kamler, 1992; Post and Lee, 1996). The change takes the value of b from close to 1.0 (isometry), to a value around 0.8. For rainbow trout, a comparison of active and routine metabolic rates suggested that the weight at which the value of b changed was about 5 g. For trout between about 0.01 g and 5 g, absolute scope (maximum aerobic metabolic rate minus routine metabolic rate) increased, but beyond 5 g, absolute scope decreased to approximately double that of early life-history stages. The factorial scope (maximal aerobic metabolic rate / routine metabolic rate) increased with weight from

about 2 to 6 during ontogeny. The changes in the relationship between metabolic rate and body weight may relate to changes in gill area relative to increasing body weight. These ontogenetic patterns in metabolism and scope probably relate to the life history characteristics of the species, but too few species have been studied for any generalizations to be drawn.

Total metabolic rate

If the total metabolic rate of a fish is R, with the component representing standard metabolism, R_s, the component associated with swimming, R_a, and the component for apparent specific dynamic action (including any metabolic costs of the synthetic processes required for growth), R_d, then:

$$R = R_s + R_a + R_d \qquad (4.12)$$

with all the components measured in joules per unit time or equivalent units (1 W = 1 J s^{-1}). The allocation of energy between the components R_s, R_a and R_d will have implications for the survival and reproductive success of the fish. Allocation will take place within the constraint that the sum of the three components cannot exceed R_{max} – the maximum rate at which energy can be made available by aerobic respiration – on a sustainable basis. An increase in one component may then have to be traded-off against a decrease in another.

Calculations for the brown trout and the cod suggested that the sum of the three components, $R_s + R_a + R_d$, could be greater than R_{max} so the maximum energy demand by each component could not be met in full simultaneously (Priede, 1985). Energy budgets for brown trout over a range of weights and temperatures have been estimated by Elliott (1976b. 1979) and from this information the following estimates were made (Priede, 1985). A 500 g brown trout in summer at 15 °C can expend energy on the three components at maximum rates of

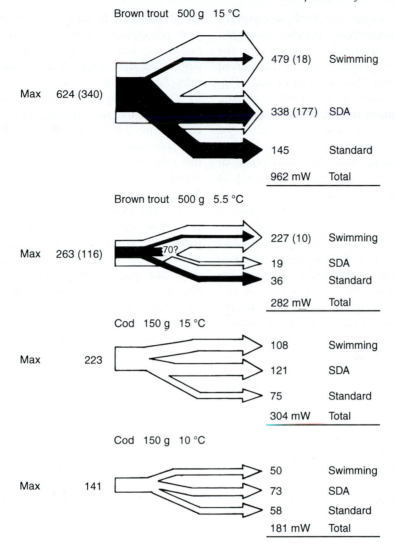

Fig. 4.3 Estimates of power capacities (open arrows) for brown trout, *Salmo trutta*, and cod, *Gadus morhua*, and power used (for trout only) (black arrows) at different temperatures. Unbracketed values are maximum capacities, bracketed values are power used. Redrawn from Priede (1985).

479 mW in swimming (R_a), 338 mW in apparent SDA (R_d) and 145 mW for standard metabolism (R_s), giving a total potential requirement for energy expenditure (R) of 962 mW. Its active metabolism (R_{max}) can supply only 624 mW, a shortfall of 338 mW (Fig. 4.3). Telemetric studies of the move-ments of free-swimming trout in a small lake suggested that trout expended only 18 mW on swimming, and fed at a rate below maximum, so were operating at well below their maximum power capacity. Priede's (1985) calculations for juvenile cod, sug-gested that in this fish also, the maximum

power expenditure on standard, locomotory and digestive metabolism together exceeded the total power capacity. In this species, the maximum power expenditure on apparent SDA could exceed that on swimming (Fig. 4.3), so the rate of energy expenditure during sustained swimming underestimated R_{max}.

Priede (1985) used an analogy with the performance of electrical or mechanical equipment to argue that the probability of dying increases the longer a fish spends at rates of energy expenditure (R) that are close to R_{max} or R_s. Consequently, a fish should regulate its swimming so that it can carry out the normal functions of foraging and patrolling with a low probability of exceeding the limits of its metabolic scope. Telemetric studies of the movements of pike and brown trout indicated that the metabolic cost of locomotion was indeed kept low (Diana *et al.*, 1977; Priede, 1985). A second advantage of a reduction in the power required for swimming is that the energy saved can be expended on growth or gametes. A reduction in swimming shown by roach, *Rutilus rutilus*, over the period July to December represented a saving in metabolizable energy of 1485 kJ per kg of fish in a period when 364 kJ of gonadal tissue per kg of fish was synthesized (Koch and Wieser, 1983).

During the early, fast-growing, life-history stages, the problem of energy allocation may be particularly acute (Wieser, 1989; Wieser and Medgyesy, 1990; Rombough, 1994). For larvae and small juveniles, there seem to be great advantages in increasing in size as rapidly as possible (Miller *et al.*, 1988) (Chapter 10). Growth takes the small fish out of the hydrodynamic regime in which viscous forces dominate (Chapter 2). With an increase in size, absolute speed of swimming increases, allowing faster escape from predators and faster approach to prey. There is an increase in sensory capacities with size (Chapter 2). With the increase in gape size, larger prey and a wider range of prey sizes can be eaten. However, growth exerts a

metabolic cost. In larger fish, there is a positive correlation between growth rate and metabolic rate. In roach larvae, estimates suggested that there was threshold growth rate. Above this threshold, further increases in growth rate were not matched by an increase in the rate of respiration. During the embryonic and larval development of chinook salmon, *Oncorhynchus tshawytscha*, the rate of growth was not correlated with metabolic rate (Rombough, 1994). If the cost of high rates of embryonic and larval growth is not met by an increase in aerobic respiration, how is it met? One possibility is that increased expenditure on growth is met by a reduction in expenditure on other activities. A second possibility is that the efficiency of growth is high during these early stages and so the cost, in terms of metabolic demand, is low. The mortality rate during the larval stage is typically high (Chapter 10). Does this high mortality partly reflect the burden that the high growth rates place on the metabolic capacity of the developing fish?

The allocation of energy between R_s, R_d and R_a has profound implications for the survival and reproductive success of a fish (Wootton, 1994a). Unfortunately there is little information on how this allocation is regulated in relation to the past history and present environment of the fish. Telemetric techniques are being developed that monitor individual, free-living fish (Lucas *et al.*, 1993). These techniques provide measurements of movement and of physiological variables correlated with rates of energy expenditure. These data will allow an assessment of the role of changes in the time spent swimming at different speeds by free-living fish in the regulation of their energy allocation.

4.4 EFFECTS OF ENVIRONMENTAL FACTORS ON METABOLISM

An important constraint on a fish's capacity to regulate the allocation of energy to maintenance, activity and feeding is the effect of changes in the abiotic environment. To minimize the effects of external environmental change, the fish must use some energy to regulate its internal environment (Fry, 1971).

Classification of effects

Abiotic environmental factors may be classified by their identity – for example temperature, oxygen or salinity – or by their effects on the fish. Fry (1971) developed a valuable classification on the latter basis. Five types of factor were recognized: lethal, controlling, limiting, masking and directive. The value of Fry's classification is that it provides a conceptual framework within which the effects of environmental variables on the capacities of individuals to function can be assessed in metabolic terms, that is in terms of the energy expenditure (Kerr, 1980). It provides a mechanistic approach to the relationship between environmental identities and the lifetime reproductive success of individuals. It is related to the concept of the niche (Chapter 9) that is used to define the range of environmental factors within which the reproductive success of individuals is sufficiently high for the population to persist (Chapter 7). However, the niche concept does not concern itself with the metabolic basis of that success.

A lethal factor imposes conditions in which the fish dies because its metabolic machinery no longer functions. Two properties of a lethal factor can be defined. The first is the incipient lethal level (ILL), the limiting value for that factor beyond which the fish can no longer live for an indefinite period. The second is the effective time (ET), which is the time it takes for the factor to kill the fish. Within the boundaries defined by the ILL, the fish is within its zone of tolerance: the life span is not directly affected. Outside the ILL, the fish is in the zone of resistance, in which events can be measured by the time taken to die. Note that these definitions are all related solely to mortality, not to other

processes such as food consumption, growth or reproduction. The boundaries for these processes lie within the zone of tolerance (Elliott, 1981) (Chapters 3, 6 and 7).

A controlling factor governs the metabolic rate of the fish by its influence on the state of molecular activation of the components of the metabolic chain. It places two bounds on the rate of metabolism. It permits a maximum and demands a minimum below which the rate is no longer sufficient to maintain the integrity of the fish. Temperature is the overriding controlling variable for ectothermic fish because of its pervasive effect on metabolism.

A limiting factor restricts the supply or removal of materials in the metabolic chain. It reduces the rate below that allowed by the controlling factors. An example will be provided by oxygen. A masking factor, illustrated by salinity, modifies the effect of other factors. It is a factor that imposes an energetic cost for its regulation. Finally, a directive factor is an identity such as temperature or salinity that elicits a response by the fish to a gradient in the factor. The gradient may be spatial or temporal.

Temperature as an environmental identity

Because, with a few exceptions, fish are obligate ectotherms, temperature is an important abiotic factor. Industry and power stations frequently discharge warm water. Consequently the effects of temperature on fish have been analysed extensively in attempts to define the impacts of thermal pollution.

Temperature as a lethal factor

Fish can only survive over a range of temperatures bounded by the upper and lower incipient lethal temperatures (UILT and LILT). The region between these lethal limits is described by a tolerance polygon (Fig. 4.4). For some eurythermal fishes, this can cover

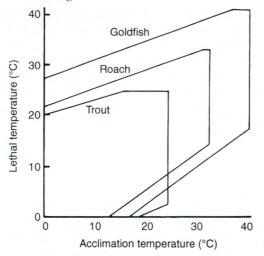

Fig. 4.4 Temperature tolerance polygons for gold-fish, *Carassius auratus*, roach, *Rutilus rutilus*, and brown trout, *Salmo trutta*. Zone of tolerance is defined by area within a polygon (see Fig. 4.7). Redrawn from Elliott (1981).

most of the range over which animal life is found. For example, goldfish, with proper acclimation can tolerate a range from 0 to 40 °C (Fry, 1971). Stenothermal fish have a narrow zone of tolerance. Antarctic fish die at temperatures a little above 5 °C (Wohls-chlag, 1964; Macdonald *et al.*, 1987). The temperature to which a fish has become acclimated by being held at that temperature for several weeks affects incipient lethal levels. UILT and LILT tend to increase with an increase in acclimation temperature (Fig. 4.4). Such static pictures of the thermal tolerances of species are limited for they refer to fish of a given size in a given physiological state.

The tolerance can change with changes in fish size and age, and also seasonally. At low acclimation temperatures, juvenile bullhead *Cottus gobio*, were less tolerant of high temperatures than adults (Elliott and Elliott, 1995). In the minnow, *Chrosomus eos*, the UILL at a given acclimation temperature was

lower for fish collected in winter than for fish collected in summer (Tyler, 1966). The zone of thermal tolerance shown in Fig. 4.4 should be thought of as a cross section through a volume, the third axis of which represents time. The area of the cross section changes through time because of ontogenetic and seasonal changes in tolerance. The size of the area provides a quantitative measure of thermal tolerance. A greater area indicates a greater tolerance.

The high specific heat of water means that changes in temperature are relatively slow, so fish can often move into more favourable conditions and avoid the lethal effect of temperature. The exception to this occurs if the fish cannot escape. The escape may be blocked, for example by a dried-out section of stream, or the lethal conditions may be present over a distance that exceeds the ability of the fish to escape. Severe cold weather which persists for a long period has been implicated in the deaths of several species of fish in the North Sea and adjacent fresh waters (Cushing, 1982). Marine species affected include the sole, *Solea solea*, plaice, *Pleuronectes platessa*, and dab, *Limanda limanda*, all demersal species. The effects of high temperatures are complicated by the lowered solubility of oxygen at such temperatures, so the direct lethal effect of temperature may be difficult to disentangle from any effects of low oxygen concentrations.

Temperature as a controlling factor

A fish experiencing temperatures outside the limit, defined by the incipient lethal level, has no pattern of allocation of energy or time that will allow it to survive for an indefinite period unless it can find a refuge from the lethal regime. Its homeostatic capacity is exceeded. The effects of temperature in its role as a controlling factor on metabolic rate are more directly relevant to bioenergetics. An increase in temperature increases the velocity at which a chemical transformation takes place. In the absence of limiting factors, a rise in temperature will increase the metabolic rate of fishes (Hochachka and Somero, 1984).

An increase in standard metabolism (R_s) with temperature is seen in fishes from geographic areas as different as the Antarctic and the tropics (Fig. 4.5). There is evidence suggesting that polar species maintain a higher metabolic rate at lower temperatures than would be expected by extrapolation from the rates shown by temperate species (Brett and Groves, 1979). Such comparisons within and between species have to be made with care to avoid the confounding effects of differences in activity, food levels and physiological state (Holeton, 1980; Clarke, 1993). However, polar species may have evolved a compensatory adaptive response to the low temperatures of their environment (Macdonald *et al.*, 1987). Correspondingly, although to a lesser extent, tropical species have a rate lower than expected by extrapolation from temperate species; nevertheless their standard metabolic rates do tend to be relatively high.

How does temperature affect the scope for activity of fishes, that is the difference between R_{max} and R_s? The effect of acclimation temperature on these two levels of metabolism in sockeye salmon is shown in Fig. 4.6 (Brett, 1964, 1971b). R_s increased with temperature approximately according to the exponential function:

$$R_s = e^{bT} \qquad (4.13)$$

where T is temperature and b is the rate constant. R_{max} initially increased exponentially with temperature, but at a faster rate than R_s, up to 15 °C, then declined with a further increase in temperature. The scope for activity increased with temperature up to 15 °C, then declined, suggesting that 15 °C represented an optimum for activity for sockeye salmon. Brown trout showed a similar

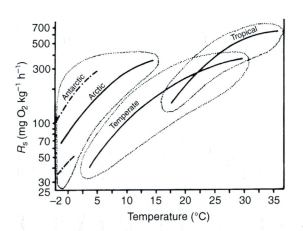

Fig. 4.5 Schematic diagram of effect of temperature on standard metabolic rate, R_s, (log scale), for polar, temperate and tropical fish (solid lines) Dotted lines indicate range of variability within each zone. Redrawn and simplified from Brett and Groves (1979).

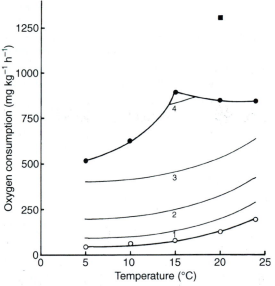

Fig. 4.6 Effect of temperature on active metabolic rate, R_{max}, and standard metabolic rate, R_s, for sockeye salmon, *Oncorhynchus nerka*. Symbols: ○ R_s; ●, R_{max}; ■ R_{max} with enhanced oxygen. Numerals give sustained swimming speed in body lengths per second. Redrawn from Brett (1964).

response to temperature, with an optimum temperature for scope for activity of about 18 °C (Elliott, 1976b). Sockeye provided with supplementary oxygen at temperatures greater than 15 °C did not show decreases in R_{max} and scope (Fig. 4.6), which suggests that the decrease at higher temperatures may result from oxygen limitation (Brett, 1964).

Pumpkinseed sunfish displayed seasonal changes in the relationship between R_s and acclimation temperature (Burns, 1975; Evans, 1984). R_s, measured at a constant acclimation temperature throughout a year, declined over the autumn and winter, then increased sharply in spring, before declining over the late summer. These changes tracked changes in daylength (Evans, 1984), which suggested they resulted from physiological changes under the control of the endocrine system.

Temperature also affects swimming performance. There is a large literature on the structural, biochemical and physiological changes that occur in muscle in response to temperature changes (Johnston, 1993), but less information on the effect of the changes on swimming performance. The maximum speed of prolonged, aerobic swimming increases with temperature up to a maximum, but then declines with any further increase in temperature (Beamish, 1978). In sockeye salmon, energy expenditure on prolonged swimming at a given speed increased with temperature (Fig. 4.6). Wardle (1980) found that the speed of contraction of white (anaerobic) muscle decreased with a decrease in temperature. In rainbow trout, maximum swimming speed during burst swimming increased with temperature over the range 5 to 15 °C but then changed little between 15 and 25 °C (Webb, 1978b). In goldfish acclimated to 10 °C, maximum speed during burst swimming increased only slightly over the range 10 to 30 °C (Johnson and Bennett, 1995). Burst swimming is likely to be important in escape from a predator or attacking a prey. In the marine cottid, *Myoxocephalus scorpius*, the strike success was 23.2% at 5 °C, but increased

to 73.4% at 15 °C (Beddow *et al.*, 1995). Atlantic salmon juveniles switch from diurnal feeding in summer to nocturnal feeding in winter (Fraser *et al.*, 1993). This change may reduce the risk of predation from endothermic piscivores such as birds at low water temperatures, when the burst swimming performance of the salmon is reduced (Chapter 5).

Metabolic expenditure associated with feeding, R_d, may also be affected by temperature. In *Tilapia rendalli*, the energy cost of processing a meal of a given size increased with temperature. This is probably because a higher proportion of the food consumed is assimilated as temperature increases (Caulton, 1978).

The three components of metabolism, R_s, R_a and R_d, are all affected by temperature, as is the scope for activity. This temperature dependency has implications for the allocation of energy to the three components. At low temperatures, brown trout have a depressed appetite (Chapter 3), so R_d is reduced. Priede (1985) estimated that at 15 °C, the maximum possible rates of energy expenditure on the swimming, feeding and maintenance in a brown trout exceed the maximum income by over 54% of the latter, but at 5.5 °C, the excess is only 7%. This suggests that the allocation problem is likely to be less acute at lower temperatures.

The effect of temperature on the rate of metabolism has an important consequence for the egg stage of a fish's life history. The time between fertilization and hatching decreases sharply with an increase in temperature (Elliott *et al.*, 1987). For example, in the threespine stickleback, the time taken to hatch increases from about 6 days at 25 °C to about 40 days at 8 °C (Wootton, 1976). Thus at low temperatures, a fish spends longer in what is often the most vulnerable stage of its life history.

Temperature as a directive factor

Seasonal changes in thermal tolerance and experimental studies on the effect of temperature on metabolism indicate that there are biochemical and physiological systems which modulate the effect of temperature (Chapter 1). Behavioural thermoregulation provides a second mechanism by which fish can control the effect of temperature on metabolism. Temperature then acts as a directive factor.

Fish placed in a temperature gradient move through it, but after 2 h or less, spend most of their time within a relatively narrow temperature range. This range defines the acute thermal preferendum of the fish. It depends on the acclimation temperature experienced by the fish before being placed in the gradient. If the fish is maintained in the gradient for several days, the preferred temperature changes progressively until the fish occupies its final preferendum, which is not a function of the prior acclimation conditions (Reynolds and Casterlin, 1979, 1980). The relationship between acute and final preferenda for bluegill sunfish and its thermal tolerance is shown in Fig. 4.7. The temperature preference of fish can also be tested in a shuttlebox (Jobling, 1994). The fish regulates the water temperature by moving between chambers. Movement into a chamber initiates either a cooling or warming of the water, depending on which chamber the fish enters.

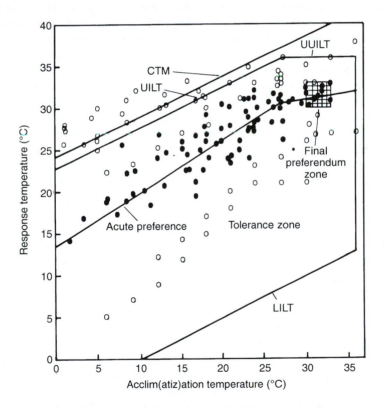

Fig. 4.7 Acute temperature preference of bluegill sunfish, *Lepomis macrochirus*, in relation to zone of tolerance. UILT, upper incipient lethal temperature; UUILT, ultimate upper incipient lethal temperature; LILT, lower incipient lethal temperature; CTM, critical thermal maximum; ● preference; ○ avoidance. Zone of resistance is defined by area between incipient lethal temperature and critical thermal maximum. Redrawn from Reynolds and Casterlin (1979).

The relationship between the behaviour of fish in a temperature gradient and their natural behaviour can be illustrated by *Tilapia rendalli* (Caulton, 1981). In a gradient from 24 to 40 °C, juvenile *T. rendalli* preferred temperatures near 36 °C, which is only slightly cooler than their UILT. In Lake McIlwaine (Zimbabwe), juveniles spent the night in deep (>1.5 m) homothermal water. They moved to shallower waters as these warmed up after daybreak so by midday large schools of juveniles were in water as shallow as 10 cm. In the evening, as the shallow waters cooled, the juveniles moved back into the warmer, deeper, homothermal water. The temperature in water 10 cm deep fluctuated between about 16 °C at night and 28 °C at midday, well below the temperatures selected by the fish in an experimental gradient. The final preferendum of the bluegill sunfish of about 30 °C is also close to the UILT and higher than temperatures the fish would usually naturally experience for long periods of time (Reynolds and Casterlin, 1979).

The ability of a fish to regulate its body temperature behaviourally and the effects of temperature on metabolism raise the question of what pattern of movement, in relation to water temperatures and food availability, a fish should adopt to maximize its lifetime production of offspring (Crowder and Magnuson, 1983). A fish that moves into warm, food-rich water may increase its rate of food consumption (page 56), but it will also have a higher rate of metabolism. A shuttlebox system can be used experimentally to record the choice of fish when presented with different food levels and temperatures. The fish, by moving through connections from one tank to another, experience differences in food level and temperature. In such an apparatus, bluegill sunfish tended to select temperatures close to their thermal preferendum, rather than the combinations of temperature and food level that would have maximized their growth rate (Wildhaber and Crowder, 1990).

Oxygen as an environmental identity

Oxygen as a lethal factor

Oxygen is possibly the most important abiotic lethal factor for fish. Fishes vary widely in their tolerance to low oxygen levels. Salmonids are typical of cool, well-oxygenated waters and are intolerant of hypoxic conditions. Brown trout require water that is about 80% saturated with oxygen (Elliott, 1994). An adequate supply of well-oxygenated water is a major factor in the design and running of facilities for rearing salmonids (Smart, 1981). Other fishes may regularly encounter a low oxygen concentration and are more tolerant of hypoxia than the salmonids. The ability to tolerate hypoxic waters may determine the species composition of fish assemblages in some localities. In Missouri, headwater prairie streams can become hypoxic in midsummer. The species that are found in these streams are typically tolerant of low oxygen concentrations (Smale and Rabeni, 1995).

The ability of some cyprinids to survive anoxia for long periods by relying on anaerobic respiration to supply their energy requirements has already been mentioned (page 69). Mathur (1967) recorded *Rasbora daniconius* as surviving for 81 days in a sealed bottle at about 30 °C!

Environments that become hypoxic or even anoxic are described in Chapter 2. Winter kills of fish in small ice-covered lakes or summer kills in unusually warm conditions are regularly reported. Although some of these deaths may be due to other causes such as the build-up of hydrogen sulphide, the low solubility of oxygen in water makes water an inherently riskier medium for active, predominantly aerobic organisms to live in than air. Fish are sensitive to episodes of pollution that cause deoxygenation of water, such as the release of organic wastes from farms or food-processing plants.

Early life-history stages such as eggs and

larvae are particularly at risk because they cannot swim out of hypoxic conditions. In Chesapeake Bay, on the east coast of North America, the recruitment of naked goby, *Gobiosoma bosc*, can be disturbed by intrusions of hypoxic water during the breeding season (Breitburg, 1992). The young life-history stages of the goby are less able to escape hypoxic conditions by swimming into shallower water than larger juveniles and adults.

Parental fish commonly show behavioural adaptations that minimize the risk: eggs are spawned in well-oxygenated flowing waters or near photosynthetically active plants, or a parent ventilates the eggs (Chapter 7).

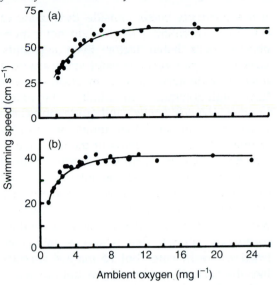

Fig. 4.8 Effect of oxygen on sustained swimming speed of (a) coho salmon, *Oncorhynchus kisutch*, and (b) largemouth bass, *Micropterus salmoides*. Redrawn from Beamish (1978).

Oxygen as a limiting factor

Oxygen is a limiting factor, because the maximum rate at which aerobic respiration can occur at a given temperature will be governed by the rate at which oxygen can be supplied (Fry, 1971). The low solubility of oxygen increases the possibility that under natural conditions a fish will experience levels such that the power capacity of the fish is limited by oxygen rather than food (Kramer, 1987).

The effects of oxygen on metabolic rate were illustrated by experimental studies on brook trout, *Salvelinus fontinalis* (summarized in Fry, 1971). As the concentration of dissolved oxygen decreased, the standard metabolic rate remained stable until the concentration dropped to about 50% of saturation. Below this, R_s increased. This increase probably represented the work done in increased ventilation, as the fish attempted to compensate for the low ambient oxygen conditions. But with a further decline in dissolved oxygen, R_s declined rapidly and the fish eventually died because the rate of metabolism fell below the level needed to sustain life.

Active metabolism, R_{max}, in the brook trout was more sensitive to reduced oxygen concentration than was R_s. R_{max} declined as the oxygen concentration declined from saturation, so that the scope for activity ($R_{max} - R_s$) decreased rapidly. The concentration of oxygen at which $R_{max} = R_s$, called the level of no excess activity, is approximately the lethal concentration of oxygen (Fry, 1971).

As the scope for activity decreases with a reduction in ambient oxygen concentration, there is a decline in the sustained or prolonged swimming speed that can be maintained (Fig. 4.8) (Beamish, 1978).

Salinity as an environmental identity

Salinity as a lethal factor

Most fishes live either in fresh water or in sea water and experience only restricted changes in the osmotic and ionic characteristics of their environment. It is unlikely that fishes normally move outside their zone of tolerance for salinity, so they do not experience it as a lethal factor. Even cyprinids, which are primary freshwater species with a long evolutionary history in fresh waters, have high tolerances to salinity. The roach, bream, *Abramis brama*, and gudgeon, *Gobio gobio*, have upper lethal limits of 15–17‰ (Hynes, 1970). Along parts of the coastline of the Baltic Sea in Northern Europe, typical freshwater species such as pike and perch migrate in summer from rivers into the sea, where they feed actively in salinities exceeding 3‰ (Hansson, 1984). Some secondary freshwater species, that is species now confined to fresh water but of marine ancestry like the Cichlidae, can tolerate full sea water at least for short time periods (Lowe-McConnell, 1975). Juvenile coho salmon living in fresh water in midsummer survived for at least 7 days in a salinity of 19‰, and by the following February they were tolerating 25‰ though they were still in fresh water (Otto, 1971).

Salinity as a masking factor

For species that live in a relatively constant salinity, the costs of osmotic and ionic regulation can be considered as part of the inherent cost of living measured by R_s. For those species that do live in waters with a fluctuating salinity, or that migrate between water bodies with different salinities, then salinity acts as a masking factor. At some salinities, the extra costs of regulation may reduce the energy available for production unless the fish can compensate by increasing its rate of feeding.

The energy costs of osmotic and ionic regulation in the rainbow trout were estimated by comparing the metabolic rates of fish swimming at imposed speeds in water of different salinities (Rao, 1968; Fry, 1971). The lowest rates at a given swimming speed were at a salinity of 7.5‰ and the highest at 30‰. For the euryhaline cichlid, *Oreochromis*

niloticus, the lowest metabolic rate at a given swimming speed was at a salinity of 11.6‰ and the highest at 30‰ (Farmer and Beamish, 1969); metabolic rate was also higher at salinities of 0, 7.5 and 22.5‰ than at 11.6‰. If a salinity of 11.6‰ is approximately iso-osmotic with the blood plasma, the increase in oxygen consumption at higher and lower salinities indicates the energy costs of regulation. Thus, for *O. niloticus*, the proportion of total oxygen consumption required for osmoregulation was 29% at 30‰ and 19% at 0, 7.5 and 22.5‰.

Salinity as a directive factor

Salinity can also act as an important directive factor for migratory fishes that move between freshwater and marine environments (Chapter 5). Movement along a salinity gradient will change the energy costs of osmotic regulation, but can also change the conditions for feeding so that higher rates of food consumption may more than compensate for any extra costs of regulation (Chapter 5). Euryhaline species show a salinity preferendum that may change with the physiological state of the fish. In midsummer, coho salmon juveniles in a salinity gradient showed a bimodal response, with zones of preference corresponding to fresh water and to a salinity of 4–5‰ (Otto and McInerney, 1970). Later in the year, as the time approached when they would normally transform into smolts and migrate downstream to the sea, the juveniles lost their preference for fresh water, and the preferred salinity increased to 6–7‰. There was also an increase in their tolerance to salinity (Otto, 1971).

Other abiotic factors

Fishes may be excluded from some unusual and localized natural habitats by lethal levels of abiotic factors such as pH (hydrogen ion concentration), carbon dioxide or soluble toxic materials. However, in most habitats such factors do not normally reach such levels that they affect the fishes' pattern of energy allocation by acting as lethal, limiting or masking factors. But in polluted waters, or during intensive fish culture, they can become important.

Fish populations in parts of North America and Europe have declined because of the acidification of poorly buffered natural waters caused by precipitation polluted with the oxides of sulphur and nitrogen from industrial and urban sources (Baker and Schofield, 1985; Haines and Baker, 1986). In addition to a direct, lethal effect on fish, low pH also mobilizes toxic metals such as aluminium. The early life-history stages, eggs, larvae and fry are often more vulnerable to acidification than the adults, so too few fish reach sexual maturity for the abundance of the population to be maintained (recruitment failure). Several countries have established programmes to mitigate the effects of acidification by liming (calcium carbonate) the affected waters or adjoining land. In a liming study in south-west Scotland, a target pH of 6.0 to 6.5 was set to allow self-sustaining populations of brown trout to become established (Howells and Daziel, 1992). This pH was judged sufficiently high to mitigate the effects of both long-term and episodic events of low pH, together with any effects of toxic metals. Effects of acidic water on patterns of energy acquisition and allocation in fish have received inadequate attention, although the disruptive effects on osmoregulation and the effective functioning of the gills are reported (Howells, 1983).

Ammonia is the main nitrogenous excretory product of fishes, but it is highly toxic. Because ammonia is highly soluble, it poses no problem to fish living at natural densities in unpolluted waters, but it can become important when fish are housed at high densities in intensive fish culture. Un-ionized ammonia, NH_3, is more toxic than the ionized form, NH_4^+, and the proportion of the un-ionized form increases as the pH of the water increases. For rainbow trout a concentration of un-ionized ammonia of 0.07 mg l^{-1} NH_3 represents a safe upper limit (Smart, 1981).

Interactions of environmental factors

Although it is convenient to describe individually the effects of environmental factors on survival and metabolism, fish do not experience factors individually but in interacting combinations. The solubility of oxygen, for example, decreases as the temperature increases and is lower at higher salinities. The effects of low pH interact with the effects of metals mobilized at low pH and the concentration of calcium, the latter tending to ameliorate the toxic effects of the pH and metals. Experimental analyses of the effects of combinations of environmental factors on metabolic rates and energy allocation have hardly begun. Yet these analyses will be required if models are to be developed which predict how a fish should behave in its natural habitat in response to changes in such factors.

4.5 EXAMPLES OF ENERGY BUDGETS

Empirical studies of the energy budgets of fishes have attacked three problems. The first is to assess whether the components of an energy budget can be estimated with sufficient accuracy that a balanced budget can be drawn up. The second is to determine the quantitative effects of environmental factors on the energy budget. The third problem is to determine the pattern of changes in the allocation of energy over time, both ontogenetic and seasonal changes. These studies have provided the basis for the development of bioenergetics models used to estimate growth or food consumption in natural populations (Chapter 6). An interesting extension of such models is their use in modelling the uptake of pollutants including organics, such as polychlorinated biphenyls (PCBs), and heavy metals.

Balanced energy budgets

Solomon and Brafield (1972) described an early attempt to obtain a balanced energy budget for perch. Individual fish were held in a flow-through respirometer for periods of 28–54 days. The rates of oxygen consumption, faecal production and nitrogenous excretion were measured throughout the experimental period, and each fish was fed a known ration. Growth was measured by the change in weight over the time the fish spent in the respirometer and was then converted to energy units. This procedure gave estimates of C, R, F, U and P_s. The results are summarized in Table 4.1. The input was estimated as the energy of the food plus any energy derived from the body tissue (degrowth). The output was estimated from the sum of faecal, excretory and respiratory losses plus any positive growth. The estimated output ranged from 84.1% to 166% of the estimated income (in one experiment that

lasted 54 days the imbalance was 249%). Deviations from 100% may reflect changes in the body composition of the fish while they were in the respirometer (Brafield, 1985).

Effect of environmental factors

A study of the energy budget of the brown trout illustrated the effect that temperature has on the allocation of energy (Elliott, 1976b, 1979). As Fig. 4.9 illustrates, standard metabolic rate increased with temperature over a range from 3.8 to 21.7 °C, but the other components of the energy budget showed sharp maxima at about 18 °C. Above that temperature, the rate of food consumption and the metabolic costs of swimming and feeding (which were not separated) declined sharply. The scope for activity increased with temperature up to about 18 °C, then declined. For trout fed maximum rations, the scope for activity increased from 12% of the energy income at 3.8 °C to 76% at 17.8 °C. These effects of temperature emphasize two points. The first is the importance of temperature. The second is that a hetero-

Table 4.1 Energy budget of perch, *Perca fluviatilis*, held in flow-through respirometer*

| Experiment | Duration (days) | Budget component† | | | | | Energy imbalance‡ |
		C (kJ)	F (kJ)	U (kJ)	R (kJ)	P_s (kJ)	
1	28	41.17	6.44	3.39	24.31	8.41	103
2	28	58.91	9.58	4.48	32.05	12.05	99
3	28	9.46	1.42	2.97	17.78	−3.89	166
4	28	19.83	3.01	2.30	16.36	−0.75	105
5	28	4.94	0.75	1.38	10.88	−4.10	144
6	28	28.58	4.18	2.97	21.25	1.84	106
7	29	37.03	5.48	2.93	20.92	1.80	84
8	30	17.03	1.97	2.22	15.90	−2.09	105
9	31	6.95	0.92	1.84	15.06	−7.07	127

*Source: Solomon and Brafield (1972).
†C, food consumed; F, faeces, U, nitrogenous excretion; R, respiration; P_s, growth.
‡Energy imbalance is the sum $F + U + R + P_s$ as a percentage of C.

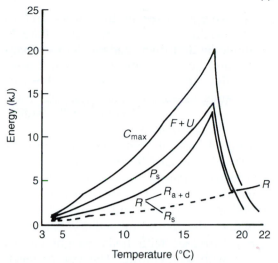

Fig. 4.9 Effect of temperature on components of energy budget of 50 g brown trout, *Salmo trutta.* C_{max}, maximum food consumption; $F + U$, faecal and excretory losses; P_s, growth; R, metabolic losses; R_{a+d}, metabolic losses due to activity and feeding; R_s, standard metabolism (shown by dashed line). Redrawn from Elliott (1976b).

thermal environment presents a fish with the opportunity to regulate behaviourally its pattern of energy allocation (Brett, 1983; Crowder and Magnuson, 1983).

Seasonal changes in energy allocation

Seasonal changes in the pattern of energy allocation were illustrated by a study of females of two populations of the threespine stickleback in Mid-Wales (Wootton *et al.*, 1980; Wootton, 1994a). Few fish in these populations survive beyond their second summer of life, so the study, which ran from September to April inclusive, covered more than half the usual life span. Components of an energy budget for an average-sized female were estimated from laboratory studies on feeding, growth and respiration and from field studies on growth rate. The estimates suggested that in late autumn and

early winter, the energy income was insufficient to cover the losses in faeces, excretion and metabolism. During this period, the fish showed a decline in their total energy content, although that of the ovaries was maintained or even increased slightly. In September and again in late winter and early spring, the energy income was sufficient to support an increase in the total energy. In spring, this increase partly consisted of a rapid increase in the energy content of ovaries prior to spawning. This study showed that when the energy provided in the food is less than the expenditures, there may still be organs that grow at the expense of decreases in the energy content of other parts of the body.

Summary energy budgets

A survey of fifteen energy budgets for carnivorous fishes (Brett and Groves, 1979) yielded a mean energy budget for young, growing fish of

$$100C = (44 \pm 7)R + (29 \pm 6)P_s + (27 \pm 3)E \qquad (4.14)$$

where $E = (F + U)$ and the means are given with 95% confidence intervals. For a few herbivorous fishes the mean budget was:

$$100C = 37R + 20P + 43E \qquad (4.14)$$

which reflects the low assimilation efficiencies of fish feeding on plant material. Such average budgets give a useful first view of energy allocation by fishes with different diets, but obscure the dynamic nature of energy allocation in relation to environmental changes and in relation to life history (Chapter 6).

4.6 APPLICATION OF BIOENERGETICS TO EXOTOXICOLOGY STUDIES

Fish can take up pollutants either directly from the water or in food (Kooijman, 1993).

Uptake from water is related to the rate of respiration, uptake in food is related to the rate of consumption, while loss may be through egestion or excretion (Post *et al.*, 1996). From a knowledge of water temperature and the rate of growth of the fish, a bioenergetics model can be used to predict the rates of respiration, consumption, egestion and excretion and hence the rates of accumulation of the pollutant. Using this approach, Post *et al.* (1996) modelled the uptake of mercury by juvenile yellow perch, *Perca flavescens*, in three Canadian lakes. Their model suggested that the total uptake and the relative importance of uptake from water or in food varied seasonally in relation to changes in water temperature, prey availability and body size of the perch.

Concern over the accumulation of PCBs in piscivorous lake trout, *Salvelinus namaycush*, in Lake Michigan led to the development of predictive models based on three submodels (Madenjian *et al.*, 1993). The submodels were a bioenergetics model of growth and consumption (Chapter 6), a predator–prey encounter model (Chapter 3) and a PCB accumulation model. The model predicted the growth and accumulation of PCB of individual trout, an example of an individual-based model (IBM) (Chapter 1). The model was used to evaluate the likely causes of the variation in PCBs found in the lake trout, including the effect of variations in PCB levels in the prey fish.

4.7 SUMMARY AND CONCLUSIONS

1. The food energy of individual fish is either lost in waste products and heat (maintenance costs) or incorporated as chemical energy in new tissue as growth and gametes (production). The sum of maintenance costs and production equals the energy income.

2. The components of an energy budget for a fish can be estimated using direct or indirect calorimetry.

3. The energy losses in the faeces and as nitrogenous waste products (largely ammonia and urea) depend on the quantity and quality of the food eaten. Losses are also affected by temperature. For carnivorous fishes, the total losses as waste products are about 10–30% of the energy of the food consumed.

4. Three levels of metabolic rate can be defined: standard (resting), routine (spontaneous swimming) and maximum active metabolic rates. The difference between maximum active and standard rates defines the scope for activity. Metabolic rate increases with body weight of fish and with temperature. The rate of metabolism also increases when food is consumed (apparent specific dynamic action).

5. The effects of abiotic factors on the activities of individual fish can be classified as lethal, controlling, limiting, masking or directive. Temperature can act as a lethal factor, but its main importance is as the major controlling factor through its effect on the rate of metabolic processes. It also acts as directive factor, with fish selecting a preferred temperature from within a gradient. Oxygen, because of its low solubility, can be an important lethal factor in environments prone to hypoxia. It can also be important as a limiting factor. The metabolic costs of osmo- and ionoregulation make salinity a masking factor for those species that experience a change in salinity. For species that migrate between waters of differing salinities, it can also act as a directive factor.

6. Energy budgets illustrate the effects of environmental factors on energy allocation between maintenance and production. They are also used to describe seasonal and lifetime changes in the pattern of allocation.

7. Bioenergetic principles can also be used in models that predict the dynamics of uptake and elimination of organic and inorganic pollutants.

5

USE OF TIME AND SPACE

5.1 INTRODUCTION

Fishes use the dimensions of time and space in many ways. Individuals of some species spend all their lives close to where they were spawned. In other species, individuals make long migrations covering hundreds or thousands of kilometres. There are species that show vertical migrations, with individuals moving up and down in the water column. The ecological significance of many of these patterns of movement is unexplained. The reproductive advantage gained is not always obvious. In several species, for example the European trout, *Salmo trutta*, some fish migrate, covering long distances, yet other fish remain resident in one area and show only restricted movements (Jonsson and Jonsson, 1993). A single watershed frequently holds both migratory and resident *S. trutta*. Why is it advantageous for some individual trout to migrate, while other coexisting individuals remain as residents in the watershed?

Rigorous studies of the ecological significance of the use of time and space by fishes are in their infancy. The allocation of time by an individual can be described by a time budget:

$$T_{tot} = T_1 + T_2 + \ldots + T_n \qquad (5.1)$$

where T_{tot} for a daily period would be 24 hours, T_i is the time allocated to activity i, and n is the total number of activities the fish can perform. The energy allocation to activity i, EA_i, can be estimated as:

$$EA_i = e_i T_i, \qquad (5.2)$$

where e_i, is the average rate of energy expenditure per unit time while performing activity i. A pattern of time allocation can thus be translated into a pattern of energy allocation.

Time budgets of parrotfish on a Caribbean coral reef revealed that during daytime over 90% of the time was spent feeding, including time spent swimming between feeding spots (Videler, 1993). Swimming accounted for 50% of the daytime budget. Most of this swimming was powered by the pectoral fins. Only about 5% of the budget was spent swimming using the caudal fin (Chapter 2).

The time (and energy) spent performing an activity will have associated with it both benefits and costs (Fig. 5.1). For example, the time spent feeding yields a benefit in the energy and nutrients acquired, but the costs may include an increased risk of predation, a reduction in the time available for other activities such as mating and the energy expended on foraging. In principle, both benefits and costs could be measured as the increase (benefit) or decrease (cost) in the lifetime production of offspring, that is in units of fitness. An individual will increase its fitness if it allocates its time in a way that increases the net benefit, defined as the difference between benefit and cost (Fig. 5.1) (Krebs and McCleery, 1984). As the environmental conditions change, for example the change in light intensity during the 24 h diel cycle, the difference between benefit and cost may also change. A fish may be able to increase its fitness by moving or switching to a different activity. The movement or change in activity will impose some cost, so for the

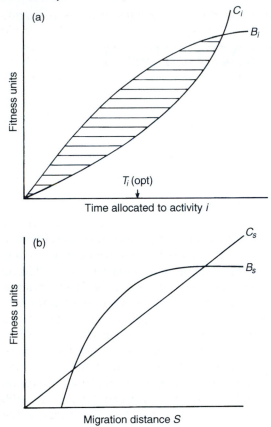

Fig. 5.1 Diagrams of cost–benefit curves. (a) Hypothetical cost–benefit curve for time spent in activity i, T_i; B_i, benefit curve for activity i; C_i, cost curve for activity i; shaded area, net benefit; $T_i(\text{opt})$, time at which difference between benefits and costs is maximized. (b) Hypothetical cost–benefit curve for migration distance, S.

behavioural change to be adaptive the net benefit must increase. The ecological problem is to determine why the behavioural change leads to an increase in net benefit.

The technical difficulties of measuring the lifetime production of offspring by individuals that show differences in their time and energy allocations are formidable. Some insight into the possible costs and benefits may be obtained from descriptive studies of the temporal and spatial patterns shown by

fish in different environments and from quantitative analyses of the energetic costs of spatial movements. Individual-based models that include rules governing the movement of individuals between habitats can generate predictions of lifetime reproductive success if the models can be based on good biological data (Tyler and Rose, 1994).

5.2 METHODS OF STUDYING THE USE OF TIME AND SPACE

The techniques used to study the allocation of time and the pattern of movements of fish range from direct observation to the use of complex electronic tracking equipment. The development of free-diving systems for use in shallow waters opened up the possibility of direct observation of fishes in their natural habitats such as coral reefs or shallow lakes. The advantage is that the species and their behaviour are observed directly and such information does not have to be decoded from indirect evidence such as sonar or radio pulses. A disadvantage of direct observation is that often the fish can be observed only for short periods and their behaviour may be disturbed by the presence of the observer.

The miniaturization of electronic equipment has made it possible to build ultrasonic acoustic or radio tags that can be attached either externally or internally to larger fish and so permit the tracking of fish movements for periods of hours or days (Weatherley and Gill, 1987; Lucas *et al.*, 1993). The behaviour of fish living in abyssal ocean depths is being studied by surrounding the tag with food as bait to encourage the fish to swallow the tag (Priede and Smith, 1986), or by attaching bait to a camera (Priede and Merrett, 1996). Electronic tags may also transmit data on physiological variables such as heartbeat and on environmental variables such as temperature. The disadvantages of electronic tags include the requirement for receiving stations that remain within the range of the transmitter, the lack of information on the behaviour of the tagged fish and any effects that the presence of the tag may have on the behaviour of the fish. The present size of electronic tags restricts their use to relatively large fish.

An alternative electronic method, passive integrated transponders (PIT tags), has been developed for use with small, stream fish (Armstrong *et al.*, 1997). A tag, with a unique code, is implanted in the body cavity. An antenna located on the stream bed detects when a fish carrying a tag crosses. Decoders and an associated computer monitor the movements of known individuals.

Less sophisticated tags or marks, such as those used in mark–recapture methods of estimating population abundance (Chapter 10), will also give information on the movement between the place at which the fish is tagged and the place at which the fish is recaptured. No information on the behaviour of the fish between the times of tagging and recapture is available unless the tagged fish are observed directly.

Echosounding devices used by fisheries to locate shoals of fish can also provide information on the shape and pattern of movement of shoals, including any vertical migrations (MacLennan and Simmonds, 1992; Misund, 1997). This technique suffers from the disadvantage that the identity of the fishes under observation is not known for certain. Samples obtained by fishing may minimize the danger of misidentification, and refinements of both the acoustic signal and its analysis are improving species identification (Misund, 1997).

5.3 TEMPORAL PATTERNS OF FISH MOVEMENT

The movement of fishes is strongly influenced by the temporal patterns to which the environment is subject. The dominant temporal patterns are the diel pattern of light and dark and the annual cycle of seasonal change. Other important patterns include the ebb and flow of the tides in the littoral zone of seas. At high latitudes, the seasonal cycle is marked by large changes in the mean daily temperature and the photoperiod. In subtropical and tropical regions, changes in these two abiotic factors are usually small but freshwater systems, and even adjacent coastal habitats, may be strongly influenced by the succession of wet and dry seasons (Payne, 1986; Lowe-McConnell, 1987).

Patterns of movement during the diel cycle

In shallow waters, the ability of fish to forage, detect predators or attract mates may vary with the daily cycle of light and dark. Species are usually active only during limited periods within the diel cycle: diurnal species in daytime, nocturnal species at night and crepuscular species at dawn and dusk, periods of rapidly changing light intensities. There is some regularity in the proportion of species that fall into these three categories, both in tropical and in temperate shallow-water assemblages of fishes (Helfman, 1978, 1993). Typically, one-half to two-thirds of the species are diurnal, one-quarter to one-third are nocturnal and about one-tenth are crepuscular. A complication is that a species that normally feeds during the day may have periods of reproductive activity at other times. Yellow perch, *Perca flavescens*, in fresh waters of North America, is a diurnal feeder but breeds at night (Helfman, 1981). On nights that are brightly moonlit, some diurnal feeders may continue to be active (Allen and Wootton, 1984).

There can be seasonal changes in the pattern of activity. During the summer, juvenile Atlantic salmon in streams feed mostly during the day. In winter they switch to feeding at night. The switch occurs when the water temperature drops below about 10 °C (Fraser *et al.*, 1993). This switch may make the juveniles less susceptible to predation during winter, when low temperatures reduce burst swimming speeds (Chapter 4), but there is a cost. The foraging efficiency of the juveniles at night is only about 35% of their efficiency during the day (Fraser and Metcalfe, 1997).

The effect of the diel cycle has been most fully studied for the fish assemblages of coral reefs. These assemblages are rich in species (Chapter 12). The fishes live in shallow, warm water and experience a photoperiod in which the lengths of day and night are similar while the transition between the two is rapid. This transition is accompanied by a change in the assemblage of active species.

During the day, the fishes active on or close to the reef include many brightly coloured species, the so-called poster-coloured forms including butterflyfishes (Chaetodontidae) and some damselfishes (Pomacentridae). Almost all the herbivorous species are diurnal; there are both solitary and shoaling herbivores. Diurnal shoaling zooplanktivore species are also abundant. Except for the herbivores, many diurnal species have nocturnal analogues, for example the squirrelfishes (Holocentridae) are nocturnal zooplanktivores (Hobson, 1972; Helfman, 1993).

Within the diel cycle on a Hawaiian reef, a sequence of events proceeds during the transition between day and night which is reversed at dawn (Hobson, 1973). As sunset approaches, the diurnal fishes make vertical or horizontal movements that take them from feeding areas to areas of shelter. Just before sunset, these fishes take shelter, some in holes in the coral, others resting in depressions or under coral heads. Some species of parrotfishes (Scaridae) secrete a mucous envelope in which they rest. Then follows a period of about 20 minutes in which there is little general movement on the reef, but when large predators are active, striking upwards at any fish that are slow to leave the water column and are silhouetted in the dying, down-welling light. This quiet if lethal period ends as light levels fall still further and the nocturnal fish emerge from shelter.

In the final phase of this transition, the nocturnal fish swim to their feeding areas. A comparison of the rods and cones of the retinas of diurnal or nocturnal prey with those of their predators suggested that the twilit periods are times when the piscivores have a visual advantage (McFarland and Munz, 1976; Hobson *et al.*, 1981; McFarland, 1991).

In temperate lakes, there is a comparable transition between day and night activity in fish assemblages, but the timing and the distinctions between diurnal and nocturnal species are more blurred. Helfman (1978, 1981, 1993) described the diel cycle in North American lakes. The diurnal species include zooplanktivores, carnivores like the yellow perch feeding on benthic and phytophilic invertebrates, and piscivores such as pike. Nocturnal species include coregonids feeding on zooplankton, benthivores such as the eel and piscivores including the bowfin, *Amia calva* (Amiidae), a holostean. When comparing the sequence of events during twilight in temperate lakes with that on a coral reef, Helfman (1993) listed the lack of a quiet period, a more prolonged change-over, greater intraspecific and ontogenetic variability in timing and a greater apparent overlap between species in change-over time. Temperate diurnal species are less likely to seek shelter at night than to rest in barren or sparsely vegetated sites or in clearings in vegetation. In response to the greater unpredictability of their abiotic environment, temperate species show a behavioural flexibility within the diel cycle that has not been recorded for coral reef species.

Although behavioural changes in response to the diel cycle may partly depend on the immediate environmental conditions, especially light levels, the timing may also be partially controlled by an endogenous circadian rhythm (Thorpe, 1978). Freshwater species that live at high latitudes, where for a period in the summer the sun does not set, can show a breakdown in the circadian rhythm and become arhythmic during mid-summer (Muller, 1978). This is a further illustration of the physiological and behavioural plasticity that is an important component of the response of some fishes to highly variable environments.

Patterns of movement during the tidal cycle

Fishes living in the marine littoral zone encounter an additional temporal pattern, the tidal cycle, which alternately immerses and exposes their habitat. Some species avoid exposure to air by moving off shore during low water. Juvenile flatfish (Pleuronectiformes) move up and down sandy beaches as the tide ebbs and flows. Other fishes including blennies (Pholidae), sculpins (Cottidae) and gobies (Gobiidae) are morphologically, physiologically and behaviourally adapted to remaining in the intertidal zone (Gibson, 1969, 1993). A component of this adaptation may be an endogenous tidal rhythm in locomotory activity such as that exhibited by the shanny, *Lipophrys (Blennius) pholis*. Individual fish in the laboratory display cyclical changes in activity level, even under constant environmental conditions. The periods of high activity coincide with the predicted times of high tide on the shore from which the fish are removed (Gibson, 1993). This endogenous rhythm will be modified in the natural environment by the responses of the fish to the immediate environmental conditions. A second component of the adaptation to an intertidal life is the ability to find shelter because of an intimate knowledge of the topography of the local area (page 96).

5.4 SPATIAL DISTRIBUTION

The spatial distribution of fish will reflect the response of individuals to several factors. Fish will respond to the intrinsic qualities of a location, including physico-chemical factors such as current speed, temperature, salinity and the physical structure of the locality (Chapter 4). A second factor is the presence of other fish. During the breeding season, distribution can reflect the presence of suitable mates (Chapter 7). Predators and interspecific competitors can also affect distribution (Chapters 8 and 9). In this chapter, the emphasis is on the effect of non-breeding conspecifics. The effect of food on the spatial distribution of fish will depend partly on the effect of the intrinsic qualities of the locality on the availability of food and partly on the effect of other fish (Chapter 3).

Habitat selection

This is choice behaviour in which the individual responds to a more or less complex set of environmental stimuli by showing a preference for living in an area that can provide that set. It is assumed that the choice tends to increase the fitness of the individual. The results of habitat selection may be seen at different spatial scales (Kramer *et al.*, 1997). Individuals of a species may be found in a river rather than a lake. Within the river they may be found in pools rather than riffles, and within the pool, near the bottom rather than at the surface. This hierarchy of scales may reflect a hierarchy of choices by the fish, or may simply reflect choices of particular stimuli on a localized scale.

Sale (1969) described the stimuli that are important in habitat selection by juvenile manini, *Acanthurus triostegus*, a surgeonfish found on Hawaiian reefs. The juveniles live in a shallow, rocky environment. Choice tests indicated that the most important stimuli were the presence of a substratum, the presence of cover and food algae, and a suitable depth of water.

Within a habitat that contains a variety of microhabitats, the balance between the costs and benefits experienced within the various microhabitats may differ for different age groups in a population. The black surfperch, *Embiotoca jacksoni*, lives in the rocky sublittoral zone of the Californian coast. Although adults and juveniles live in the same habitat, the adults prefer to forage for invertebrates in areas covered with an algal turf, while the juveniles pick invertebrates off foliose algae (Holbrook and Schmitt, 1984). If different species coexisting in a habitat show intrinsic preferences for moving into different microhabitats, this habitat selection potentially provides a mechanism that allows the coexistence of two or more species (Chapters 9 and 12).

The correlations between habitat characteristics and the presence or absence of fish have led, particularly for riverine species, to the development of methods for assessing habitat quality. One example is the instream flow incremental methodology, which uses the microhabitat variables of water depth, velocity, and nature of substratum to assess flow requirements for riverine fishes in North America (Allan, 1995). In Wales (UK), a more detailed set of habitat variables, including type of substrate and type of cover, has been used to predict the density of young salmonids in streams (Milner *et al.*, 1993). The disadvantage of such models is that they are not derived from an understanding of the behavioural response of fish to habitat features, but rather from correlations between fish density and habitat variables (Elliott, 1994).

A causal explanation of microhabitat selection was illustrated by a study of the current speeds selected by rainbow trout and rosyside dace, *Clinostomus funduloides*, in a stream in North Carolina, (Hill and Grossman, 1993). A bioenergetics model was developed for each species that predicted the energy gain and energy expenditure at a given current speed. Both species fed on invertebrates drifting in the water column. The total energy content of prey increased with current velocity, but at high speeds, the success of fish in capturing the drifting prey declined. Energy expenditure by fish holding a position also increased with current speed.

The model suggested that the current speeds selected by the fish were often close to the speeds at which the net energy intake was maximized.

There may also be age-related changes in use of habitats. The grayling, *Thymallus thymallus*, studied in French rivers, provides an example (Sempeski and Gaudin, 1995; Sagnes *et al.*, 1997). The yolk-sac larvae emerge from gravel nests (Chapter 7) in the main channel of the river, but move to shallow water by the banks of the river, characterized by slow water currents. The larvae, 15–20 mm in length, prefer current velocities of less than 200 mm s^{-1}, and live close to the surface. As they grow, they move away from the banks and closer to the main channel, with water velocities of 200–300 mm s^{-1} and the juveniles become more benthic. Finally, at a length of about 60 mm, they move into the main channel, the adult habitat. During these habitat changes, the fish show ontogenetic changes in morphology, in particular the development of a streamlined body form suitable for the hydrodynamic conditions of the main river channel.

In rivers, habitat units on a scale of metres can be defined by geomorphic features and flow characteristics (Rabeni and Jacobson, 1993). This classification is independent of the distribution of fish. It includes habitat units such as riffles, runs and backwater pools. Common Centrarchidae (bass and sunfish) showed changes in distribution between these habitat units with age in a stream in Missouri. There were also differences between the species in their use of the habitat units (Chapter 9).

Where there are ontogenetic changes in habitat selection, individuals must be able to move from habitat to habitat at the relevant age (or perhaps size). Modifications of the environment that restrict the possibility of these (size- or) age-related movements will have adverse effects on the fish population. In assessing environmental quality, it is crucial to consider habitat selection over the whole life history of a species.

Effects of conspecifics

The presence of conspecifics may affect the spatial position of an individual because they adversely affect the quality of the habitat for that individual. However, the individual may react to the presence of other fish by conforming its movements to theirs. This social response will result in the formation of shoals (page 94). If a fish avoids or repels other fish, the consequence may be some form of territoriality. Another possibility is that an individual is indifferent to the presence of other fish so that its pattern of movement and use of space is uninfluenced by them.

Density-dependent habitat selection

In the absence of conspecifics, an individual fish can select the best habitat from those accessible. Conspecifics may modify the quality of that habitat by exploiting its resources. As the density of conspecifics increases, the quality of the habitat declines. In this density-dependent context, an individual may do as well by selecting a habitat that is intrinsically of lower quality, but which also holds a lower density of conspecifics. The behaviour of the individual depends on the behaviour of other individuals, but the individual is reacting to the consequences of the presence of conspecifics rather than to social interactions with other fish. If individuals are free to move from habitat to habitat and can accurately assess the quality of a habitat, the resulting distribution of individuals across habitats is called the ideal free distribution (IFD). A characteristic of the IFD is that no individual can do better by moving to another habitat. At a given population density of fish, the frequency distribution of the population across habitats reflects the intrinsic quality of the

habitats (Sutherland, 1996; Kramer *et al.*, 1997). Chapter 3 (page 48) gives examples of the application of the IFD model when food is the resource being depleted.

MacCall (1990) has developed an ambitious extension of the concept of density-dependent habitat selection for fish occupying large geographical ranges such as pelagic or demersal marine fish. His model suggests that at low population densities, the fish occupy only the best habitats. If population abundance increases, the population occupies a wider and wider geographical range as intrinsically poorer habitats are occupied, because of the density-induced declines in quality in the better habitats. Population densities reflect the intrinsic qualities of the habitats, with higher densities in the better habitats and lower densities in poorer habitats. However, the fitness of fish is the same across the whole range. If valid, the model has important implications for fisheries management. The densities in the better habitats will remain high, even if fishing reduces overall abundance. By concentrating its fishing effort in the better habitats, a fishery can continue to obtain good catches even as fish abundance continues to fall. The range occupied by the population contracts as the poorer-quality habitats are abandoned. The conservation of the population would call for a restriction of fishing in the high-quality habitats to maintain a core population, while allowing heavy exploitation in poorer habitats that are replenished by density-dependent processes from the core. MacCall (1990) illustrated his model by applying it to the distribution of spawning northern anchovy, *Engraulis mordax*, off the coast of California.

The evidence for the model is equivocal. The geographical range of cod in the southern Gulf of St Lawrence, Canada, expanded as abundance increased (Swain and Kramer, 1995). The spatial distribution of the cod in relation to temperature altered. With an increase in abundance, the cod tended to select colder waters than at low abundance.

This change in temperature preference may have a bioenergetics interpretation. Experimental studies suggest that in fish, as food levels decrease, the temperature at which growth rate is highest also decreases (Chapter 6). If the increase in density of cod reduced their food supply, they could protect their growth rate by selecting colder waters. In contrast, for American plaice in the southern Gulf, there was no increase in the geographical range with an increase in abundance (Swain and Morin, 1996).

Ideal free distribution models assume that the rule governing fish movement is that an individual moves if, by doing so, it can do better in terms of fitness than by staying at its present locality. This leaves open the question of what fitness-maximizing rule is being used. One possibility is that the rate of food intake, and so growth, is maximized (Chapter 6). Another possibility is that the risk of mortality is minimized (Chapter 10). A more subtle rule is that the ratio of mortality to growth is minimized. This latter rule was developed to interpret the size-dependent movements of fish in relation to the distribution of food (Chapter 3) and predators (Chapter 8) (Werner and Gilliam, 1984).

An individual-based simulation model that used quantitative values for juvenile yellow perch feeding on zooplankton suggested that the rule that maximized survival over 150 days depended on the spatial distribution of zooplankton and predators, and on the density of the juveniles (Tyler and Rose, 1997). This study illustrated the potential power of individual-based models to relate fish movement to growth and survival, but also warned of the potential complexity of these relationships.

Shoaling

In some circumstances, the benefits of associating with other fish may outweigh the costs. A shoal is a group of fish that remain together for social reasons (Pitcher and Parrish, 1993). School describes a shoal in which the fish are swimming in the polarized and synchronized pattern that characterizes many pelagic species like the herring and mackerel. An aggregation occurs when fish group because individuals are responding to some environmental cue such as temperature or current speed rather than to social cues.

Shoaling may have several functions (Pitcher and Parrish, 1993). The relationship between shoaling and foraging has already been discussed (page 52). An individual may also reduce its risk from predation by joining a shoal (page 186). In the context of the present chapter, two other possible functions are also relevant. Shoaling may improve the accuracy of migration, thereby minimizing the risk to an individual of failing to find a suitable habitat and perhaps also reducing the energy cost of the migration by reducing the time spent swimming in an inappropriate direction. A fourth function applies only to schools. An individual might gain some hydrodynamic advantage by being a member of a school and so reduce its energy expenditure on swimming. This last function cannot be substantiated on present evidence.

The relative importance of the functions of shoaling behaviour may change during the life cycle of the fish. In some species, the benefits of shoaling may exceed the costs for all or most of an individual's lifetime. Such species are sometimes called obligate shoalers because individual fish are rarely, if ever, found away from a shoal. Species that live in the essentially unbounded environment of the pelagic zone of seas or large lakes are frequently obligate shoalers. These are environments in which there are no physical structures in which to shelter. Other species are facultative shoalers: individuals join shoals at some times, but live as solitary fish at other times. Juvenile salmon show territorial behaviour while feeding in streams, then join shoals during their downstream migration to the sea. The analysis of shoal structure and function now emphasizes the balance between costs and benefits for the individuals that make up a shoal (Pitcher and Parrish, 1993; Krause, 1993, 1994).

Dominance, territoriality and home range

The ideal free distribution (IFD) assumes that individuals are unconstrained in their movements by the presence of conspecifics. Shoaling behaviour violates this assumption, with individuals consorting together. The assumption is also violated when movement is inhibited by the presence of conspecifics. Dominance and territorial behaviour impose such inhibition (Grant, 1993; Huntingford, 1993). The effect of dominance is to allow the dominant individuals to obtain a disproportionate share of resources at the expense of subordinate individuals. Territoriality occurs when the dominance is associated with site attachment. The territorial individual defends a locality and sequesters the resources therein. The resource that is defended may be food, a spawning site, a refuge from predators or a combination of resources. In comparison with the IFD, the effect of the presence of dominant individuals is to lower the quality of the habitat to any individual trying to settle compared with its quality to residents. This circumstance is modelled by the ideal despotic distribution (Sutherland, 1996).

The resources should be defensible by virtue of their spatial or temporal distribution and predictability (Grant, 1993). The benefits obtained by monopolizing the resources must exceed the costs of their defence. The size of the defended area will depend on the balance between the costs and benefits of holding a territory of a given size (Davies and Houston, 1984). As this balance changes, the size of the territory can change. Hixon (1980) developed a theoretical analysis of the benefits and costs associated with a feeding territory in relation to territory size in terms of the time devoted to foraging and

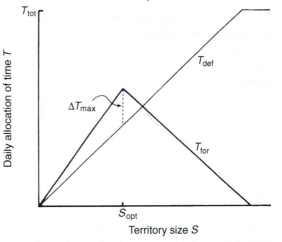

Fig. 5.2 Simplified diagram illustrating Hixon's (1980) model of the relationship between territory size, S, and daily allocation of time, T. T_{tot}, total time available; T_{for}, time spent foraging in territory; T_{def}, time spent defending territory; ΔT_{max}, maximum time available for foraging; S_{opt}, optimum territory size at which time available for foraging is maximized.

defence (Fig. 5.2). As territory size increases, the time devoted to defence increases. The time devoted to foraging at first increases because a larger territory provides more food, but then declines because the time available for both foraging and defence is limited. Territory size of juvenile coho salmon decreased with an increase in food availability (Dill *et al.*, 1981), a result predicted by Hixon's model. But well-fed, mature male threespine sticklebacks expanded their territories at the expense of neighbours that had received less food (Stanley and Wootton, 1986). For juvenile salmonids the territory is solely a feeding territory (Frost and Brown, 1967). For a male stickleback, the territory provides an area in which mature females can be courted and mated as well as an area in which to forage (Wootton, 1984a).

A study of the behaviour of the medaka,

Oryzias latipes, showed how the use of space can change in relation to the presence of other conspecifics and the spatial distribution of food (Magnuson, 1962). With limited food spread throughout a tank containing several fish, the medaka established a dominance hierarchy with the dominant fish obtaining more food than the subordinates. With the food in short supply but also localized, the dominant fish became territorial, each defending an area around the food. When food is dispersed, defence of an area is not economical (Grant, 1993). Territoriality also broke down when there was a high density of intruders into the area where food was located, which suggests that the high cost of defence made territoriality no longer economical.

The interaction between bioenergetics and behaviour was illustrated by a study of Arctic grayling, *Thymallus arcticus* (Hughes and Dill, 1990; Hughes, 1992). When groups of four fish were introduced into a pool, they adopted positions such that the profitability of the position in terms of energy returns correlated with the dominance status of the individual. The highest-ranking fish adopted the most profitable site and the lowest-ranking fish the least profitable. When the highest-ranking grayling was removed, its position was taken by the next-ranking grayling and the least profitable position was abandoned as each remaining fish moved into the site vacated by the fish immediately above it in rank.

In such a social system, the location of individuals may change as the profitability of locations changes. However, the dominance relationships continue to regulate the access to profitable sites (Jenkins, 1969).

Territories frequently have some focal point or points around which the fish spends a high proportion of its time. *Ophioblennius atlanticus* is a tropical blenny living in shallow water on Caribbean coral reefs. Each fish maintains a permanent territory within which it grazes algae. A time-budget study of individual blennies showed that about 50% of their time was spent in about 15% of the area of the territory (Nursall, 1977, 1981).

Most field studies on the territorial behaviour of fishes are carried out on coral reefs or in small streams and pools, environments in which direct observations can be made. In larger or more turbid waters, direct observation of fishes is not usually possible and so territorial defence cannot be seen. Gerking (1959) reviewed evidence that in many populations of freshwater fishes, individual fish remain in restricted areas for long periods of time. He suggested that they either have home ranges, which are not defended, or are territorial. This view was supported by some studies in which the movements of individual freshwater fish were tracked using radio or sonic telemetry (O'Hara, 1993). However, a study of movement of cyprinids in rivers in eastern England suggested that fish were not living in home ranges but formed mobile populations influenced by the flow characteristics of the river (Linfield, 1985). Within a population, some individuals may be relatively mobile while others move less, remaining in restricted areas. Juvenile brook trout observed in pools in streams tended to spend either a high or a low portion of their time moving while searching for food (McLaughlin *et al.*, 1992). The degree of mobility shown by fish in populations has important implications for their management, so detailed studies on the typical patterns of movement by fish in rivers and lakes are urgently needed (O'Hara, 1993).

There can be benefits of site attachment in the form of a home range even if resources are not sequestered. Fishes that restrict their activities to a limited space may learn its characteristics in intimate detail. Littoral species frequently show restricted movement and tend to return to the same site at low tide. Studies in which such fish were displaced short distances showed that the fish had a good knowledge of their immediate environment and could return to their home site after the displacement (Gibson, 1993). A spectacular example of this ability is provided by the tropical goby, *Bathygobius soporator*. If this fish is disturbed at low tide, it will jump accurately into an adjacent pool. The fish learns the topography of the area by exploring it at high tide (Aronson, 1951).

The size of home ranges and territories is related to body size. A general survey of freshwater species found that the relationship between estimated area of home range, *S*, and body weight, *W*, took the form:

$$S = aW^b \qquad (5.3)$$

where the value of b was 0.58 (Mins, 1995). Home ranges in lakes were significantly larger than in rivers. In a survey of the juveniles of stream salmonid species, the relationship between territory size and body weight took same form, but the value of b was 0.86 (Grant and Kramer, 1990). Atlantic salmon in a stream in New Brunswick had a larger value of 1.12 (Keeley and Grant, 1995). The allometric form of the relationship between home range (territory) size and body size recalls the allometric relationships of food consumption and metabolic rate with body size (Chapters 3 and 4).

Because of its effect on the spatial distribution of fish, territorial behaviour may have effects on population density, a topic taken up in Chapter 10.

5.5 MOVEMENT PATTERNS

Strong attachment to a site will not always be appropriate behaviour. Resources are not always localized or predictable. As fish grow, their requirements may change, necessitating a change in location. Fish may also move location as the characteristics of their environment change, for example seasonal changes in water temperature or level.

Weakly directional movements

Particularly when feeding, fish may move from site to site and not stay in any focal area. There may be a diffuse home range, but in few cases can an individual fish be tracked for a sufficiently long period to show that it is moving within a definable area. The movements of six pike, *Esox lucius*, in Lac Ste Anne in Alberta showed no evidence of home range behaviour (Diana *et al.*, 1977; Diana, 1980). The pike were tagged with ultrasonic transmitters and tracked for periods ranging from 5 to 51 days. Although inactive for much of the time, some pike ranged over distances exceeding 0.5 km. By contrast, most marked 0+ and 1+ pike in a small lake in the Netherlands were recaptured a year later within 100 m of the point of release (Grimm and Klinge, 1996). On their feeding grounds in the North Sea, plaice change position by entering mid-water at night and taking advantage of the tidal currents. The timing of their entry into mid-water is such that their movements lack a strong directional component, although at other times of the year the plaice do make strongly directional movements to and from their spawning grounds (Arnold and Cook, 1984). The movement patterns shown by the pike in Lac Ste Anne and the feeding plaice suggest that they are exploiting resources that are not easily defensible and that familiarity with their immediate environment may not offer high benefits. However, such patterns of movement are poorly studied and understood. They may grade into home range or territorial behaviour in appropriate circumstances, and longer-term studies may indicate regular patterns of visits to different localities.

Migrations

Many commercially important species – including salmonids, tunas, cod, plaice and herring – show migratory behaviour. Much attention has been paid to the mechanisms that migrating fish use to orientate their movements (Harden Jones, 1968; Hasler and Scholz, 1983; McCleave *et al.*, 1984;

McKeown, 1984; Smith, 1985). Less attention has been paid to the ecological aspects of the migrations (Northcote, 1978, 1984).

A convenient definition of migration is those movements that result in an alternation between two or more separate habitats, occur with a regular periodicity and involve a large proportion of the population (Northcote, 1978, 1984).

Horizontal migrations

Species such as tunas, herring or cod, the migrations of which take place entirely within the sea, are called oceanodromous. If species' migrations take place entirely within fresh waters, for example migratory fishes of floodplain rivers, they are called potamodromous. Those interesting species having migrations that take them between the sea and fresh water are diadromous (McDowall, 1987). Anadromous species spawn in fresh water but spend a significant proportion of their life in the sea, for example many salmonids and shad (Clupeidae). Catadromous species spend much of their lives in fresh water but migrate to the sea to breed. The best-known examples are the American and European eels. Amphidromous species migrate between the sea and fresh water but this movement is not directly linked to reproduction.

The scale of the movements can range from tens of metres to many hundreds or even thousands of kilometres. The periodicity may be that of the diel cycle or it may be the length of the life span of the fish. Are there any features common to the various patterns of directed movements?

A potentially fruitful model of migration assumes that the balance between the fitness benefits and costs of residence at a particular locality changes at different stages in the life cycle. If the benefits that accrue from moving to another locality outweigh the costs of making the move, migration is adaptive. One cost is the energy expenditure associated with swimming. At optimum swimming speeds, the cost of transport decreases with an increase in body size (Videler, 1993) (Chapter 4). Roff (1988) has argued that migration is correlated with increased size at maturity, and that small individuals do not migrate as far as large individuals. Other costs are those associated with moving into an unfamilar environment.

Three types of habitat can be recognized: one suitable for reproduction, one suitable for feeding and one suitable as a refuge in periods of unfavourable abiotic or biotic conditions. Individual fish can maximize their fitness if they move between these habitats at the appropriate times in their life span (Fig. 5.3).

The pattern of migration may be more complicated if the habitats suitable for feeding or refuge are different at different stages in the life cycle. Many species have nursery areas in which the juvenile fish feed or take refuge, but these are not revisited once sexual maturity has been reached (Harden Jones, 1968; Northcote, 1978, 1984). The availability of suitable nursery areas can be an important factor in determining the abundance of a population. In many rivers, shallow, well-vegetated margins and backwaters are important nursery areas for larval and juvenile fishes (Mills and Mann, 1985). River modifications to improve drainage or navigation that involve straightening and dredging may destroy such nursery areas to the detriment of the fish populations (O'Hara, 1993). A comparison of unregulated and flow-regulated rivers in Alabama found that the regulated river had a lower abundance of larval fish in nursery habitats and the larval assemblage contained a lower proportion of typical riverine species (Scheidegger and Bain, 1995). In tropical and subtropical regions, coastal mangrove forests are cut down to provide space for the installation of shrimp ponds, although the forests can be important nursery areas for coastal, marine fishes.

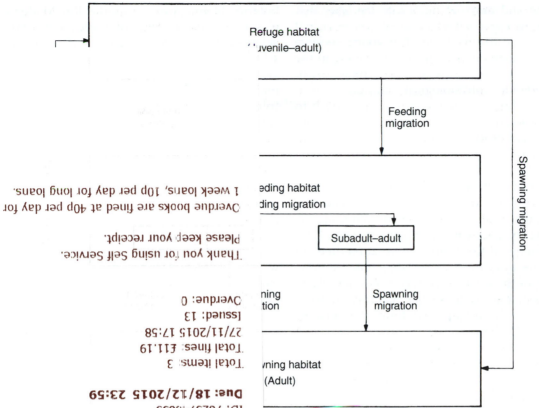

Refuge habitat
(Juvenile–adult)

Feeding
migration

Feeding habitat
...ding migration

Subadult–adult

Spawning migration

...ning ...tion

Spawning
migration

...wning habitat
(Adult)

...een the three basic habitats utilized by migratory fishes.

its open nature is probably also an area of danger from predators once light intensities are sufficiently high (page 90). Membership of a shoal may improve the accuracy of the migration of individual grunts. Transplanted individuals learned the migration routes and the refuge sites from the local residents (Helfman and Schultz, 1984).

The Skeena River system in British Columbia, Canada, provides important breeding grounds for sockeye salmon (Foerster, 1968; Brett, 1983, 1986). Adults spawn in the autumn in the upper reaches of the inlet and outlet streams of the lakes in the system. The fry emerge from the gravel in mid- to late spring and move at night into the lakes that form the nursery feeding areas. The principal nursery is Babine Lake. The young sockeye usually spend a year in the lake, living in the open, limnetic waters. Next they smolt, undergoing physiological changes which prepare them for the transition from fresh to sea water. The smolts aggregate in schools to move out of the lake in well-timed and well-orientated migrations to the sea. Once in the sea, the sockeye move first north-westerly and then south-westerly to enter the massive Alaskan Gyre (Fig. 5.4). Most fish spend two or three years at sea before returning as adults to spawn in the Skeena system. After spawning all the adults die, so at no time are parents and progeny coexisting. Interestingly, some of the sockeye young do not migrate to the sea, but reach sexual maturity in fresh water as the land-locked form, the kokanee (Foerster, 1968).

A final example is provided by the potamodromous migrations of large characins including *Brycon* and *Prochilodus* in such tributaries of the Amazon as the Rio Madeira (Goulding, 1980). The tributaries are characterized by a cycle in water level, with the adjacent, low-lying forest floor flooding each year. Early in the annual flood, the adults migrate down side tributaries into the main tributary where they spawn in the turbid waters. After spawning, the fish move back up the side tributaries and disperse across the flooded floor of the forest, which provides a rich but temporary feeding area. After several months of feeding, as the water levels begin to fall, the fish move back into side tributaries and down to the main tributary and then upstream. The main tributary provides a refuge during periods of low water whereas the forest floor provides a feeding habitat. Seasonal migrations that take advantage of feeding grounds opened by flooding are also shown by species living in floodplain rivers in Africa and Asia (Lowe-McConnell, 1975, 1987; Welcomme, 1979).

The mechanisms by which fish orientate themselves during migration are not considered in detail given the discussions by Leggett (1977), McCleave *et al.* (1984), McKeown (1984) and Smith (1985). A major point at issue in these discussions is the precision of that orientation. There is evidence, particularly from studies of migratory salmonids, that the fish can obtain directional information from the sun, polarized light and even from geomagnetic fields. Water movements, including the unidirectional flow in rivers and the tidal-stream currents in the sea, may also be important for directional movement. However, simulation studies have suggested that migratory fish may need to show only a small bias towards movement in the required direction for the homing to be successful eventually. Fish have sophisticated sensory and central nervous systems, so it may be more economical in time and energy expenditure if they use good-quality directional information rather than a biased random search.

For some anadromous salmonids, there is

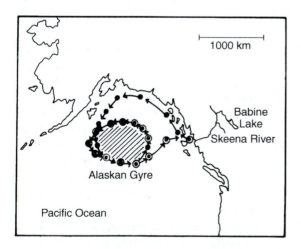

Fig. 5.4 Coastal and offshore migration of Skeena sockeye, *Oncorhynchus nerka*, during 2 years of ocean life. Sockeye occupy whole of shaded area anually. Encircled points indicate positions after first year at sea. Redrawn from Brett (1983).

convincing evidence that as many as 95% of spawners return to the stream in which they spent their early life-history stages (Harden Jones, 1968; Hasler and Scholz, 1983). In a classic series of studies, Hasler showed that the adult salmon recognizes its home stream by olfaction. Each stream may have its own bouquet naturally generated by the unique combination of abiotic and biotic factors that characterize that stream. In those salmonid species in which the adults migrate into streams that contain juveniles that have still to migrate seawards, pheromones produced by the juveniles may attract the adults (Nordeng, 1977; Smith, 1985). The fish become imprinted on the odour of their home stream as smolts before they begin their seaward migration and retain the memory throughout their life in the sea. Hasler showed that coho salmon, stocked in the North American Great Lakes, could be imprinted as smolts to a synthetic chemical, morpholine. When the coho returned to the lakeside on their spawning migration, the imprinted fish moved into a stream into which morpholine had been released (Hasler and Scholz, 1983).

A problem with determining the functions of migration and homing for fish of a given population is to find or establish control groups of fish in the same population which do not show these behaviour patterns. Likely candidates are species that have both migratory and non-migratory forms inhabiting the same watershed (Jonsson and Jonsson, 1993). Examples include the sockeye and kokanee forms of *Oncorhynchus nerka*, the steelhead and rainbow trout forms of *Oncorhynchus mykiss*, the sea and brown trout forms of *Salmo trutta* and the trachurus and leiurus forms of the threespine stickleback, *Gasterosteus aculeatus*. The first-named form of each pair is anadromous and the second permanently resident in fresh water. Ideally, the survivorship, growth and fecundity schedules of the anadromous and resident forms of a species in a watershed should be com-

pared, but comparative demographic studies of fish populations are still in their infancy. Some comparisons can be made. The anadromous form usually reaches a larger adult size than the resident form. This has been observed in *O. nerka*, *S. trutta* and *G. aculeatus* (Foerster, 1968; Frost and Brown, 1967; Hagen, 1967). This larger adult size is reflected in a higher fecundity (Gross, 1987; Jonsson and Jonsson, 1993) (Fig. 5.5). A cost of migration may be a higher mortality rate as the migrants encounter new habitats (Jonsson and Jonsson, 1993).

Even within resident freshwater populations, movement from a natal stream into a lake can lead to an increased growth rate. A good example is provided by the brown trout (Frost and Brown, 1967; Swales, 1986). These comparisons provide evidence that the downstream migrations of these species do take the fish to richer feeding sites. A plausible advantage of the upstream spawning migration is that the eggs are laid in an environment in which there are fewer predators on the eggs and young compared with downstream and many marine habitats (Kedney *et al.*, 1987). Quantitative studies on the predation rates on eggs and young fish in different habitats would help in the evaluation of this suggested advantage. Head waters are often highly oxygenated and free from silt, both conditions advantageous to immobile eggs.

A comparison of the global distribution of anadromous and catadromous species provided circumstantial evidence that migration is from areas of low production (poor feeding) to areas of high production (rich feeding) (Gross, 1987; Gross *et al.*, 1988). Catadromous species are more common at low latitudes, where primary production in fresh waters tends to be higher than in the seas. Anadromous species are more common at high latitudes where it is the marine environment that has the higher rates of primary production.

All adults of the Pacific salmon (*Oncor-*

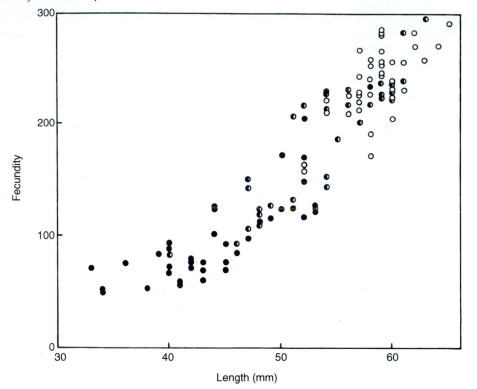

Fig. 5.5 Differences in fecundity between three plate morphs of threespine stickleback, *Gasterosteus aculeatus*, from Little Campbell River, British Columbia: ○ fully plated morph (migratory); ◐ partially plated morph (migratory); ● low plated morph (resident). Redrawn from Wootton (1984a).

hynchus spp.) die after spawning and their decaying bodies can form a significant source of nutrients in head waters which would be of advantage to the progeny. Northcote (1978, 1984) has suggested that this is a further advantage of migration in this group of anadromous fishes. But this enrichment is simply a fortuitous by-product of the life-history strategy. In the terminology of Williams (1966), it is an effect rather than a function of migration. Other suggested functions of migration, for example that it aids colonization by a population, also assume that the advantage is to the population rather than to the individuals in the population. Some form of group selection is assumed. Selection at the level of populations rather

than individuals is of controversial importance and should be invoked as a mechanism only if supported by strong evidence (Maynard Smith, 1976). Such evidence is lacking in studies of migration.

Homing in spawning migrations brings the fish back to an environment for which the fish is itself evidence that the environment is suitable for reproduction, at a time when other sexually mature fish will also be present. A consequence of homing is that gene flow is restricted largely to within the population of fish that home to a given area (Carvalho, 1993). Genotypes within that population may become highly adapted to the specific environmental conditions experienced by the population. With the increase in the farming of Atlantic salmon along the coasts of Ireland, Scotland and Norway, there is concern that farmed salmon, which escape from their holding cages, are inter-breeding with local, wild salmon, altering the genetic composition of the local populations.

A migratory species may become divided into populations reproductively isolated from each other by their tendency to home to specific breeding sites. Only strays that fortuitously find their way to a different spawning site will maintain any gene flow between the populations. Straying may also be a factor in determining population abundance (Sinclair, 1988) (Chapter 10).

Stocks of sockeye salmon that use Babine Lake as their nursery show local adaptation. When the time comes for the smolts to leave the lake system, the different stocks have to follow different migratory routes to the outlet. Groot (1965) suggested that the patterns of preferred direction shown by the different stocks are genetically based and depend on accurate homing which maintains the genetic differences by limiting gene flow between stocks.

Vertical migration

Some species show diel vertical migrations. Diurnal zooplanktivores associated with coral reefs move upwards in the water column to feed during the day, but then descend in the evening to night-time locations just above the reef (Helfman, 1993). These migrations probably represent a simple alternation between a feeding and a refuge habitat.

Many pelagic species show extensive vertical migrations. For example, the herring, *Clupea harengus*, stays near the bottom during the day, but moves towards the surface at sunset, then around midnight sinks in the water column, rising again at dawn before finally sinking to the daytime depth (Harden Jones, 1968). In Babine Lake, sockeye salmon juveniles show a comparable vertical migration during the summer when the lake is thermally stratified. During the day they sink into the cold hypolimnion at a depth of 30–40 m. At dusk they rise into the warm upper epilimnion where they feed for about 2 h. They then sink into the upper zone of the thermocline but at dawn rise again. As the light increases they sink to the hypolimnion. At this time, the temperature of the hypolimnion is 4–6 °C, while the epilimnion ranges from 16 to 18 °C. This vertical migration ceases when the thermal stratification of the lake breaks down with the autumn turnover (Brett, 1983).

Three functions have been suggested for the vertical migration of pelagic species (Brett, 1971b). The first is that the fish are following the vertical movements of their prey, the zooplankton. The second is that during daylight the fish are moving into darker water to reduce predation. The third suggestion is that the fish are maintaining a homeostatic control over their energy expenditure by moving, after they have fed, into cooler waters where their rate of energy expenditure is reduced. Estimates of energy expenditures and experimental studies both suggest a thermoregulatory effect of the vertical migrations of sockeye fry (Fig. 5.6) (Brett, 1983). However, Clark and Levy (1988) suggested that no single explanation is adequate. They developed a model, which predicted that for the sockeye juveniles there are brief antipredation windows at dawn and dusk. These correspond to times when the light intensities are such that the sockeye can still successfully detect their zooplankton prey, but the sockeye are at reduced risk from their predators. There is a trade-off between food intake and risk of predation, which are both likely to be higher at daytime light intensities. In other habitats, for example coral reefs, predation risk is highest at dawn and dusk. Pitcher and Turner (1986) found that minnows detected a stalking pike at shorter distances at light intensities typical of twilight and so are probably then at greater risk.

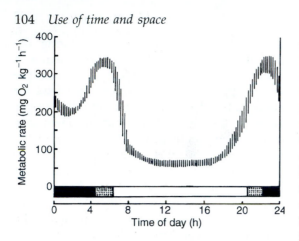

Fig. 5.6 Schematic representation of daily mid-summer metabolic rates of young sockeye, *Oncorhynchus nerka*, in Babine Lake, British Columbia. Black bars, night; dotted bars, dawn (left) and dusk (right); open bar, daytime. Redrawn from Brett (1983).

Energy cost of migration

An attack on the problem of determining the benefits and costs of migration has been made by measuring the energy costs of migratory behaviour and the possible consequences of these costs on other activities of the fish.

In a bioenergetics *tour de force*, Brett (1983, 1986) described the lifetime energetics of sockeye salmon from the Babine Lake catchment. His analysis was based on experimental studies on the effect of temperature on sockeye bioenergetics (Brett, 1971b). The sockeye's lifetime can be divided into three phases: freshwater, marine and spawning. The freshwater phase runs from when the young fry emerge from the gravel in which the eggs were spawned to when the downstream migration of the salt-tolerant smolts brings them to the sea. For almost all this period, the sockeye fry are feeding pelagically in Babine Lake on zooplankton. Brett concluded that during this period, the sockeye are food limited. Starting life as an egg containing 1.55 kJ, the smolt reaches the

sea some 17 months later weighing about 5 g, with a total energy content of 26.0 kJ. In the sea, the fish grow at a rate that suggests they are feeding at maximum rations, consuming mostly euphausiids. A female salmon that spends two years at sea returns to the mouth of the Skeena weighing 2270 g and has a total energy content of 17 620 kJ, having consumed some 66 950 kJ. During the migration upstream to the spawning streams and during the spawning period, the fish do not feed. Migration and spawning consume more than half of the total energy content of the fish that entered the river. For a 2270 g female, the upstream migration reduces her body energy to 7950 kJ, while spawning further reduces it to 3890 kJ. Males are reduced to 7110 kJ after spawning. All adults then die.

The American shad, *Alosa sapidissima*, is an anadromous clupeid that makes spawning runs into rivers that drain the eastern seaboard of North America from Florida to the St Lawrence River in Canada. During the freshwater phase the shad do not feed, but metabolize reserves laid down while at sea. In the northern populations there are significant proportions of repeat spawners. These are shad that spawn in the river, return to the sea and in future years re-enter the river and spawn. In the most southerly populations all the shad die after spawning (Leggett and Carscadden, 1978). Figure 5.7 shows estimates of the energy cost of the river migration for shad in the Connecticut River in New England. There is an inverse correlation between the proportion of total energy reserves used in the migration and the proportion of fish that are repeat spawners. Fish that run into the Connecticut late in the season encounter higher temperatures and consequently have high metabolic rates. They use a higher proportion of their energy reserves (Fig. 5.7(a)) and probably suffer higher mortality rates (Glebe and Leggett, 1981a). Shad in a southern population (St Johns River, Florida) use 70–80% of their

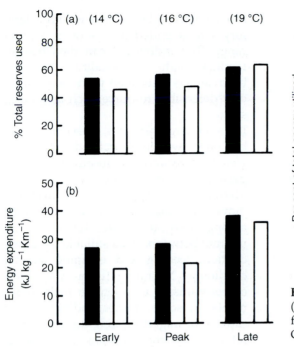

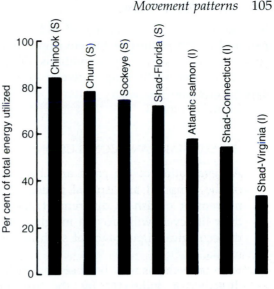

Fig. 5.8 Energy use by representative semelparous (S) and iteroparous (I) spawning anadromous fishes during freshwater migration. Redrawn from Glebe and Leggett (1981b).

Fig. 5.7 (a) Proportional and (b) absolute energy usage during freshwater migration of shad, *Alosa sapidissima*, in the Connecticut River and associated average water temperatures. Black columns, males; open columns, females. Early, peak and late relate to timing of migration within spawning season. Redrawn from Glebe and Leggett (1981a).

total energy reserves during a spawning run, whereas shad in the Connecticut population use only 40–60% (Glebe and Leggett, 1981b).

Data from anadromous populations of the Pacific salmon *Oncorhynchus*, in which there are no repeat spawners, suggested that they use more than 70% of their stored reserves during the spawning run. The Atlantic salmon, *Salmo salar*, is physiologically able to spawn more than once, although the proportion of re-spawners is usually low. A survey of Norwegian salmon populations found that average post-spawning survival declined from about 22% when 60% of stored reserves was used to 2% when 70% was used (Jonsson *et al.*, 1997). These data suggest that

there is a relationship between the energy expenditure during the migration and spawning, and the survival rate after spawning (Fig. 5.8) (Glebe and Leggett, 1981b).

Females of the South American characin, *Prochilodus mariae*, may illustrate a situation in which the energy expended on migration is traded-off against investment in egg production. Females that remained resident in a lagoon devoted about five times as much energy to egg production as did females that undertook an up-river migration (Saldana and Venables, 1983). Such a large difference in investment in eggs is surprising and suggests that migration confers important benefits in survival that compensate for the lower investment in egg production.

These studies on the energetics of migration, when coupled with demographic studies of migratory and resident populations, should provide a more firmly based account of the functions of migration. This account will emphasize the costs and benefits in relation to the size and age of individual fish and in the context of the environmental conditions encountered by individuals. Other aspects of the relationship between energy expenditure and life-history patterns are discussed in Chapter 11.

5.6 SUMMARY AND CONCLUSIONS

1. An evaluation of the costs and benefits of the temporal and spatial patterns of movement of fish at different stages in their life cycle will provide insights into the ecological significance of such movements.
2. Such movements may be observed either directly, or indirectly by the use of acoustic or radio transmitters or other tags carried by the fish.
3. Some movements relate to temporal patterns in the environment. This is illustrated by the diel movements of fish on coral reefs and in temperate lakes. These diel movements probably relate to the need to forage and to avoid predators.
4. Movements are also influenced by physical characteristics of the environment and by the presence of other fish. Density-dependent habitat selection occurs when conspecifics lower the quality of a habitat through their exploitation of resources.
5. In some species, individuals form shoals in which the movements of individuals are related to the movements of other fish in the shoal. Individuals may reduce their risk of predation by joining a shoal. Shoaling may also improve foraging or allow the learning of migration routes.
6. Movement can be located within a territory – a defended area – or in a home range. The individual can then become familiar with the locality. Territorial behaviour is associated with the presence of a defensible resource such as food or shelter.
7. In other circumstances, movement may be only weakly localized and only weakly directional if resources are dispersed and not spatially predictable.
8. Migrations are strongly directional movements that result in an alternation among two or more habitats. Fish migrate between areas suitable for reproduction, feeding and escaping adverse conditions. Migrating fish use a variety of mechanisms to orientate their movements. In some species, individuals are capable of migrating long distances, then returning to their starting points accurately. This homing can result in populations evolving adaptations to local conditions because it reduces gene flow between populations.
9. The vertical migration shown by some species in seas and lakes may function as a behavioural thermoregulatory mechanism. But it may also be a pattern of movement that yields the best trade-off between foraging success and predator avoidance.
10. Migrations have energy costs. In some anadromous species, the energy expended on the spawning migration is related to post-spawning mortality. As the proportion of stored energy reserves spent on the migration increases, the proportion of fish surviving to spawn again decreases.

6

GROWTH

6.1 INTRODUCTION

If its rate of food consumption is sufficiently high, a fish can, in addition to meeting the energy costs of maintenance, synthesize new tissue. This tissue may be either retained within the body as growth, including any storage products, or disseminated as gametes. Growth constructs the framework and metabolic machinery necessary to synthesize and to protect the gametes until their release. The process of natural selection will lead to the evolution of patterns of growth that tend to maximize the lifetime production of offspring. Growth and reproduction are complementary processes, but both depend on the limited resources of energy and nutrients made available by the foraging behaviour of the fish. This chapter discusses the factors that affect the growth of fishes and explores the methods that are used to predict growth.

6.2 DEFINITION AND MEASUREMENT OF GROWTH

Definition of growth

If growth, food consumption and the losses associated with maintenance are all measured in energy units, then growth is defined as:

$$\text{Growth} = \text{In} - \text{Out} \qquad (6.1)$$

(Ursin, 1979). In terms of the energy budget described in Chapter 4. and assuming a period over which no gametes are released, then:

$$P_s = C - (F + U) - R \qquad (6.2)$$

Although this bioenergetic definition provides a useful starting point for the development of predictive models of growth, most studies of growth measure it in units of either length or weight. It is not valid to apply mass-balance equations when growth is measured in this way, because the outputs and input cannot be measured in a common currency such as energy units.

Measurements of growth as length quantify axial growth; measurements as weight quantify growth in bulk. These two categories of growth are usually highly correlated. But a fish can change in weight without changing in length, or vice versa. For example, measurements of the length and weight of Norway pout, *Trisopterus esmarkii*, during the second and third year of life showed that between January and April the weight dropped sharply but length stayed constant (Ursin, 1979). As a fish grows, the energy content per unit mass of tissue may also change. In the African catfish, *Clarias lazera*, there was an allometric relationship between energy content and body weight (Hogendoorn, 1983). The energy content per gram of tissue increased as the fish grew heavier.

The relationship between weight and length for fish in a given population can be analysed either by measuring the weight and length of the same fish repeatedly throughout their life span, or by measuring the weights and lengths of a sample of fish taken at a particular time. The relationship between weight, W, and length, L, typically takes the form:

$$W = aL^b \qquad (6.3)$$

or in the linear form:

$$\log W = \log a + b \log L$$

where a and b are constants, usually estimated by regression analysis.

An example of the relationship between weight and length is shown in Fig. 6.1. If the fish retains the same shape, it is growing isometrically and the length exponent, b, takes the value 3.0 (assuming that the specific gravity of the fish does not change). A value significantly larger or smaller than 3.0 indicates allometric growth (Tesch, 1971; Ricker, 1979). A value less than 3 shows that the fish becomes lighter for its length as it grows; an exponent greater than 3 indicates that the fish becomes heavier for its length as it increases in size.

The relationship between length and weight provides a simple index frequently used by fisheries biologists to quantify the state of well-being of a fish. This index is the condition factor, K:

$$K = W / L^3 \qquad (6.4)$$

(Tesch, 1971; Weatherley, 1972). Fish with a high value of K are heavy for their length, while fish with a low value are light for their length. In effect, the K value for a given fish measures its deviation from some hypothetical ideal fish of that species growing isometrically. Changes in the K value of fish may indicate gonadal maturation or changes in feeding intensity. The condition factor of fish of the same species from different populations can also be compared (Weatherley, 1972). A second index of condition is the relative condition factor (K_n) defined as:

$$K_n = W / W_{pred} \qquad (6.5)$$

where W is the observed weight of the fish and W_{pred} is its predicted weight from the weight-length relationship (LeCren, 1951). K_n measures the deviation from the weight predicted for a fish of a given length for that population. Because K_n is specific to a given population, it can be used for quantifying changes in condition within that population but not for comparing the condition of fish from two or more populations. A third index is relative weight, W_r, defined as:

$$W_r = 100(W / W_s) \qquad (6.6)$$

where W is the observed weight and W_s is its predicted weight from a standard weight-length relationship (Cone, 1989). Although condition factors are frequently used, a preferable method of comparing weight-length

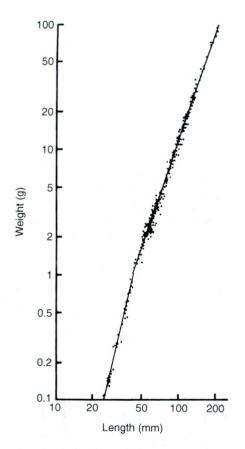

Fig. 6.1 Weight–length relationship for brown trout, *Salmo trutta*. Note presence of two growth stanzas, with transition at 42 mm. In this example, the lines have been fitted by eye. Redrawn from Tesch (1971). Logarithmic scales.

relationships is covariance analysis. This technique provides a method of statistically comparing the intercepts and slopes of the relationship both within and between populations (Wootton and Mills, 1979; Cone, 1989).

Growth is measured over a defined time period, for example between times t_1 and t_2. If growth is measured in terms of weight (W), then the absolute growth rate is:

$$(W_2 - W_1)/(t_2 - t_1)$$

where W_1 and W_2 are weights at times t_1 and t_2. The instantaneous growth rate is:

$$dW/dt.$$

A disadvantage in using the absolute growth rate is that it depends strongly on the size the fish has reached. In absolute terms, a fish growing from 1000 g to 1001 g over a period of a week has the same growth rate as a fish growing from 1 g to 2 g over the same period, yet the latter has doubled its weight! A more appropriate measure of growth rate is specific growth rate, g. This is the instantaneous rate of growth per unit weight, that is:

$$g = dW / Wdt. \qquad (6.7)$$

Over a defined period of time, the specific growth rate is calculated as:

$$g = (\log_e W_2 - \log_e W_1) / (t_2 - t_1) \qquad (6.8)$$

Frequently it is expressed as % per unit time, that is

$$G = 100g. \qquad (6.9)$$

The weight at the end of the time period is given by:

$$W_2 = W_1 \exp (g(t_2 - t_1)). \qquad (6.10)$$

This assumes that the fish grows exponentially during the defined time interval (Fig. 6.2), although the real pattern of growth within the interval will be unknown. If the time intervals over which the specific growth

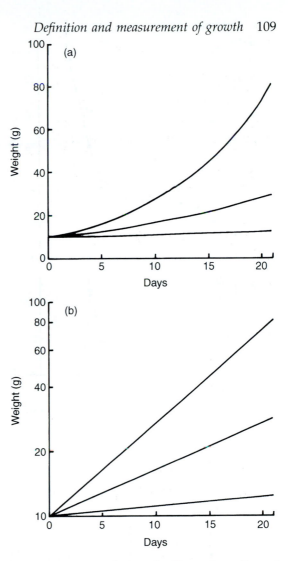

Fig. 6.2 Exponential growth, illustrating effect of daily specific growth rates of 1%, 5% and 10% (bottom, centre and top curve respectively in each graph) for fish of initial weight 10 g. (a) Arithmetic scale; (b) logarithmic scale.

rate is measured are short, detail is obtained about the pattern of growth shown by a fish during its lifetime (Fig. 6.3) (Ricker, 1979).

Although absolute and specific growth rates are described for changes in weight, the concepts can also be applied to changes in length, L or changes in total energy content, E:

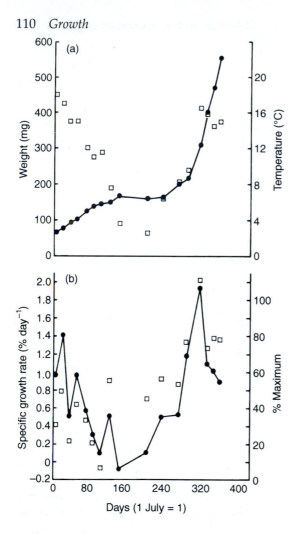

Fig. 6.3 Seasonal growth in first year of life of threespine stickleback, *Gasterosteus aculeatus*, in a mid-Wales reservoir, Llyn Frongoch (note: any size-selective mortality has been ignored). (a) Changes in mean weight (●) and surface water temperature (□); (b) changes in mean specific growth rate (●) and in growth rate as a percentage of predicted maximum rate (□). Redrawn from Allen and Wootton (1982b).

$$g(L) = (\log_e L_2 - \log_e L_1) / (t_2 - t_1) \qquad (6.11)$$
$$G(L) = 100g(L)$$

$$g(E) = (\log_e E_2 - \log_e E_1) / (t_2 - t_1) \qquad (6.12)$$
$$G(E) = 100g(E)$$

Ageing fish

In laboratory studies, the time period over which the observed growth takes place is known accurately. If fish in a natural population are tagged with marks that are specific to each individual, then the interval between tagging and recapture is also known accurately. However, in most studies of natural populations, estimates of growth rates depend on methods that allow fish to be aged.

An annual pattern in which a season of slow growth is followed by a season of fast growth may leave a record in the calcified parts of the body such as scales, ear otoliths, bones or spines. These can then be used to age the fish.

The structure most frequently used for ageing fish is the scale (Bagenal, 1974). Scales grow by accretion at their margins as the fish grows (Van Oosten, 1957). As a scale grows, a series of ridges or circuli are laid down. Usually the circuli are concentric (Fig. 6.4), but in clupeids they are almost transverse. It is the pattern of the circuli that is the essential characteristic for the purpose of ageing. Some feature of this patterning must signal the end of a year's growth: this feature is the annulus. The form the annulus takes varies between species and families. The common forms of annuli include (Tesch, 1971):

1. a zone of closely spaced circuli that is followed by a zone of widely spaced circuli
2. a clear zone devoid of circuli;
3. discontinuous circuli;
4. circuli that appear to cut across other circuli (called cutting over);
5. waviness of the circuli.

The number of annuli that are identified on the scale indicates the age of the fish. Widely spaced circuli may be laid down when fish that are physiologically competent to show fast growth encounter good feeding conditions, but the effect of environmental factors on the pattern of circuli is still not well understood (Barber and Walker, 1988). In some species the annuli are distinct; in others scale reading requires skill and experience. Scales can yield details of the life history of the fish in some species: the anadromous salmonids provide a good example (Fig. 6.4).

Otoliths are calcareous bodies in the inner ear. Each otolith consists of a matrix of proteinaceous material within which crystals of calcium carbonate are deposited. The largest of the three otoliths in each ear, the sagitta, is usually used for ageing. A feature of an otolith that allows it to be used for ageing is a sequence of alternating opaque and translucent (hyaline) concentric zones. Ideally one opaque zone and one translucent zone represent one year's growth. The opaque zone is usually deposited in a period of rapid growth and the translucent zone in a period of slow growth (Blacker, 1974; Beckman and Wilson, 1995). An example of such a sequence is provided by the otoliths taken from threespine sticklebacks in a small reservoir (Table 6.1), the seasonal growth pattern of which is shown in Fig. 6.3.

Otoliths also show daily increments, which are particularly clear in the first few months of life. These daily increments allow an accurate estimation of the age of young fish in days. (Pannella, 1971; Campana and Neilson, 1985; Secor *et al.*, 1995). Each daily growth increment consists of an incremental and a discontinuous zone (Fig. 6.5). The latter consists primarily of the proteinaceous matrix, whereas the former has a higher calcium content. The formation of these daily growth increments depends on an endogenous, circadian rhythm of deposition, synchronized with the diel cycle by the photoperiod. This daily cycle may be masked by environmental factors such as fluctuations in temperature or in food supply, which can lead to the formation of subdaily increments or to the cessation of otolith growth. A disadvantage of using otoliths rather than scales for ageing is that the fish has to be killed.

The analysis of otolith microstructure is a powerful tool in the interpretation of patterns of growth, particularly in the first few months of life of a fish. Examination of the otoliths of juvenile chinook salmon, *Oncorhynchus tshawytscha*, allowed a reconstruction of the events in the first spring and summer of their life in an Oregon river (Neilson *et al.*, 1985). The age at which the fish moved from fresh water into the estuary was identified by an increase in the width of the daily growth increments, while the growth rate in the estuary was estimated from these widths. Daily growth increments in the otoliths of

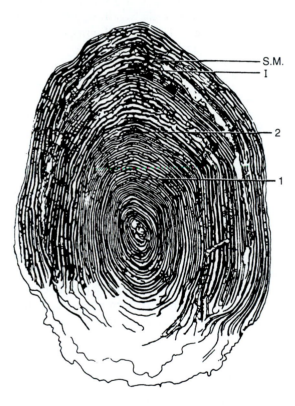

Fig. 6.4 Scale of anadromous trout, *Salmo trutta*, illustrating circuli, annuli and a spawning check: 1, annulus marking completion of first year of growth in fresh water; 2, annulus marking completion of second year of growth in fresh water; I, annulus marking completion of first year of growth in sea; SM, spawning check. Reproduced with permission from Fahy (1985).

Table 6.1 Otolith characteristics in a population of threespine sticklebacks, *Gasterosteus aculeatus*, sampled at monthly intervals from Llyn Frongoch, Mid-Wales*

Month	Percentage of sample in each otolith class†					
	CO	COT	COTO	COTOT	COTOTO	COTOTOT
August	15	10	70	2.5	0	2.5
September	2.5	45	42.5	5	5	0
October	0	95	2.5	2.5	0	0
November	0	93	0	7	0	0
December	0	98	2	0	0	0
January	0	90	3	7	0	0
February	0	75	15	7	0	3
March	0	70	15	10	2.5	2.5
April	0	70	18	7	5	0
May	0	68	20	7.5	2.5	2.5
June	0	75	20	5	0	0
July	0	30	65	5	0	0
Mean total length of fish (mm)	18.8	31.5	38.7	43.9	48.6	49.2
Number of fish in sample	3	326	110	26	6	5

*Source: Allen and Wootton (1982b)

†C, otolith nucleus; O, opaque zone; T, transparent zone observed with reflected light).

the juveniles of coral reef fishes were used to estimate the age at which juveniles settle on a reef after the pelagic phase of their life cycle (Victor, 1983) (Chapter 12). Counts of daily rings can also provide an estimate of the date on which an individual hatched. The dates of hatching of young fish that have survived to a given age can be compared with the dates on which spawning in a population took place. This technique was used to reconstruct the hatch dates of bluegill sunfish (Cargnelli and Gross, 1996).

Daily growth increments may also be valuable in ageing some species of tropical fishes that do not form distinct annuli in calcareous structures because of a lack of seasonality in their growth patterns. Ralston and Miyamoto (1983) developed a technique for ageing the Hawaiian snapper, *Pristipomoides filamentosus*, from measurements of growth increments at different distances from the

focus of the otolith. Their method assumed that the increments are laid down daily. Oxytetracycline, when injected into fish, is laid down in calcareous structures and can be visualized by illumination with ultraviolet light. Juvenile snappers injected with oxytetracycline showed a high correlation between the number of growth increments and the number of days that elapsed between injection and the day the fish were killed.

Calcareous structures other than scales or otoliths are also used to age fishes. Perch can be aged using the opercular bone, which shows alternating opaque and transparent bands (Le Cren, 1947). Vertebrae and fin spines may also show annuli (Tesch, 1971; Beamish and McFarlane, 1987).

The use of calcareous structures for ageing depends on the assumption that periodic features such as annuli or growth increments are laid down at a constant frequency. If the dis-

Freshwater
Growth ⟶ ⟵ Estuarine
Growth ⟶

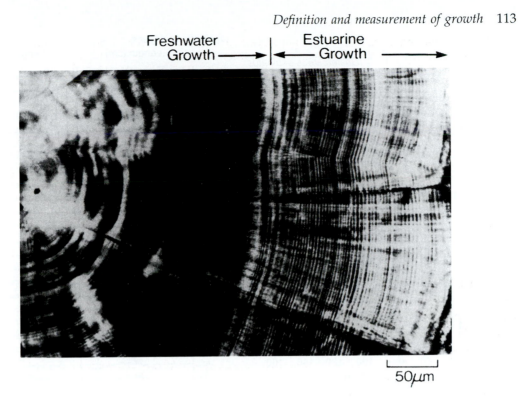

50μm

Fig. 6.5 Polished section of sagittal otolith of juvenile chinook salmon, *Oncorhynchus tshawytscha*, showing daily growth increments and microstructural differences associated with movement from fresh water to estuary. Reproduced with permission from Neilson *et al.* (1985).

tance between successive periodic features is related to the growth rate, then it is possible to reconstruct the growth pattern of a fish. The length of the fish at younger ages can be back-calculated from its current length and the record of its previous growth fossilized in the scale or otolith (Tesch, 1971; Bagenal, 1974). The first step in the back-calculation of lengths in a particular population is to determine the relationship between the size of the scale or otolith and the length of the fish over a wide range of fish lengths. Specifically, consider the relationship between the radius of a scale (R) and fish length (L), where R is a function of L. That is:

$$R = f(L). \tag{6.13}$$

Then the radius at some earlier age (R'), measured as the distance from the centre of the scale to the annulus used to identify that age, is a function of the length of the fish at that younger age, $f(L')$. In the simplest case, length and scale radius are directly proportional so that:

$$L' = R'L / R. \tag{6.14}$$

In many fishes the relationship between scale radius and fish length is more complex than simple proportionality and so an appropriate expression must be used for predicting L' from L, R and R' (Tesch, 1971; Francis, 1990, 1995).

Lee's phenomenon refers to the situation in which back-calculated lengths at an age are smaller the older the fish from which the lengths are back-calculated (Tesch, 1971; Ricker, 1979). If the appropriate function relating scale radius to fish length has been

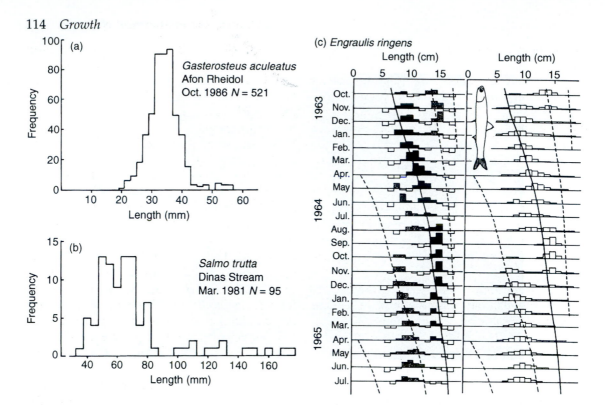

Fig. 6.6 Examples of length-frequency distributions. (a) Threespine stickleback, *Gasterosteus aculeatus*, from Afon Rheidol, Mid-Wales – note single dominant mode of young-of-year fish. (b) Brown trout, *Salmo trutta*, from Dinas Stream, Mid-Wales – note difficulty of identifying separate modes. (c) Peruvian anchovy, *Engraulis ringens* (northern/central stock). ELEFAN method of analysis used to fit growth curve. Original length-frequency data on right, restructured data showing peaks (black or grey) and troughs (white) on left. Redrawn from Longhurst and Pauly (1987).

used, then a source of the phenomenon can be size-selective mortality in which the smaller fish of a given age have better survival. Such size selection may be caused by natural sources of mortality, or by a fishing technique that tends to select the faster-growing, hence larger, fish of any cohort (Ricker, 1979).

Ageing by identifying annuli (or daily increments) in calcareous structures must be validated. Such structures frequently show false annuli or checks that can be mis-interpreted as true annuli. The patterns of deposition in scales, otoliths and other structures used in ageing may be disrupted during periods of stress caused for example by inadequate feeding or during reproduction. In some species, spawning is marked by a reabsorption of the edge of the scale and when growth is resumed this is recorded on the scale as a spawning check.

The pattern of growth increments in the otoliths may be more stable than that on scales, so otoliths may provide a more representative record of growth than scales (Campana and Neilson, 1985). However, attempts to identify changes in growth rates in larval and juvenile fish from changes in the width of the daily increment have found that the pattern of growth of the otolith may differ from the pattern of overall growth in length or weight. Otolith growth may be partly de-coupled from overall growth, particularly when overall growth is slow (Wright *et al.*, 1990; Moksness *et al.*, 1995). This recalls observations that changes in growth rate during early life history stages are not correlated with changes in the overall metabolic rate (Chapter 4).

As understanding of the dynamics of otolith growth in relation to the physiological state of the individual and environmental conditions increases, the reconstruction of growth rates of young fish will become a powerful tool for interpreting events during the crucial early weeks and months of life (Chapter 10).

As fish get older, their growth rate slows, and in old fish, successive annuli in a scale become difficult or impossible to recognize so the age of the fish is underestimated. Beamish and McFarlane (1987) suggest that in many long-lived species, ages determined by scale reading are too low and ages determined from otolith sections are more reliable. They describe a rougheye rockfish, *Sebastes aleutianus*, with an estimated age of 140 years.

Methods for validating the ageing include (Blacker, 1974): the use of tank or pond experiments in which the structures from fish of a known age can be analysed; the use of marked fish; the observation of the same population for several years; and the use of samples taken throughout the year so the time when annuli are formed is identified accurately.

An ageing scheme based on calcareous structures may be corroborated by ages estimated from the size-frequency distribution of the population. This latter method depends on the fish having a seasonal reproductive cycle so that recruitment to the population occurs at intervals separated by approximately 1 year. All the fish born at approximately the same time form a cohort. Each cohort recruited in a given year has a 1 year growth advantage over the next cohort to be recruited. Cohorts should form distinct modes in the size-frequency distribution of

the population (Fig. 6.6). Several processes may obscure the distinction between the modes representing successive cohorts. The faster-growing fish in the younger cohort may catch up with the slower-growing fish of the previous cohort. A protracted season may lead to the production of several distinct cohorts within a single year. The relatively slow growth that often characterizes sexually mature fish may result in the older cohorts becoming indistinguishable, and males and females may have different growth patterns. Graphical and statistical techniques that improve the analysis of size-frequency data are available (Cassie, 1954; MacDonald and Pitcher, 1979). This method for ageing fish is valuable because fish need not be killed or mutilated and it provides an independent method of corroborating an ageing scheme that is based on calcareous structures.

If length-frequency data are obtained for a population at known intervals of time, the growth of fish in a particular cohort can be estimated by the shift in the position of the mode for that cohort. The shift occurs because of the increase in the mean length of the fish in the cohort over time. If the data are collected over a sufficiently long time, the full lifetime growth curve for the cohort can be obtained. The ELEFAN system for the analysis of such length-frequency data has been developed to provide estimates of growth rates (Pauly, 1987). This system was originally designed to obtain growth curves for tropical marine species. Such species are often difficult or impossible to age from scales.

The methods available for ageing fish provide a powerful tool for the study of growth patterns in natural populations. But the tools need to be used with care and with a critical awareness of their limitations. Validation and corroboration are required if reliable ages are to be assigned, particularly to older, slow-growing individuals (Beamish and McFarlane, 1987; Francis, 1995).

Other methods of estimating growth rates

Glycine uptake by scales

Collagen is the main structural protein in fish scales. The amino acid glycine is a major component of collagen. Scales removed from a live fish and incubated with glycine labelled with ^{14}C incorporate the amino acid at a rate that is correlated with the specific growth rate of the fish (Ottaway and Simkiss, 1977; Adelman, 1987). This relationship allows the prediction of growth rates from measurements of the rate of incorporation.

RNA:DNA ratios

Protein synthesis depends on RNA because of the role of this nucleic acid in the transcription of information encoded in DNA, as a component of the ribosomes where protein synthesis takes place and as a transfer molecule for amino acids. High growth rates are often correlated with high RNA:DNA ratios, changes in this ratio, or in the concentration of tissue RNA, provide information on changes in the growth rate of fish in a population over time (Bulow, 1987).

Liver-somatic indices and condition factors

Indirect, but sometimes useful indicators of changes in growth rates are provided by indices of condition, such as the condition factor, or the ratio of liver weight to total weight (liver-somatic index, LSI) (Heidinger and Crawford, 1977). In immature European minnows kept under controlled conditions and fed known rations, the specific growth rate was highly correlated with the LSI and less strongly, but still positively with the condition factor (Cui and Wootton, 1988). Such indices also change with changes in the reproductive status of the fish and care has to be used in their interpretation as indicators of growth rates.

It is important to validate these indirect methods of estimating growth rate in experiments under controlled conditions to define the contexts in which they provide good predictors of growth or nutritional status.

6.3 PATTERNS OF GROWTH

Plasticity of fish growth

These methods of ageing and estimating growth rates in fishes allow the description of both short-term and long-term growth patterns. If the time interval between successive measurements of size is long, for example a year, then the growth curve will be relatively smooth. If the time interval is short, the growth curve will reflect short-term variations in food supply, other environmental factors and the physiological state of the fish and will be more irregular.

The growth of most fish is indeterminate. Sexually mature individuals do not have a characteristic adult size. This is in contrast to animals with a determinate growth pattern exemplified by many insects, birds and mammals. Given suitable environmental conditions, most fish continue to grow throughout life, although the rate of growth does tend to decline with age. Beverton (1992) has argued that it is better to describe the growth pattern of fish as asymptotic in contrast to those vertebrates whose growth stops abruptly when sexual maturity is reached. Some mammals, such as whales, show asymptotic growth.

A second major characteristic of fish growth is its flexibility. The same species may show different patterns of growth in different environments, with sexual maturity being reached at different sizes or at different ages (Fig. 6.7). The weights of five-year-old brook trout from lakes in the Canadian Rocky Mountains ranged from 65 g in Temple Lake to 1751 g in Lake Patricia (Donald *et al.*, 1980). These differences were correlated with the density of amphipods, a

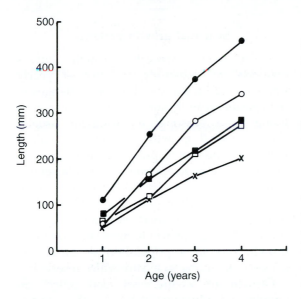

Fig. 6.7 Interpopulation differences in growth of brown trout, *Salmo trutta*, in UK. Mean lengths (mm) at age from: ● Llyn Alaw; ○ Loch Leven; ■ Windermere; ○, Llyn Frongoch; ×, Yew Tree Tarn. Data from Swales (1986).

Flexible growth patterns provide a further mechanism for an adaptive phenotypic response to a changing environment (Chapter 1). Fish can achieve sexual maturity at sizes that are unusually small or large in comparison with those found in other populations of the same species (Chapter 11).

Lifetime growth patterns

When the length or weight of fish in a population is measured at intervals of a year, the pattern of growth is usually sigmoidal (Fig. 6.8). The point of inflection on the growth curve marks the time at which the long-term absolute growth rate starts to decline. For some species, growth is divided into a series of stanzas. Within each growth stanza, the pattern is typically sigmoidal. Each stanza is characterized by a different weight–length

major food item. Within a population, fish born in different years can show different growth patterns. In Windermere, UK, perch born in 1959 grew more slowly and had a smaller asymptotic size than perch born in 1968 (Craig, 1987).

An extreme example of this indeterminate growth is the phenomenon of stunting. In stunted populations, the fish reach sexual maturity at an unusually small size. Some species of African cichlid, such as *Tilapia zillii*, normally reach sexual maturity at an age of 2–3 years and a length of 200–300 mm. They can reach sexual maturity when a few months old and less than 100 mm in length when kept in small ponds (Fryer and Iles, 1972). This precocious maturation can result in dense populations of small fish, unsuitable for harvesting. Much effort has been devoted to developing aquaculture techniques that avoid stunting (Pullin and Lowe-McConnell, 1981).

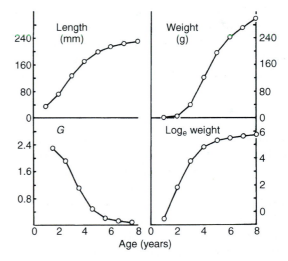

Fig. 6.8 Growth in length and weight and specific growth rate, G (% year^{-1}), in a population of bluegill sunfish, *Lepomis macrochirus*, in Spear Lake, Indiana, illustrating lifetime pattern. Note inflections in curves for length and weight and approach to an asymptotic size. Redrawn from Ricker (1975).

relationship (e.g. Fig. 6.1). A change from one growth stanza to the next occurs when there is some major change in the pattern of growth. A list of such changes would include (Ricker, 1979):

1. metamorphosis, for example the change from the bilaterally symmetrical larvae to the laterally flattened juveniles shown by flatfish (Pleuronectiformes);
2. an abrupt change in body form less extensive than metamorphosis, for example from the larval to the juvenile body form;
3. physiological changes, for example anadromous salmonids have a freshwater growth stanza and a saltwater stanza;
4. a sudden change in growth rate, for example when a new and more profitable food becomes available. In perch, the onset of piscivory may mark the start of a new stanza.

A common but not universal pattern is for the exponent, b, in the weight–length relationship, $W = aL^b$, to increase at some point during the larval stage (Müller and Videler, 1996). The rapid growth in length relative to weight indicated by a relatively low value of b may reflect the advantage of growing quickly out of the hydrodynamic environment characterized by a viscous flow regime (Chapter 2). The low b value indicates a long, thin body at a given weight. Such a shape maintains a high surface-to-volume ratio at a stage in life history when cutaneous respiration is important because gills cannot function in a viscous flow regime. The shape is also appropriate for the anguilliform mode of locomotion appropriate in the hydrodynamic regime. Once the larvae have grown into the inertial flow regime, they change shape, with the body adopting the juvenile/adult form. During this period, b can take values greater than 3.0 (Kamler, 1992). In older juveniles and in adults, growth is frequently isometric, with b approximately equal to 3.0.

Seasonal growth patterns

The relatively smooth growth curves obtained by measuring the fish at yearly intervals hide what may be major changes in the specific growth rate within the year. Fish living in a seasonal environment usually show seasonal growth patterns, with alternating periods of rapid and slow growth. Fish that live in temperate or subpolar environments usually grow slowly or not at all during the winter months, but rapidly during the spring and summer. Fig. 6.3 shows the seasonal changes in size and growth rate in a population of threespine sticklebacks living in a Mid-Wales reservoir.

Growth in weight can also reflect an annual reproductive cycle. As the gonads ripen, they usually increase in weight and there is a correlated increase in the total weight of the fish. This increase in weight is lost when the gametes are discharged. These changes in weight are not accompanied by changes in length. In some species, a cycle of storage and mobilization of lipids is associated with the reproductive cycle (Shul'man, 1974). This cycle may lead to changes in the energy content that are only weakly correlated with changes in body weight and not correlated at all with changes in length. In North Sea herring, fat reserves are accumulated during a period of intensive feeding during spring and early summer, then mobilized as the fish cease feeding with the onset of sexual maturation (Iles, 1984).

This cycle of storage and mobilization links a feeding cycle with the reproductive cycle and so allows sexual maturation and spawning to take place at a time that is favourable to the progeny but bioenergetically unfavourable to the adults (Wootton, 1979; Reznick and Braun, 1987).

6.4 FACTORS AFFECTING GROWTH RATES

The factors that determine the rate at which a fish can grow can be divided into exogenous factors imposed by the environment and endogenous factors related to genotype and physiological condition of the fish.

Exogenous (environmental) factors: 1. Food

As the energy-budget equation emphasizes, growth is an output based on an input, food consumption. Both the quantity and quality of the food are relevant to the growth performance. Food quality can differ both in the energy and nutrient content and in the size of the food particles.

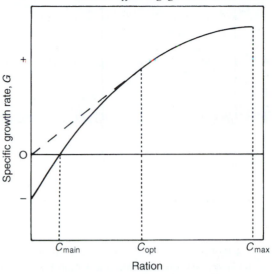

Fig. 6.9 Schematic diagram of relationship between growth rate and ration. Solid line, idealized relationship; broken line, tangent from origin, identifying optimum ration. C_{main}, maintenance ration; C_{opt}, optimum ration; C_{max}, maximum ration. Redrawn from Wootton (1984a).

Relationship between growth and quantity of food

The relationship between growth and quantity of food consumed can be analysed by determining the shape of the relationship between specific growth rate, g or G, and ration, C (Brett, 1979). An idealized form is shown in Fig. 6.9. At a given temperature, three pivotal ration levels are defined, C_{main}, C_{opt} and C_{max}. At the maintenance ration, C_{main}, the fish neither gains nor loses weight (or energy). At rations less than C_{main}, the fish loses weight (or energy). At the optimum ration, C_{opt}, the gross conversion or growth efficiency is at a maximum. This ration is defined by a line from the origin of the relationship between G and C that forms a tangent to the G–C curve (Fig. 6.9). C_{max} is the maximum ration that the fish will consume. By analogy with the scope for activity described in Chapter 4, scope for growth is defined as:

$$C_{max} - C_{main}$$

Three growth efficiencies are defined (Brett, 1979): gross growth efficiency (K_1), where:

$$K_1 = 100(P_s / C) \qquad (6.15)$$

or in terms of assimilated energy, A':

$$K_2 = 100(P_s / A') \qquad (6.16)$$

and net growth efficiency, K_3 which measures the conversion efficiency of the food

consumed in excess of maintenance requirements and is calculated as:

$$K_3 = 100[P_s / (C - C_{main})] \qquad (6.17)$$

where P_s is the increase in weight (or total energy content) in a defined time interval, C is the weight (or energy content) of food consumed during the interval, A' is the weight (or energy) of food assimilated, and C_{main} is the maintenance ration for the interval. By definition, when the ration is C_{main}, growth efficiency is zero.

The G–C relationship over the range from $C = 0$ to C_{max} is usually a negatively accelerated curve (Fig. 6.10). This curve describes the relationship for several species including sockeye salmon (Brett *et al.*, 1969), brown trout (Elliott, 1975c), English sole, *Solea solea* (Williams and Caldwell, 1978), threespine stickleback (Allen and Wootton, 1982a), African catfish (Hogendoorn *et al.*, 1983) and

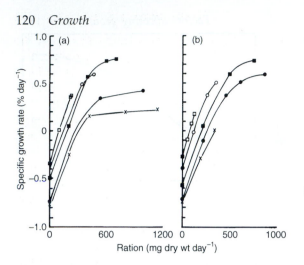

Fig. 6.10 Effect of temperature on relationship between specific growth rate and ration for brown trout, *Salmo trutta*. (a) ●, 7.1 °C; ○ 10.8 °C; ■ 12.8 °C; ● 16.2 °C; ×, 17.8 °C. (b) ●, 5.6 °C; ○ 9.5 °C; ■ 13.6 °C; ● 15.0 °C; ×, 19.5 °C. Redrawn from Elliott (1975c).

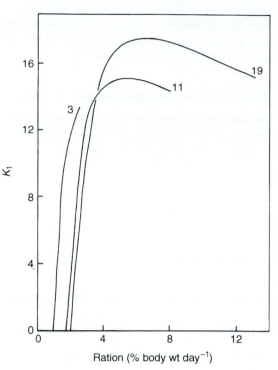

Fig. 6.11 Effect of daily ration on gross growth efficiency for threespine stickleback, *Gasterosteus aculeatus*, at 3, 11 and 19 °C. Redrawn from Wootton (1984a).

common carp, *Cyprinus carpio* (Goolish and Adelman, 1984). In sockeye salmon juveniles at high temperature, the G–C relationship is sigmoidal because of a positively accelerated phase at rations below C_{main} (Brett *et al.*, 1969). A similar positively accelerated phase at low rations and high temperature is shown by brown trout (Elliott, 1975c). The G–C relationship may be close to linear over a narrow range of rations, for example in the cod (Soofiani and Hawkins, 1985) and the Alaskan pollack, *Theragra chalcogramma* (Yoshida and Sakuri, 1984). The relationship is also linear for the herbivorous grass carp feeding on duckweed (Cui *et al.*, 1994).

With a curvilinear G–C relationship, the ration at which gross growth efficiency is maximized, C_{opt}, is less than the maximum ration, C_{max}. A fish does not maximize both its growth rate and its growth efficiency simultaneously. Growth efficiency must be zero at C_{main}, so the relationship between growth efficiency and ration is a domed curve, with a maximum at C_{opt} (Fig. 6.11).

Between C_{opt}, and C_{max}, growth efficiency declines. In developing a model of fish growth, Paloheimo and Dickie (1965) suggested that the relationship between gross growth efficiency and ration can be described by a decreasing linear function of the form:

$$\log K_1 = a - bC \qquad (6.18)$$

which they called the K-line. This relationship can only apply for rations greater than C_{opt}. In the brown trout, at least, this is at relatively high rations and high growth rates (Elliott, 1975c).

Do fish in natural populations forage at a rate that maximizes their growth rate or their growth efficiency? An analysis of swimming speeds of young bleak, *Alburnus alburnus*, feeding on zooplankton suggested that they swim at a speed that maximizes their growth rate rather than their growth efficiency (Ware, 1975a). This is compatible with optimal foraging theory, described in Chapter 3, which assumes that fish forage in a way that maximizes their net rate of energy intake. Some of the advantages of foraging to maximize growth rate are discussed in Chapters 8 and 11. However, studies on compensatory growth (page 138) suggest that a distinction has to be drawn between short-term and long-term values for C_{max}. The value of C_{opt} may only be of relevance in the context of aquaculture and not in natural populations.

Relationship between growth and quality of food

The effect of food quality on growth is illustrated by comparing the growth of omnivorous fish when fed different diets. Omnivorous roach, 2–3 years old, provided with either grass or meal-worms in sufficient quantities to ensure that the fish fed at close to maximum rates, grew at different rates (Hofer *et al.*, 1985). Fish fed meal-worms grew by an average of 89 kJ in three weeks, the fish fed grass by 6.3 kJ. Roach eating grass consumed a weight of food more than twice that consumed by the roach feeding on meal-worms, yet the gross growth efficiency of the former was only 8.9% compared with 46.2%. *Barbus liberiensis*, a cyprinid of West Africa, had a maximum gross growth efficiency of 25.7% when fed on shredded beef muscle, but only 5.5% when fed groundnuts (Payne, 1979). In both these examples, fish eating an animal diet had the advantage of a higher growth rate and growth efficiency.

In Lake Sibayi (Republic of South Africa), *Oreochromis mossambicus*, although fast-growing as juveniles, grew slowly as adults (Bowen, 1979). The fish ate detritus, which was abundant, but at the depths at which the adults fed, it provided insufficient protein to prevent them from experiencing malnutrition. With an increase in depth, the energy content of the organic component of the detritus did not change, but the ratio of digestible protein to digestible energy decreased. This example illustrates the need to augment analyses of growth using energy budgets with analyses of nitrogen budgets. Several species, including both carnivores and herbivores, show a linear relationship between daily protein requirement per unit body weight and specific growth rate (Tacon and Cowey, 1985).

Nutritional studies on fish are motivated primarily by the need to formulate diets for use in intensive aquaculture (Milliken, 1982; Cowey *et al.*, 1985; Weatherley and Gill, 1987). These studies help to define the requirements of a few species for protein, lipid, vitamins and minerals, and to define the ratio of protein to lipid that yields the best growth performance. The extent to which fish in natural populations experience diets that are nutritionally suboptimal, and the consequences for survival, growth and reproduction in such populations have received little attention. Pike and white bass, in Lake Erie, Canada, display thyroid hyperplasia (goitre) which was attributed to an iodine deficiency (Ketola, 1978), but the ecological implications of this were not described. Populations of Pacific salmon, *Oncorhynchus* spp., have become established in the Great Lakes of North America. These populations also exhibit high levels of thyroid hyperplasia. Leatherland (1994) suggested that this was not caused by an iodine deficiency, but was related to poor environmental quality, including the presence of goitrogens in the water.

Relationship between growth and size spectrum of food

The size spectrum of prey may also influence the pattern of growth of the predator. Both the mechanistic and functional models of prey size selection described in Chapter 3 predict that there is a limited range of prey sizes that a fish will take, and that as a fish grows the mean size of prey taken will tend to increase. The profitability of a prey of a given size depends on its total energy content in relation to the time and energy needed for the fish to detect, pursue and handle a prey of that size. Laboratory experiments in which juvenile Atlantic salmon were fed on food pellets of different diameters showed that growth rate was closely related to pellet size (Wankowski and Thorpe, 1979). For most of the year, the pellet diameter that yielded maximum growth was 0.022–0.026 times the fork length of the fish. The pellet size giving maximum growth increased in direct proportion to fish length. Manufacturers of food pellets for salmonid culture recognize the relevance of food particle size for growth. They provide pellets of different diameters, together with recommendations of the appropriate diameter to be used for fish of a given size (Wankowski and Thorpe, 1979).

If the environment does not provide a size spectrum of prey that allows a fish to take larger and larger prey as it grows, growth may cease (Mittelbach, 1983). The fish is unable to select prey which yield an energy return sufficient to support a further increase in size. This mechanism may explain stunting in some populations of perch (Craig, 1987). Perch can become piscivorous if they grow to a size at which they can capture and handle fish prey. If the size spectrum of the available invertebrate prey is deficient in larger prey, the perch will not reach the threshold size at which they can start to take the profitable fish prey (Chapter 9). The growth of brown trout stocked in separate sections of a stream in Virginia demonstrated the effect on growth rate of a switch from eating invertebrates to piscivory (Gorman and Nielson, 1982). Two size categories of trout were stocked (> 280 mm and < 280 mm in total length). The larger trout fed consistently on fish,

while smaller trout rarely ate fish. Over a period from May to November, the average weight of the large trout increased by 21%, whereas that of the small trout increased by only 4.6%.

Effect of exogenous factors on growth: 2. Temperature

Temperature has a pervasive controlling effect on the rates of both food consumption and metabolism (Chapters 3 and 4), and so has effects on growth. The observation that the growth rates of fishes living at moderate or high latitudes are usually reduced during the winter is circumstantial evidence for an effect of temperature (Fig. 6.3). Interpretation of such seasonal patterns in natural populations is made difficult because any effects of temperature are confounded with those of daylength and other seasonally varying factors.

In analysing the effects of abiotic factors such as temperature on growth, two categories need to be distinguished. Firstly, the abiotic factor may affect the rate of food consumption. Secondly, it may affect the growth efficiency at a given rate of food consumption.

Experimental studies illustrate the effect that temperature has on growth rate and demonstrate that the effects are relatively similar for different species. An extensive study was that on brown trout (Elliott, 1975c,d, 1976b, 1979, 1994). Trout fed on maximum rations, C_{max}, over a temperature range from 3.8 to 21.7 °C, showed an increase in the specific growth rate with temperature up to a maximum at about 13 °C; at higher temperatures the rate declined (Fig. 6.12). The temperature at which specific growth rate was maximized was lower than that at which the rate of food consumption was maximized (Fig. 3.26). The G–C relationships shown by the trout over the temperature range (Fig. 6.10) revealed the effect of temperature on C_{main}, C_{opt} and C_{max}. The main-

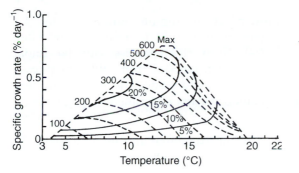

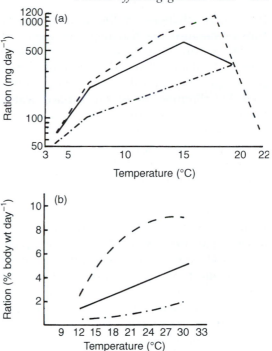

Fig. 6.12 Effect of ration and temperature on specific growth rate (broken lines) and gross growth efficiency (solid lines) for brown trout, *Salmo trutta*. Numbers, ration in mg day^{-1}; percentages, growth efficiencies. Redrawn from Elliott (1975c).

Fig. 6.13 Effect of temperature on maximum (dashed line), optimum (solid line) and maintenance ration (dotted line) of (a) brown trout, *Salmo trutta*, and (b) carp, *Cyprinus carpio*. Redrawn from Goolish and Adelman (1984).

tenance ration, C_{main}, increased with temperature over the whole range tested. C_{opt} was indistinguishable from C_{max} at temperatures less than about 7 °C, but then increased with temperature at a slower rate than C_{max}, reaching a peak at 15 °C before declining (Fig. 6.13). Similar effects of temperature on C_{main}, C_{opt}, and C_{max}, were found for the threespine stickleback (Allen and Wootton, 1982a) and juvenile carp (Goolish and Adelman, 1984). In the brown trout, specific growth rate declined with an increase in temperature for a given ration when that ration was less than C_{max}. The lower the ration, the lower was the temperature at which the trout showed maximum growth rate. Gross growth efficiency was highest at about 10 °C and intermediate ration levels (Fig. 6.12). This was a lower temperature and ration level than those at which growth rate was maximized.

The causal bases for these interrelationships among consumption, temperature, growth rate and growth efficiency still have to be clarified. Experimentally determined relationships can be used to predict growth rates in natural populations. An analysis based on the experimental results for the brown trout suggested that the observed differences in mean growth rates in trout populations in streams at various locations and of varying characters in the UK were explained largely by differences in the temperature regimes (Edwards *et al.*, 1979). The experimental studies also showed that if food is not limited and so the voluntary rate of consumption by the fish can equal C_{max}, the growth rate will increase up to the optimum temperature for growth, then decrease. If food is limited, an increase in temperature will cause a decrease in growth.

A complication in the study of the effect of temperature on growth rate is that in many environments the fish may experience cyclical changes in temperature, either because of changes in the water temperature or because the fish move between different thermal regimes. Juvenile sockeye salmon in Babine Lake show a diel pattern of vertical migration that takes them from a cold hypolimnion to warm surface waters (Chapter 5). In an experimental study, juvenile sockeye were exposed either to a cyclical temperature regime that mimicked the changes experienced during vertical migration or to constant temperatures. When fed moderate rations, sufficient to sustain growth but less than C_{max}, the fish on the cyclical regime had growth rates equal to or greater than fish on constant regimes (Biette and Geen, 1980). At a low ration, growth was greatest at the lowest constant temperature (6.2 °C), while at a high ration, growth was greatest at the highest constant temperature (15.9 °C). This experimental study supports the suggestion of Brett (1971b, 1983) that the vertical migration of sockeye juveniles is a behavioural mechanism that allows them to maximize their growth rate in a thermally heterogeneous environment in which food is limited (Chapter 5). In other species, including rainbow trout and the striped bass, *Morone saxatilis*, a cyclical temperature regime also enhanced growth rates, particularly when the average temperature of the cycle was below the temperature at which the maximum possible growth rate was achieved (Hokanson *et al.*, 1977; Cox and Coutant, 1981). The mechanism by which enhanced growth is achieved will be revealed only when the patterns of energy allocation of fish experiencing cyclical temperatures are accurately described.

A second complication is that the relationships between temperature and growth may differ between populations, especially in those species such as brown trout, which have a wide geographical range. Experimental and some field studies in Norway (Forseth and Jonsson, 1994) on brown trout have suggested an optimum temperature for growth at maximum rations 2 to 3 °C higher than the 13 °C recorded by Elliott (1994). Such discrepancies can be understood if fish derived from different populations are reared under the same conditions (common garden experiments) (page 128).

Effect of exogenous factors on growth: 3. Oxygen

Oxygen acts as a limiting factor, in the sense defined in Chapter 4, for growth. In some environments, oxygen may at times be a more important limiting factor than food (Kramer, 1987). Typically, there is a critical oxygen concentration, below which growth rate declines with a decrease in oxygen concentration even when food is in excess (Brett, 1979; Brett and Blackburn, 1981). For three species of temperate regions, the largemouth bass, the carp and the coho salmon, the critical concentration is about 5 mg O_2 l^{-1} (Brett, 1979). The guppy, *Poecilia reticulata*, a tropical poeciliid, showed no effect of low oxygen concentration on growth as long as the fish had access to the surface layer (Weber and Kramer, 1983). When access was prevented, there was a progressive decrease in growth rate at concentrations lower than 3.0 mg O_2 l^{-1}, which was associated with a decline in feeding rate. This is another illustration of the importance that the oxygen-rich surface layer can have for fish that live in environments that are likely to become hypoxic (Chapters 2 and 4).

The concentration at which oxygen becomes a limiting factor for growth decreases as the food ration is decreased simply because the effect of food as a limiting factor tends to exceed the effect of oxygen (Brett, 1979). At high rations, fish will have a high demand for oxygen because of apparent specific dynamic action (Chapter 4), so intensive aquaculture systems require large quantities of well-oxygenated water if the growth potential of the fish is to be achieved.

Effect of exogenous factors on growth: 4. Salinity

Although most species live in environments that show little change in salinity, for fish that experience fluctuating salinities, the energy costs of osmotic and ionic regulation will mean that less energy can be allocated to growth (Brett, 1979). Salinity acts as a masking factor for growth. Such an effect should show up as a decrease in the growth efficiency at a given ration.

Juvenile coho salmon show an increase in their salinity tolerance between emergence from the gravel in spring and the following January. Otto (1971) described an experimental study of the effect of salinity on growth over this period for coho kept at 10 °C. In summer and early autumn, food consumption, growth rate and growth efficiency did not vary significantly over a salinity range of 0‰ to 10‰, but then declined sharply at higher salinities. In January, growth rate and growth efficiency were reduced at 0‰ and at salinities greater than 10‰. A complex interaction between salinity, temperature and ration influenced the growth rate and growth efficiency of the estuarine flatfish the hogchoker, *Trinectes maculatus* (Peters and Boyd, 1972). Salinity had little effect on voluntary food consumption, which was mainly determined by temperature. At 15 °C, the growth efficiency and growth rate increased with an increase in salinity from 0 to 30‰. At 35 °C, both these growth variables decreased with an increase in salinity. The highest growth rate was achieved at 30‰ and 25 °C, but the highest growth efficiency was at 30‰ and 15 °C. A complex interaction between the effects of temperature and salinity was also seen in the euryhaline desert pupfish, *Cyprinodon macularius* (Kinne, 1960). When fed *ad libitum*, the pupfish showed the highest growth rates at 30 °C and a salinity of 40‰, but the highest growth efficiencies were at 20 °C and 15‰.

Many anadromous species, including migratory salmonids display, higher growth rates after they move from the fresh water in which they hatched into the sea. This suggests that increased availability of food more than compensates for any penalties arising from the energy costs of osmoregulation in a hyperosmotic medium and paid in the form of a reduced growth rate. The development of salmonid farming in sea lochs and fjords and other sheltered coastal localities has led to a greater need to understand the effects of salinity on growth, so that the salinities that provide the best growing conditions can be selected (Thorpe, 1980). In rainbow trout, a species that is farmed, growth declined at salinities greater than 20‰, but there was no significant effect on growth over the range 0–20‰ (McKay and Gjerde, 1985).

Effect of other abiotic factors on growth

Less attention has been paid to the effects of other abiotic factors on growth. Growth is inhibited at high levels of ammonia, but such conditions are only likely to occur in intensive fish culture. Concern at the effects of water acidification caused by industrial and urban emissions of the oxides of sulphur and nitrogen has led to studies on the effect of pH on growth. Fish from naturally acidic waters are frequently relatively small, a phenomenon shown by populations of brown trout in the British Isles (Frost and Brown, 1967). Several causes for the slow growth of trout in acid waters are proposed, including a direct effect of water quality on growth, temperature regime, a shortage of suitable food in such waters and high population densities relative to food availability because such waters often have good spawning habitats (Campbell, 1971). Experimental studies have not yielded consistent results, partly because of a failure to distinguish between the effects of water quality on voluntary food consumption and on growth efficiency. In threespine sticklebacks, voluntary food consumption declined at pH lower than 5.5 and there was a corresponding decline in growth rate (Faris, 1986). In contrast, no effect of pH on the growth rate of brown trout was found down to a pH of 4.4 as long as the calcium concentration in the water was greater than 7 μmol l^{-1} (Sadler and Lynam, 1986).

Current speed may act as a masking factor for growth, because the energy the fish has to expend maintaining its position in or swimming against a current is not available for growth. Two factors may confound this effect. Firstly, fish experiencing moderate currents have a better physiological condition than those in still water – a training effect (East and Magnan, 1987). Secondly, competitive interactions with conspecifics may be more intense in still water (page 127). Brook trout showed higher growth at a current speed of 0.85 body lengths per second (L s^{-1}) than at either $0\,L$ s^{-1} or $2.50\,L$ s^{-1} (East and Magnan, 1987).

Investigations of the effects of abiotic factors on growth must ask the question: what is the effect of the factor on the shape of the G–C relationship (Brett, 1979)? By using this approach, the effects on appetite, measured as C_{max}, can be disentangled from those on growth efficiency. An abiotic factor may produce a reduction in growth rate by causing a reduction in the voluntary rate of food consumption or by reducing the growth efficiency achieved at a given rate of consumption, or by a combination of both effects. A definition of the effects of abiotic factors on the pivotal ration levels of C_{main}, C_{opt}, and C_{max} and on growth efficiencies will provide insights into the causal basis for the effects of the factors on growth and will provide the basis for predictive, quantitative models of growth.

Effects of social interactions on growth

A factor that complicates the experimental study of the effects of exogenous factors is the effect that social interactions between fish has on their growth. Many experimental studies use groups of fish as their basic unit of replication, so any effects of social behaviour are confounded with the effects of the abiotic factors. The effects of social interactions can be classified as competitive or beneficial. In experiments on the effect of exogenous factors, the design should take account of the social interactions likely to occur. A good example of such design comes from the study of the effect of food particle size on the growth of juvenile Atlantic salmon (Wankowski and Thorpe, 1979). In preliminary experiments, socially dominant fish effectively denied subordinate fish access to food. When cover was provided for the fish, this overt dominant–subordinate effect was eliminated.

Competitive effects on growth occur when behavioural interactions between fish cause an unequal distribution of a resource that is related directly or indirectly to growth. Food and space are the resources unequally distributed most frequently by such interactions (Chapters 3 and 5). A symptom of competition is an increase over time in the coefficient of variation of size of a group of fish when compared with the variation in a similar number of fish kept on their own (Purdom, 1974). (The coefficient of variation gives a measure of the variability in the sizes of the fish in a group and is calculated as 100(SD) / $\overline{X}$, where SD is the standard deviation of fish size and $\overline{X}$ is mean size.) The process is called growth depensation. The dominant fish consistently sequester an unequal proportion of the resource and so grow faster than they would if the resource was distributed equally amongst all the fish, consequently subordinate fish grow more slowly. The faster growth of the dominant fish usually enhances their dominance and so allows them to sequester resources even more effectively and further increases the difference in growth rate. In the pygmy sunfish, *Elassoma evergladei*, an increase in the density of fish significantly decreased their mean growth rate, but only slightly affected the growth rate of the fastest-growing fish – the best competitors (Rubenstein, 1981). Such growth depensation is recognized in the intensive culture of salmonids, eels and other species. At regular intervals, the growing fish are size graded so that only fish of approximately the same size are reared together.

In some species, growth depensation occurs only when food is in short supply. When food was restricted, large medaka chased small fish away from the food and grew faster, but when food was present in excess the large fish had no competitive advantage over the smaller fish (Magnuson, 1962). In groups of rainbow trout, restricted feeding led to increased size differences between fish in a group. But when switched to high rations, the fish showed a reduction in these differences (Jobling and Koskela, 1996). In other salmonids, growth depensation can occur when food is not a limiting factor (Jobling, 1985b; Jobling and Baardvik, 1994).

The behavioural interactions have a direct effect on either the rate of food consumption or the rate of energy expenditure. A study of the effect of density on the growth of juvenile rainbow trout in an artificial stream suggested that metabolic rates increased with an increase in density (Li and Brocksen, 1977). Both the mean growth rate and mean gross growth efficiency of the trout declined as density increased. Three processes are thought to cause these effects: starvation of some subordinate fish, increased exercise by subdominant fish forced into regions of fast water and a general increase in the level of excitation. At each density, the dominant trout grew faster and more efficiently than subordinate fish. In an artificial stream, the most dominant coho salmon took up those feeding stations that were most profitable, so growth rate was correlated with dominance rank (Fausch, 1984) (Chapter 5).

A contrast to growth depensation is seen in some species in which the growth of individuals kept in isolation is inhibited. In species that normally shoal, isolated fish often show erratic, abnormal behaviour. Yearling sea bass, *Dicentrarchus labrax*, when isolated remained motionless unless disturbed. They had poor appetites, poor growth and reduced growth efficiencies (Stirling, 1977). In contrast, when they were in pairs they were active and fed readily. Studies on the effect of shoaling behaviour on the division of time between feeding and vigilant behaviour in the minnow (Chapter 3) suggest that growth rates might be higher in fish in shoals because they devote more time to feeding. Yamagishi (1969) developed a model that synthesized the potential effects of social behaviour on growth (Fig. 6.14). The model suggested that shoaling tends to reduce the individual variation in size, in contrast to situations in which shoaling does not develop or does not persist.

Studies of the effect of social interactions on the shape of the *G–C* curve need careful experimental design, but are required if the confounding effects of such interactions are to be understood. It is even more difficult to determine the importance of social interac-

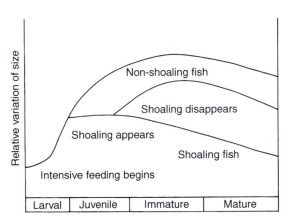

Fig. 6.14 Yamagishi's (1969) schematic model of trends in relative variation in size in fish in relation to social behaviour and state of development. Redrawn from Brett (1979).

tions for growth rates in natural populations, although such effects may be of importance if they result in a decrease in growth rates as population density increases. Such density-dependent effects on growth could result in a decrease in the reproductive success of individuals (Chapter 7) and an increased risk of predation (Chapter 8). Examples of the effects of intraspecific and interspecific interactions on growth are given in Chapter 9.

6.5   ENDOGENOUS CONTROL OF GROWTH RATES

Most studies emphasize the importance of environmental factors such as food availability and temperature for the pattern of growth shown by fish. An equally important, but less studied factor is the intrinsic control of growth. Given the capacity of fish to maintain homeostatic control over many aspects of their metabolism (Chapters 1 and 4), a trait as important as growth will be under regulatory control through the neuroendocrine system.

Genetic control of growth

Although the flexibility of the growth of fishes means that individuals of the same species may show different growth patterns, there is usually a range of sizes characteristic of sexually mature fish (Fig. 6.7). These species-typical sizes are evidence that a genetic component has a role in determining the patterns of growth. The analysis of the genetics of growth in fish is made difficult by the flexibility of their growth patterns and the sensitivity of growth to myriad environmental influences, including social interactions (Purdom, 1974, 1979, 1993). Heritability gives an estimate of the contribution of genetic factors, relative to environmental factors, to the variation shown by a phenotypic trait (such as growth) in a population at a given time. In its narrow sense, heritability is the proportion of the total variance

in the trait that is due to additive genetic variance (Falconer, 1989). It quantifies the extent to which phenotypes are determined by genes transmitted from the parents. It takes values from 0 (no genetic contribution) to 1 (no environmental contribution). Heritability for body weight is low in juvenile Atlantic salmon and rainbow trout, about 0.1, but higher in adults (0.2–0.4) (Gjerde, 1986). The high fecundity of fish in comparison with other domesticated animals (Chapter 7) allows a high intensity of selection for a desired trait. High responses to artificial selection for growth rate in rainbow trout, Atlantic salmon and channel catfish, *Ictalurus punctatus*, have been obtained despite low heritabilities (Gjerde, 1986). The success of programmes of artificial selection on growth traits indicates that natural selection for similar traits can occur in natural populations.

The Atlantic silverside, *Menidia menidia*, provides an important example of population differences in growth performance that have a genetic basis (Conover and Present, 1990; Conover, 1992). The silverside lives along the eastern seaboard of North America from the Gulf of St Lawrence to northern Florida, typically in salt-marshes, estuaries and bays. The species experiences a steep gradient in mean water temperature from north to south, with the gradient much steeper in winter than summer (Chapter 1). The typical life span is only a year, with spawning taking place in spring or early summer. Northern populations spawn later than southern populations. The summer growth season is much shorter for northern than for southern populations, yet there is little difference in mean size at the end of the growing season. Field collections provided evidence that overwintering survival is size dependent, with larger fish surviving the winter better.

Experimental studies showed that at a given temperature, larvae from northern populations grew faster than larvae from southern populations when food was in excess. This was true for larvae hatched from eggs collected in the field, or raised from eggs laid by parental fish raised in the laboratory. Even at high temperatures, the growth rates of the northern populations were higher than for the southern populations, although the southern populations typically experience the higher temperatures. This provides an example of countergradient variation (Conover, 1990; Schultz *et al.*, 1996). The higher intrinsic growth potential in the northern populations compensates for poorer environmental conditions.

The interpretation of these results is that the high growth rates in the northern populations allow the young to reach a size that improves their chances of surviving their first winter. The high growth rates compensate for the short growing period experienced by the northern populations. The genetic basis for the differences in growth rate between the populations was demonstrated by the use of larvae obtained from parental stocks raised in the laboratory under controlled conditions, but derived from northern and southern populations. The experiments are examples of 'common garden' experiments. Genetic differences between populations are revealed by raising progeny of the populations under identical conditions.

Effect of body size on growth rate

Although most fish species show an indeterminate growth pattern, the specific growth rate typically declines throughout the life span. The early stages are characterized by high growth rates. Threespine sticklebacks in a Mid-Wales reservoir had a specific growth rate in terms of weight of about 10% day^{-1} in the first months of life, but in adults the maximum rate declined to about 2% day^{-1} (Allen and Wootton, 1982b). Similar ontogenetic changes in growth rate occur in other species (Brett, 1979). In general, the relation-

ship between specific growth rate, g, and body weight, W, for fish fed maximum rations takes the form:

$$g = a'W^{-b}, \qquad (6.19)$$

or in the linear form:

$$\log g = a - b \log W \qquad (6.20)$$

where the weight exponent, b, is approximately -0.4 (Brett, 1979; Jobling, 1983b). With an increase in body size, the increase in maximum rate of food consumption, C_{max}, is less than the increase in the maintenance ration, C_{main}, so the scope for growth, $C_{max} - C_{main}$, declines with size.

The physiological basis for these size-dependent relationships is not understood. Pauly (1981, 1994) attributed the effects to the allometric growth of the gill surface area (Chapter 2). Another possibility is that the increase in surface area over which food is absorbed becomes limiting. Ware (1978) suggested that in pelagic fish, the decline in specific growth rate is a consequence of the allometric relationships between length and the optimal speed of a foraging fish and hence volume of water searched. The result is a decrease in the ration per unit weight as fish weight increases. Another factor, which will be described in Chapter 7, is the need to synthesize and release gametes. If food consumption does not increase sufficiently to meet the costs of gamete production, then the rate of somatic growth must decrease simply because of a limitation of resources.

Endogenous patterns of growth

Fish kept under constant environmental conditions and supplied with excess food might be expected to grow in a smooth, if weight-related, pattern. But this may not occur. In a pioneering experimental study on the growth of brown trout, Brown (1946, 1957) maintained two-year-old trout at a constant temperature, 11.5 °C, and a constant photoperiod (12L:12D). Under these constant conditions the trout displayed two cyclical patterns of growth. There was a short-term cycle in which the fish showed alternating two-week periods of rapid and slow growth in weight. Growth in length was correlated with growth in weight in the preceding two weeks, so the growth cycles in weight and length alternated. A longer cycle was superimposed on this short-term cycle. Growth rate decreased to a minimum in October, reached a maximum in February, then fell gradually throughout the summer until August. A sharp drop in autumn was associated with the onset of sexual maturity. Under a constant temperature (4 °C) and photoperiod (12L:12D) maintained for 13 months, Arctic charr showed seasonal changes in food consumption and growth rate, with rates highest in midsummer (Saether *et al.*, 1996).

Juvenile rainbow trout showed cyclical changes in weight gain that had a period of 3–4 weeks (Wagner and McKeown, 1985). Coho salmon juveniles also showed a cyclical growth pattern when maintained at a constant temperature and photoperiod (Farbridge and Leatherland, 1987). In coho, the length of the growth cycle was about 14 days and periods of low growth tended to coincide with new and full moons. The control and significance of such cyclical patterns is not known. Fish may respond to external cues not detected by the investigators, or they may have a genetically controlled, endogenous growth cycle which is synchronized by photoperiodic, lunar or other external but reliable cyclical phenomena which act as *zeitgebers* (timing signals) (Thorpe, 1977; Farbridge and Leatherland, 1987).

Further evidence of endogenous control of growth comes from studies that find that changes in growth rate are independent of changes in environmental temperature. Experimental studies show that temperature should be a major environmental factor influencing growth. However, the growth increments of the lake whitefish, *Coregonus*

clupeaformis, were more closely correlated with daylength than with water temperature (Hogman, 1968). Intensive feeding and growth of North Sea herring start when temperatures are low and cease when temperatures are still favourable and food availability is high (Iles, 1984).

Juvenile Atlantic salmon reared under identical conditions often divide into two distinct size groups by the first winter after hatching. This bimodal size distribution emerges because fish in the lower mode almost cease growing for up to 6 months, while fish in the upper mode continue to grow, though at a reduced rate (Thorpe, 1977; Higgins and Talbot, 1985). This difference results from a decline in the feeding motivation of fish in the lower mode that is independent of temperature, food availability and the presence of competitors (Metcalfe *et al.*, 1986, 1988). The suppression of appetite seems to be endogenous and results in a characteristic growth pattern. It also has consequences for the life-history pattern, because fish in the lower mode migrate to sea at least a year later than fish in the upper mode. These differences in growth patterns and life-history cannot be predicted from the assumption that growth rate is solely under the control of exogenous, environmental factors. Fish that were smaller and subordinate in midsummer were more likely to show the decline in appetite (Metcalfe *et al.*, 1989). Respirometry experiments in spring, about the time the young salmon start feeding for the first time, suggested that fish with relatively high standard metabolic rates tended to become dominant (Metcalfe *et al.*, 1995). This result establishes a causal link between a metabolic characteristic and a life-history trait, the age at which the juvenile salmon migrate to sea.

Another mechanism by which a fish might regulate its growth rate is by changes in its locomotory activity. The energy costs of swimming are related to swimming speed (Chapter 4), while the rate of acquisition of food in many predatory species will also be a function of swimming speed (Ware, 1975a). Roach may reduce their locomotory activity during a period of gonadal growth (Koch and Wieser, 1983). A fish may also achieve a behavioural regulation of its growth rate by its selection of thermal regimes (Brett, 1971b, 1983; Crowder and Magnuson, 1983).

These studies imply that the fish is more independent of a direct effect of abiotic factors such as temperature on growth than the short-term experimental studies discussed earlier might suggest. Fish have a genetically determined programme or programmes of growth defining a framework within which external factors can cause modifications. The timing of the execution of the programme will depend on the neuroendocrine system, which may use *zeitgebers* such as photoperiod to synchronize the growth pattern with the temporal patterns that characterize the external environment and, in part, to anticipate such external patterns. The neuroendocrine control of growth of fishes is poorly understood (Matty, 1985) and growth regulation in fishes is a concept that requires more attention.

6.6 MODELLING THE GROWTH OF FISH

The ability to predict the growth of fish is a valuable tool for the successful management of a natural fishery or of an aquaculture facility. The catch that can be taken from a fishery will depend partly on the growth of fish in the exploited population (Pitcher and Hart, 1982; Hilborn and Walters, 1992), while the growth of farmed fish largely determines the success of the enterprise. Experimental studies on the effects of abiotic and biotic factors are providing an empirical framework within which prediction of growth can be made (Brett, 1979). Empirical models of growth based on regression analysis have been developed for the threespine stickleback (Allen and Wootton, 1982a,b). But for most species such information is not available. An

alternative, though complementary approach has been the development of growth models on theoretical grounds (Ricker, 1979). Two types of model are used. In the first, growth is assumed to be some function of the size the fish has achieved. In the second approach, bioenergetic principles are used to formulate the model.

Models based on achieved size

The basic assumption of all models based on achieved size is that the specific growth rate at some time, t, is a function of the size of the fish at that time. Thus in terms of weight, W:

$$dW_t / W_t dt = g_t = f(W_t) \qquad (6.21)$$

The models differ in the choice of the function, $f(W_t)$. The models that have been suggested for fish are listed in Table 6.2. By a judicious choice, a suitable fit to an observed growth curve may be obtained by applying one of these models. The chosen model, with its parameters, then provides a compact description of the observations. Unfortunately, it does no more. It provides no insight into the causal processes that have generated that growth pattern, nor does it suggest how the pattern will change if circumstances change. Any model of fish growth that is to be used for prediction rather than mere description must include a component that is related to the rate of food consumption.

A more empirical approach to modelling growth based on achieved size has been developed by Elliott (1975c,d, 1994), initially to predict growth of the brown trout on maximum rations over a range of temperatures and body sizes (Elliott *et al.*, 1995). The model is based on data obtained from feeding experiments on trout. The model takes the relationship (page 129)

$$G = a'W^{-b} \qquad (6.21)$$

and incorporates an effect of temperature, giving

Table 6.2 Growth models based on achieved size, where specific growth rate, $g = dW/Wdt$, is a function of weight, i.e. $g = f(W)$*

Model	Assumption of relationship	$f(W)$†
Logistic	g linear function of W	$k(1 - W/W_\infty)$
Gompertz	g linear function of $\log_e W$	$k(\log_e W_\infty - \log_e W)$
Monomolecular	g rectangular hyperbolic function of W	$k[(W_\infty/W) - 1]$
Richards	g linear function of W^n	$[1 - (W/W_\infty)^n]\, k/n$

*Source: Causton *et al.* (1978).
† W_∞, asymptotic weight; k and n are constants.

$$G = a'W^{-b}(T - T_{\lim})/(T_m - T_{\lim}) \quad (6.23)$$

where T is temperature, T_m is the optimum temperature for growth, $T_{\lim}$ is T_l if T is less than T_m, or T_u if T is greater than T_m. T_l and T_u are the lower and upper temperatures at which growth is zero (Fig. 6.12). This model has provided a good fit to growth data from brown trout and Atlantic salmon.

The von Bertalanffy model of fish growth

This model has dominated the description of growth patterns of fish. It is based on bioenergetics principles and takes as its starting point the assumption that the rate of growth, dW/dt, is equal to the difference between the rate at which tissue is synthesized, the rate of anabolism, and the rate at which tissue is broken down, the rate of catabolism. Both the rate of anabolism and the rate of catabolism are assumed to be functions of the weight of the fish. That is:

$$dW/dt = hW^n - kW^m \quad (6.24)$$

where h and k are coefficients of anabolism and catabolism respectively, and n and m are the weight exponents for anabolism and catabolism.

Bertalanffy (1957) assumed that the rate of anabolism is proportional to the surface area over which oxygen can be absorbed, S, and

that the relationship between S and W takes the form:

$$S = W^{\frac{2}{3}} \quad (6.25)$$

while the rate of catabolism is directly proportional to body weight, so that the growth equation takes the form:

$$dW/dt = hW^{\frac{2}{3}} - kW. \quad (6.26)$$

From this, by integration, the weight of the fish at time t is given by:

$$W_t = W_\infty\{1 - \exp[K(t - t_0)]\}^3 \quad (6.27)$$

and the length at time t is given by:

$$L_t = L_\infty\{1 - \exp[-K(t - t_0)]\} \quad (6.28)$$

where W_∞ and L_∞ are the asymptotic size, that is the size reached when the rate of catabolism equals the rate of anabolism:

$$hW^n - kW^m = 0 \quad (6.29)$$

and t_0 and K are parameters that together with the asymptotic size determine the shape of the growth curve (Beverton and Holt, 1957; Pauly, 1981): K defines the rate at which the curve approaches the asymptote and t_0 is the hypothetical time at which the size of the fish is zero. Estimates of the three parameters are required to fit a curve to data. Although the model can be used to describe the growth of an individual, more often it is applied to a population. In the latter situation, W_t, W_∞, L_t and L_∞ refer to mean values for the population (Francis, 1995).

A Walford plot is often used in association with the von Bertalanffy growth model. The length of the fish at time $t+1$ is plotted against its length at time t (Ricker, 1979). If the fish has ceased growing, $L_{t+1} = L_t$, and the length at which this occurs can be regarded as an estimate of L_∞. (Fig. 6.15). Some methods for fitting the von Bertalanffy model to a set of data are described in Ricker (1979). The use of non-linear regression methods to fit the model is now facilitated by suitable computer programs (Ratkowsky, 1986).

The popularity of the von Bertalanffy model is due partly to its derivation from physiological principles, but mainly to its use in an influential theory of fisheries management developed by Beverton and Holt (1957). There are numerous examples of its use as a description of the growth in length or, less frequently, weight in a population of fish. Usually, it is used to describe the changes in size at 1 year intervals (Fig. 6.16).

Despite its popularity, inspection of the form of the von Bertalanffy model reveals a major weakness. Although it is derived from general physiological principles, in its final form it has no terms that relate in a simple way to the rate of food consumption. It cannot be used as a predictive model, only as a descriptive model of a set of growth data already collected. In this sense it is as limited as the models based on achieved size. Indeed, Schnute (1981) showed that the von Bertalanffy model and several of the achieved size models are special cases of a more general model. This more general model assumes that the variable, relative rate

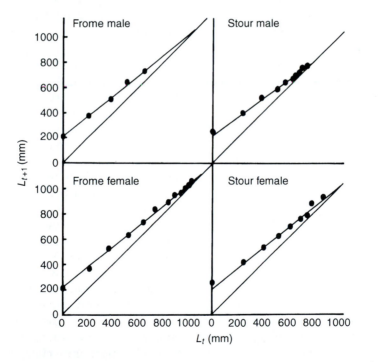

Fig. 6.15 Walford plots for two populations of pike, *Esox lucius*, from southern England. Upper graphs, males; lower graphs, females. L_{t+1} and L_t, lengths at ages $t+1$ and t. Intersection of regression line with diagonal ($L_{t+1} = L_t$) gives estimate of L_∞. Redrawn from Mann (1976a).

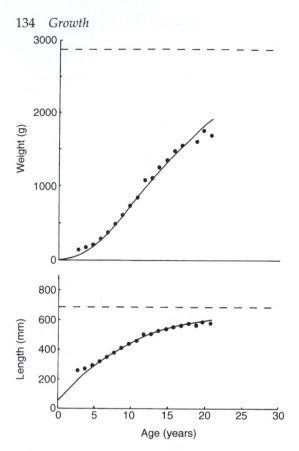

Fig. 6.16 Von Bertalanffy growth model fitted to weight- and length-at-age data for North Sea plaice, *Pleuronectes platessa*. Weight model: W_∞ (broken line), 2867 g; K, 0.095; t_0, −0.815 year. Length model: L_∞ (broken line), 685 mm; K, 0.095; t_0, −0.815 year. Redrawn from Beverton and Holt (1957).

of the specific growth rate, dg/gdt, is a linear function of specific growth rate, that is:

$$dg / gdt = - (a + bg) \qquad 6.30)$$

where a and b are parameters. This strange-looking function expresses an interesting biological idea. It suggests that changes in the specific growth rate over time, dg/dt, depend on the current rate of growth, g (page 137). This idea may be relevant to the regulation of growth because the neuroendo-

crine system may be able to monitor changes in the level of substances such as amino acids that reflect the current growth rate. On the basis of this information, the growth rate could be modulated by changes in levels of hormones such as growth hormone and the thyroid hormones (Matty, 1985; Weatherley and Gill, 1987).

Pauly (1981) proposed a generalized version of the von Bertalanffy model which assumes that the substances needed for anabolism enter the body of the fish across a surface whose area, S, is related to fish length by the relationship:

$$S = pL^a \qquad (6.31)$$

and that catabolism is directly related to fish weight, W, with a relationship between W and L of the form:

$$W = qL^b \qquad (6.32)$$

where p, a, q and b are parameters. If D = a−b, the generalized model for length takes the form:

$$L_t = L_\infty\{1 - \exp[-KD(t - t_0)]\}^{1/D}. \qquad (6.33)$$

This model contains more parameters than the original model and so will probably yield a better fit to any data set. Furthermore, Pauly gives a biological interpretation for his model. He argues that fish do not have large stores of oxygen and that the growth of the gill surface area is slower than growth in body weight (Chapter 2). The relative gill area decreases as the fish increases in weight, restricting the oxygen availability and leading to a reduction in growth efficiency as the fish gets bigger. This hypothesis suggests that in fishes it is oxygen rather than food supply that restricts growth. This is a hypothesis about the evolution of growth patterns and body size in fishes. As a causal explanation, it cannot be used to account for relatively short-term changes in patterns of growth in environments in which oxygen is not a limiting factor. As Chapter 11 discusses, the evolution of growth patterns and other life-history traits will be constrained by what is physiologically possible.

A different extension of the original von Bertalanffy model led to the development of predictive models of fish growth by explicitly including a rate of feeding in the models (Ursin, 1979; From and Rasmussen, 1984). The basic model took the form:

$$\text{Growth} = \text{Anabolism} - \text{Catabolism}. \quad (6.34)$$

Anabolism was assumed to be a function, H, of the rate of food consumption, $H(dC/dt)$. Catabolism was assumed to be a function, J, of two components: starving catabolism, which is a function of body weight, W_t, and feeding catabolism which is a function of the consumption rate, dC/dt. The catabolic rate was then expressed as $J[W_t, H(dC/dt)]$. Growth, dW/dt, is given by:

$$dW/dt = H(dC/dt) - J[W_t, H(dC/dt)] \quad (6.35)$$

More realism was introduced into the model by assuming that the rates of both anabolism and catabolism are temperature dependent.

The basic model was expanded into a form in which the parameters were estimated experimentally from studies on the effects of temperature and fish size on rates of consumption, digestion, excretion and respiration. In a test of the model, the parameters were estimated for rainbow trout fed moist pellets and were then used to predict the growth, over short periods of time (8 and 16 days), of trout fed dry pellets (From and Rasmussen, 1984). Predicted final weights differed from observed final weights by between −2.1 % and +9.1%.

Bioenergetics models of growth

A related set of growth models start from the balanced energy budget (Chapter 4):

$$C = F + U + R + P \quad (6.36)$$

or in its fuller version:

$$C = F + U + R_s + R_d + R_a + P_s + P_r, \quad (6.37)$$

where C is the energy of the food consumed; F and U are the energy lost in faeces and excretion; R_s, R_d and R_a are standard metabolism, digestive metabolism and activity metabolism respectively; P_s is somatic growth (including storage) and P_r, is gamete production (Fig. 4.1) (Chapter 4). Then:

$$P_s + P_r = C - F - U - R_s - R_d - R_a \quad (6.38)$$

If, for the moment, production in the form of gametes is ignored, the growth is:

$$P_s = C - F - U - R_s - R_d - R_a \quad (6.39)$$

Each component on the right-hand side can be expressed as functions of present body weight, W_t, temperature and other relevant factors (Chapters 3 and 4). In principle the value of each component can be estimated, and so P_s can be calculated and the new size of the fish calculated as:

$$W_{t+1} = W_t + P_s E \quad (6.40)$$

where E is the factor required to convert the increase in energy, P_s, into an increase in weight. In most applications of this model, a time step of one day is used, so that the model predicts the daily increase in weight.

Because of difficulties in estimating natural rates of food consumption (Chapter 3), the model is often used to deduce food consumption from observed growth, rather than growth from observed food consumption.

Kitchell and his colleagues developed this bioenergetics model to describe the growth and food consumption of several species including yellow perch, largemouth bass and lake trout (Kitchell *et al.*, 1977; Kitchell and Breck, 1980; Rice *et al.*, 1983; Stewart *et al.*, 1983; Rice and Cochran, 1984). The development of a computer package based on this form of a bioenergetics model (Hewett and Johnson, 1989) has popularized its use.

The original bioenergetics model for the yellow perch (Kitchell *et al.*, 1977) provided the basis for subsequent developments. The subcomponents of this model took the forms:

1. consumption (C): $C = C_{max}pr_c$, where C_{max} is the maximum daily rate of consumption at the optimum temperature, p is a proportionality constant taking values between 0 and 1.0, and r_C is a term that adjusts for temperatures away from the optimum temperature;

2. metabolism (R): $R = R_{max}Ar_r + SC$, where R_{max} is the maximum weight-specific standard metabolic rate, A is an activity parameter specifying metabolic rates above the standard rate (a value of A = 2 indicates twice the standard metabolic rate), r_r is a temperature-dependent adjustment and S is the apparent specific dynamic action coefficient defining the proportion of C utilized in food processing;

3. faecal (F) and nitrogenous (U) excretion: F or $U = CaT^be^{cP}$, where a, b and c are constants, p is the proportionality constant and T is temperature (Chapter 4).

Estimates of the parameters were obtained from the literature on the biology of *Perca flavescens* and *P. fluviatilis* or by extrapolation from studies on non-percid species. The effect of temperature on the specific growth rate and gross growth efficiency predicted by the model is shown in Fig. 6.17. This can be compared with the equivalent figure based on empirical data for the same relationships in the brown trout (Fig. 6.12). The resemblance suggests that the model did generate biologically realistic predictions. The model was used to predict the growth trajectory of yellow perch in Lake Erie for the first 5 years of life (Fig. 6.18).

The techniques of sensitivity analysis can be used to explore how sensitive the predictions of the model are to alterations in the values of the parameters (Bartell *et al.*, 1986; Cui and Wootton, 1989). These techniques indicate which parameters should be estimated with the greatest accuracy to ensure

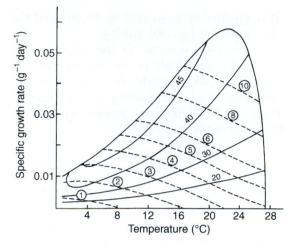

Fig. 6.17 Predicted relationships between growth rate at fixed rations (broken lines) (rations as % body weight per day shown by encircled numbers) and gross growth efficiencies as percentage (solid line) from bioenergetics model for yellow perch, *Perca flavescens*. Compare with Fig. 6.12. Redrawn from Kitchell *et al.* (1977).

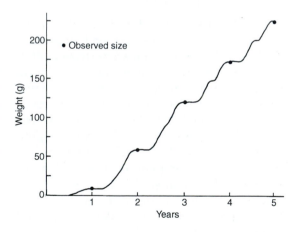

Fig. 6.18 Growth trajectory between observed weights-at-age (●) predicted by bioenergetics model for yellow perch, *Perca flavescens*, in Lake Erie. Redrawn from Kitchell *et al.* (1977).

precise predictions and so help to guide future research programmes. A development of the bioenergetics model extends it to take account of the biochemical transformations involved in feeding, metabolism and growth (Machiels and Henken, 1986).

An impressive feature of the model developed by Kitchell's group has been the extensive use made of it in the context of the management of fish populations. A bioenergetics model of the lake trout in Lake Michigan was used to estimate the impact that increasing the rate of stocking of trout into the lake might have on the rate of exploitation of alewife, an important prey (Stewart *et al.*, 1983). A similar model was used to assess the relative importance of thermal effluent and food availability on the growth and condition of largemouth bass in a reservoir in South Carolina (Rice *et al.*, 1983). Bevelhimer *et al.* (1985) used bioenergetic principles to assess the suitability of three species of *Esox* for stocking into waters with different temperature regimes in Ohio.

An application of the bioenergetics model of growth has used it to assess the importance of spatial and seasonal variation in abiotic characteristics and prey density on consumption and growth of fish. This approach was used to model the growth of striped bass, *Morone saxatilis*, in Chesapeake Bay on the Atlantic seaboard of the USA (Brandt and Kirsch, 1993). The model took into account spatial variations in prey sizes, prey densities and temperature. Growth rates predicted using the spatially explicit model were lower than those obtained using mean temperatures and prey densities. This finding emphasizes the potential importance of spatial and temporal variation for growth in fish populations (Chapter 5)

As the use of bioenergetics models increases, there is more focus on their potential weaknesses. The models are usually used to predict growth from consumption or consumption from growth in natural populations. However, there is usually little or no information on the levels of activity shown by the fish being modelled. A field study of yellow perch from 12 lakes in Quebec found that estimates of food consumption based on

stomach contents (Chapter 3) were a poor predictor of summer growth rates estimated for three age classes (Boisclair and Leggett, 1989b). One interpretation of this surprising result is that energy expenditure on activity varied from population to population in relation to the prey taken and the numerical density of the fish assemblage in each lake (Boisclair and Leggett, 1989a,c). A second problem is that the energy expenditure will vary with the temperature regime experienced by the fish (Chapter 4). Information on levels of activity and temperature regimes requires a better understanding of the behaviour of fish in natural populations, information that may be obtained from techniques such as telemetry and acoustic surveys (page 88) (Hansen *et al.*, 1993). Video analysis of free-swimming fish has provided estimates of the energy expenditure of free-swimming brook trout (Krohn and Boisclair, 1994).

Models based on bioenergetics principles are taking the modelling of fish growth from an era of curve-fitting to an era of predictive, causal models. These models are being embedded in individual-based models that predict growth and mortality rates in cohorts of fish of the same age (Chapter 10). Development of bioenergetics models now requires a fuller, quantitative understanding of the physiological bases for effects of body size and environmental factors on the processes of consumption, digestion, excretion and metabolism. At the moment, although based on bioenergetic theory, the models rely heavily on empirically determined relationships between the processes and the factors.

Modelling endogenous control of growth

Causal models of growth, including bioenergetic models, must also incorporate the concept of the regulatory control of growth.

In a discussion of the growth pattern of isopods, Hubbell (1971) described one possible framework by which such control can be built into a growth model. He suggested that growing organisms have a 'desired' growth rate that represents their optimum growth trajectory. This is the pattern of growth that will maximize the lifetime production of offspring. The physiological nature of the 'desired' growth rate is not specified in the model, which is developed in terms of control theory. At each moment, achieved growth rate is compared with this 'desired' growth rate, and foraging behaviour and metabolism are adjusted to minimize the difference between achieved and desired growth rate (Fig. 6.19).

The phenomenon of compensatory (or catch-up) growth provides some experimental evidence for this model. When fish are deprived of food for a period of time, they lose weight. If then provided with unlimited food, the fish show growth rates higher than those shown by fish continually on unlimited rations (Jobling, 1994). These increased growth rates enable the fish to wholly or partially catch up with the size of fish continually on unlimited food. The previously deprived fish can regain the growth trajectory of the well-fed fish. The increased growth rate can be a result of increased food consumption and higher growth efficiencies (Miglavs and Jobling, 1989; Russell and Wootton, 1992). Several species show compensatory growth, including salmonids and cyprinids, although the conditions in which compensatory growth will be shown are poorly defined.

In some situations, the compensatory response is not shown to growth rate but some other indicator of physiological status. Some juvenile Atlantic salmon lose appetite as winter approaches but deplete their lipid reserves during the overwintering period. When juveniles were starved, accelerating the depletion of their lipid reserves, the fish

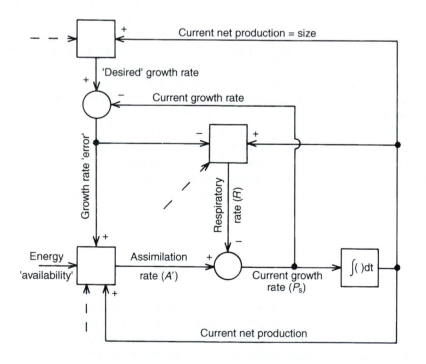

Fig. 6.19 Simplified flow diagram for regulation of growth rate based on Hubbell's control-theory model. Broken arrows signify environmental disturbances. Redrawn from Calow (1976).

showed a temporary increase in appetite when subsequently provided with food (Metcalfe and Thorpe, 1992). This increase caused a recovery of the lipid reserves to levels characteristic of unstarved fish. A model of compensatory growth in salmonids suggested that fish maintain a ratio of non-mobilizable to mobilizable tissue, and that changes in appetite seen in compensatory growth are adjustments to changes in that ratio (Broekhuizen *et al.*, 1994).

These experimental studies are uncovering the regulatory capacities of fish in the context of energy allocation, growth and storage. The importance of these capacities for natural populations has still to be defined. The compensatory responses would allow fish to buffer the effects of variations in food availability on growth trajectories and levels of reserves.

6.7   SUMMARY AND CONCLUSIONS

1. Growth is a change in size of an individual, usually an increase, measured in units of length, weight or energy. It is best measured as specific growth rate. The relationship between weight and length provides an index of the condition of the fish.

2. Estimating growth rates in natural populations usually depends on adequate methods of ageing fish. Calcareous structures including scales and otoliths may show distinct annual patterns that allow fish to be aged. Otoliths also show daily growth rings that allow a finer resolution of age, especially in young fish. Length-frequency data can also be used to estimate the number of age classes present in a population. Care must be taken to adequately validate the methods of ageing used.

3. With a few exceptions, growth in fishes is indeterminate and flexible, so fishes frequently respond to environmental changes with changes in their growth rate. Superimposed on a lifetime growth pattern are seasonal variations in growth rates that may reflect the effects of both environmental (exogenous) factors and endogenous factors. The growth pattern of a fish is the result of an interplay between a potential for growth that depends on the genotype of the fish and the complex of environmental conditions it encounters as it grows.

4. Growth depends on the energy and nutrients provided in food. The relationship between growth rate and ration size is usually curvilinear, approaching an asymptote at high rations. Consequently growth (conversion) efficiency is at a maximum at an intermediate, optimum ration. Growth rates also depend on food quality and on the size of food particles avilable.

5. At maximum rations, growth rate increases with an increase in temperature up to a peak, but then decreases with a further increase in temperature. The peak defines the optimum temperature for growth. At low rations, growth rate declines with an increase in temperature.

6. At low concentrations oxygen is a limiting factor for growth. Salinity acts as a masking factor. Other abiotic factors affecting growth rates include pH and current speed.

7. The presence of conspecifics can lead to depensatory growth, in which larger fish are more successful in sequestering resources than smaller fish, leading to an increase in the relative size differential. The effects of fish density on growth rates may have important consequences for the survival and fecundity of individuals.

8. Growth also depends on endogenous factors including the genotype, body size and physiological condition of the individual. Endogenous changes in appetite may override the effects of abiotic factors.

9. Two main approaches to modelling growth are used. The first assumes that specific growth rate is a function of achieved body size, i.e. $dW/Wdt = f(W)$. The second is based on bioenergetic principles. An early bioenergetics model, the von Bertalanffy curve, is frequently used to describe lifetime growth. More promising as predictive models are those based on a description of the allocation between maintenance and growth of the energy supplied in the food. Such models are increasingly being used as management tools.

10. Experimental studies on compensatory growth are defining the capacity of fish to regulate energy intake and allocation.

7

REPRODUCTION

7.1 INTRODUCTION

If a fish is to be represented genetically in the next generation, at some time in its life it must begin to allocate time and resources to reproduction. A study of the ecology of reproduction will include analyses of the effects of environmental factors on when and where spawning takes place and what resources are allocated to reproduction as opposed to maintenance and growth. The problem of timing raises two sets of questions. The first set asks at what age does a fish becomes sexually mature and what factors determine this age? The second set asks what factors determine when in the year reproduction takes place? The problem of allocation also has two basic components: what portion of available resources is allocated to each reproductive attempt, and of the material resources that are allocated to reproduction, what portion is allocated to each individual offspring? This chapter explores each of these questions, where possible in relation to the effect that environmental factors have on their resolution.

Strategies and tactics of reproduction

Fishes show a profusion of patterns in where they reproduce and in their allocation of time and resources to reproduction (Breder and Rosen, 1966; Balon, 1975). Most species are gonochoristic, that is the sexes are separate. But there are species that are sequential hermaphrodites in which an individual functions as one sex and then becomes transformed to function as the other sex (Warner, 1978; Shapiro, 1984). There are also a few species that are simultaneous hermaphrodites in which an individual has functional ovaries and functional testes concurrently (Warner, 1978). Most species are oviparous, but both ovoviviparity and full viviparity occur. A few species are parthenogenetic, but these parthenogenetic forms require sperm from males of a closely related species to activate the development of the eggs. Some of these unusual reproductive strategies are considered later in this chapter, but the typical teleostean reproductive mode is of separate sexes with cross-fertilization. Compared with elasmobranchs, reptiles and birds, the fishes lay clutches of numerous, small eggs. Parental care of eggs is not uncommon, but in the majority of species the eggs are not tended. Compared with terrestrial vertebrates or the elasmobranchs, most teleosts specialize in the quantity rather than the quality of the progeny they produce.

Each individual fish has a suite of reproductive traits that are determined by its genotype and hence by the evolutionary history of the gene pool of which the fish is a member. The combination of reproductive traits characteristic of individuals belonging to the same gene pool can be regarded as the reproductive strategy of those individuals (Fig. 7.1). An understanding of reproductive strategies rests on identifying the selective processes that have led to the evolution of the observed strategies. The causal factors responsible for the selection are sometimes called ultimate factors. The genotype of an individual will also define the phenotypic range that a trait may express (Fig. 7.1).

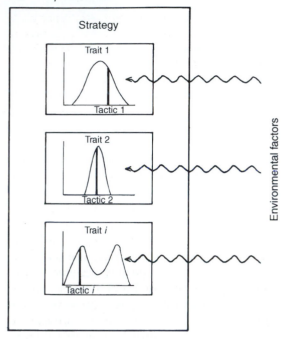

Strategy

Trait 1

Tactic 1

Trait 2

Tactic 2

Trait *i*

Tactic *i*

Environmental factors

Fig. 7.1 Reproductive strategy combines traits 1,2, ... *i*; each trait has a range of phenotypic expression illustrated by curves. From this range the tactic expressed, indicated by thick vertical line, depends on the environmental conditions experienced by the fish.

Some traits may be plastic so an individual can exhibit a wide range in their expression (Chapter 11). Other traits may be inflexible and show little phenotypic plasticity. The environment experienced by the fish will determine the expression of the trait, and the variations in expression of a trait in response to environmental changes are the tactical responses of the individual to those changes (Wootton, 1984b). Such tactics represent a homeostatic response of the fish that minimizes the cost of the changes (see also Chapter 1). An understanding of tactical variations depends on identifying those environmental factors that cause variations within the lifetime of the individual. These are often called proximate factors.

The most important ultimate and proximate factors shaping the reproductive ecology of fishes probably include: the harshness and variability of the abiotic environment; the availability of food for parental fish and their offspring; the presence of predators on parental fish and on the offspring; and the nature of the habitat of parental fish. The profusion of reproductive patterns shown by fishes offers a richness of material for comparative analysis of the possible effects of these environmental variables. However, this profusion has probably slowed the application of experimental and quantitative methods to studies of reproduction in comparison with studies of metabolism or growth.

7.2 TIMING OF REPRODUCTION

Age at first maturity

The onset of sexual maturity represents a critical transition in the life of an individual. Before, the allocation of time and resources is related only to growth and survival. After, there is a potential conflict between the allocation of time and resources to reproduction or to survival and growth. The resolution of this potential conflict should be a pattern of allocation that maximizes the number of offspring produced in the lifetime of the individual under the prevailing environmental conditions (Chapter 11). In some species, adults breed only once: the age of first reproduction defines the life span. All fish reaching sexual maturity die either while breeding or soon after. This is a semelparous life history. The pattern is exemplified by Pacific salmon of the genus *Oncorhynchus*. After the completion of their anadromous migration to the spawning grounds in the head waters of the Pacific seaboard rivers, the Pacific salmon spawn and die (Chapter 5). Species in which some adults in the population survive to breed again are called iteroparous. The salmonid genus *Salmo* is iteroparous, although the proportion of adults surviving to spawn again is often small (Jonsson *et al.*, 1991). Rainbow and steelhead trout (*Oncorhynchus mykiss*) are interoparous, although now classified in the same genus as the Pacific salmon.

A fuller discussion of this problem of allocation is postponed until Chapter 11 because it involves the synthesis of information on growth rates (Chapter 6), reproduction (this chapter) and mortality rates (Chapter 10).

Fishes display a wide range of ages at first reproduction. At one extreme, cyprinodonts, particularly those found in temporary pools in tropical and subtropical areas, can reach sexual maturity at the age of a few weeks (Miller, 1979a; Simpson, 1979). At the other extreme, flatfish of the genera *Hippoglossus* and *Hippoglossoides* may not become sexually mature until 15 years old (Roff, 1981). Even within the flatfish order, Pleuronectiformes, the age at first maturity ranges from 1 year (*Cynoglossus semifasciatus*) to 15 years or more (*Hippoglossoides platessoides*). In flatfishes, the reproductive life span after the onset of sexual maturity is correlated with the age at maturity (Roff, 1981). Species that mature at a young age have only a short reproductive life span, whereas those species that mature late have a long reproductive span.

Within a species, the age at maturity may vary. In American plaice, *Hippoglossoides platessoides*, from the sea around Newfoundland including the Grand Banks area, age at which 50% of the females in a population are mature has varied from 7.8 to 15.2 years (Pitt, 1966). For males, the range is 5.3–7.5 years. Similar interpopulation variation is found in many other species. Part of this variation certainly reflects genetic differences. Experiments with brown trout showed that when fish from different populations were reared under similar conditions, they still had differences in their age and size at maturity (Alm, 1959). However, part of the variation also reflects proximate, environmental effects.

The effect of an environmental change on the age (and size) at maturity depends on the effects of the change on the growth and mortality rates (Chapter 11). A common response to a change that increases the growth rate of the fish is a decrease in the age of maturity. This may be accompanied by a change in the size at which the fish matures, but not always (Alm, 1959; Stearns and Crandall, 1984; Stearns and Koella, 1986). Age at first maturity for American plaice females from a Grand Banks population declined from about 14 years in the period 1961–65 to about 11 years in 1969–72, but the length at maturity did not change significantly (Pitt, 1975). This change occurred when heavy fishing pressure reduced the abundance of the plaice. Their growth rate increased, probably because there was more food for surviving plaice. Experimental evidence shows that differences in food supply can alter the time at which fish become mature. In brown trout, threespine sticklebacks and the Pacific herring, *Clupea harengus pallasi*, fish receiving a higher food ration matured earlier (Bagenal, 1969; Wootton, 1973; Hay *et al.*, 1988).

In a few species, social control of the timing of maturation has been demonstrated experimentally. The presence of large or fast-growing males inhibits the onset of sexual maturity in smaller or slower-growing juvenile male platyfish, *Xiphophorus maculatus* (Borowsky, 1973; Sohn, 1977).

Seasonal timing of reproduction

As conditions change during the year with the change in season, the suitability of the environment for the vulnerable early life-history stages will vary. A fish should reproduce at that time of year that will tend to maximize its lifetime production of offspring. The larval fish should hatch into a world that can provide appropriate food, protection from predators and benign abiotic conditions. Most freshwater fish species in

Canada, for example, spawn between April and July, although the salmonids provide a minor peak of spawning in the autumn (Wootton, 1984b). Hatching or emergence of the larvae in the Canadian waters is concentrated into a short period between April and July.

Timing at high latitudes

In high latitudes, the major seasonal changes are in temperature and hours of daylight. Correlated with the changes in these abiotic factors, there are usually changes in the abundance and quality of food. Such changes are clearly illustrated by production cycles in the sea (Cushing, 1975, 1995). In one such cycle, the abundance of phytoplankton is low in winter, but increases rapidly during the spring. Over the summer the abundance falls, but there is frequently a secondary peak in the autumn preceding the winter decline. In parallel, there are changes in the abundance of the zooplankton that graze the phytoplankton. A similar production cycle occurs in lakes (Wetzel, 1983). Larval fish, because of their small size and poor locomotory and sensory capacities, can only capture less-evasive prey that fall into appropriate size categories (Frank and Leggett, 1986; Miller *et al.*, 1988). The timing of reproduction must ensure that the larvae are hatching at a time of year when the size spectrum and abundance of the prey are appropriate. The eggs and young larvae are also the life-history stages most vulnerable to predation, both by other fish and by invertebrate predators (Chapter 9). Rapid development of the eggs and subsequent rapid growth of the larval fish into size classes that are less vulnerable to predation will also be advantageous.

Fish in high latitudes do show strong seasonal patterns of reproduction. In several species the timing of spawning is precise. The dates of peak spawning in some populations of sockeye salmon, herring, plaice and cod have a standard deviation of less than a week (Cushing, 1969). Seasonality in the timing of reproduction can be illustrated by species living in northern fresh waters, in which three basic patterns of timing can be recognized (Fig. 7.2).

Many salmonid species are autumn and early winter spawners. Gametogenesis takes place during the summer so that ripe eggs and sperm are available in the autumn. The

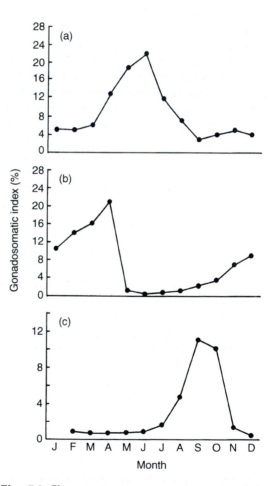

Fig. 7.2 Changes in mean gonadosomatic index shown during annual reproductive cycles of three northern freshwater species illustrating (a) summer (*Phoxinus phoxinus*), (b) spring (*Perca fluviatilis*) and (c) autumn (*Salvelinus fontinalis*) spawners. Redrawn from Mills (1987), Craig (1974) and Wydoski and Cooper (1966).

eggs are large by teleost standards, typically 4–6 mm in diameter. The spawning female buries the eggs in gravel. At the low temperatures of winter and early spring, the rate of development of the eggs is low, so the larvae do not hatch until spring. The larvae remain in the gravel while they absorb their yolk. Only then do they emerge to take up an active predatory life. On emergence they are large in comparison with most other fish larvae and are able to feed on relatively large food particles (Fahy, 1979).

A second seasonal pattern is shown by species in which gametogenesis takes place in autumn and through the winter. Spawning takes place in spring or early summer with more-northerly populations later in this period. Pike show this pattern. The female lays all the eggs for that year over a short time period, so the breeding season is short. The eggs are laid in vegetated shallow backwaters or littoral areas where water temperatures increase quickly as the air temperature increases. The larvae hatch early in that period of the year within which almost all the fish species produce larvae. The eggs of the pike are relatively large, about 2.5 mm in diameter, and by the time all the contents of the yolk sac have been absorbed the larvae are about 15 mm long. Initially the larval pike feed on zooplankton, but growth is rapid and the young fish soon feed on insect larvae and amphipods, and then on fish, and reach a length of 100 mm or more by the first autumn of life (Mann, 1976a; Balvay, 1983; Craig, 1996). Other carnivorous species including the perch show a similar timing of reproduction (Craig, 1987).

Summer spawners illustrate a third spawning pattern. In these species, though functional sperm may be present in the testis by late autumn, the development of the ovaries is slow during the autumn and winter. Only in spring does the rapid maturation of the ovaries occur and spawning takes place in late spring and summer. Examples of this pattern include the stickle-backs, the European minnow and several other cyprinid species (Wootton, 1979). In many species with this pattern, the female can produce several batches of eggs during a spawning season, which is extended over two or three months. The eggs are typically 1–2 mm in diameter and they develop rapidly in the warmer, summer waters.

The consequences of time of spawning and egg size were illustrated by a study of the growth of nine species in Lake Opinicon, Ontario, in their first year of life (Keast and Eadie, 1984). The species have different feeding habits and life-history strategies. The young of species that spawned early in the summer reached a larger size by September than the young of species that spawned later. There was a positive correlation between egg size and the weight the fish reached by the end of the summer. Over the summer, the patterns of growth of the nine species were similar, with an almost linear increase in length in June and July. Growth almost ceased by September as water temperatures declined. Although such studies show the consequences of different times of spawning and different egg sizes for the progeny, the adaptive and ecological significance of such life-history traits require more rigorous, quantitative study.

Timing at low latitudes

At low latitudes, seasonal changes in temperature and daylength have much smaller amplitudes than in high latitudes. Often the major seasonal event in fresh waters is the change in water level with the sequences of wet and dry seasons (Lowe-McConnell, 1975, 1987; Payne, 1986; Munro, 1990). Seasonal patterns of reproduction in freshwater fishes at low latitudes frequently relate to the effect of the changes in water level. Coastal waters may also be influenced by seasonal changes in the volume of freshwater run-off from adjacent land.

Species that are typical of the large floodplain river systems such as the Amazon, the Zaire and the Mekong, spawn just prior to or during the period of flooding (Lowe-McConnell, 1975, 1987; Munro, 1990). As the river floods, inundating adjacent land, it opens up large areas, which provide abundant food and cover for the newly hatched fish. The spawning females typically are highly fecund and they shed all their eggs over a short period of time. In an Amazonian genus, *Prochilodus*, which spawns as the flood waters begin to rise, the ovaries account for up to a quarter of the total weight of the fish and about 300 000 eggs are shed (Schwassmann, 1978).

In small, tropical streams, many species spawn in the dry season (Lowe-McConnell, 1979; Townshend, 1984). This may be an adaptation that protects the larval fish from being washed away by spates. The species in streams are frequently able to spawn several times during the breeding season and many show some form of parental care. Stream cichilids such as *Cichlasoma* spp. are biparental, with both male and female guarding and tending the eggs and newly hatched larvae. Communication between parents and offspring would probably be much more difficult in a turbid, fast-flowing spate stream. However, several different seasonal breeding patterns may coexist in the same small stream. Six small, egg-laying, characin species (Characiformes) in a small Panamanian stream revealed a range of seasonal patterns (Kramer, 1978). Two species spawned over a short period at the start of the rains, two species spawned for 2–4 months during the dry season, and two species spawned throughout the year. A comparable study of three viviparous poeciliid species in two Costa Rican streams found evidence of breeding throughout the year, though with some seasonal variation particularly in the females (Winemiller, 1993). Livebearers do not require special spawning or larval nursery habitats and so may be less affected by changes in water level. However, explanations for these differences in seasonality would require more information on patterns of growth, survival and resource use than is currently available.

In tropical lakes, which are less influenced by changes in water level than rivers, a variety of seasonal patterns of breeding is observed (McKaye, 1977, 1984; Lowe-McConnell, 1987; Munro, 1990). Some species have well-defined breeding seasons. In other species, fish in reproductive condition are found throughout the year, but there are seasonal peaks in the proportion of fish in breeding condition. In a small crater lake in Nicaragua, Lake Jiloa, eight or nine species of cichlids bred in shallow, rocky areas (McKaye, 1977, 1984). All spawned on the substrate, with both parents guarding the eggs and young. There was a bimodal seasonal breeding pattern, with peaks in the dry season around April and in the wet season around August. Competition for breeding sites in the shallow water was intense. The species that dominated the cichlid biomass, *Cichlasoma citrinellum*, bred in the wet season when it excluded most of the other species from breeding sites by its aggression. In the dry season, *C. citrinellum* did not breed and other species were able to move into the spawning area. The failure of the dominant cichlid to breed in the dry season was probably caused by a shortage of food. So in this system, the seasonality of breeding of the cichlids was partly caused by a seasonal cycle of food abundance, but this effect was amplified by the behavioural interactions between the species competing for a limited spawning area. The presence of a predatory fish, *Gobiomorus dormitor*, which lives in deeper waters, may be the factor that forced the cichlids into the shallow water.

Tropical marine fishes vary from species in which some fish are spawning throughout the year to species in which there are well-defined breeding seasons (Lowe-McConnell, 1979, 1987; Longhurst and Pauly, 1987). Even in the former group, there are often clear seasonal variations in the proportion of fish in a population that are reproductively active.

Control of timing

To ensure that a fish is in breeding condition at the appropriate season, a physiological mechanism must control the timing of gonad maturation. This timing mechanism probably has two components: an endogenous cycle of gonadal development and a mechanism that synchronizes this cycle with environmental cues. These factors have been the subject of many experimental studies (Wootton, 1982; Lam, 1983; Bye, 1984; Munro *et al.*, 1990). The cues used by fish at low latitudes to ensure the correct timing of their breeding effort are less clear. They may include changes in the abundance of food, in water chemistry, in meteorological conditions and subtle changes in temperature or daylength (Lowe-McConnell, 1979; Burns, 1985; Munro, 1990).

Many species live over a wide range of latitudes and so encounter different temperature and photoperiod regimes. This can result in major differences in the timing of reproduction. In the south of its range, the threespine stickleback breeds in early spring, but in more northerly populations breeding takes place in midsummer (Wootton, 1984a; Baker, 1994). The maturation of the gonads in the stickleback is under photoperiodic control but the final stages of maturation occur more quickly at warmer temperatures (Baggerman, 1980, 1990). Some species have been transplanted with relative ease over wide geographical distances, for example brown trout have been transplanted from Europe to the highlands of East Africa and to New Zealand (Frost and Brown, 1967). This suggests that the timing mechanisms for reproduction are flexible and can function over a range of environmental circumstances. The neuroendocrine system, which is responsible for the timing of reproduction, forms an important component of the homeostatic systems of the fish (Munro *et al.*, 1990). By exercising a choice over the thermal and light regimes they experience (Chapter 4), fish can, within limits, control behaviourally the temporal pattern of stimulation from such external cues (Scott, 1979).

Diel and other temporal patterns of spawning

In some species, the act of spawning only takes place during a restricted time of the day. Of twelve species studied off the coast of Long Island, New York, eight spawned mainly in the evening or at night, two spawned in the afternoon and into the night, and one spawned in the morning (Ferraro, 1980). Only one of the twelve, the mackerel, *Scomber scombrus*, spawned throughout the day. Many coral reef species that have pelagic larvae spawn in the evening (Helfman, 1993). This timing may reduce predation on the eggs because of the low light levels and because the numbers of fish feeding on zooplankton are reduced at that time of the day, though it may increase the risk to the spawners (Chapter 5).

Fishes that lay their eggs in the intertidal zone may have times of spawning related to the lunar cycle, because of its relationship to the tidal cycle. The Californian grunion, *Leuresthes*, lays its eggs in sand high up the beach. The fish make spawning runs soon after the highest daily tide, 2–4 days after both the new and full moon phases (Moffat and Thomson, 1978). The eggs, buried in the sand, develop until uncovered by the high tides of a subsequent moon phase. The Atlantic silverside spawns in the upper intertidal zone and deposits the eggs on vegetation, including detrital mats (Middaugh, 1981). Spawning runs only take place during the day at a time that coincides with high tide. The heaviest runs occur near the time of the new and full moons.

The damselfishes (Pomacentridae) of coral

reefs show a variety of temporal patterns of spawning (Thresher, 1991). Some species have a lunar cycle, others a semilunar cycle. Other species have no identifiable periodicity.

Although descriptive and physiological studies of seasonal, lunar and daily patterns of reproduction are common, there are few quantitative studies of the ecological significance of these temporal patterns. What are the consequences for the survival and growth of progeny if reproduction takes place at unusual times? By manipulation of environmental conditions or treatment with exogenous hormones, it is now possible to bring several species into reproductive condition at unusual times of the year (Bye, 1984). This means that experiments are now feasible in which eggs and larvae are exposed to natural conditions at unusual times, to test quantitatively the common-sense hypothesis that eggs are laid and larvae hatch at times that maximize their survival.

7.3 THE SITE OF REPRODUCTION

Eggs and the young free-living stages of fishes are vulnerable both to unfavourable abiotic conditions and to attack by invertebrate and vertebrate predators. The lack of mobility of these early life-history stages means that they can display little or no behavioural response to such hazards. Consequently, the spawning site largely determines the intensity and nature of the hazards that the eggs and larvae encounter. Probably the most important potential hazards for the eggs and larvae are a lack of oxygen in the surrounding water, a danger of being smothered by silt, infection by microorganisms and predation. However, spawning sites that are well concealed so that the dangers of predation are reduced are likely to be sites that become deoxygenated, prone to siltation and colonization by microorganisms (Keenleyside, 1979). The site of spawning will also determine whether appropriate nursery areas

for larvae and juvenile fish are accessible (Sinclair, 1988).

Balon (1975, 1981) has proposed a comprehensive classification of reproductive styles in fish based largely on the site of spawning, the adaptations of the eggs and embryos to the site and the degree of parental care (Table 7.1).

A basic division is between pelagic and demersal eggs. Pelagic eggs are characteristic of offshore marine species (Russell, 1976). Most coral reef fishes also have pelagic eggs (Sale, 1980; Thresher, 1984) (Chapter 12). In such species, spawning usually takes place at sites that ensure the dispersal of the eggs away from the reef on the ebbing tide. Some freshwater fishes also spawn pelagic eggs (Balon, 1975). Examples are provided by some cyprinids of large Chinese rivers including the four important cultured species, the grass carp, *Ctenopharyngdon idella*, bighead, *Aristichthys nobilis*, silver carp, *Hypophthalmichthys molitrix* and black carp, *Mylopharyngodon piceus*. Their eggs drift downstream in the turbid river water as they develop. After hatching, the larvae make their way to the river margins into backwaters and side-lakes for further growth. Pelagic eggs, kept in motion by water movements, are in little danger of experiencing anoxic or silty conditions, but can suffer high mortalities by predation (Ware, 1975b).

Demersal eggs are characteristic of most freshwater species and many inshore marine fishes. Even the pelagic Atlantic herring lays demersal eggs on traditional spawning grounds. Eggs spawned on the substratum may be in danger of siltation unless the spawning site is in a water current sufficiently strong to keep the eggs clean and well aerated. Eggs that are buried in the substratum are protected from predation, but the nature of the substratum must be such that a water current bathes the eggs. Salmonids choose gravel through which water flows, when selecting where to dig nests in which their eggs are laid. When eggs are

Table 7.1 Reproductive guilds of teleost fishes based on Balon's (1975) classification (see also Moyle and Cech, 1996)

I.   Non-guarders of eggs and young

   A. Open substrate spawners

      1. Pelagic spawners, e.g. *Mola mola*

      2. Benthic spawners

         a. Spawners on coarse substrates (rocks, gravels, etc.)

            (1) Pelagic free embryo and larvae, e.g. *Morone saxatilis*

            (2) Benthic free embryo and larvae, e.g. *Phoxinus phoxinus*

         b. Spawners on plants

            (1) Non-obligatory, e.g. *Rutilus rutilus*

            (2) Obligatory, e.g. *Esox lucius*

         c. Spawners on sandy substrates, e.g. *Gobio gobio*

   B. Brood hiders

      1. Benthic spawners, e.g. *Oncorhynchus nerka*

      2. Cave spawners, e.g. *Anoptichthys jordani*

      3. Spawners on invertebrates, e.g. *Rhodeus amarus*

      4. Beach spawners, e.g. *Leuresthes tenuis*

      5. Annual fishes, e.g. *Nothobranchius guentheri*

II. Guarders

   A. Substrate choosers

      1. Rock spawners, e.g. *Chromis chromis*

      2. Plant spawners, e.g. *Pomoxis annularis*

      3. Terrestrial spawners, e.g. *Copeina arnoldi*

      4. Pelagic spawners, e.g. *Ophiocephalus* spp.

   B. Nest spawners

      1. Rock and gravel nesters, e.g. *Ambloplites rupestris*

      2. Sand nesters, e.g. *Cichlasoma nicaraguense*

      3. Plant material nesters

         a. Gluemakers, e.g. *Gasterosteus aculeatus*

         b. Non-gluemakers, e.g. *Micropterus salmoides*

      4. Bubble nesters, e.g. *Betta splendens*

      5. Hole nesters, e.g. *Cottus aleuticus*

      6. Miscellaneous materials nesters, e.g. *Lepomis macrochirus*

      7. Anemone nesters, e.g. *Amphiprion* spp.

III. Bearers

   A. External bearers

      1. Transfer breeders, e.g. *Oryzias latipes*

      2. Forehead breeders, e.g. *Kurtius gulliveri*

      3. Mouthbrooders, e.g. *Oreochromis mossambicus*

      4. Gill-chamber brooders, e.g. *Typhlichthys subterraneus*

      5. Skin brooders, e.g. *Bunocephalus*

      6. Pouch brooders, e.g. *Syngnathus abaster*

   B. Internal bearers

      1. Ovi-ovoviviparous, e.g. *Glandulocauda inequalis*

      2. Ovoviviparous, e.g. *Sebastes marinus*

      3. Viviparous, e.g. *Poecilia reticulata*

immobile on a surface, parental care with one or both parents guarding and ventilating the eggs can reduce predation and the problems of oxygen supply and siltation (page 154). Such care may extend to the parental fish modifying the substratum, for instance by cleaning the site, digging a pit or building a nest.

The presence of cues from suitable spawning sites can be important in stimulating the final stages of reproduction (Stacey, 1984; Munro *et al.*, 1990). Ovulation in females goldfish is stimulated by the presence of aquatic vegetation.

In thirteen families of teleosts, the eggs are fertilized internally and develop in the ovarian lumen or the follicles. The young are born free-swimming. Reproduction is emancipated from the requirement of a suitable external spawning site (Wourms, 1981). In these viviparous fish, the female is responsible for supplying oxygen and disposing of the waste products from the developing eggs and larvae. In the pipefishes and sea horses (Syngnathidae), the male cares for the eggs, with the female transferring eggs to the parental male. In some species the male holds the developing eggs in a brood pouch on the abdomen. Oral brooding, in which the fertilized eggs develop in the mouth of the parent, occurs in about six families. In the sea catfishes (Ariidae), it is usually the male that broods the unusally large eggs. In the Cichlidae, female oral brooding is common. All these modes of parental care reduce or eliminate the risk of predation at the egg stage of development.

7.4 ALLOCATION OF RESOURCES TO REPRODUCTION

Resources must be allocated to reproduction to meet the costs of the development of any secondary sexual characteristics, of reproduction behaviour, and of the gametes (Miller, 1979b; Wootton, 1985). The pattern of allocation of time and material resources will be related to the breeding system (Turner, 1993). The problem of relating breeding systems, a partial classification of which is given in Table 7.2, to environmental characteristics presents a major challenge to fish biologists.

Allocation to secondary sexual characters

In many species, secondary sexual characters develop with the onset of sexual activity. Such characters can take the form of breeding colours such as the red throat and belly of the male threespine stickleback (Wootton, 1976) or the brilliant colours of some African Great Lakes cichlids (Fryer and Iles, 1972). Secondary sexual characters may also be morphological traits. Adult male salmon develop a hooked jaw called a kype. Some cyprinids develop epidermal tubercules during the breeding season (Breder and Rosen, 1966; Smith, 1974). Male swordtails, *Xiphophorus*, develop the elongated tail from which they take their name (Turner, 1993). In sexually mature male sticklebacks, the kidneys become enlarged and start to secrete a mucus. The energy requirements for the development of such characteristics are not known but in some cases could be high. An increased risk of predation will be another cost if the secondary sexual characteristics increase the conspicuousness of the fish to predators or reduce its ability to escape (Chapter 8). The benefits depend on the nature of the characteristics. In the male stickleback, the mucus secreted by the kidneys is used to construct a nest. The kype of the salmon is used in fighting with rival males. Conspicuous colours and other ornamentation will allow species recognition and may also be important for mate choice during courtship. Variation in the development of secondary sexual characteristics may have important consequences for the lifetime reproductive success of the individuals (Andersson, 1994).

Table 7.2 Components of teleost breeding systems

1. Sex determination
   (a) Genetic, e.g. most teleosts
   (b) Environmental, e.g. *Menidia menidia*

2. Gender system
   (a) Gonochoristic, e.g. most teleosts
   (b) Hermaphroditic
       (i) Simultaneous hermaphrodite, e.g. *Rivulus*, *Serranus*
       (ii) Protandrous hermaphrodite, e.g. *Amphiprion*
       (iii) Protogynous hermaphrodite, e.g. *Thalassoma*
   (c) Parthenogenetic
       (i) Hybridogenetic, e.g. *Poeciliopsis*
       (ii) Gynogenetic, e.g. *Poeciliopsis*

3. Mode of fertilization
   (a) External, e.g. most teleosts
   (b) Internal, e.g. *Poecilia*
   (c) Buccal, e.g. some *Oreochromis*

4. Mating system
   (a) Promiscuity – both sexes have multiple partners within a single breeding season, e.g. *Clupea*, *Gasterosteus*
   (b) Polygamy
       (i) Polygyny – male has multiple partners within a breeding season (includes lekking species), e.g. *Anthias*, *Oreochromis*
       (ii) Polyandry – female has multiple partners within a single breeding season, e.g. *Amphiprion* (in some ecological circumstances)
   (c) Monogamy – mating partners remain together, e.g. *Cichlasoma*, *Serranus*

5. Secondary sexual characteristics
   (a) Monomorphism – sexes indistinguishable even when sexually mature, e.g. *Clupea*
   (b) Sexual dimorphism
       (i) Permanent dimorphism – sexes always distinguishable after onset of sexual maturity, e.g. *Betta*
       (ii) Temporary dimorphism – sexes distinguishable only during breeding season, e.g. *Gasterosteus*
   (c) Sexual polymorphism – one or both sexes represented by more than one distinguishable form, e.g. primary and secondary males of *Thalassoma*, precocial and adult male salmonids

6. Spawning site preparation
   (a) Site undefended, e.g. *Salmo*
   (b) Site defended, e.g. *Gasterosteus*, *Oncorhynchus*

7. Parental care
   (a) No care, e.g. majority of teleosts
   (b) Male parental care, e.g. *Gasterosteus*, *Syngnathus*
   (c) Female parental care
       (i) Oviparity with post-spawning care, e.g. *Oreochromis*
       (ii) Ovoviviparity, e.g. *Sebastes*
       (iii) Viviparity, e.g. *Poecilia*, *Embiotoca*
   (d) Biparental care, e.g. *Cichlasoma*
   (e) Juvenile helpers, e.g. *Lamprologus*

Allocation to reproductive behaviour

Reproductive behaviour can include several components: movement to and choice of a breeding site, preparation of a spawning site, defence of a spawning site, courtship and mating, and parental care (Keenleyside, 1979). Not all these components need occur, but those that are performed require the allocation of time and some may be energetically costly (the energy costs of long migrations to spawning grounds are discussed in Chapter 5). Time and energy spent on reproductive behaviour are not available for activities that would directly increase the investment in gamete production or contribute to somatic growth and maintenance. During the breeding season, both male and female pike in Lac Ste Anne (Alberta) showed significant depletion of their somatic energy stores (Diana and MacKay, 1979). There was little gonadal growth during this period, so the losses were probably the result of spawning-related activities exacerbated by a fast that occurs in pike.

Reproductive behaviour will reflect environmental factors such as: the presence of stable, two- or three-dimensional structures which can act as focal points for courtship and spawning; the availability of food for parents and offspring; and the presence of predators of parent and offspring (Loiselle and Barlow, 1978). The differences in size between the parents and their eggs and larvae that is typical of fishes means that the effects of food and predation on parents and offspring are often related to different sets of food organisms and predators. Conspecific juveniles and adults can be major predators of the eggs and larvae. In the freshwater mottled sculpin, *Cottus bairdi*, over 90% of females collected in the breeding season had fed on sculpin eggs (Downhower *et al.*, 1983). Predation including cannibalism will favour breeding systems that include protection of eggs and larvae, or rapid dispersal of eggs and larvae, or synchronized breeding which swamps potential predators with an excess of food (Loiselle and Barlow, 1978; Turner, 1993). Major advantages that accrue from a behavioural investment in reproduction probably include an increase in the survival rate of eggs and larvae. Even in breeding systems that do not involve parental care, courtship gives individuals the opportunity to exert some choice of their partner in mating (Andersson, 1994).

Allocation to spawning site preparation

In some fishes, for example pelagic spawners, no preparation of a spawning site occurs, while at the other extreme some species construct elaborate nests in which the eggs are spawned (Keenleyside, 1979). Most salmonids select a spawning site in the gravel bed of shallow head waters and rivers. The female energetically digs a pit in the gravel by strong flexions of her body. In the Atlantic salmon, the female may dig several nests in a restricted area. The area of disturbed gravel containing the nests forms a redd. A female may use more than one redd, and may move as far as 0.5 km to dig a new one (Fleming, 1996). Large male sunfish of the genus *Lepomis* excavate a nest pit by mouth and by sweeping movements of the tail. The male of the littoral and sublittoral wrasse, *Crenilabrus melops*, builds a nest of algae in a crevice. Different algae are used at different sites within the nest, one form being preferred where the eggs are laid and a different alga preferred in the outer, protective layer (Potts, 1984). A male stickleback builds a nest from bits of vegetation bound together with mucous threads (Wootton, 1976). Other examples of spawning sites are listed in Table 7.1.

Allocation to territorial defence

A spawning site or nest is a resource that may be economically defensible, that is the benefits exceed the costs of defence (Chapter

5). The female salmonid defends the area in which she is digging the nest, and a large male salmonid will defend an area which may contain the redds of several females. In the semelparous Pacific salmon, the female defends the nest site after spawning. The male stickleback vigorously defends the area around his nest from both conspecific and heterospecific intruders. In biparental, substrate-spawning cichlids, both male and female defend the area around the cave or hollow in which the eggs are spawned. In cichlids of the genus *Oreochromis*, the eggs are orally brooded by the female Spawning takes place at nests dug in the substrate by the male and defended against other males (Fryer and Iles, 1972).

Although defended, the nests or territories may be contiguous, so a breeding colony is formed. This is found in the threespine stickleback and the bluegill sunfish. The clumping of territories of parental males may increase the ability of the territorial males to defend their nests against intruders, which in some stickleback populations take the form of shoals of females that raid the nests of males and devour the eggs (Whoriskey and FitzGerald, 1985). In the orally incubating *Oreochromis*, the males do not have a parental role and so the colony of nesting males forms a lek (Loiselle and Barlow, 1978). A female ready to spawn visits the lek to have the eggs fertilized by one or more of the nest-holding males. The presence of several reproductively active males in a small area potentially provides the female with an opportunity to select the males with which she is going to mate.

The energy costs of territorial defence in the pupfish, *Cyprinodon*, have been estimated (Feldmeth, 1983). Dominant male pupfish defend small territories within which females spawn. Part of this defence is directed against intruding satellite males that attempt sneak fertilizations inside the territories of the dominant males. Dominant males spend a high proportion of the daylight period

swimming through their territories, defending them against intruders. Mature females, in contrast, form slow-moving shoals away from the hectic activity of the breeding territories. The energetic cost of territorial defence was about twice the cost of merely holding station in the territory (0.32 kJ against 0.16 kJ per 16 h day). Satellite males had about the same rate of energy expenditure as territorial males, because satellites showed high levels of activity. Territorial defence reduced the growth of male threespine sticklebacks (Wootton, 1994a) and male *Oreochromis mossambicus*, a cichlid that nests in colonies (Turner, 1986).

Allocation to courtship and mating

No estimates of the energy costs of courtship are available. In most cases it seems unlikely that they are a significant component of the energy budget of a fish, but time allocated to courtship is not available for foraging and the courtship interaction may make the fish more vulnerable to predation. Courtship in fishes ranges from a temporary association of ripe males and females to a long-term interaction between monogamous pairs (Keenleyside, 1979; Turner, 1993). Courtship probably fulfils several mutually compatible functions: species recognition, mutual orientation to the spawning site, and synchronization of the activities of the male and female so that gametes are released at the appropriate time. It may also serve to minimize aggressive interactions between the potential mates. This can be particularly important if the spawning site is strongly defended by one of the pair (Keenleyside, 1979; Wootton, 1984a).

Courtship tends to be more elaborate, and hence more costly in terms of time, energy and risk of predation, in fishes in which there is preparation of the spawning site and some degree of parental care. In species in which mating is promiscuous and there is no parental care, courtship may be poorly developed. In the Atlantic herring, the spawning fish congregate on the spawning ground and there is no easily detectable difference between the sexes. The release of eggs is probably triggered by a pheromone released with the milt of the males (Turner, 1993). Courtship in polygamous species, in which there is either no parental care or care only by one parent, may be elaborate but occupy only a short period of time. In monogamous, biparental species, the courtship can be elaborate and time-consuming (Baerends, 1986).

Courtship also provides an opportunity for mate choice (Turner, 1993; Andersson, 1994). For females, choice of a suitable male will be important because eggs are probably more costly to produce than sperm (page 163). In breeding systems in which the male takes the parental role, choice of the females by the male is important because the male may show little or no courtship after he has started parental care (Turner, 1993). Secondary sexual characteristics, which make one or both of the sexes conspicuous, may play a role in determining mate choice, although the ultimate factors that have led to the evolution of such characteristics are still a subject of debate. This debate focuses on the question of whether the secondary characteristics signal information about the genetic quality of the carrier, or whether they simply make the carrier more likely to be chosen by a potential mate (Bradbury and Andersson, 1987; Andersson, 1994).

The female mottled sculpin lays a single clutch of eggs in a breeding season. The eggs are laid in a shelter, which is defended by the parental male. One male can spawn with several females within the same shelter. The male defends the eggs until the yolk sac is resorbed. Females prefer to mate with large males. This preference increases the reproductive success of the female because egg mortality is lower in shelters guarded by large males. However, small females avoid the largest males because of the danger that the male will eat them. Males prefer to mate with large females, which are more fecund

(Downhower and Brown, 1981; Downhower *et al.*, 1983). Observations on mate choice by sockeye salmon in a small inlet stream of Babine Lake showed that fish tended to mate with similar-sized fish (Hanson and Smith, 1967). Female threespine stickleback prefer to mate with males that have well-developed red breeding colours rather than with drabber males, perhaps because brighter males are more effective at defending their nests from raiders (Semler, 1971). In some species in which the male cares for eggs in a nest, females prefer males already guarding eggs to males with empty nests. This preference was shown by females of *Aidablennius sphynx*, a blenny of rocky Mediterranean shores (Kraak and Videler, 1991). The survival of eggs may be higher in nests containing many eggs.

Other examples are described by Turner (1993). Evidently, at least some species of fish can exert some selectivity in the choice of mate, rather than rely simply on the random fusion of gametes released into the water.

Allocation to parental care

Most fish do not show any post-fertilization care of the eggs beyond, in some cases, concealing them. No parental care is found in 78% of teleostean families, care by the male is found in 11%, care by the female in 7% and biparental care in 4% (Sargent and Gross, 1986). Some form of parental care is more prevalent in freshwater families (57.4%) than in marine families (15.4%) (Baylis, 1981). This and other evidence suggests that parental care has tended to evolve in fishes occupying environments that are spatially and temporally unpredictable for zygotes and young. The selection, guarding and tending of a spawning site by the parental fish reduces the danger to the zygotes of that unpredictability (Clutton-Brock, 1991; Krebs and Davies, 1993). The prevalence of male parental care in fishes may be a consequence of the strong relationship between body size and fecundity in female fish (Sargent and Gross, 1986). If parental behaviour is energetically costly and so reduces growth, the reduction in the future production of offspring is likely to be greater for females than for males. Parental care by the male may also be favoured in fish because external fertilization reduces the chance that the male is showing parental care to offspring it has not fathered (Blumer, 1979) or because the male can defend a spawning site that is in short supply (Baylis, 1981).

Care may include guarding the eggs and larvae from predators, ventilating the eggs to provide a flow of oxygenated water while also clearing away any accumulations of silt, and the removal of dead eggs. Parental care can be costly both in the allocation of time and in the energy expenditure (Smith and Wootton, 1995). While tending eggs in his nest, the male stickleback spends a high proportion of his time vigorously fanning a current of water through the nest (van Iersel, 1953; Wootton, 1976). Biparental cichlids take turns to fan the developing eggs, although the female performs most of the fanning. The amount of fanning shown by parental female convict cichlids, *Cichlasoma nigrofasciatum*, decreased when they were fed reduced rations (Townshend and Wootton, 1985b), which suggests that parental care is energetically costly.

Parental care by females is shown by many African cichlids, with the eggs and larvae being brooded in the mouth of their mother (Fryer and Iles, 1972). Even after the young fish become free swimming, they return to their mother's mouth if danger threatens. These species are characterized by relatively low fecundities but large eggs (Welcomme, 1967), suggesting a trade-off between the behavioural investment in parental care and a cytoplasmic investment in more numerous eggs. Theoretical analyses have suggested that parental care favours the evolution of relatively large eggs (Sargent *et al.*, 1987).

Flexibility in reproductive behaviour

Descriptions of reproductive behaviour usually record the typical form of the behaviour for a species, but details of the behaviour may change in response to environmental changes. A strategic evolutionary change is illustrated by the ninespine stickleback, *Pungitius pungitius*. Throughout most of the circumpolar range of this species, the male builds a nest in rooted vegetation, off the substrate (Morris, 1958). But in some populations of *P. pungitius* in North America, the nest is built on the substrate (Keenleyside, 1979). *Cichlasoma panamense* is a stream-dwelling cichlid of Central America that illustrates a tactical change. Typically it is a biparental substrate spawner, with both male and female guarding the eggs and fry (Townshend, 1984). However, in a stream in which the levels of predation on the fry were judged to be low, some males deserted their partner and offspring, possibly to pair with another female. In contrast, in a stream in which predation was high, desertion by the male was rare (Townshend and Wootton, 1986). Male desertion in *Cichlasoma* may also be related to the abundance of caves suitable as spawning sites. If caves are in short supply, male deserters may have a poor chance of spawning again in the present breeding season (Wisenden, 1994).

Individuals of the same sex, usually the male, showing different reproductive behaviours may coexist in the same population at the same time. These alternative mating patterns are described in more detail on page 169.

Allocation of resources to gonads

A distinction must be made between the investment represented by the size and energy content of the gonads at any one time and the rate of turnover of gonadal material. A gonad that is small in comparison with total body size may represent a major investment of resources if the production and release of gametes is rapid. A common method of describing the relative size of the gonads is the gonadosomatic index (*GSI*) which is defined as either:

$$GSI = 100 \text{ (weight of gonads / total body-weight)} \quad (7.1)$$

or:

$$GSI = 100 \text{ (weight of gonads / somatic body weight)} \quad (7.2)$$

where somatic weight is the weight of the body excluding the gonads. The *GSI* may be expressed in terms of wet or dry weight or even in energy units (Wootton, 1979). Although the *GSI* provides a simple index with which to describe changes in the relative size of the gonads over time, it suffers from the disadvantage of all such indices. A change in *GSI* may result from a change in the weight of either the gonads or the soma, or may arise because there is an allometric relationship between gonadal and somatic weights. The index should be replaced, wherever feasible, by the appropriate regression analyses of the relationship between gonad size and body size.

The growth rate of gonads can be described either as the absolute change in weight (or energy content) per unit time, or as their specific growth rate (Allen and Wootton, 1982c). As for somatic growth (Chapter 6), efficiencies of production can be measured. The efficiency of egg production (gross reproductive efficiency) is:

100 (energy content of eggs produced over a defined time period / energy content of food required for production of those eggs).

Allocation to ovaries

Compared with a sperm, each egg represents a massive cytoplasmic investment. The egg provides the yolk, which the zygote utilizes while developing to the stage at which it can assume an independent life and feed exogenously. Vitellogenesis, the provisioning of the eggs with yolk, takes place in the ovaries, although the precursors of the yolk are synthesized in the liver (Tyler and Sumpter, 1996). Changes in the absolute and relative size of the ovaries during the reproductive cycle largely reflect the growth of the developing oocytes as they accumulate yolk. Examples of changes in the *GSI* during the annual reproductive cycle of an autumn spawner, an early spring spawner and an early summer spawner are shown in Fig. 7.2. There is a wide interspecific range for the *GSI* of a ripe female. In some species, including salmonids and some cyprinids, the ovaries form 20–30% of the total body weight just before the female spawns (Wootton, 1979). In others, even when ripe, the ovaries form less than 5% of the body weight. The ovary of a ripe female cichlid, *Oreochromis leucostictus*, represents only 3% of the body weight (Welcomme, 1967). Even within a single suborder, the gobies (Gobioidei), the *GSI* ranges from about 5% to 30% (Miller, 1984).

These differences partly reflect differences in the temporal pattern of egg development and spawning. The females of some species shed all the eggs that they will mature in that breeding season over a short time interval. Such species are called total spawners: examples include salmon and trout, pike and many of the cyprinids of floodplain rivers that spawn at the beginning of the flooding (Lowe-McConnell, 1975). In total spawners, the *GSI* is usually high just prior to spawning. Other species are batch spawners: these may spawn several times during a breeding season. Batch spawners may have a relatively low *GSI*, but they may also have a high rate of egg production if they produce many batches in a season. Small egg-laying cyprinodonts such as *Cyprinodon nevadensis* can spawn nearly every day (Shrode and Gerking, 1977). The *GSI* of spawning females in this species varies from 2% to 14%. However, the threespine stickleback, which can spawn at intervals of three to five days, has a *GSI* just prior to spawning of about 20% (Wootton, 1984a).

Fecundity

The number of eggs that a female spawns over a defined time period depends on the number of eggs per spawning and the number of spawnings (Bagenal, 1978; Wootton, 1979). When describing the fecundity of fish it is necessary to distinguish among batch fecundity, breeding season fecundity and lifetime fecundity.

Batch fecundity, F_{batch}, that is the number of eggs produced per spawning, is a function of body size (Fig. 7.3). Typically the relationship with length (L) takes the form:

$$F_{batch} = aL^b \qquad (7.3)$$

or:

$$\log F_{batch} = \log a + b \log L. \qquad (7.4)$$

Because batch fecundity will be related to the volume of the body cavity available to accommodate the ripe ovaries, geometry suggests that the length exponent, b, would be 3.0. A survey of 62 species found that the exponent ranged from about 1.0 to 5.0, and lies most commonly between 3.25 and 3.75 (Wootton, 1979). There is a tendency for the value to be higher in marine species than in freshwater species, but the ecological significance of interspecific variation in the length exponent is still obscure.

The volume occupied by the ripe eggs can be calculated as the product of batch fecundity and the mean egg volume just before spawning. For a sample of 238 species, including both marine and freshwater fishes, the relationship between total volume (*TV*) and fish length (*L*) took the form (Wootton, 1992b):

$$TV = aL^{3.09} \qquad (7.5)$$

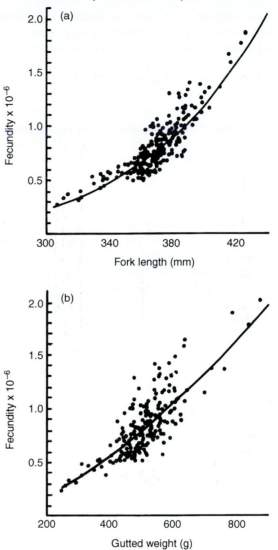

Fig. 7.3 Relationship between fecundity (millions of eggs) and and (a) fish length and (b) fish weight illustrated by Atlantic mackerel, *Scomber scombrus*, a pelagic marine species. Lines are fitted regressions: $\log_{10} F = -8.346 + 5.544 \log_{10} L$ and $\log_{10} F = 1.721 + 1.5547 \log_{10} W$, where F is fecundity, L is length (mm) and W is gutted weight (g). Redrawn from Morse (1980).

The length exponent of 3.09 indicates that total volume increases isometrically with body weight. The strong correlation ($r = 0.88$, $P < 0.01$) between total volume and body size suggests that there is limited scope for the independent variation of egg size and fecundity. Large eggs can be produced only at the cost of a reduced batch fecundity (Duarte and Alcaraz, 1989; Elgar, 1989; Wootton, 1992b). Fig 7.4 illustrates the relationship between total volume of eggs and body length for Canadian freshwater fishes. For this regional sample, the length exponent is significantly lower than 3.0 (Wootton, 1984b).

Although there is usually a good correlation between batch fecundity and fish size, individual females of the same size do differ in their batch fecundities and even the same female can have different batch fecundities within a breeding season (Bagenal, 1978; Wootton, 1979). Furthermore, the mean batch fecundity at a given size can vary from year to year in a population, or can differ between populations of the same species (Bagenal, 1978). When fish of the same size and age were compared, the fecundity of plaice in the German Bight area of the North Sea in 1980 was 1.44 times as great as in 1979 (Horwood *et al.*, 1986). Plaice fecundity in the period 1977–1985 was 30–100% higher than in the period 1947–49, although there was a much

smaller difference in ovary size between the two periods (Rijnsdorp, 1994). In the North Sea, the fecundity of plaice at a given length and age declines northwards from the eastern English Channel to the German Bight. A survey of threespine stickleback populations revealed significant interpopulation differences in the relationship between batch fecundity and length (Baker, 1994). The predicted fecundity for a female with a standard length of 55 mm ranged from 10 to 250 eggs. An important ecological question is whether variations in batch fecundity are predictable or are simply the consequence of random events during the maturation of the ovaries and essentially unpredictable. Predictable variation in batch fecundity could arise from genetic differences between females, or from the effect of environmental factors on oogenesis or from an interaction between genetic and environmental factors. Although fecundity is a critical biological trait, comparatively little effort has been devoted to the experimental study of the factors that cause variations in batch fecundity.

For total spawners such as brown trout, breeding season fecundity and batch fecundity are the same, but for batch spawners, the breeding season fecundity will depend on the number of times the female spawns during the season. The importance of the number of spawnings for breeding season and lifetime fecundity has stimulated the development of methods for estimating spawning frequency based on the presence of hydrated eggs just prior to spawning or on histological evidence of a recent spawning (Hunter and Macewicz, 1985; Lasker, 1985). If, for example, 20% of batch-spawning females in a sample show evidence that they have spawned within 24 h, the mean frequency of spawning per female for that population is once every 5 days. Such techniques suggested that the northern anchovy in the Southern California Bight spawns about 20 batches per year (Hunter and Leong,

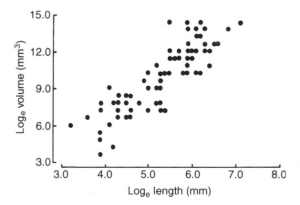

Fig. 7.4 Relationship between total volume of eggs and body length of Canadian freshwater fishes. Each point represents one species. Redrawn from Wootton (1984b).

1981). An earlier estimate had suggested only 1–3 batches per year.

For batch spawners, a distinction is made between species with determinate or indeterminate fecundity (Hunter *et al.*, 1992). In species with determinate fecundity, the potential breeding season fecundity is fixed before the start of spawning. As successive spawnings occur, the stock of oocytes remaining to be spawned in that breeding season declines. During the breeding season, some yolked oocytes may be resorbed through the process of atresia, so the achieved fecundity is lower than the potential fecundity. In species with indeterminate breeding season fecundity, there is recruitment to the stock of spawnable oocytes during the breeding season, so the achieved fecundity depends on the balance between rates of recruitment, spawning and atresia.

Environmental effects on fecundity

The most relevant environmental factor is food, because the critical transformation is food into progeny (Fig. 1.2). The effect of food raises two questions. What is the relationship between the quantity of food and fecundity? When, during the reproductive cycle, is fecundity sensitive to the quantity of food consumed? Because of the relationship between fecundity and fish size (Fig. 7.3), any effect of food on fecundity that reflects the effect of food on growth and hence body size (Chapter 6) needs to be distinguished from direct effects of food on fecundity, independent of body size.

Rainbow trout is a total spawner, with eggs typically produced in autumn or early winter. Ration size affects fecundity both through its effect on trout growth and through a direct effect on fecundity (Bromage, 1995). Even when fish of the same size were compared, trout that had received high rations in the first part of the year had a higher fecundity than fish receiving a high ration later in the year (Bromage, 1995). In Atlantic cod, *Gadus morhua*, a batch spawner with determinate fecundity, the effect of food on fecundity differed between females spawning for the first time and older repeat spawners (Kjesbu *et al.*, 1991; Kjesbu and Holm, 1994). In first-time spawners, potential fecundity was determined by the size of the fish at the beginning of the spawning period. In repeat spawners, potential fecundity was related to food availability during the period in autumn when the oocytes were accumulating yolk (vitellogenesis), as well as to body size.

Laboratory studies on the threespine stickleback demonstrated an effect of food availability on batch and breeding season fecundity (Wootton, 1977; Fletcher and Wootton, 1995). Batch fecundity was determined primarily by the size of the fish. The effect of food supply on batch fecundity was mediated largely through the effect of food on growth rather than a direct effect on fecundity. However, the number of spawnings during a breeding season was related to food availability during the breeding season. Better-fed fish spawned more often and with shorter intervals between successive spawnings (Fig. 7.5) (Wootton, 1977; Fletcher and Wootton, 1995). Female convict cichlids had lower batch fecundity and fewer spawnings when fed low rations (Townshend and Wootton, 1985a).

The guppy is viviparous, producing a litter at intervals of about 28 days. Low rations were associated with small litters, and an increase in the length of the interval between successive broods (Hester, 1964; Reznick and Yang, 1993).

Other environmental factors may also affect fecundity. In the desert pupfish, *Cyprinodon n. nevadensis*, both the proportion of days on which spawning took place and the number of eggs per spawning were significantly related to temperature. There was a decline in both variables when the temperature was lower than 24 °C or higher than 32 °C (Shrode and Gerking, 1977). Other

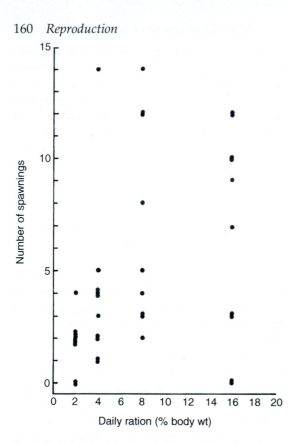

Fig. 7.5 Relationship between number of spawnings in breeding season and ration as % body weight per day for female threespine stickleback, *Gasterosteus aculeatus*. Source: Fletcher and Wootton (1995).

factors that stress female fish, such as acidic water, the presence of pollutants or abrupt changes in water level, may also cause reduced fecundity (Gerking, 1980), in some cases because of an increased resorption of oocytes through atresia (Billard *et al.*, 1981). Thus the number of eggs spawned during a breeding season can be determined both by factors such as food availability that enhance the recruitment of oocytes into the vitellogenic phase and by factors that reduce the stock of vitellogenic oocytes by inducing atresia.

Some caution is needed when applying the results of laboratory studies on the effects of

environmental factors on fecundity. Under laboratory conditions, female convict cichlids bred repeatedly, with the number of spawnings positively correlated with ration size (Townshend and Wootton, 1985a). In contrast, a field study of convict cichlids in streams in Costa Rica found that most females bred only once in a breeding season (Wisenden, 1995). A physiological potential may not be fulfilled ecologically.

A study showing that environmental effects are important compared the number of spawnings in a breeding season by the freshwater cottid, *Cottus gobio*, from two contrasting environments (Mann *et al.*, 1984). Female *Cottus* from an infertile, upland stream in northern England spawn once a year, but females in a fertile chalk stream in southern England spawn several times. Northern fish transplanted to a southern site spawned the number of times that is typical of southern fish, while southern fish at a northern site spawned only once. These reciprocal transplant experiments showed that the differences in spawning characteristics were primarily the result of environmental differences rather than genotypic effects (see also Chapter 11).

Lifetime fecundity depends on breeding season fecundity, and life span and will be considered in Chapter 10 in the context of the population dynamics of fishes.

Bioenergeties of ovarian maturation

In terms of the energy budget equation (Chapter 4):

$$C = P_s + P_r + R + F + U \qquad (7.6)$$

attention is focused on production in the form of gametes, P_r, in relation to the energy made available through feeding, C. The bioenergetics of ovarian maturation are illustrated by three examples.

The threespine stickleback usually reaches sexual maturity at the age of 1 year. During this year, both the adult somatic framework and the ripe ovaries are synthesized. Estimates are available for the rates of food consumption, somatic growth and ovarian growth in a Mid-Wales population during the period September to May (Wootton *et al.*, 1978, 1980). In the first few months of life before September, the somatic growth rate was high, so by September the total energy content of the soma was about 2 kJ but that of the ovaries only about 2% of this. In winter the ovaries grew only slowly (Fig. 7.6), at about 0.2–0.3 J day^{-1}, which represented about 1% of the daily energy income.

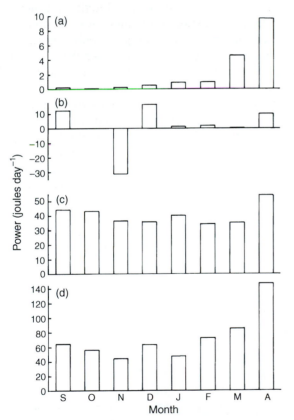

Fig. 7.6 Seasonal changes in estimated energy budget of an average-sized female threespine stickleback from Afon Rheidol in Mid-Wales, illustrating growth of ovaries. (a) Ovarian growth; (b) somatic growth; (c) routine metabolism; (d) consumption. Source: Wootton *et al.* (1980).

But, during this time the energy income was barely sufficient to meet the daily costs of maintenance so somatic growth was negligible. From September to when breeding started in May, ovarian growth represented about 3% of the total energy income and somatic growth about 4%. During the spawning season, May to July, the soma showed virtually no change in its total energy content. Experimental studies showed that if a spawning female's soma is not changing in energy content, she is investing 15–20% of the energy income on egg production (Wootton, 1977, 1979).

For female pike in Lac Ste Anne (Alberta), the annual investment of energy in reproduction increased from 0% in the first year of life to 11–16% over the second, third and fourth years (Diana, 1983a). Somatic investment declined from 42% to 5–8%. In contrast to the stickleback, most of the ovarian growth in the pike occurs during the winter months, and represents a high proportion of the energy income over this period (Diana and MacKay, 1979; Diana, 1983a). Both the stickleback and the pike illustrate a tendency for periods of somatic and ovarian growth to alternate during the year, at least in fish from strongly seasonal environments.

The northern anchovy is a pelagic, marine species with an extended breeding season. The percentage of the energy income that females invest annually in reproduction was estimated to increase from 8% for 1–2-year-olds to 11% for 3–4-year-olds (Hunter and Leong, 1981).

In some species, a temporal alternation of somatic and ovarian growth is linked to a cycle of storage and depletion that allows ovarian growth to occur during a period when the rate of energy income through feeding could not support it (Shul'man, 1974; Wootton, 1979; Reznick and Braun, 1987). In such species, reproduction is dependent on a capital of resources built up during the period of feeding rather than a resource income from feeding during the reproductive period (Henderson *et al.*, 1996). Many clupeid fishes show this cycle clearly. In the North Sea herring, intensive feeding, somatic growth and accumulation of fat reserves take place in spring and early summer. In late summer and autumn, feeding ceases and the fat reserves are mobilized as a source of energy while protein is translocated into the ovaries (Iles, 1984). Alewife, *Alosa pseudoharengus*, in Lake Michigan accumulate lipid reserves during a period of intensive feeding in autumn. These reserves are mobilized for energy during the winter and during ovarian growth and maturation (Flath and Diana, 1985). Lipid accumulated in reserves is used primarily as a fuel for metabolism rather than as a component to be translocated into the ovaries. The high energy content of lipid and its low specific gravity make it a better storage material than either protein or carbohydrate. Cycles of storage and depletion form part of the normal annual cycle of many species of fish. The depletion does not normally represent a pathological condition.

The priority that ovarian growth has over somatic growth can be assessed by observing the effects of reduced ration levels. In the threespine stickleback, periods of food deprivation that lasted 21 days did not alter the pattern of ovarian growth during the period October to April, although the growth rate of the liver and the carcass declined (Allen and Wootton, 1982c). In contrast, food deprivation in the winter flounder, *Pseudopleuronectes americanus*, reduced both the proportion of females that had ovaries that contained vitellogenic oocytes and the percentage of vitellogenic oocytes per ovary (Tyler and Dunn, 1976).

During the breeding season, female sticklebacks on low rations invested a higher proportion of their energy income on the production of eggs than females on high rations. At low rations, the females showed a significant depletion of the energy content of their soma, which suggests that the soma subsidized egg production (Wootton, 1977). In the medaka, *Oryzias latipes*, the proportion of the energy income invested in egg production declined with an increase in ration at 29 °C and 27 °C, but increased at 25 °C (Fig. 7.7) (Hirshfield, 1980). At the lowest temperature, the rate of food consumption, even at maximum rations, was low compared with that at the higher temperatures. Both the stickleback and the medaka are small,

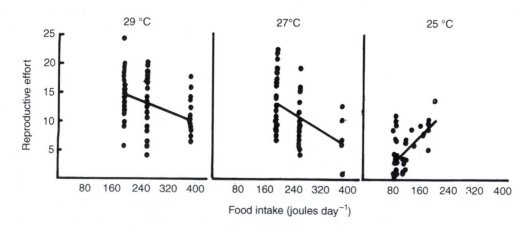

Fig. 7.7 Relationship between reproductive effort (energy content of eggs spawned as % of energy of food consumed) and ration in female medaka, *Oryzias latipes*, at three temperatures. Redrawn from Hirshfield (1980).

short-lived fish. In longer-lived fish, the pattern of allocation of energy between the ovaries and the soma probably will favour the soma more. The experiments with the winter flounder, a long-lived flatfish, certainly suggested this (see also Chapter 11).

Successful spawning will also depend on the allocation of nutrients and minerals to the developing oocytes (Wootton, 1979). In the winter flounder, significant amounts of zinc and copper are transferred to the developing ovaries even after feeding has ceased (Fletcher and King, 1978). These translocated minerals probably play an essential role in the ovaries during oogenesis, but also the developing embryonic fish derives these essential elements from the ovum rather than from the water.

Allocation to testes

The testes frequently represent a much lower proportion of the body weight than the ovaries. While the ovaries of a female stickleback about to spawn form over 20% of her total body weight, the testes of the male she is mating form less than 2% of his weight (Wootton, 1984a). In mature Atlantic salmon, the ovaries form about 20–25% of body weight, while the testes of mature, anadromous males form 3–6% (Fleming, 1996). The testes of sexually mature males of the cichlid, *Oreochromis mossambicus*, represent only about 0.2% of body weight (Hodgkiss and Mann, 1978). A sexual dimorphism in gonad size is not always present. In the Arctic cod, *Boreogadus saida*, the testes of mature males form 10–27% of the body weight (Craig *et al.*, 1982). The size of the testes may relate to the mode of fertilization. If the male and female are in close proximity at the time of spawning or if the eggs are laid in an enclosed space, the quantity of sperm required to fertilize the eggs may be lower than when the eggs and sperm are more widely disseminated. Large testes may also be advantageous if several

males are competing to fertilize eggs, so production of a large volume of sperm will increase the chances of fertilizing at least some eggs.

The allocation of energy to the testes has been studied in detail in pike in North America. In pike populations of Alberta and Michigan, 6–18 times as much energy was deposited in the ovaries as in the testes (Diana, 1983b). Growth of the testes was completed by the autumn, whereas ovarian growth continued throughout the winter. In Lac Ste Anne male pike, the annual allocation of energy to reproduction increased from 0% in the first year of life to 4–5% in the next three years of life (Diana, 1983a). The male pike had a lower rate of food consumption than the females, although both sexes had similar rates of somatic production. This difference in consumption between the sexes points to the importance of the endogenous control of appetite in relation to the temporal patterns of somatic growth and reproduction (Chapter 6). The interactions between somatic growth, gonadal growth and appetite have received totally inadequate attention in analyses of the annual cycles of fishes.

For many fish species, the energy costs of sperm production are probably lower than those of egg production, but the testes do not provide a limitless supply of sperm. There was a steady decrease in the percentage of eggs fertilized by male lemon tetras, *Hyphessobrycon pulchripinnis*, with an increase in the frequency of daily spawning acts (Nakatsuru and Kramer, 1982). This decrease could be caused by a decline in either the quantity or the quality of the sperm. After 30 spawning acts, the reproductive success of the male was almost entirely limited by the fertilization rate rather than the supply of eggs. Such sperm depletion could be important in species with high rates of mating by males, but is unlikely to be of any significance in species in which the frequency of matings is low (Wootton, 1984a).

7.5 ALLOCATION TO INDIVIDUAL PROGENY

The contribution, other than genetic, of the fertilizing sperm to the zygote can be ignored. In addition to her genetic contribution, the female governs cytoplasmic allocation to individual progeny. This maternal allocation has two components, its size and its composition. These two components, together with the genetic component, define the quality of the egg, that is its potential to produce viable fry (Kjorsvik *et al.*, 1990; Bromage, 1995).

Egg size

A characteristic of teleosts compared with elasmobranchs, is the production of large numbers of small eggs. A theoretical analysis suggested that this strategy is adaptive when the resources required by the young stages are patchily distributed over a large spatial scale (Winemiller and Rose, 1993). This pattern of resource distribution is, for example, characteristic of the size classes of plankton that form the prey of the early life history stages of pelagic fishes.

Compared with the range of adult body sizes in fish, from about 10 mm to several metres, the range in egg size is restricted. The smallest egg is about 0.25 mm in diameter and the largest about 20 mm (Miller, 1979a; Wootton, 1992b). The largest eggs are those of mouthbrooding sea catfishes (Ariidae). Surveys of egg diameters of marine and freshwater species showed that the frequency distribution is skewed towards the smaller diameters (Fig. 7.8) (Wootton, 1979; Kamler, 1992; Winemiller and Rose, 1992). An interpretation of this pattern is that there is an evolutionary tendency in fish to minimize egg size. There is a trade-off between egg size and fecundity, given that the volume of the abdominal cavity that can accommodate ripe eggs is limited (page 157). Fecundity can be maximized by minimizing egg size. However, the optimal egg size is

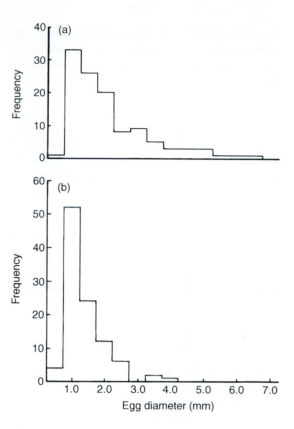

Fig. 7.8 Frequency distribution of egg diameters for species from north temperate (a) fresh waters (*N* = 33) and (b) sea waters (*N* = 101). Redrawn from Wootton (1979).

that which maximizes the number of offspring surviving to become reproductively active, so it is the size at which the product of fecundity and survival is a maximum (Sibly and Calow, 1986). An increase in egg size is likely to increase juvenile survival because bigger eggs tend to produce bigger larvae (Ware, 1975b; Pepin, 1991). Larger larvae will be able to take a wider range of prey sizes (Chapter 3), can probably survive periods of food shortage better (Blaxter and Hempel, 1963), and have fewer predators (Chapter 8) (Miller *et al.*, 1988). Thus, an increase in fecundity achieved by a reduction in egg size is likely to be counterbalanced by a decrease in survival. The optimal egg size will relate to the size-dependent mortality rates experienced by eggs, larvae and juveniles (Ware, 1975b; Wootton, 1994b) (Chapter 10). In turn, these rates will depend on the size and abundance of the prey and predators of the young fish.

Large eggs are likely to have an adaptive advantage if the food supply for the larvae is sparse or variable, or if the period spent in the egg stage is long or relatively unpredictable. Of two species of grunion, *Leuresthes*, that spawn in beach sand at high tide, the northern species, *L. tenuis*, has significantly larger eggs than the more southerly species, *L. sardina* (Moffatt and Thomson, 1978). This difference is probably an adaptation to the more irregular tides experienced by *L. tenuis*. The larger egg allows the embryo to stay alive in the sand for a longer period.

Egg size is related both to latitude and to mode of spawning. In marine families that spawn demersal eggs, the size of the egg tends to increase with latitude. But the same trend is not shown strongly by pelagic spawning families (Thresher, 1988). At low latitudes, the mean egg size of demersal and pelagic spawners is similar, but at higher latitudes, demersal spawners tend to have larger eggs than pelagic spawners (Russell, 1976; Thresher, 1988).

Egg size varies within a species as well as interspecifically (Fig. 7.9) (Bagenal, 1971; Ware, 1975b). In the eastern North Atlantic, herring populations that spawn at different times of the year have different egg sizes. Egg size is largest in populations that spawn in early spring, smallest in populations spawning in late spring and summer, and intermediate for late autumn and winter spawners (Blaxter, 1969). Fecundity is highest in populations with the smallest eggs. These interpopulation differences in egg size probably relate to seasonal changes in the density, size spectrum and spatial distribution of the planktonic prey of the herring larvae. Such interpopulation differences are

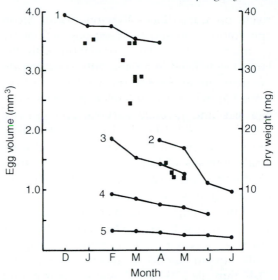

Fig. 7.9 Seasonal changes in mean egg volume of: 1, *Pleuronectes platessa*; 2, *Trigla gurnardus*; 3, *Melanogrammus aeglefinus*; 4, *Merlangius merlangus*; 5, *Limanda limanda* and (■) seasonal changes in mean egg weight of eastern Atlantic *Clupea harengus*. Redrawn from Bagenal (1971).

not confined to marine species. The eggs of the small cobitid, *Barbatula* (*Noemacheilus*) *barbatulus*, from a Finnish population have a volume 2.5 times as great as eggs of a population from a river in southern England (Mills and Eloranta, 1985). By contrast, the mean egg weight of the anadromous coho salmon declines with an increase in latitude in north-west America, although there is considerable interpopulation variation (Fleming and Gross, 1990). Mean fecundity (standardized to a common body length) tends to increase with latitude, illustrating the trade-off between fecundity and egg size. Fleming and Gross (1990) suggested that egg size is adapted to the local population conditions, while the variations in mean fecundity reflect the interpopulation variations in egg size.

Even within a geographically localized area, egg size can differ between populations. Isolated populations of brook trout occupy rivers on Cape Race, Newfoundland. Mean egg volume differed between populations by as much as 35%. The population with the largest eggs lived in a river with a low density of stream invertebrates, the staple food of the trout. On the basis of the relationship between egg size and juvenile survival and growth, Hutchings (1991, 1993a) calculated that female fitness was maximized when she produced small numbers of large eggs under low food conditions, and large numbers of small eggs under high food conditions.

Within a population, mean egg size may vary from year to year. The eggs of the dace, *Leuciscus leuciscus*, in a southern English chalk stream varied in mean volume from 1.647 mm^3 to 1.994 mm^3 over a 7 year period (Mann and Mills, 1985). In six of the seven years, fecundity and mean egg size were inversely correlated, but in a seventh year both fecundity and egg size were low. This exception is probably because the previous year had been exceptionally warm and the dace had made an unusually high investment in somatic tissue.

Egg size may vary even in the same year. Many populations of pelagic fishes show a progressive decline in egg size over a spawning season (Fig. 7.9) (Bagenal, 1971). This decline may be due partly to larger fish spawning earlier in the season, but the same phenomenon is shown by batch spawners (Ware, 1975b). This seasonal change in egg size may be an adaptive response to the change in conditions that confront the eggs and larvae. Ware (1975b) developed a model predicting that the optimal egg size will decrease as the incubation period (the time from fertilization to hatching) decreases. Incubation time decreases with an increase in temperature, so as sea temperature increases over the summer, the optimal egg size should decrease. The change in egg size will also allow the size of larvae to match the size spectrum of prey that are available (Jones and Hall, 1974; Ware, 1975b). A comparison of the seasonal change in egg size in a mack-

erel population in the Gulf of St Lawrence with the seasonal change in the size of the plankton showed a close correlation between mean egg diameter and mean particle size in the plankton (Ware, 1977).

Some species show no relationship between egg size and female body size, and in interspecific comparisons, body size was a poor predictor of egg size (Duarte and Alcaraz, 1989; Wootton, 1992b). But in some species, egg size does vary with the size of the female. In many salmonids, mean egg size increases with female length (Thorpe *et al.*, 1984). Dace show a similar relationship (Mann and Mills, 1985). It is not clear why egg size should vary with female size. It might be expected that the optimal egg size would be independent of the size of the parent, in that it is related to the environmental conditions experienced by the egg and larvae. However, the relationship between egg size and fitness may itself be a function of the size of the female (Bernardo, 1996). In coho salmon, egg size increases with body size. Fleming and Gross (1990) conjectured that this is related to the quality of the gravel in the nests dug by large and small females. Where there is competition between recently hatched siblings, theoretical studies suggested that egg size should increase with female size (Parker and Begon, 1986). Such competition may occur where eggs are laid in clutches in a nest or other confined space.

In the small Texan percid, *Etheostoma spectabile*, offspring from larger eggs were larger at hatching and took longer to starve (Marsh, 1986). There was no effect of egg size on the growth rate of feeding fry. A similar lack of effect was found in well-fed Atlantic salmon and rainbow trout fry in hatcheries (Thorpe *et al.*, 1984; Springate and Bromage, 1985). Differences in egg size, which result in differences in the size of the larvae at hatching, are probably of more significance if food is in short supply or if the size spectrum of the potential prey is restricted than if food is freely available (Hutchings, 1991).

The food supplied to the female can also

affect egg size, but the effect varies from species to species. In the brown trout (Bagenal, 1969) and the Pacific herring (Hay and Brett, 1988), better-fed fish produced smaller eggs. In the threespine stickleback, the better-fed fish produced slightly larger eggs (Fletcher and Wootton, 1995). It remains to be shown that this food-induced effect on egg size has any effect on the survival or growth of larval or juvenile sticklebacks.

Egg composition

The yolk provides the developing embryo with the reserves of energy and nutrients it requires until it starts feeding exogenously. The energy content of eggs per unit weight tends to be higher than that of somatic tissue. A mean for 50 species was 23.5 J mg^{-1} dry weight (Wootton, 1979), with total energy content per egg largely, but not entirely, determined by egg size (Kamler, 1992). The main constituents of eggs are protein (55–75% of dry weight) and lipid (10–35% of dry weight) (Kamler, 1992). Experimental studies on the threespine stickleback (Fletcher and Wootton, 1995) and the rainbow trout (Springate *et al.*, 1985) found that changes in the quantity of food supplied had no significant effect on the chemical composition of the eggs. Red sea bream, *Pagrus major*, fed diets with different compositions showed little difference in the levels of protein, fat, water or minerals in their eggs, although the fatty acid content of the eggs reflected that of the diet (Watanabe et al., 1984). These studies on sea bream have identified the importance of highly unsaturated fatty acids and carotenoids for egg quality. Fish maintain a quality control over eggs, which buffers the effect on egg composition of changes in the composition of the diet, at least over reasonable ranges of diet composition. The demands of the aquaculture industry for high-quality eggs with predictable fertility and survival characteristics are acting as a stimulus for experimental studies that define the effect of

maternal diet on egg quality (Bromage and Roberts, 1995).

Individual variation in egg quality

Many studies on egg size and composition have focused on interspecies comparisons, for example the trade-off between egg size and fecundity (Wootton, 1992b). However, selection operates at the level of the individual, so it is important to know the consequences of variation in egg size and composition within and between individuals in a population (Chambers and Leggett, 1996). For individual capelin, *Mallotus villosus*, females, initial yolk volume was correlated with length of the larvae at hatching, but not with posthatching life span (Chambers *et al.*, 1989). Posthatching life span was correlated with volume of the oil globule in the egg at hatching. Egg traits differed between females, with initial yolk volume and oil globule volume contributing to these differences. The analysis also suggested that the nutritional status of the female influenced egg traits. Initial yolk volume was correlated with indices of condition and lipid content of females. In a survey of North Atlantic marine fishes, Chambers and Leggett (1996) provided other examples of variation between females in egg size for several species. Their analysis suggested that within a species, these differences in egg size could translate into differences in the size of the young fish for months after hatching. This has implications for mortality rates suffered by these young fish (Chapter 10). If egg characteristics do depend on the nutritional status of the female, environmental effects on the female may have consequences well into the offspring generation.

Viviparity

In some species in which fertilization is internal and the eggs are retained, the female does not provide any further nutrients to the developing eggs, which rely entirely on their yolk. Until the larva is born and starts feeding, there is a progressive loss of material as the yolk is utilized. In other species, the mother provides nutrients for the developing eggs and there is a gain in the weight of the embryo while it is in the ovary. In the ovoviviparous rockfish, *Sebastes marinus*, the eggs lose over 30% of their dry weight during development, but in the viviparous poeciliid *Heterandria formosa*, the fertilized egg weighs 0.017 mg, yet at birth the embryo weighs 0.68 mg, a gain of 3900% (Wourms, 1981). Some viviparous Poeciliidae show superfetation, simultaneously carrying more than one brood at different stages of development (Wourms, 1981). Embryos of the poeciliid, *Poeciliopsis occidentalis*, show a 32% increase in their total energy content during their development. The parental female *P. occidentalis* alternately allocates energy to either an older brood or a younger brood depending on their stages of development (Constantz, 1980).

The effect of amount of food available to the female on the size of her newly hatched young can depend on mode of viviparity (Reznick *et al.*, 1996a). In the poeciliids, *Poecilia reticulata*, *Priapichthys festae* and *Heterandria formosa*, only the third provisions the developing embryo. Under low-food conditions, *P. reticulata* and *P. festae* produced fewer but larger young, with larger fat reserves, whereas *H. formosa* produced fewer and smaller young. Reznick *et al.* (1996a) suggested that the production of larger young with fat reserves was an adaptive response to low-food conditions.

Diapause in eggs

Diapause, an arrest in the development of eggs, occurs in annual fishes (Wourms, 1972). These are cyprinodont fishes that inhabit temporary ponds in South America and Africa. They are characterized by small body size and short life span. When the pools dry out, the adults along with any juveniles die.

A population in a pond may persist because the eggs survive in the mud and hatch when the pond refills. However, there is a danger that the eggs will hatch at inappropriate times if the pond only fills for a period too short for the fish to complete their life cycle. This risk is reduced by a trait that generates a developmental unpredictability, counteracting the environmental unpredictability.

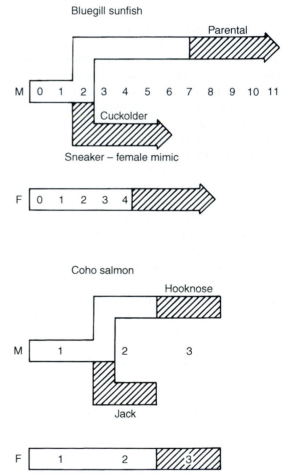

Fig. 7.10 Alternative life histories in male bluegill sunfish, *Lepomis macrochirus*, and male coho salmon, *Oncorhynchus kisutch*. M, male; F, female; numerals, age in years; shaded, sexual maturity. Redrawn from Gross (1984).

During the development of eggs, there are three diapause stages in which development is arrested. These stages may be obligate or facultative and are of prolonged but variable length. The consequence is that a clutch of eggs consists of subpopulations that hatch at different times depending on the pattern and length of diapause the eggs experience. This temporal pattern of hatching decreases the risk that all the eggs in a clutch hatch at inappropriate times.

7.6 ALTERNATIVE REPRODUCTIVE STRATEGIES

Although in many, perhaps most, species there is a typical reproductive strategy for each sex, in some species one sex, often the male, may have alternative reproductive strategies. Thus within a population, there can be subpopulations that mature at different ages and sizes and display different reproductive behaviour (Fig. 7.10).

Some insight into the situations in which alternative reproductive strategies might evolve comes from the application of games theory to evolutionary problems (Maynard Smith, 1982). Maynard Smith introduced the concept of the evolutionarily stable strategy (ESS). A population that has adopted an EES cannot be invaded by a rare mutant adopting an alternative strategy. It is a pure ESS if all the population uses the same strategy and a mixed ESS if portions of the population display different strategies. A mixed strategy can be stable if there is negative, frequency-dependent selection. This means that the fitness obtained by playing a particular strategy is a function of the frequencies of strategies in the population, and the fitness of each strategy declines as the proportion of the population adopting that strategy increases. At equilibrium, the strategies deliver the same fitness. Evolutionary game theory is related to the cost–benefit model described earlier (page 87), but is applicable to situations in which the costs

and benefits are determined by what other individuals in the population are doing (Chapter 5).

In many salmonid populations, some males become sexually mature at an early age. In Pacific salmon, sexually mature small males are called jacks and the larger, later-maturing males hooknoses. In a population of the coho salmon that spawned in Deer Creek Junior, a small stream in Washington State, the jacks spent only 5–8 months at sea before returning to spawn, whereas other males, hooknoses, spent 17–20 months (Gross, 1985). Jacks had a length range of 250–390 mm while hooknoses had a range of 390–720 mm. The two types of males differed in the behavioural tactics they used to get close to a female about to spawn. The hooknoses fought for a position close to the female but the jacks used stealth, sneaking into position. On average, the lifetime production of offspring by jacks was similar to that by large hooknoses. Large jacks were at a disadvantage because there were fewer hiding places for them and small hooknoses were at a disadvantage because they lost fights.

Precocious males in the Atlantic salmon become sexually mature as parr (the name for the juvenile phase in migratory salmon), without spending a period of time at sea. These mature parr compete with older, large, anadromous males for matings with anadromous females (Jones, 1959; Hutchings, 1993a). Modelling studies suggested that the fitness of precocious males decreased as their frequency increased in a population, and as age at maturity increased (Hutchings and Myers, 1994). The probability of a male maturing as a parr seemed to depend partly on the environment and partly on the genotype of the male (Chapter 11). Fast juvenile growth favoured precocious maturation, but different genotypes may differ in the growth rate that has to be achieved for precocious maturation to occur. The incidence of precocious males in populations can be as low as 1%, but can reach 100% (Hutchings, 1993a; Fleming, 1996).

Male bluegill sunfish show alternative mating strategies (Gross, 1982, 1984, 1991). In Lake Opinicon, Ontario, bluegill males become sexually mature either at 2 years or at about 7 years of age (Fig. 7.10). The late-maturing, parental males dig nest pits, which they defend and in which the females spawn. The precocious males, called cuckolders, do not defend nests but try to sneak into a nest and release their sperm just as a female is spawning. In cuckolder males the testes represented 4.6% of body weight, but in parental males only 1.1% (Gross, 1982). In some bluegill populations, a third male strategy occurs. Larger, cuckolder males mimic adult females and this allows them to get close to a parental male as it is about to spawn with a true female (Dominey, 1980). In breeding colonies of bluegill sunfish in Lake Opinicon, the pairing success of cuckolder males was density dependent. At high densities, their success declined (Gross, 1991). Estimates for the population of the pairing success of cuckolders compared with that of parental males suggested that at a high frequency of cuckolders, their success decreased. This suggests that negative frequency-dependent selection maintained the mixed strategy in the population. Within colonies, the density at which cuckolders were most successful was a function of the cover available to the cuckolders.

These alternative mating strategies illustrate an important principle. The behaviour that a fish must adopt to maximize the number of offspring it produces may depend on environmental factors extrinsic to the population, but also on the behaviour of other fish in the population.

7.7 UNUSUAL REPRODUCTIVE STRATEGIES

Although the majority of teleosts are gonochoristic and probably have a genetic sex-determination mechanism, some unusual reproductive strategies are helping to throw light on the adaptive significance of reproductive traits.

Environmental sex determination

In many species, the sex of an individual depends entirely on its genotype and so is independent of the environment in which the individual develops. In fish, there are a variety of mechanisms for genotypic sex determination (GSD) (Purdom, 1993). However, in some species, sex depends irreversibly on the environment experienced during early development, a phenomenon called environmental sex determination (ESD). The extent and distribution of environmental sex determination in fish is not known.

The best known example of ESD in fish is the Atlantic silverside. This species is distributed down the eastern seaboard of North America (page 128). In most populations of silversides, there is an inverse relationship between the temperature of rearing and the proportion of fish that develop as females. Low rearing temperatures favour the development of females (Conover and Kynard, 1981). The critical period for this effect of temperature is when the larvae are between 8 and 20 mm in total length. The importance of genetic and temperature effects in determining sex in silversides varies geographically (Lagomarsino and Conover, 1993). In a northern population from Nova Scotia, rearing temperature had little or no effect on sex, which is genetically determined. In a southern population from South Carolina, rearing temperature had a major effect on the sex ratio. The proportion of females produced dropped sharply between 15 and 21 °C (Conover and Heins, 1987).

An adaptive interpretation is possible for these patterns of sex determination in silversides. Their breeding biology is such that large body size is likely to be more of an advantage to females than males. Low water temperatures occur early during the prolonged breeding season in spring and summer. Females produced early in the breeding season will have a long growing period before the low temperatures of winter bring growth to a halt, allowing them to achieve a relatively large body size, and consequently a high fecundity when they spawn at 1 year old. Males will be produced later in the breeding season when temperatures are warmer. Taken over the whole breeding season, the sex ratio, the proportion of females to males, will be close to the 1:1 expected on theoretical grounds. In northern populations, the length of the growing season is too short for ESD to be advantageous and sex is determined genetically (Conover and Heins, 1987). Silversides from a range of populations were exposed to either high or low temperatures, during the critical period in larval development for sex determination, over several generations in the laboratory. Over successive generations, the sex ratio in the laboratory populations approached the theoretically expected 1:1 value (Conover *et al.*, 1992)

ESD is predicted to be adaptive when the environment that the offspring enter has an effect on fitness that depends on gender. In the example of the silversides, the early breeding season is more favourable to females because it permits a longer growing period. Careful rearing studies are required to reveal the extent of ESD in other species.

Hermaphroditism

Hermaphroditism has evolved in several fish lineages (Warner, 1978). Sequential is more frequent than simultaneous hermaphroditism. In the former, some species show protogyny in which the fish first functions as a female and then as a male. Protandric species have males that transform to females (Warner, 1978; Shapiro, 1984).

Simultaneous hermaphroditism is found in the Aulopiformes, which are bathypelagic. In this group, bisexuality may be related to low population densities, so that any conspecific encountered is a possible mate (Warner, 1978). A freshwater cyprinodont, *Rivulus marmoratus*, and some species in the perciform group Serranidae, which occur in shallow marine waters, are also simultaneous hermaphrodites.

Sequential hermaphrodites are found in 13 families. They include species that are herbivores, zooplanktivores or piscivores; species that are monogamous, polygamous or promiscuous, large or small species and species with a wide range of social systems (Shapiro, 1984). The evolution of sequential hermaphroditism seems to depend more on the social environment than the abiotic environment. Indeed the change of sex is usually induced behaviourally, either by the disappearance of an individual of the dominant sex, or by a change in the sex ratio of a social group (Shapiro, 1984). It is favoured if a fish can reproduce more effectively as one sex in one set of circumstance, but under different circumstances can reproduce more effectively as the other sex (Shapiro, 1984). The benefits of the sex change must outweigh the costs incurred in the reorganization of the gonads and of any secondary sex characteristics.

Protogynous labrids (Labridae) provide a good example of sequential hermaphroditism (Warner and Downs, 1977; Robertson and Warner, 1978; Warner and Robertson, 1978). In this group, size is the critical factor in determining male reproductive success. A large male, through its territorial dominance, can have a daily spawning rate 10 to 40 times greater than that of a smaller male. Individuals avoid an unproductive period by being a female when small and a male when they are large enough to compete for a territory. In some labrids, for example the blue-head wrasse, *Thalassoma bifasciatum*, there are small males that sneak fertilizations or engage in group spawnings, in which several males consort with one female. The proportion of small males in a population increases with population abundance, probably because the large males have increasing difficulty in defending a territory and there is a greater supply of ripe females (Warner and Hoffman, 1980).

Parthenogensis and hybridogenesis

Asexually reproducing species are rare in the fishes. In principle, if only females are produced, a population should have a higher rate of increase than a population of sexually reproducing fish. If the fecundities are equal, for every female produced by a sexually reproducing fish, an asexually reproducing fish will produce two and so its population will grow in abundance at twice the rate of the sexual form (Moore, 1984). The potential disadvantages of asexual reproduction include the lack of genetic variation among the offspring and the accumulation in the genome of lethal mutations, which cannot be purged by recombination. The best-known examples of asexual reproduction are viviparous cyprinodonts in south-western North America, and occur in the genera *Poecilia* and *Poeciliopsis* (Moore, 1984; Vrijenhoek, 1984).

In *Poeciliopsis* in north-eastern Mexico, two types of all-female populations occur, hybridogenetic and gynogenetic. Both involve hybridization between females of the sexually reproducing species *P. monacha* and males of a closely-related species. Hybridogenetic populations arise from diploid hybrids. Only female hybrids are viable and these produce eggs that carry only the maternal genome. These eggs are then fertilized by sperm from the other species, again producing female hybrids. By this mechanism, the female *P. monacha* genome is conserved in the hybrid line. Gynogenetic populations are triploid. Although the development of the eggs produced by gynogenetic females has to be activated by sperm from a sexual species, the sperm contribute neither genotypically nor phenotypically to the offspring. In contrast, in hybridogenetic lines, the genotype of the father is expressed in the offspring, but those offspring do not transmit any of their paternal genes to their offspring (Fig. 7.11). Both hybridogenetic and gynogenetic all-female populations of *Poecilopsis* depend parasitically on the sperm of males from closely related sexually reproducing species.

The ecological circumstances that allow

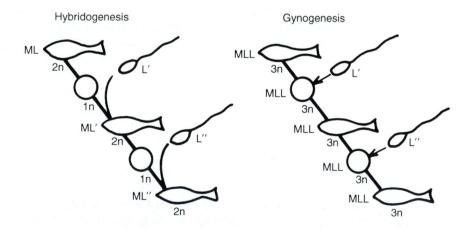

Fig. 7.11 Two modes of parthenogenetic reproduction in *Poeciliopsis*. See text for further details. M, *monacha* genome; L, *lucida* genome; primes represent different allelic markers of *lucida* genomes. Redrawn from Vrijenhoek (1984).

this coexistence of closely related sexual and asexual populations are not fully understood (Moore, 1984; Vrijenhoek, 1984). The 'frozen niche-variation' hypothesis assumes that many clones of parthenogenetic forms coexist in a population, having arisen on several occasions through hybridization between sexual species. As these new clones arise, they 'freeze' and maintain genotypes that affect their use of food and spatial resources, with each clone expressing a different phenotype. Selection between clones will then produce a structured assemblage of clones each relatively specialized for a particular combination of environmental qualities. In other words, the clones differ in their niches (Chapter 9) (Vrijenhoek, 1989). The model assumes that the sexual forms can coexist with these clones for two reasons. The genetic diversity of the sexual forms allows them to use a wider range of resources than the clones, and the clonal forms depend on the presence of the sexual forms for sperm. Field studies on *Poeciliopsis* provided evidence of a differential use of habitats by different clones, especially in relation to water current and diet. Some experimental evidence suggested that a sexual form can use a wider range of resources, i.e. that it has a wider niche (Weeks, 1995).

This cannot be a general hypothesis. The cyprinids *Phoxinus eos* and *P. neogaeus*, through hybridization, give rise to parthenogentic lines by gynogenesis. In a drainage system in Minnesota, only a single clone was found coexisting with the sexual species. The drainage contained a variety of habitats that varied both spatially and temporally because of the damming activities of beaver (*Castor canadensis*). The clone was distributed over a range of habitats, but reached its highest frequency compared with the sexual species in the marginal habitat of meadow formed by the collapse of lowland beaver pools (Elder and Schlosser, 1995).

These unusual fishes potentially provide rich material for studying the ecological and evolutionary consequences of asexual and sexual reproduction and hybridization.

7.8 SUMMARY AND CONCLUSIONS

1. The reproductive success of an individual depends on where and when it breeds and the resources it devotes to reproduction. Typically in fishes, the sexes are separate and the females produce a high number of small eggs. Within the reproductive strategy characteristic of a species, individuals show tactical variations in their expression of reproductive traits in response to environmental circumstances.

2. There is wide interspecific variation in the age at maturity. Within a species there may be considerable inter- and intrapopulation variation, reflecting both genetic and environmental influences.

3. Breeding is often restricted to a particular period of the year. Reproductive seasonality is well defined at high latitudes, but is also observed at low latitudes in some freshwater and marine species. The seasonal timing depends on an endogenous reproductive cycle synchronized to external cues such as photoperiod and temperature.

4. A major division is between pelagic spawners depositing their eggs in the water column, and demersal spawners laying their eggs on or near the substrate. The spawning site defines the nature of the hazards to which the eggs and newly hatched larvae are exposed.

5. Resources allocated to reproduction must meet the costs of any secondary sexual characteristics, of reproductive behaviour and of the production of gametes.

6. Reproductive behaviour can include the components: movement to a spawning site, preparation and defence of a spawning site, courtship and parental care. The combination of components observed for any species is related to the

physical nature of the breeding site, the abundance and type of food resources available and the abundance and range of predators on eggs, juveniles or adults.

7. The number of eggs produced by a female over a breeding season depends on her batch fecundity and the number of times she spawns. Batch fecundity increases with an increase in body size. In some species it may also be a function of food ration. For batch spawners, the number of spawnings is affected by the food supply and abiotic factors such as temperature. The proportion of the energy income that is invested in eggs can vary with ration, temperature and other environmental factors.

8. In many species, the testes are smaller than the ovaries and it is likely that the energy allocated by males to sperm production is lower than that allocated to eggs by females.

9. Egg size varies both inter- and intraspecifically. Larger eggs may give larvae an advantage if larval food is sparse, but an increase in egg size is usually traded-off against a reduction in batch fecundity. A few fishes bear live young, being either ovoviviparous or truly viviparous.

10. Less usual reproductive strategies include alternative mating strategies, environmental sex determination, sequential and simultaneous hermaphroditism and parthenogenesis. The ecological circumstances that favour these unusual strategies are now being studied.

8

BIOTIC INTERACTIONS: I. PREDATION AND PARASITISM

8.1 INTRODUCTION

Abiotic factors such as temperature and oxygen set limits on where individual fish can survive and these factors influence the patterns of allocation of time and material resources by individuals (Chapter 4). Within the limits set by abiotic factors, encounters with other organisms can also change an individual's risk of dying, or may change its pattern of allocation of time and resources with consequences for its growth and reproduction. Encounters that result in the fish feeding or mating have already been described (Chapters 3 and 7). Other types of encounter are the subjects of this chapter and the next. Interactions with other organisms are of a fundamentally different nature from interactions with abiotic factors because the fitnesses of all the interacting organisms may be affected. In contrast, abiotic factors, although they may be capricious, are indifferent to the outcome. The outcome of a biotic interaction may have consequences for the distribution, abundance and genetic composition of the populations of which the interacting organisms are members and for the species composition of the assemblage present in a location (see Chapter 12 for a discussion of assemblages). But even if the consequences of the interactions are more easily observed through changes at the population or assemblage level, these consequences result from the effects of the interactions on the individual participants.

One classification of interactions between organisms is by the net effects on the fitnesses of the organisms involved (Table 8.1). Another dimension for classification is whether the interaction is intraspecific or interspecific. Predation and parasitism are interactions in which one participant gains in fitness but the other loses (interactions of the form + − in Table 8.1).

Such a simple classification cannot reflect all the complexities of the interactions. For example, a predator usually benefits at the expense of its prey, but not always. If a predator makes an unsuccessful strike at a prey, the prey may learn from that experience and become more difficult to attack in the future. After an infestation by a parasite, the host fish may become immune to further infestations by that parasite.

Biotic interactions take place in a context defined by the abiotic environment. As that environment changes, the nature and effects of the biotic interaction may change, giving rise to abiotic–biotic interactions. The effects of biotic interactions will mainly be described as though they were taking place against a static abiotic background, but its dynamic presence should not be forgotten. This chapter discusses the effects of predation and parasitism on fishes, while Chapter 9 considers competition and mutualism (interactions of the form + +, + 0, − − and − 0 in Table 8.1). The effects of biotic interactions on the abundances of fish populations and on the composition of fish assemblages also form topics in Chapters 10 and 12.

Table 8.1 Classification of biotic interactions*

Effect of presence of individuals of species i on fitness of individuals of species j	Effect[‡] of presence of individuals of species j on fitness of individuals of species i		
	+	0	−
+	+ +	+ 0	+ −
0	0 +	0 0	0 −
−	− +	− 0	− −

*Mutualism: + +, + 0, 0 +; competition: − −, − 0, 0 −; predation/parasitism: + −, − +; neutral, 0 0.

[‡] +, positive effect on fitness; −, negative effect on fitness; 0, no effect on fitness.

8.2 PREDATION

Components of predation

The role of fish as predators has been described in Chapter 3; here the emphasis is on fish as prey. A fish may be eaten by a cannibalistic conspecific or by a heterospecific. The chances that an individual will be attacked will depend on the density of the prey, the density of the predators and often on the size of the predator relative to the prey. The number of prey taken by predators over a defined time is determined by the functional, numerical and developmental responses of the predator–prey interaction (Fig. 8.1) (Taylor, 1984). The functional response is the relationship that describes the change in the number of prey taken as a function of the prey and predator densities. The numerical response describes the change in predator density as a function of prey and predator densities. This change in density may be caused by a behavioural response by the predator, for example if predators tend to aggregate in regions of high prey density. A numerical response may also reflect demographic changes in the predator population. High food levels may increase the survival rates or the fecundity of the predators (Chapter 7). Behavioural responses can cause

a rapid numerical response, but demographic changes will be slower. The developmental response is the change in size of the predator as a function of prey and predator densities (Chapter 6). This change in size is important because bigger predators usually have higher absolute rates of prey consumption than smaller predators and can take larger prey (Chapter 3). For an individual, the combined effects of the functional, numerical and developmental responses define the risk of being attacked by a predator.

A piscivorous duck, the common merganser, *Mergus merganser*, feeding on salmonid juveniles in small coastal streams of Vancouver Island, Canada, illustrated these responses (Wood, 1985, 1987a,b; Wood and Hand, 1985). During spring and summer, large numbers of salmonid smolts move down the streams on their seaward migration (Chapter 5). As Fig. 8.2(a) shows, the functional response of the mergansers was the typical, negatively accelerated curve (compare with Fig. 3.21), which means that the risk of being attacked declines as the density of the prey increases. But the mergansers increased in abundance on the streams at times when high numbers of smolts were migrating (Fig. 8.2(b)). This aggregation was a behavioural response, but there was also a demographic numerical

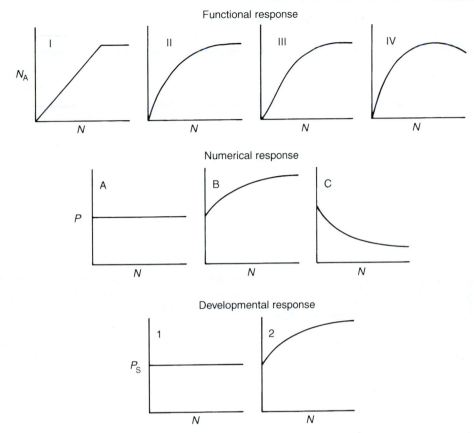

Fig. 8.1 Idealized functional (I, II, III and IV), numerical (A, B and C) and developmental (1 and 2) responses in predation. N_A, number of prey attacked; N, density of prey; P, density of predators; P_S, size of predator.

response. The mergansers bred during the summer, and the ducklings also fed on salmonid juveniles. The developmental response was complex. As the ducklings grew, their daily rate of food consumption expressed as a percentage of body weight declined but their body weight increased rapidly.

Types of predation

Cannibalism

Cannibalism takes two forms: intracohort and intercohort (Smith and Reay, 1991). In intracohort cannibalism, the cannibals are consuming conspecifics born at approximately the same time as themselves. Two examples are provided by piscivorous freshwater species. Larvae of walleye, *Stizostedion vitreum*, showed a cannibalistic phase during the short period between 6 and 16 days after hatching (Li and Mathias, 1982). The cannibals were too similar in size to their prey to eat them completely: they could not swallow the heads. Although the intensity of the cannibalism decreased as the supply of an alternative food increased, it was not totally suppressed. Cannibalistic walleye larvae could achieve twice the size of non-canniba-

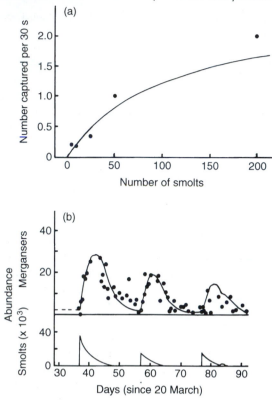

Fig. 8.2 Functional and numerical response of mergansers, *Mergus merganser*, feeding on salmonid smolts. (a) Partial functional response (excluding digestive pause) of megansers feeding on coho, *Oncorhynchus kisutch*, smolts in stream enclosure. Redrawn from Wood and Hand (1985). (b) Abundance of mergansers (each point represents a pair of birds, females with broods excluded) in relation to releases of salmonid smolts and fry in Rosewall Creek, Vancouver Island. Line is trend in merganser abundance predicted by an aggregation model. Redrawn from Wood (1985).

listic larvae. Intracohort cannibalism also occurs in pike fry (Giles *et al.*, 1986; Wright and Giles, 1987; Craig, 1996). Mortality attributable to cannibalism increases with the density of the pike fry.

Intercohort cannibalism is the consumption of younger and smaller conspecifics by fish in older age classes. Egg cannibalism is common. The eggs of conspecifics occur in the stomachs of pelagic fish such as the northern anchovy, demersal spawners including coregonids and in species in which eggs are guarded including sticklebacks and cottids (Chapter 7). Early larvae are quickly digested (Hunter, 1981), which can make detection of intercohort cannibalism difficult. Predation of older pike on 0+ conspecifics can play a major role in reducing the size of fluctuations in the abundance of pike populations, with cannibalism acting as a density-dependent regulatory process (Craig, 1996) (Chapter 10).

Cannibalism should not be seen as a pathological behaviour. Cannibalistic individuals can gain advantages (Polis, 1981; Smith and Reay, 1991). Cannibalistic fish consume a high-quality diet, the chemical composition of which closely resembles that of their own flesh. This allows them to grow quickly through stages when they are at risk from other predators, and may also allow them to increase their fecundity (Chapters 6 and 7). Unless the cannibals are eating their own offspring, or perhaps fish that are closely related to them, they are increasing their fitness at the expense of their conspecifics.

Interspecific predation

Taxonomically, predators of fish range from invertebrates such as coelenterates and molluscs to vertebrates, including agnathans (lampreys), fishes, reptiles, birds and mammals. The impact of different predators changes as prey fish grow and move to different habitats. As fish grow, they probably become potential prey to a narrower range of predators. Large piscivorous fish are absent or rare in the shallow head waters of rivers, where birds and mammals may be more important as predators. Further downstream, predatory fish become common. In coastal waters the risk of predation may be high because of the presence of piscivorous elasmobranchs and fishes, dense colonies of piscivorous birds such as gulls, terns and auks, and piscivorous mammals, especially pinnipeds and cetaceans. In the pelagic zone of the open sea, fish predators include the tunas and jacks, fish adapted to chasing down their prey (Chapter 2). The deeper waters of the seas contain some piscivorous fishes that are equipped with huge mouths, including the deep-sea angler fishes (Lophiiformes) and the aptly named swallowers (Saccopharyngidae) and gulpers (Eurypharyngidae). Many aspects of the biology of fishes can be interpreted as adaptations to this galaxy of natural enemies.

Piscivorous invertebrates feed mainly on larval and young juvenile stages. For example, 40–45 mm final instar dragonfly nymphs (*Aeshna* spp.) given threespine sticklebacks from three size classes, 15–25 mm, 30–40 mm and 50–60 mm, took 97.8% of the smallest fish but only 4.4% of the medium size and 1.1% of the largest fish (Reimchen, 1980). Piscivorous fishes and some piscivorous birds usually swallow their prey whole. This imposes an upper size limit on the size of prey that can be taken by a predator of a given size and frequently means that the predator has to reach a critical size before it becomes piscivorous (Chapter 3). Those birds and mammals that rip and tear their fish prey into pieces are less constrained by the size of their prey. Thus the size of a fish determines the nature of the predation risk to which it is exposed.

Modes of predation

Given the taxonomic diversity of predators on fish, it is helpful to impose an ecological classification (Bailey and Houde, 1989). Invertebrate and vertebrate piscivores can be classified in terms of how they hunt, detect and capture their prey (Chapter 3). Predators may ambush their prey or cruise in search of them. Predators may detect prey by direct contact, by vision, mechanoreception or chemoreception or by some combination of

these. The prey may be filtered, entangled, or seized. The mode of predation will influence the vulnerability of the prey. Vulnerability of individual prey depends on the rate of encounter between prey and predator and the susceptibility of the prey (Bailey and Houde, 1989). Both of these are likely to depend on the size of the prey (Chapter 3). As prey get larger, their swimming speed increases (Chapter 5), tending to increase the rate of encounter, especially for ambush predators. With an increase in size, the prey become easier to detect visually or by mechanoreception. However, as the prey increase in size, they become less susceptible to both size-limited and gape-limited predators. Consequently, vulnerability may first increase but then decrease with an increase in prey size, or may simply decrease as prey size increases.

An analysis of data on predation on larval fish suggested that their vulnerability to predatory crustaceans and to predatory medusae decreased as the size of the fish prey relative to the predator increased. For predatory fish and predatory ctenophores the relationship between vulnerability and prey size was dome-shaped (Paradis *et al.*, 1996). The larvae were most vulnerable when their length was about 10% that of the predator.

Prey can reduce their vulnerability by reducing the rate of encounter or by reducing their susceptibility to attack even if detected (Fuiman and Magurran, 1994) (page 184).

Consequences of predation

Density dependence

Predators have direct effects: killing or injuring their prey. Predators may also have indirect effects. The presence of predators may restrict prey to habitats in which they are less likely to be attacked but which are less profitable for foraging (Chapter 3). Poorer foraging will lead to slower growth rates (Chapter 6). Consequently, the prey may be susceptible to size- or gape-limited predators for longer. Prey may also accumulate smaller reserves and so have lower survival under severe conditions such as an overwintering period (Tonn *et al.*, 1992). Reproductive activities may be inhibited if there is a risk of predation.

The risk of attack for an individual prey fish may be related to the density of the prey population (see also Chapter 10). There are three possibilities. The risk may be independent of the prey density i.e. density independence. The risk may increase with an increase in prey density i.e. direct density dependence. An increase in the mortality caused by predation compensates for the increase in prey density. The third possibility is that the risk declines as the prey density increases i.e. inverse density dependence. Such mortality is depensatory, because as the prey density increases, the mortality caused by predation decreases.

The asymptotic nature of the functional response curve (Figs 8.1 and 8.2(a)) means that for an individual predator, there is a density of prey such that even if prey density increases further, the predator cannot consume more prey. Unless there is a strong numerical response, so that the number of predators increases rapidly at high prey densities, predatory mortality is depensatory at high prey densities. The risk to each prey is lessened by the presence of other prey.

Depensatory mortality can occur during fish migrations when the migrating fish pass through an array of predators over a short time period. During the downstream movement of Pacific salmon smolts they are attacked by piscivorous birds and mammals. With high numbers of migrants, predators can be swamped, but with low numbers of migrants, the predators take a greater proportion of the migrants (Ricker, 1962). The mortality resulting from merganser predation on smolts in the Vancouver Island streams was clearly depensatory (Wood, 1987a).

There was a reduction in the risk to a smolt of being eaten if it timed its downstream run to coincide with the time when most other smolts in the population were also migrating. As the young salmonids move offshore, they may also have to run a gauntlet of piscivorous adult salmon returning to the coastal waters prior to their upstream spawning migrations.

Cannibalism can impose two forms of direct density-related mortality on a population. Intracohort cannibalism can impose a risk of being eaten, which is a function of the density of the cohort. As the density changes, the level of cannibalism changes almost immediately. Intercohort cannibalism will be related to the densities of the cannibalistic cohorts as well as the prey cohort. A large cannibalistic cohort may have a major effect on the abundance of a prey cohort, even when the latter is not abundant. Subsequently, when in its turn the original prey cohort becomes cannibalistic, its low abundance will mean that it can only impose light predatory pressure even if the prey cohort is now large. Some of the potential consequences of cannibalism for the population dynamics of fish are recognized (Chapter 10), but the effects of cannibalism on the dynamics of real populations and its role in the determination of age-class strengths are still obscure (Smith and Reay, 1991).

North Sea cod illustrate a situation in which intercohort cannibalism could play a role in determining cohort abundance. The young-of-the-year cod (0-group) live in shallower water than older cod. The young have to pass through an array of larger, cannibalistic conspecifics as they move from the spawning grounds to their nursery grounds (Hislop, 1984). Ricker (1962) has suggested that intercohort cannibalism is a factor in generating a two-year cycle of abundance in many populations of pink salmon, *Oncorhynchus gorbuscha*, that spawn in the rivers of the Pacific North-west. Although in some populations, density-dependent cannibalism may tend to dampen any fluctuations in abundance (Chapter 10), this is a fortuitous effect of a behaviour that has evolved because of the advantages that accrue to the cannibalistic individuals.

Quantitative effects of predation on mortality of prey

The risk to individual fish of being eaten by a predator can be estimated from the proportion of a population that is eaten. If subgroups within a population are either more or less at risk from predation, the mortality rates for the subgroups have to be estimated.

There are relatively few examples of quantitative estimates of the mortality caused by predators. A study of salmonid populations in the Au Sable river in Michigan provided estimates of the intensity of predation in a cold-water stream (Alexander, 1979). In the river, brook and brown trout after their first autumn of free-swimming life suffered annual mortalities of 70–90%. Piscivorous birds, brown trout, mink and otter were the major natural predators and together with anglers accounted for 50–90% of the trout mortality (Table 8.2). In Vancouver Island streams, some salmonid fry remain over the summer, migrating as smolts in the following year. Merganser predation on the fry was estimated as being equivalent to 24–65% of the smolt production (Wood, 1987b). Other studies showed that the survival of stream salmonids can be improved by the removal of predatory birds including mergansers and kingfishers (Elson, 1962). An increase in the rate of predation by other natural predators could also result (Alexander, 1979), as could an increase in intra- and interspecific competition (Chapter 9).

The mortality of lake trout in Lake Ontario

Table 8.2 Estimated percentage mortality of brown trout, *Salmo trutta*, and brook trout, *Salvelinus fontinalis*, caused by natural predators and angling in Au Sable River*

Fish species	Predator		Age class		
			Eggs – 0	0 – I	I – IV
Brook trout	American merganser		0.0	3.6	10.5
	Great blue heron		0.1	7.7	3.3
	Belted kingfisher		0.9	4.0	0.1
	Brown trout		2.8	58.0	5.1
	Mink		0.1	5.0	4.8
	Otter		tr	0.6	1.2
	Angler		0.0	6.6	43.7
		Total	3.9	85.5	68.7
Brown trout	American merganser		0.0	10.7	12.8
	Great blue heron		0.0	6.8	14.1
	Belted kingfisher		0.2	3.2	0.0
	Brown trout		0.0	16.1	0.2
	Mink		0.0	7.7	11.4
	Otter		0.0	0.2	5.4
	Angler		0.0	6.8	46.0
		Total	0.2	51.5	89.9

*Source: Alexander (1979); tr, trace (less than 0.1%).

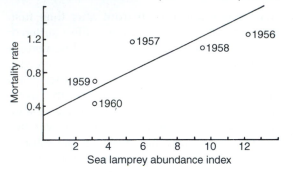

Fig. 8.3 Instantaneous annual mortality of lake trout, *Salvelinus namaycush*, in relation to an index of sea lamprey, *Petromyzon marinus*, abundance in Lake Ontario during the 1950s. Redrawn from Christie (1974).

between 1956 and 1960 was strongly correlated with the density of sea lampreys, *Petromyzon marinus* (Fig. 8.3) (Christie, 1974). In the Volga Estuary, fishes were estimated to take over 10% of a generation of the vobla, *Rutilus rutilus caspicus* (Popova, 1978). For Drizzle Lake on the Queen Charlotte Islands in the Pacific North-west, Reimchen (1994) estimated that invertebrate, fish, bird and mammal predators took a total of 562 000 juvenile and adult threespine sticklebacks per year, with the adult stickleback population estimated at 75 000. In this population about 13% of live sticklebacks sampled had injuries, probably caused by predatory attacks that the fish had survived. These injuries included fractured spines, skin lacerations and the imprints of the beaks of piscivorous birds (aviscars).

Discussion of the effects of predation on the abundances of fish populations is continued in Chapter 10.

Spatial distribution of predators and prey

Mobile prey such as fish should avoid localities where predators are abundant and, where possible, seek out refuges inaccessible to predators. On the other hand, the pre-

dators should seek out localities where prey are abundant. An echosounding study of the density and distribution of cod, the predator, and capelin, the prey, in the Gulf of St Lawrence, Canada, illustrated the spatial patterns that may result (Rose and Leggett, 1989, 1990). The study also illustrated the importance of the scale at which the distributions of prey and predator are compared. Capelin had a temperature refuge, being found at temperatures both higher and lower than were cod. At a spatial scale of tens of kilometres, cod distributions were positively correlated with high densities of capelin within the temperature range 1–9 °C. Outside this range, the densities were not correlated. At smaller spatial scales, there was a negative correlation between cod and capelin densities, which probably reflected the avoidance by prey of predator aggregations. At low densities of capelin, there was no correlation between densities of prey and predator.

Indirect effects of predation on foraging

In the presence of predators, prey fish may modify their behaviour in a way that reduces the risk of being eaten (Lima and Dill, 1990; Milinski, 1993). Such modifications of behaviour involve trade-offs with other behaviours including habitat selection and foraging. If the habitats in which the prey fish find it more profitable to forage are also places in which they are at greater risk from predators, the prey may have to balance the profitability against the risk. A fish that has its foraging success restricted by the presence of predators will have a reduced growth rate (Chapter 6) and may have a reduced breeding success (Chapter 7). Slower growth can also lengthen the time over which a fish is at risk from size- or gape-limited predators. Evidence for indirect effects of predation comes from field and experimental studies. The trade-off between foraging and predation risk has been analysed theoretically (Houston *et al.*, 1993).

Some streams, for example in the forests of Panama and in the prairie margins of Oklahoma, have ribbons of green along their margins formed by stands of algae, although in deeper water the standing crops of algae are low (Power, 1987). Around patches of coral, there is sometimes a halo of bare substrate surrounded at a greater distance from the reef by stands of vegetation (Randall, 1965). The effects of predators on the distribution of their prey, herbivorous fishes, generate these patterns (Power, 1987). The size distribution of fishes in streams and lakes may also reflect the effects of predation. Smaller species or the juveniles of larger species are frequently found in shallow waters where there are hiding places amongst the vegetation or between stones.

In a Panamanian stream, algal-grazing, armoured catfish (Loricariidae) had a distinctive bigger–deeper distribution (Power, 1987). The small catfish were restricted to shallow waters by large piscivorous fish that occupied the deeper pools. The larger catfish, because of their size, were at less risk from fish predators, but in shallow water were at risk from avian predators, most of which concentrated their hunting in shallow waters. The small catfish found refuge from the birds by hiding in small crevices and holes. The large catfish in the Panamanian stream heavily grazed the algal in the deeper waters but did not exploit the algae stands along the margins of the stream. This suggests that the risks of predation outweighed the gains to be made from foraging in the margins.

In Oklahoman streams, herbivorous minnows, *Campostoma anomalum*, were largely excluded from deeper pools by largemouth bass, *Micropterus salmoides*, and spotted bass, *Micropterus punctulatus* (Power, 1987). In an experimental stream in Illinois, juvenile cyprinids, percids, catostomids and centrarchids preferred to occupy deeper pools, but if the pools contained adult catostomids or centrarchids, the juveniles were largely restricted to riffles or raceways that

formed refugia from the predators (Schlosser, 1987a).

In experimental ponds, small bluegill sunfish remained in fringing vegetation when piscivorous largemouth bass occupied adjacent open water. However, the open water was probably a more profitable foraging habitat for the sunfish because, in the absence of the predatory fish, the small bluegills did move into it (Chapter 3) (Werner *et al.*, 1983a,b). Bluegills that were too large to be prey of the bass moved into the open water when it became profitable to do so, even in the presence of bass.

The effect of a predator on the use of habitats by prey can depend on the size and species of the predator. This was illustrated by a study of four lakes in Sweden, of which two contained perch and piscivorous pike, whereas two others held only cannibalistic perch. Pike grow larger than perch and can take larger prey. In lakes with only perch, the perch used both the epilimnion and hypolimnion. In lakes with pike, the perch were found mainly in the littoral zone. In the latter lakes, the abundance of perch was lower but the growth rates of individual perch higher than in perch-only lakes (Persson *et al.*, 1996).

The effect of piscivorous perch on crucian carp, *Carassius carassius*, in shallow lakes in Finland demonstrated the consequences of an interaction between the direct and indirect effects of predation (Tonn *et al.*, 1992). Some lakes contained only carp because the waters became anoxic in winter. The carp survived this condition by using anaerobic respiration based on glycogen reserves laid down in late summer and autumn (Chapter 4). These lakes contained dense populations of stunted carp that exploited both the littoral and pelagic habitats. In lakes containing piscivorous perch, the carp populations consisted of relatively few, but large, fish. A field experiment found that the presence of perch reduced recruitment of young-of-the-year carp by 90%. The young carp were entirely confined to the vegetated littoral zone, and had a reduced growth rate. The few carp that survived probably benefited from reduced intraspecific competition (Chapter 8), and could eventually grow to sizes that were invulnerable to perch predation.

These examples pose the question of what rules prey fish might follow when faced with a choice between habitats that are profitable but hold dangerous predators and those that are less profitable but safer. One such rule is that an individual should minimize the ratio of mortality rate to growth rate (Werner and Gilliam, 1984) (Chapter 5). Gilliam and Fraser (1987) suggested that in some circumstances, foraging fish should use the habitat that has the lowest ratio of mortality risk to gross foraging rate. Foraging juvenile creek chub, *Semotilus atromaculatus*, switched from a region of lower predation risk to one with a higher risk but a higher density of food. The density of food at which they switched was approximately that predicted by assuming that they were minimizing the mortality-to-foraging-rate ratio (Gilliam and Fraser, 1987).

These studies on sunfishes, perch and carp are good illustrations of the way in which laboratory experiments, field experiments and comparative, observational studies can be integrated to provide insights in to the consequences of the direct and indirect effects of predation.

Indirect effects of predation on reproduction

Mating often involves conspicuous courtship behaviour and secondary sexual characteristics (Chapter 7), with the participants paying attention to their mates. The potential trade-off between reproduction and risk of predation has received less study than the trade-off with foraging (Sih, 1994). *Rivulus hartii* and *Hoplias malabaricus* live in streams in Trinidad. *Hoplias* is a piscivore. In an experimental stream, the presence of *Hoplias* led to a 50% reduction in daily total egg pro-

duction by *Rivulus* compared with when the piscivore was absent (Fraser and Gilliam, 1992). Studies on guppies from Trinidadian streams suggested that increased risk of predation reduced non-random mating, with males adopting less conspicuous courtship behaviour and the females less able to reject male copulation attempts (Sih, 1994; Magurran *et al.*, 1995).

Reduction of predation risk

The pervasive effect of predation is probably best illustrated by the adaptations that have evolved to protect fish from their enemies. Defences against predation can be classified as primary or secondary (Edmunds, 1974; Keenleyside, 1979; Fuiman and Magurran, 1994). Primary defence reduces the possibility of detection by a predator, whereas secondary defence reduces the chance of the prey being eaten even after detection. The sequence of events in an encounter between prey and predator was described by Lima and Dill (1990) (Fig. 8.4).

Primary defences

Several mechanisms may in combination or separately make prey less conspicuous (Edmunds, 1974; Keenleyside, 1979). Of course, the same mechanisms may also make the predator less conspicuous to the prey. These mechanisms include cryptic coloration, cryptic behaviour and the use of refuges. The effectiveness of these will depend on the physical characteristics of the environment and the mode of feeding of the predators. The light-transmitting properties of natural waters and the background colour vary greatly (Chapter 2), and this will influence what is conspicuous or cryptic to a predator dependent on vision. In green waters, red and its complementary colour blue-green are most conspicuous, whereas in blue waters yellow and the complementary indigo blue are most visible (Lythgoe, 1979).

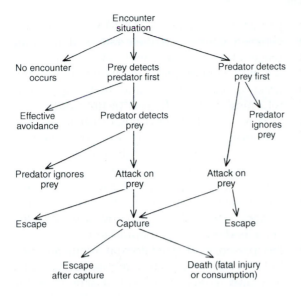

Encounter situation

No encounter occurs

Prey detects predator first

Predator detects prey first

Effective avoidance

Predator detects prey

Predator ignores prey

Predator ignores prey

Attack on prey

Attack on prey

Escape

Capture

Escape

Escape after capture

Death (fatal injury or consumption)

Fig. 8.4 Diagram of possible outcomes of encounter between prey and predator. Simplified from Lima and Dill (1990).

In shallow, fresh and coastal waters, many fishes are cryptically coloured with bodies that are drab browns and greens, often with darker blotches. The colour patterns may also break up the outline of the fish, making it more difficult to distinguish from its background. Good examples are provided by the sculpins, blennies and gobies found in the rock pools on many shores. In some species, the body shape resembles an inedible object. An extreme example of this is the seadragon, *Phyllopteryx eques*, which has been described as "more like a complex mass of seaweed" (Keenleyside, 1979).

Benthic fishes such as the Pleuronectiformes associated with featureless substrates rely on camouflage, cryptic behaviour and in some species a remarkable ability to change the body coloration to match the colour of the substrate (Keenleyside, 1979).

In open waters, structural refuges are not available. Pelagic species such as mackerel and herring are usually countershaded with a silver ventral surface and a bluish back,

which reduces their visibility to predators above or below them. Because of the mirror-like way In which they reflect the downwelling and scattered light, the silvery flanks of many pelagic species also make a normally orientated fish difficult to see from the side (Lythgoe, 1979). Juvenile salmonids become silvery as they become smolts in preparation for the downstream migration to the sea (Chapter 5). The smolts lose the blotchy-brown coloration characteristic of their stream phase. In deeper, pelagic waters some fishes have light-producing organs, photophores, along the abdomen and lower flanks. These may match the fish to the faint, downwelling light and so make it difficult for a predator to see them from below (Lythgoe, 1979; Williams, 1996).

Secondary defences

These include morphological characteristics that provide some protection from predators. The spines of sticklebacks make it difficult for predators such as small pike to swallow them (Hoogland *et al.*, 1957; Reimchen, 1994). When presented with sticklebacks, together with spineless species such as minnows, small pike prey preferentially on the minnows. Some fish defend themselves with poisonous spines formed by spiny fin rays that have become associated with a venom gland, for example the weevers (Trachinidae) of the sandy, marine littoral, or the beautiful stonefish, *Pterois*, of coral reefs. Other fish including the porcupinefish (Diodontidae) and puffers (Tetraodontidae) can swell up by swallowing water so that their body size is greatly increased and the sharp spines on their body stand erect.

A behavioural secondary defence is flight into shelter. Stands of vegetation, crevices and caves in rocky or coral reefs, and undercut river banks all provide refuges from predators. Species differ in the their use of types of shelter. When juvenile roach and perch were exposed to piscivorous perch and pike, the juvenile perch used both stands of vegetation (artificial) and plastic pipes on the bottom as refuges. The juvenile roach used only the vegetation (Eklov and Persson, 1996). The effectiveness of refuges is also dependent on the species of prey and predator involved in the encounter. In small outdoor pools, largemouth bass captured small bluegill sunfish at a lower rate as the density of vegetation increased. Density of vegetation had no effect on the rate at which bass captured fathead minnows. When the predator was a pike, vegetation density had no effect on the capture rate for either prey species, with the minnows caught at a faster rate than the bluegills (Savino and Stein, 1989).

Flatfish, living on a level substrate, which has few hiding places, bury themselves in the substrate by rocking and settling movements, further protected by their cryptic coloration. Some fishes dive into soft silts (Keenleyside, 1979).

The availability of numerous shelters into which prey can rapidly retreat at times of danger may reduce the importance of cryptic coloration. Many of the most brightly coloured fishes are found associated with coral or rocky reefs.

Ontogenetic effects on response to predators

Within the lifetime of a fish, experience may play an important role in the behavioural response to a potential predator. Guppies that had experience of being chased by conspecifics as juveniles had a greater likelihood of surviving an attack by a predatory fish than inexperienced guppies (Goodey and Liley, 1986). Threespine sticklebacks that had prior experience of a stalking pike subsequently reacted to the presence of a pike at a greater distance, were more likely to retreat into weed and were less likely to approach the pike than naive sticklebacks. In some populations of the threespine stickleback, the development of an effective response to a

predator depended on the interaction between young and their father at the time when the young fish were starting to leave the nest (Huntingford *et al.*, 1994). At this time, a male chases any straying young, returning them to the nest in his mouth.

In some populations where predatory fish are present, crucian carp develop relatively deep bodies compared with carp in predator-free populations. This deeper body shape may lower the risk from gape-limited predators such as pike or perch (Bronmark and Miner, 1992). Well-fed carp also develop deep bodies. Lower population densities in the presence of predators may mean high food levels for the surviving carp. The latter may also reduce their level of activity in the presence of predators, potentially allowing more energy to be diverted into growth (Chapter 6) (Holopainen *et al.*, 1997).

Shoaling and predation

Shoaling can also be an effective defence against predation for fishes living in open waters. Experimental studies show that shoaling can reduce the attack success of a predator (Fig. 8.5). Shoaling behaviour could reduce predation by several mechanisms. Some theoretical studies argue that shoaling reduces the chance of a predator detecting the prey, because the predator has to search through a greater prey-free volume than if prey are scattered. However, the probability of detection for a given individual remains the same, whether it is isolated or in a shoal (Pitcher and Parrish, 1993). Estimates of shoal sizes of roach and pike densities in a river in the Midlands of England suggested that the shoals were never out of range of predatory pike (Pitcher, 1980). For a visually hunting predator, a shoal may be more visible than individual prey simply because of the size of the shoal. However, when in a shoal, an individual has a lower probability of being the fish attacked. An individual will gain an advantage by joining a shoal if the

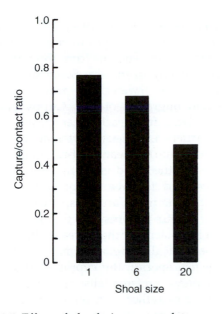

Fig. 8.5 Effect of shoal size on predatory success measured by the capture/contact ratio of perch, *Perca fluviatilis*, attacking guppies, *Poecilia reticulata*. Redrawn from Neill and Cullen (1974).

combined effects of the lowered risk of being attacked when in the shoal, together with any increased possibility of the shoal being detected, mean that the individual is less likely to be eaten than if it stays on its own. This aggregate effect of detection and attack dilution is called attack abatement (Pitcher and Parrish, 1993). Shoaling can also make it more likely that an approaching predator will be detected because there are more fish scanning the environment (Magurran *et al.*, 1985). Once the predator has detected a shoal, the presence of a number of potential prey may generate a confusion effect: the predator is unable to concentrate on the pursuit of any one fish (Milinski, 1984a). As a member of a shoal, an individual can take advantage of cooperative escape tactics performed by the group (Pitcher and Parrish, 1993). The presence of a potential predator often increases the cohesion of a shoal and results in an increase in the degree of syn-

chronization of the behaviour of the constituent individuals.

When a predator is detected, some fish may leave the shoal, approach the predator, then turn and return to the shoal. This behaviour, predator inspection, probably allows the inspecting fish to gain information about the risk presented by the predator, but at the cost of the increased danger of being attacked (Pitcher, 1992; Pitcher and Parrish, 1993). Fish that remain in the shoal also seem to obtain information about the predator from the behaviour of the inspecting fish. The conditions under which fish should run the risk of approaching a predator have been analysed in the framework of game theory. The analysis considers whether a fish should continue an inspection or retreat to the shoal in the context of what other fish are doing (Milinski, 1996).

In many species of the superorder Ostariophysi, which includes the Cypriniformes, fish in a shoal show a distinctive alarm reaction if a conspecific is injured (Smith, 1992). This reaction is triggered by a chemical, the alarm substance or *schreckstoff*, released from the damaged skin of the injured fish. Although most studies on alarm substances have used ostariophysans, other taxa, including the gobies (Gobiidae), may include species that produce alarm substances. The function of these substances in natural populations is still uncertain, because most descriptions of their effects come from laboratory studies.

In some circumstances, shoals may form not as a defence against predation but to enhance it. Shoaling allows a group of fish to overcome the defence of a resource by territorial behaviour that would be effective against one or a few intruders (Robertson *et al.*, 1976). The neotropical Pacific wrasse, *Thalassoma lucasanum*, forms shoals to exploit the egg masses guarded by parental Pacific sergeant major, *Abudefduf troschelii* (Foster, 1987). These shoals are formed only when the eggs are present, which suggests that the primary benefit gained from joining a shoal is that it facilitates the exploitation of this resource. Some piscivorous fish hunt in shoals, with the behavioural tactics of the hunters countering those of their shoaling prey. The jack, *Caranx ignobilis*, will form shoals to hunt the anchovy, *Stolephorus purpureus* (Major, 1978). Shoals of jacks break up the anchovy shoals, isolating individuals or small groups of prey which are then more easily captured before they rejoin their companions.

Other aggregations may act as secondary defences against predation. Some species, such as the Pacific herring, spawn in shoals, releasing all the eggs over a short interval of time. Such synchronous releases of eggs may swamp the capacity of predators to eat them, improving the chances of survival of individual eggs and larvae. The breeding colonies of the bluegill sunfish described in Chapter 7 may also have an antipredator function (Dominey, 1981).

Evolutionary responses to predation

The effects of predation on populations of guppies and on closely related species of sticklebacks have provided some of the best evidence in vertebrates of the process of natural selection.

Guppies live in small streams on the islands of Trinadad and Tobago and in north-eastern Venezuela. There is sexual dimorphism. Females have indeterminate growth, while the smaller males, unusually for fishes, have a determinate growth pattern. Females have a simple colour pattern, but the males have a complex colour polymorphism which involves interference colours in addition to pigments such as carotenoids (Endler, 1980, 1983).

Pioneering studies by Liley and his coworkers stimulated an extensive research programme on the influence of predation on the life-history traits, colour and behaviour of the guppy (Seghers, 1974b,c; Liley and

Seghers, 1975; Endler, 1980, 1983; Reznick and Endler, 1982; Magurran *et al.*, 1995). On Trinidad, many streams run through deep valleys directly to the sea. Some streams are broken up by waterfalls, which bar the upstream migration of fish. Guppy populations isolated from each other by these physical barriers experience different intensities of predation. In clear, fast-running waters typical of the head streams and springs, guppies tend to have a large body size at maturity. The males are conspicuous because of their colour patterns and the fish disperse themselves across the stream bed. Such streams typically contain only a single predatory fish species, the cyprinodont *Rivulus*, which preys preferentially on small, immature guppies rather than on adults. In lower sections of the streams, the guppies are smaller at maturity, males are less brightly coloured and the fish show a more vigorous avoidance response to potential predators. The guppies are frequently found shoaling along the edge of the stream. Predatory fish can include the cichlid *Crenicichla* and the characin *Hoplias*. These predators can reach a larger body size than *Rivulus* and can prey on adult guppies.

Careful mark–recapture experiments showed that mortality rates of guppies were higher in stream pools that contained *Crenicichla* than in pools holding *Rivulus* (Reznick *et al.*, 1996b). Although the causes of the mortality were not observed directly, predation was probably an important agent. The higher mortalities in *Crenicichla* pools were shown by all size classes of guppies and not just concentrated on the larger fish (Chapter 11).

Male guppies that are conspicuous against their background are more successful in courtship (Chapter 7). The size of the spots based on carotenoid pigments may also indicate males that are good foragers. But brightly coloured males are probably also more conspicuous to visually hunting predators such as *Crenicichla*. The mark–recapture study found that males had poorer survival than females or juvenile males (Reznick *et al.*, 1996b). In an experiment, guppies derived originally from many populations were exposed to different predation intensities (Endler, 1980, 1983). The fish were scored for body coloration after three and nine generations. Guppies from populations exposed to higher predation intensities were less conspicuous because of a loss of bright colours, a drop in the size of patches of colour and a decline in the number of spots (Fig. 8.6). In a field experiment, guppies were transferred from a site of presumed high predation pressure to a low-predation site. When sampled about 15 generations later, the guppies were more brightly coloured,

indicating an evolutionary response to the relaxation in predatory pressures. The experiments also showed that some life-history traits changed (Chapter, 11). Each guppy population has evolved male coloration patterns and behaviour that represent a compromise between the need to attract receptive females for copulation and the need to minimize the risk of predation. The males in a population do not have identical patterns, which suggests that there can be many individual ways of achieving an adaptive level of conspicuousness.

In the stickleback family (Gasterosteidae), the ecology of different species reflects the effectiveness of their morphological and behavioural secondary defences (Bell, 1984; Wootton, 1984a). The ninespine stickleback has shorter spines than the threespine stickleback and this makes it more vulnerable to predation (Hoogland *et al.*, 1957). However, the ninespine stickleback is more likely to enter weed beds and spends longer in them than the threespine stickleback. Threespine sticklebacks build their nests on the substrate, and though they do take advantage of available cover, will build their nests in relatively open situations. In contrast, in most populations of ninespine sticklebacks, the nest is built in vegetation and the species seems to prefer habitats characterized by well-developed stands of vegetation (Wootton, 1976, 1984a). In Swedish lakes, the ninespine stickleback is typically confined to vegetation along the shore (Svardson, 1976). After one lake had been treated with rotenone, which killed off the piscivorous species, the abundance of sticklebacks increased dramatically, with fish milling around in the pelagic zone of the lake away from the vegetated zones.

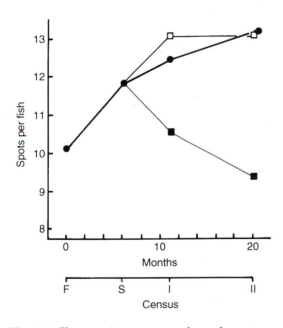

Fig. 8.6 Changes in mean number of spots per fish on male guppies, *Poecilia reticulata*, during experiment on effect of predation. Symbols: ● no predation; ○, weak predation; ■ strong predation; F, foundation population (no predation); S, start of experiment when predators added; I and II, first and second censuses, respectively 3 and 9 generations from start. Redrawn from Endler (1980).

8.3 PATHOGENS

Pathogens can be classified as micro- or macroparasites (Anderson, 1981; Dobson and May, 1987). The former include viruses, bacteria, fungi and protozoans. They are characterized by their small size, short generation times and their ability to multiply within an infected host. They are often transmitted directly and so fishes that live in dense shoals may be highly susceptible to these pathogens. Macroparasites of fishes include platyhelminths, nematodes, acanthocephalans, crustaceans and agnathans. Some of these have complex life histories in which the fish is an intermediate host and the parasite becomes sexually mature only when the infected fish is eaten by the final definitive host such as another fish, a bird or a mammal. An example is the cestode parasite of sticklebacks, *Schistocephalus solidus*, which is frequently used as a model in studies of parasite–host interactions. Sticklebacks become infested when they eat copepods, which are infested with the procercoid stage of the parasite. The parasite migrates to the body cavity of the stickleback where it grows and can cause gross distension of the abdomen. For *S. solidus* to complete its life cycle, the stickleback must be eaten by a bird, the final host in which the cestode rapidly becomes sexually mature. The cestode's eggs are defecated into water by the bird, where they are ingested by copepods. For other macroparasites, the fish may itself be the definitive host.

Endoparasites are found in a range of sites in the bodies of fish (Table 8.3). Some macroparasites are ectoparasites, attaching themselves to the surface of the fish. Species of the monogenean fluke, *Gyrodactylus*, are frequently attached to the gill filaments or fins. At the other end of the size scale, lampreys attach themselves to the body of fishes, rasping coin-sized holes in the flesh with the horny teeth of their suctorial mouths.

Table 8.3 Locations of parasites infesting herring, *Clupea harengus**

Site	Number of species at site	% Species at site
Skin	8	6.1
Fins	5	3.8
Gills	20	15.3
Operculum	1	0.8
Eyes	2	1.5
Eye lens	1	0.8
Nasal cavities	3	2.3
Mouth	2	1.5
Buccal cavity	2	1.5
Pharynx	1	0.8
Stomach	17	13.0
Pyloric caeca	6	4.6
Intestine	32	24.4
Liver	4	3.1
Gall bladder	5	3.8
Mesenteries	1	0.8
Visceral cavity	10	7.6
Kidney	2	1.5
Testes	2	1.5
Heart	1	0.8
Swim bladder	1	0.8
Musculature	5	3.8

*Source: MacKenzie (1987).

Fishes are hosts to an impressive number of protozoans and macroparasites. Over 80 species are recorded from the herring (MacKenzie, 1987) and similar numbers from the threespine stickleback (Wootton, 1976) and perch (Craig, 1987). Any individual fish carries only a subset of the parasites that infest that species. The effects of infection or infestation range from being undetectable to causing the death of the host.

An important feature of many macroparasite species is their aggregated distribution within a fish population. Most fish carry no or only a few parasites, but a few fish are heavily infested (Fig. 8.7). This type of distribution can usually be described by a negative binomial distribution (Anderson, 1981). An aggregated distribution implies that the effects of the parasite are not spread approximately equally across all members of the fish population, but that a small proportion of individuals in the population experience the consequences of heavy infestations.

Pathogens as cause of death

There are occasions when the mortality rates in a population are so high that the effects of the pathogen are observed. In 1976, disease killed over 98% of the perch population in Windermere (UK) (Craig, 1987). The pathogen was tentatively identified as an *Aeromonas* bacterium. An infestation by the ciliate, *Ichthyophthirius multifiliis*, which causes white-spot or ich, almost completely destroyed the 0+ age class in a population of threespine stickleback in a Scottish loch (Hopkins, 1959). In 1954–55, a disease caused by the fungus, *Ichthyophonus hoferi*, killed about 50% of the herring population in the Gulf of St Lawrence. There have been six recorded outbreaks of this disease in herring populations in the western North Atlantic since 1896 (Sindermann, 1966). A model of an outbreak in herring in the North Sea in 1991, suggested that herring infested with

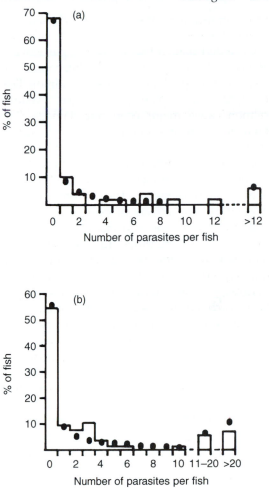

Fig. 8.7 Frequency distribution of parasites per fish for two species parasitic on threespine stickleback, *Gasterosteus aculeatus*. (a) *Echinorhynchus clavula* (Acanthocephala); (b) *Diplostomum gasterostei* (Trematoda). Filled circles (●) denote frequency predicted by negative binomial distribution. Redrawn from Pennycuick (1971).

the fungus suffered an additional mortality rate more than ten times higher than uninfested fish (Patterson, 1996).

Along the Atlantic coast of Norway, the monogenean *Gyrodactylus salaris* has inflicted massive mortality on Atlantic salmon parr, although salmon from the Baltic are relatively immune to the parasite (Johnsen and Jensen, 1991).

Crowded fish in fish farms are particularly at risk from pathological infections. In western Europe, infestations by salmon lice, *Lepeophtheirus salmonis*, of caged Atlantic salmon cause major economic damage. The high densities of salmon in cages also act as reservoir for the lice from which natural populations of salmonids can be infested.

These are spectacular examples of the effects of pathogens. The chronic risk of death from the effects of pathogens is difficult to assess for individual fish. Lester (1984) reviews methods for estimating the mortality rate due to parasitism in natural populations. They include analyses of the frequency distribution of parasites within hosts, detailed autopsies and determining the frequency of infestations known to be eventually lethal. No single method currently available provides a definitive estimate. A model of the potential impact of mortality caused by pathogens on a representative Atlantic salmon population in Scotland predicted that the effect of pathogens on the juvenile fish in fresh water (parr stage) had the greatest detrimental effect on population size (Des Clers, 1993).

Sublethal effects of pathogens

Even if the pathogen does not directly cause the death of its host, it may have detrimental effects that reduce the host's lifetime reproductive output. A variety of qualitative effects are listed by Sindermann (1987). The *Schistocephalus solidus* – stickleback model provides much quantitative evidence of sublethal effects.

Effect on energy allocation

Pathogens can affect the pattern of energy allocation in several ways. A fish that mounts a defensive response against the pathogen may increase its maintenance rate of energy expenditure. Large parasites can divert a large portion of their host's energy income to themselves. *S. solidus* grows rapidly and can reach a large body size in the body cavity of its stickleback host, making heavy demands on the supply of energy and materials of the host. Sticklebacks infested with *S. solidus* had a higher respiration rate than uninfested fish (Meakins and Walkey, 1975). As the oxygen content of water declined, parasitized sticklebacks resorted to aquatic surface respiration at higher oxygen tensions than uninfested fish (Giles, 1987). This requirement for higher oxygen concentrations by the parasitized sticklebacks may force them into the more oxygen-rich surface waters where they will be at more risk from bird predators (Lester, 1971; Giles, 1987).

Infested sticklebacks also showed evidence of being hungrier than sticklebacks not carrying the cestode (Milinski, 1985). This increased hunger of parasitized sticklebacks made them more prone to forage in areas where they were at risk from predators (Milinski, 1985). However, infested sticklebacks were poor competitors for food when in the presence of uninfested fish (Milinski, 1985).

The presence of *S. solidus* also affects the growth and reproduction of the stickleback. Parasitized fish had a similar growth efficiency to unparasitized fish, but this reflected the growth efficiency of the parasite. The growth efficiency of the parasitized fish was only about half that of the unparasitized fish when the growth of the parasite was excluded (Walkey and Meakins, 1970). The presence of the cestode inhibited the development of ripe eggs by an infested female, probably because the energy demands of the parasite prevented the female from allocating sufficient resources to the ovaries (Meakins, 1974). In a Scottish population of sticklebacks in which infestation took place in the first autumn of life, few parasitized fish reached sexual maturity in the following spring (Tierney *et al.*, 1996).

Another larval cestode, *Ligula*, which parasitizes cyprinid fishes, may inhibit the sexual maturation of the fish by interfering with the hormonal control of reproduction (Kerr, 1948). A protozoan parasite, *Eimeria sardinae*, causes parasitic castration of sardines, *Sardina pilchardus*, if they are abundant in the testes of the fish (Sindermann, 1987).

Sensory impairment

Some parasites lodge in sensory organs. Larval digeneans including species of *Diplostomum*, occur in the eye lens or retina. *D. spathaceum* lives in the eye lens of cyprinids including dace. For dace, the maximum reactive distance to prey was inversely related to the numbers of the parasite in the eyes of the fish (Crowden and Broom, 1980). The foraging success of the dace was also adversely affected by the presence of the parasites.

Other sublethal effects

Some parasites cause morphological deformations that may impede the locomotion of the fish. *S. solidus* can cause massive distension of the abdomen of sticklebacks. The protozoan microsporidian, *Glugea anomala*, causes unsightly tubercules on the body. Part of the increased rate of respiration of parasitized sticklebacks may represent the increased cost of swimming. Other parasites cause lesions of the musculature, skin erosion, liver and cardiac damage (Sindermann, 1987), and these symptoms may indicate that such parasites are a chronic cause of mortality.

Even if the parasites do not directly cause the death of the fish, they make the fish less able to survive unfavourable environmental conditions. Sticklebacks with *S. solidus* died sooner from starvation than uninfested fish (Walkey and Meakins, 1970), and were less able to tolerate water containing toxic heavy metals such as cadmium (Pascoe and Cram, 1977). Parasitized fish may have poorer overwintering survival than uninfested fish.

There is evidence that infestation by the salmon louse disrupts the typical migration patterns of anadromous salmonids. In anadromous European trout, infestations have been linked to a premature migration back into fresh water from the sea (Chapter 5), with adverse effects on growth (Birkeland, 1996).

Interactions between predation and parasitism

Parasitized or diseased fish may also have an increased risk of predation. Such fish may swim erratically or have conspicuous skin lesions that make them more conspicuous to predators. Sticklebacks with a heavy infestation of *S. solidus* had their swimming impeded and were more likely to swim near the surface where they would be at risk from birds (Lester, 1971; Giles, 1987). Infested sticklebacks also resumed foraging faster than uninfested fish after they had been disturbed, for example by a predatory bird (Giles, 1983, 1987). They approached closer to a large and potentially predatory fish to forage (Milinski, 1985). All these behaviours are likely to increase the risk to the infested stickleback.

Killifish, *Fundulus parvipinnis*, on the California coast can be infested by the trematode *Euhaplochoris californiensis*, which encysts in the brain. The final host of the parasite is a piscivorous bird. Infested fish showed a higher frequency of conspicuous behaviours than uninfested fish. In a field experiment, bird predation on parasitized fish was much higher than on unparasitized fish. Heavily infested fish were up to 30 times more susceptible to predation (Lafferty and Morris, 1996).

The effect of predators on the distribution of fish prey may change the risk of parasitic infestation. When piscivorous pike–perch, *Stizostedion lucioperca*, was introduced to a Norwegian lake, juvenile roach, which had previously lived in the pelagic zone, occupied the vegetated littoral zone. This was presumably because the risk of predation was lower in the vegetation. This habitat change was associated with an epidemic infestation by *Ichthyophthirius multifiliis* of 0+ roach, with a dramatic reduction in the abundance of the roach. The habitat shift of the roach into the littoral zone probably increased their risk of infestation because the parasite has a free-living stage associated with the substrate (Brabrand *et al.*, 1994).

Individual fish that are uninfested or carry only light parasitic burdens gain a twofold advantage. They do not suffer the debilitating effects of the presence of the parasites and they do not have the increased risk from predation.

8.4 SUMMARY AND CONCLUSIONS

1. Predation and parasitism are interactions in which some participants make a net gain in fitness at the expense of other participants, which suffer a net loss.
2. The risk of being attacked by a predator is determined by the functional, numerical and developmental responses characterizing the interaction in relation to the densities of the prey and predators.
3. Predation of conspecifics can take the form of intra- or intercohort cannibalism. Fishes are also exposed to a wide range of heterospecific predators. Invertebrate predators of fish are usually strongly size limited and take only early life-history stages. Vertebrate predators take both young and adult stages, but the risk from

fish predators usually diminishes with an increase in prey size.
4. The risk of predation can vary either directly (compensatory risk) or inversely (depensatory risk) with prey density. Estimates suggest that predation can be responsible for a high proportion of the mortality in a natural fish population.
5. Predation can also have indirect effects on the prey, influencing the time and energy budgets of the latter. Predators can confine their prey to refuges in which the foraging is less profitable. The presence of predators may also cause modifications in the reproductive behaviour of prey.
6. Fishes employ both primary and secondary defences against predation. Defences include cryptic coloration, morphological features such as spines, and behaviour, especially shoaling.
7. Studies on guppies and sticklebacks show that predation can be a potent evolutionary factor leading to changes in coloration, behaviour, morphology and life history traits of the prey populations.
8. Pathogens can be classified as micro- or macroparasites. The latter are also classified as either ecto- or endoparasites. Occasionally extremely high mortality rates occur in fish populations due to pathogens. Sublethal consequences for their fish hosts include effects on energy allocation, reproduction, behaviour and the formation of lesions.
9. Infestation can also increase the host's risk of predation. Some parasites cause the host to be less wary in the presence of a potential predator.

9

BIOTIC INTERACTIONS: II,
COMPETITION AND MUTUALISM

9.1 INTRODUCTION

Competition is an interaction between individuals in which one or more of the participants suffers a net loss of fitness and none shows a net gain compared with values in the absence of the competitive interaction. In terms of Table 8.1, competition is defined as − 0 or − −. The competition is asymmetrical if the loss in fitness suffered by some participants is much greater than that suffered by others. Mutualism is classified as + 0 or + + in Table 8.1. Some or all of the participants in the interaction show a net gain in fitness and none shows a net loss.

Although the quantitative effects of predators and pathogens are still debated by ecologists, that there are effects is not in doubt. Competition has been a more contentious concept: some ecologists have even avoided using it (Andrewartha and Birch, 1954, 1984). Other ecologists argued that competition is frequently the most important interaction both intra- and interspecifically (Giller, 1984). A sceptical analysis of the concept helped to emphasize that the importance of competition must be assessed on the basis of empirical evidence rather than assumed.

9.2 COMPETITION

Exploitation and interference competition

The loss in fitness that defines competition occurs because in the presence of competitors an individual has a reduced access to a resource that it requires to promote its survival, growth or reproduction. Such reduced access may occur because the presence of a competitor further reduces the availability of a resource that is in short supply. The competing individuals need never meet. This is competition by exploitation. A prey consumed by one predator is not available for consumption by a subsequent predator. Access to a resource may also be reduced if the behaviour of competitors interferes with the ability of an individual to acquire that resource. This is interference competition. In this interaction, there need not be an absolute shortage of the resource, but the behaviour of the competitor creates a relative shortage. Territorial behaviour may deny other animals access to a resource within a territory, although within that territory there may be an excess of the resource.

Territorial coral reef fish frequently defend their territories against heterospecific as well as conspecific intruders (Low, 1971; Choat, 1991). The intensity of such defence and the distance at which a defender reacts to an intruder is greater the more the diet of the intruder overlaps with that of the defender (Myrberg and Thresher, 1974; Ebersole, 1977). Dusky damselfish, *Stegastes dorsopunicans*, defend their territories on coral reefs against the blue tang surgeonfish, *Acanthurus coeruleus*. The territories of larger damselfish contained a higher biomass of algae than the territories of smaller damselfish. The larger damselfish were more aggressive and so were more effective at inhibiting the blue tang from cropping the algal mat (Foster, 1985a,b).

Concept of the niche

Closely associated with the concept of competition is the concept of the niche, which is used in attempts to place a population (or sometimes an individual) in its abiotic and biotic environment. It has some resemblance to the concept of the zone of tolerance developed by Fry and described in Chapter 4. Whereas the zone of tolerance defines the ranges of abiotic factors within which an organism will survive, the niche defines the ranges of both abiotic and biotic factors within which the organism or population of organisms can be reproductively successful (Kerr, 1980; Werner, 1980). The analysis of the effects of environmental factors developed by Fry (1971) (Chapter 4) provides a bioenergetic underpinning to the concept of the niche. Niche theory has provided some of the more baroque developments in theoretical ecology over the past 30 years (Roughgarden, 1979; Giller, 1984; Begon *et al.*, 1996). The concept can still provide a useful framework in which to discuss competitive interactions (Elliott, 1994), although its use does tend to focus attention on the effects of competition detected at the population or assemblage levels, rather than its effects on the interacting individuals.

One early use of the term niche was to describe the functional role of an animal. Pike and largemouth bass would be described as occupying the piscivory niche. This usage was considerably modified by Hutchinson (1958), who introduced the concepts of the fundamental and realized niche. The fundamental niche of a species (or population) is defined in terms of the niche dimensions that are relevant. Niche dimensions include abiotic factors such as temperature, oxygen and salinity, and biotic dimensions such as the size spectrum of the prey consumed. The fundamental niche defines the ranges of the relevant environmental factors within which, in the absence of natural enemies, a population will maintain itself by natural recruitment. For a given abiotic dimension, this range is usually less than the zone of tolerance (Chapter 4). The realized niche is a subset of the fundamental niche that is occupied by the population in the presence of natural enemies including predators, pathogens and interspecific competitors (Pianka, 1981; Giller, 1984).

The environment does not provide a defined and limited number of niches which populations then fill. The niche of a population is the outcome of the dynamic interaction between the individuals in that population and the qualities of the environment defined by the niche dimensions. Niche can be distinguished from habitat, because the latter can be defined in terms that are independent of the presence of a particular species or group of species. The habitats in a stream, for example, can be defined by hydraulic and geomorphic characteristics, independently of what species occupy those habitats (Rabeni and Jacobson, 1993).

As with the effects of predation and parasitism, the effects of competition may vary in intensity from mildly deleterious effects on survival, growth or reproduction to the exclusion of one of the competing species from a particular habitat – competitive exclusion. Niche theory predicts that competitive exclusion is more likely, the more the niches of the competing species overlap (Giller, 1984).

Quantitative descriptions of the properties of niches have been developed, the most important of which are niche width (or breadth) and niche overlap (or similarity) (Table 9.1) (Levins, 1968; Hurlbert, 1978; Feinsinger *et al.*, 1981). Niche width is a measure of the extent to which a niche dimension is utilized by an individual or population (Fig. 9.1). Niche overlap measures the degree to which two individuals (or more commonly two populations) utilize the

Table 9.1 Some measures of niche width (breadth) and niche overlap (similarity)*

Let: x_i = number of individuals in population X that use resource state i
X = total abundance of population X, i.e. $X = \Sigma x_i$
y_i = number of individuals in population Y that use resource state i
Y = total abundance of population Y, i.e. $Y = \Sigma y_i$
p_{xi} = proportion of population X that utilizes resource state i, i.e. $p_{xi} = x_i / X$
p_{yi} = proportion of population Y that utilizes resource state i, i.e. $p_{yi} = y_i / Y$
a_i = number of items available in resource state i
A = total number of items available, i.e. $A = \Sigma a_i$
q_I = proportion of resources available that consists of items in resource state i,
     i.e. $q_i = a_i / A$
R = number of resource states available

Measures of niche width (breadth) include:

1.

$$B = 1 / \sum_i p_{xi}^2$$

2.

$$B_n = 1 / \left(R \sum_i p_{xi}^2 \right)$$

3.

$$B' = X^2 / \left[A \sum_i (x_i^2 / a_i) \right] = 1 / \sum_i (p_i^2 / q_i)$$

4.

$$PS = 1 - 0.5 \sum_i \left| p_{xi} - q_{xi} \right|$$

Measures of niche overlap (similarity) include:

1.

$$C_{xy} = 1 - 0.5 \sum_i \left| p_{xi} - p_{yi} \right|$$

2.

$$L = (A / XY) \sum_i (x_i y_i / a_i)$$

3.

$$C_m = 2 \sum_i (p_{xi} p_{yi}) / \left[\sum_i p_{xi}^2 + \sum_i p_{yi}^2 \right]$$

same range of a niche dimension (Fig. 9.1). The greater the niche overlap or similarity between populations, the greater the possibility of contemporary, competitive interactions between individuals from those populations. However, a narrow overlap may indicate that there has been a competitive interaction in the past, and the evolutionary responses of the populations have minimized that competition by a reduction in overlap. These quantitative measures of niche properties do not provide evidence of the mechanisms or

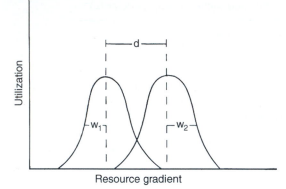

Fig. 9.1 Niches of two species represented along a single niche dimension, approximated by the resource utilization curves for two species along a resource gradient (niche dimension). Possible gradients include food particle size (Chapter 3) or temperature (Chapter 4). Symbols: w_1 and w_2, niche widths for species 1 and 2 measured as standard deviations of resource utilization curve; d, measured difference between niches; niche overlap is shown by region of gradient shared by both curves.

of the effects of competitive interactions between individuals; they merely provide preliminary evidence for the occurrence of such interactions.

9.3 INTRASPECIFIC COMPETITION

Intra- and intercohort competition

Individuals within the same population are likely to be most similar in their resource requirements and so are potentially intense competitors. Two forms of intraspecific competition are recognized (Hassell, 1976; Łomnicki, 1988). In scramble competition, the share of resources taken by an individual depends on the amount available and the population density, irrespective of the rank of the individual in the population. In contest competition, the resource is unequally divided because the share taken by high-ranking individuals is independent

of the abundance of low-ranking individuals. The depensatory growth shown in many groups of fish kept in confined volumes (Chapter 6) may indicate the existence of contest competition. Territorial behaviour is a form of contest competition if some fish that are physiologically capable of taking up territories are prevented from doing so by other territorial fish (Chapters 5 and 10).

Both intra- and intercohort competition may occur within populations. Evidence for intracohort competition came from a field experiment with brown trout fry in a small stream in the English Lake District (Backiel and LeCren, 1978). Screened sections of the stream were stocked with brown trout fry at a range of initial densities. The mean specific growth rate of the trout declined with an increase in stocking density, although there was a threshold density above which there was no further decrease in growth rate. The per capita mortality rate increased with an increase in stocking density for densities greater than 10 m^{-2}. Studies of a natural population of migratory trout in a Lakeland stream also observed density-dependent mortality in the fry, although not density-dependent growth (Chapter 10) (Elliott, 1985a, 1994). After emerging from the gravel, trout fry take up feeding territories and the density-dependent mortality may be of fish that are unsuccessful in obtaining territories in the contest for suitable areas of the stream bed (Chapter 10). The experiments with the trout showed that because of the density-dependent growth and mortality, there was a maximum rate of production that was independent of initial stocking density for densities greater than 12 m^{-2}. Experiments in which ponds were stocked with common carp also suggested that there is an optimum rate of stocking which maximizes the production of fish in the pond. If rates of stocking are too high, the density-dependent effects on growth and mortality result in reduced production (Backiel and LeCren, 1978).

Density-dependent growth in the first year occurs in several populations of marine fishes including herring, cod and haddock, *Melanogrammus aeglefinus* (Shepherd and Cushing, 1980). This suggests intracohort competition for food, although interspecific competition may also play a part. The intensity of intracohort competition during the early life-history stages may depend on the extent to which the 0-group fish congregate in nursery areas (Beverton, 1984).

Fish from different age classes (cohorts) in a population usually differ in size (Chapter 6). If the age classes share the same habitat and their diets overlap, then there is the possibility of intercohort competition for food or perhaps other resources. In behavioural interactions, larger fish usually dominate smaller fish, which suggests that the older, larger fish will have a competitive advantage over younger fish. But younger fish have a smaller absolute requirement for food (Chapter 3), so they may have a competitive advantage at low levels of food abundance.

The feeding ecology of perch in their third (2+) and fourth (3+) year of life in a small, shallow, eutrophic Swedish lake provided evidence that intercohort competition can be significant (Persson, 1983a). Fish in the population were stunted, suggesting that food was a limiting resource. In both age classes, the daily rate of food consumption increased to a peak in late summer, then declined over the autumn. The composition of the diet was similar for both age classes and the niche breadth for the food dimension showed similar seasonal changes. The niche overlap between the two classes was high for diet. After allowance was made for the differences in the rate of food consumption between the two age classes, the diet of the 3+ fish overlapped more with that of the 2+ fish than the diet of the 2+ fish overlapped with that of the 3+ fish. Persson (1983a) suggested that the two age classes were unable to avoid intercohort competition by using different habitats because the lake contained few dif-

ferent habitats. Annual mortality in the two age classes was high, 70% for 2+ and 45% for 3+ fish. This suggests that intense intercohort competition could lead to populations in which there were few age classes. Intraspecific competition would then be an important factor in the evolution of life-history traits (Chapter 11).

Intercohort competition in which individuals in the younger age class are at a competitive advantage was used to explain a 2 year cycle observed in the abundance of the planktivorous vendace, *Coregonus albula*, in Lake Bolmen, Sweden (Hamrin and Persson, 1986). The hypothesis was that an abundant 0+ year class depleted food during the summer, so individual 0+ fish were small in the autumn. Adults in the population would also show little growth and would have a low reproductive output when they spawned in November because of the depletion by the abundant 0+ class. In the following summer, the small fish from the previous year would now grow quickly and experience little competition because the new 0+ cohort was numerically small. After a summer of fast growth, they would have a high reproductive output, generating another abundant year class. Intercohort competition was also implicated in an apparent two-year cycle of abundance in a population of roach in a shallow eutrophic lake in eastern England (Townsend, 1989).

Effect of intraspecific competition on niche characteristics

Individuals in a population could avoid competition with conspecifics by minimizing the extent to which they exploit the same limiting resources. This could be achieved either by individuals being generalists that can rapidly switch to exploit those resources that are abundant or by being specialists on a restricted range of resources (see also Chapter 3). The former solution adopted in a population leads to a high within-phenotype component to the niche width, the latter solution to a high between-phenotype component (Fig. 9.2) (Giller, 1984). The distinction is between populations composed of behaviourally flexible individuals and populations composed of individuals that are relatively inflexible in their behavioural responses to the ecological conditions, but which are specialized at responding to a restricted range of conditions. It is frequently suggested that species from apparently abiotically stable, benign environments, such as the African Great Lakes cichlids or species associated with coral reefs, tend to have a high between-phenotype component. Species from harsher and more unpredictable environments such as temperate fresh waters are expected to have high within-phenotype component (Giller, 1984). The observation that some African Great Lakes cichlids do switch to exploiting locally abundant foods, even when apparently specialized to exploit other food categories (Chapter 3), shows that the characterization of a high between-phenotype component has to be made with care.

A component of phenotypic plasticity can be a change in behaviour of individual fish as a consequence of competitive interactions. When threespine sticklebacks were fed *Daphnia* of different sizes, successful competitors took a higher proportion of the larger *Daphnia* than less-successful fish (Milinski, 1982). Subsequently, when tested on their own, the less-successful fish continued to take proportionately more of the smaller *Daphnia*, suggesting that a preference acquired under competition was subsequently retained. Subordinate brown trout learned to pick artificial prey off the bottom of an aquarium, a behaviour never performed by the dominant fish (Nilsson, 1978).

Niche theory predicts that intraspecific competition can lead to the evolution of a wider niche as the tendency to avoid the competition leads to the exploitation of wider and wider ranges along some of the niche dimensions. Where the pumpkinseed sunfish coexists in lakes with the bluegill sunfish, the adult pumpkinseed typically feed on molluscs and other hard-bodied prey in the littoral, with adult bluegills feeding on zooplankton and soft-bodied invertebrates (Chapter 3). In some lakes in the Adirondack region of north-eastern USA, only pumpkinseeds occur. Two morphological forms of the pumpkinseed occur in some of these lakes (Robinson *et al.*, 1993, 1996). The littoral form retains the typical pumpkinseed diet, while a pelagic form feeds on zooplankton. The morphological differences between the forms are

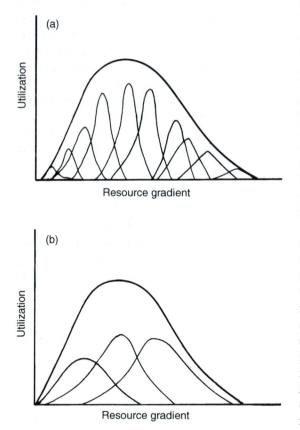

Fig. 9.2 Resource utilization curves for populations with (a) high between-phenotype niche width (specialists) and (b) high within-phenotype niche width (generalists). Outer curve represents sum of individual resource utilization curves. Redrawn from Giller (1984).

subtle, but there is evidence that the differences are biologically significant (Robinson *et al.*, 1996). Fish with extreme 'littoral' or 'pelagic' body forms had, on average, higher lipid contents and growth rates than fish with an 'intermediate' morphology.

The coexistence of littoral and benthic forms within a single lake has also been described for threespine sticklebacks and for Arctic charr and other salmonids (Robinson and Wilson, 1994; Smith and Skulason, 1996) (page 210).

Intraspecific competition in fish populations can result in poor growth of individuals (Persson, 1983b; Hanson and Leggett, 1985). This effect on growth may reduce the number of size classes present in the population and so the range of prey that can be exploited by that population. A consequence of the flexible, indeterminate growth pattern characteristic of many fishes is that intraspecific competition can lead to a reduction of the niche breadth for a population rather than an increase.

9.4 INTERSPECIFIC COMPETITION

Evidence for interspecific competition

Three important lines of evidence have been used to assess the ecological effects of interspecific competition between fish. The first is from the patterns of resource use by species when they are living in the same area, that is when they are sympatric, and when they are living in different areas and so are allopatric. Competition between them is possible only when they are sympatric. Differences between their patterns of resource usage in sympatry and allopatry may be caused by competition (Fig. 9.3). Two basic patterns are possible – selective segregation and interactive segregation. The two species may be adapted to use different resources so their pattern of resource usage is not significantly affected by the presence or absence of the putative competitor. Nilsson (1967) calls this

selective segregation. Such selective segregation could have evolved at some time in the past in response to competition, but whatever its origin, it minimizes contemporary competition as a significant interaction. Evidence that interspecific competition is an important, contemporary interaction comes from an observation that the use of resources differs when the species are sympatric although it is similar when they allopatric. Such interactive segregation may be caused by exploitation or interference competition (Nilsson, 1967).

Secondly, evidence for competition also comes from changes in the abundance of a species when another species invades or is introduced into its range. In this case, any effects of predation have to be carefully disentangled from the effects of competition. Thirdly, experimental studies provide evidence of interspecific competition.

Important analyses of interspecific competition come from studies of salmonids in streams and lakes, lake-dwelling centrarchids and the interactions between perch and roach in lakes.

Evidence of interspecific competition in stream-dwelling salmonids

Many streams in the temperate and subpolar Northern Hemisphere carry populations of two or more salmonid species. These species are characteristic of cool, well-oxygenated, running water. Morphologically and ecologically they are similar, particularly in the juvenile period before the anadromous forms make their seaward migration (Chapter 5). Because of their importance both to commercial and to recreational fisheries, species have been carried far outside their natural geographical distribution and introduced into streams alongside the indigenous salmonids. Streams on the eastern seaboard of North America may now contain brown trout from Europe, rainbow trout from north-western America, together with the indigenous non-migratory brook trout and anadromous

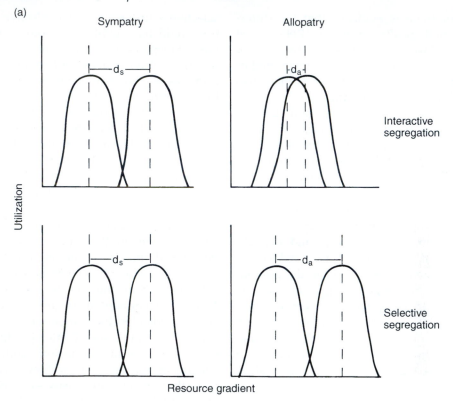

Fig. 9.3 (a) Interactive and selective segregation illustrated by resource utilization curves of species when sympatric and allopatric. (b) (Opposite) Diets of brown trout, *Salmo trutta*, and Arctic charr, *Salvelinus alpinus*, in Scandinavian lakes when sympatric (black bars) and allopatric (open bars), illustrating increase in difference along a diet dimension when species are sympatric. Diet categories: A, fish; B, small crustaceans; C, large crustaceans and molluscs; D, insect larvae; E, terrestrial insects; F, miscellaneous. Redrawn from Giller (1984).

Atlantic salmon. Competitive interactions between the introduced and resident salmonids will influence the success and consequences of the stocking programme for both the stocked and indigenous species (Hearn, 1987).

Field observations and laboratory experiments have suggested that both selective and interactive segregation play a part in the coexistence of juvenile salmonids, with the relative importance of the two depending on the species present and perhaps the physical environment.

Interactive segregation was demonstrated in an exemplary field and experimental study (Hartman, 1965). In many coastal streams of British Columbia, anadromous coho salmon and steelhead trout coexist as juveniles. Although these species live in the same habitats, their distribution differs between microhabitats. In spring and summer, when population densities are high, coho are more frequent in pools whereas steelhead occur more in riffles. In winter, densities are low and both species are mainly in the pools. The microhabitat preferences of

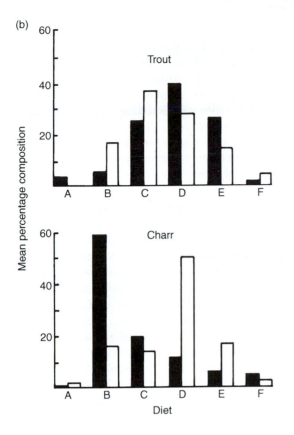

the two species were studied using an experimental stream, in which juveniles were presented with gradients in light, depth and cover or with a sequence of pools and riffles. The preferences of the two species were most similar in spring and summer. Yet this is when the species are most segregated in the river. Both species directed their aggressive behaviour intra- and interspecifically, but the level of aggression was high in spring and summer and low in winter. Steelhead juveniles defended areas in the riffles but not in the pools, whereas the coho were aggressive in the pools but less aggressive in the riffles. The mechanism for the interactive segregation is interference competition for space in different microhabitats in the stream.

A field experiment suggested that introduced brown trout aggressively excluded indigenous brook trout from preferred resting places in a Michigan stream, when such resting places were in short supply (Fausch and White, 1981).

Both selective and interactive segregation were involved in determining the distribution of juvenile Atlantic salmon and rainbow trout in a stream in Vermont (Hearn and Kynard, 1986). In autumn in the field, the salmon lived principally in riffles, and the trout in pools. In an experimental stream, 0+ and 1+ trout preferred pools whether salmon were present or absent. Salmon were distributed more evenly between pools, riffles and an intermediate habitat. Although the 0+ salmon were unaffected by the presence of trout, the 1+ salmon made more use of riffles in the presence of trout.

Studies of other, coexisting pairs of salmonids also suggested that segregation in the stream can depend on the different microhabitat preferences of the species. Examples include: chinook salmon and coho salmon (Lister and Genoe, 1970), chinook salmon and steelhead trout (Everest and Chapman, 1972), brook trout and cutthroat trout, *O. clarki* (Griffith, 1972). In some cases, the species have different times of emergence from the gravel, so they differ in mean size, and choice of habitat is related to size. Chinook fry emerge from the gravel a month earlier than coho, are larger on emergence and grow faster. At any given time, the chinook prefer locations in the stream with a higher current velocity than do coho (Lister and Genoe, 1970). These differences in the microhabitat preferences of species probably do not apply to all populations, so a pair of species in one stream that are showing selective segregation may in another stream display interactive segregation (Hearn, 1987).

Differences in microhabitat preference probably relate to the bioenergetics of life in moving water, particularly the balance between the cost of holding station and the

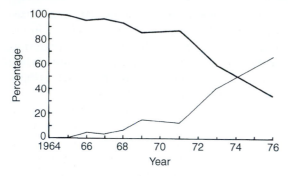

Fig. 9.4 Composition of catch in gill nets in Lake Ovne, showing effect of introduction of whitefish (thin line) on charr (thick line). Redrawn from Svardson (1976).

benefits of being close to a current which is carrying drifting prey (see also Chapter 5). Both intra- and interspecific competition in stream-dwelling salmonids are primarily for positions in the stream which provide cover, a hydrodynamic advantage and access to food (Hearn, 1987).

Evidence of interspecific competition in lake-dwelling salmonids

Pairs of salmonid species in lakes in British Columbia have provided evidence for interspecific competition from the comparison of allopatric and sympatric populations. Cutthroat trout and dolly varden, *Salvelinus malma*, occur both allopatrically and sympatrically in these lakes (Andrusak and Northcote, 1971). When allopatric, adult cutthroat trout eat a wide variety of prey from most habitats in the lakes, including chironomid pupae, zooplankton and benthic invertebrates. Allopatric populations of dolly varden also take a range of prey, including terrestrial insects, zooplankton and benthic organisms. Both species are present in Marion Lake, where in the summer, cutthroat trout live mostly in the upper 3 m and inshore, whereas the dolly varden are mostly

below 5 m with a more uniform horizontal distribution. The trout are heavily dependent on terrestrial insects taken from the surface, while the dolly varden are primarily bottom feeders. So in sympatry, the two species show a more restricted spatial distribution and a narrower diet. The segregation between the two species in Marion Lake is greatest in summer, probably the time of the highest rate of food consumption (Chapter 3).

In the laboratory, the two species showed differences in their feeding behaviour even when individual fish were observed (Schutz and Northcote, 1972). Dolly varden were the more successful at feeding on benthic prey, and were also more successful at low light levels. Cutthroat trout were better at taking surface prey. These results suggested that selective segregation related to differences in feeding behaviour is important in Marion Lake. However, the experimental fish were taken mainly from Marion Lake and so could have adopted differing feeding responses because of previous interactive experiences (page 200). The comparison of the allopatric and sympatric populations suggested that both species retain considerable phenotypic plasticity, allowing them to change diets or habitats if circumstances required (Schutz and Northcote, 1972).

A comparison of allopatric and sympatric populations of cutthroat and rainbow trout in British Columbian lakes showed that interspecific interactions do not always have outcomes that can be neatly classified (Nilsson and Northcote, 1981). Allopatric populations of rainbow trout tend to have higher mean body sizes than allopatric populations of cutthroat trout, but in sympatry the cutthroat trout tend to be larger. In the presence of rainbow trout, the cutthroat trout generally grow faster than when isolated or in the presence of other salmonid species. This increased growth rate may be related to the increased importance of fish in the diet of sympatric cutthroat trout. Behavioural observations suggested that the rainbow trout are more aggressive than the cutthroat in both intra- and interspecific interactions. When the two species are sympatric, there may be interactive segregation caused by inter-specific aggression. Forced into the littoral areas of the lake by the presence of the rainbow trout, the cutthroat trout would encounter large prey in the form of small fish such as sticklebacks. Cutthroat trout are mor-phologically and behaviourally equipped to exploit this food source. Consequently, the cutthroat show a high growth rate! This example demonstrates that simple concepts of competition and its effects may not always readily withstand exposure to the realities of natural interspecific interactions.

Studies in Scandinavia on salmonid populations of lakes have provided examples both of allopatric and sympatric comparisons and of the effect of species introductions (Nilsson, 1963, 1965, 1967, 1978; Svardson, 1976). In Scandinavian lakes, there is a general correlation between the size of the dominant cladocerans in the zooplankton and the composition of the fish fauna. The size of cladocerans decreases in a sequence: fishless lakes, brown trout only, Arctic charr only, charr + trout, charr + trout + whitefish (*Coregonus* spp.), whitefish only. Although charr are more effective planktivores than trout, the whitefish are even more effective and can exploit smaller zooplankters (Chapter 3). In lakes containing just trout and charr, the former are restricted to the lit-toral zone, while the charr exploit the lim-netic zone, which is usually more extensive. The introduction of whitefish into a lake that previously contained only trout and charr leads to the decline or even the elimination of the charr (Fig. 9.4). When whitefish are also present, they virtually eliminate the larger zooplankton on which charr forage. Charr are unable to exploit the littoral zone. They are excluded by the trout, possibly because they are less aggressive. Their only refuge is as deep-water populations.

Even within the coregonids, there are dif-ferences in the effectiveness of exploitation of zooplankton. The cisco, *Coregonus albula*, is a specialized planktivore, but in Sweden is confined to waters at altitudes below 300 m. Where cisco and other whitefish species are sympatric, the whitefish live close to the shore or bottom and only the cisco is caught in the pelagic zone. When allopatric, the whitefish are regularly caught in pelagic gill nets. Chapter 3 emphasizes the role that prey size has on the composition of the diet of fishes. The studies of the Scandinavian sal-monids suggest how interspecific differences in prey-size selection are translated into dif-ferences in the distribution and abundance of both prey and predator species.

Evidence for interspecific competition between centrarchids

In lakes in central North America, up to eight species of centrarchids (sunfishes, crap-pies and bass) may coexist. The presence together of several, closely related species in small, enclosed water bodies provokes the questions: (1) what interactions occur among the species? and (2) what are the con-sequences for the distribution and abundance of the centrarchids? These questions have been attacked in a programme of research, which explicitly used niche and optimal fora-ging theory as guides to organizing the empirical data and to suggest future research (Werner and Mittelbach, 1981). Perhaps the most important characteristic of this pro-gramme has been its attempt to determine the mechanistic basis of species interactions by quantifying the capacities of individual fish and the changes in these capacities with the growth of the fish.

In many lakes, bluegill and green sunfish (*L. cyanellus*) coexist with the largemouth bass (Werner, 1977). Usually the biomass of the bluegill is greater than that of the bass and far greater than that of the green sunfish. The three differ morphologically.

The bluegill is adapted for high manoeuvrability and the capacity to take small prey, the bass is a piscivore with a spindle-shaped body and large mouth (Chapter 3), and the green sunfish is morphologically intermediate between the other two. Measurements of the pursuit and handling time of the three species in relation to prey size (Chapter 3) showed that the optimal prey size of the bluegill is smaller than that of the bass, and that for the green sunfish is intermediate (Fig. 9.5). An analysis of the niche widths and overlaps of the three species suggested that bluegill and bass could coexist because they are sufficiently segregated along a prey-size niche dimension, but that the green sunfish could not coexist with either because of high niche overlap. The distribution of the three species in natural lakes is compatible this prediction. Adult bluegill and bass have similar distributions: they are found in deeper areas of the littoral zone.

Green sunfish are restricted to the shallowest inshore habitats. The low biomass of green sunfish in many lakes is correlated with the small proportion of lake area represented by its typical habitat.

Experiments in small ponds provided evidence that the differences in the distributions of the bluegill and green sunfish are due to interactive segregation (Werner and Hall, 1976, 1977, 1979). The ponds contained three habitats: a border of emergent vegetation (cattails), open water and the substrate. Some of the latter was exposed sediment; the rest was covered with submerged vegetation or debris. Three species of sunfish, the bluegill, green and the pumpkinseed, were stocked as year-old fish either together or separately. On its own, each species preferred to forage in the vegetation and the diets were dominated by invertebrates associated with the vegetation. When the three were stocked together, the diet of the green sunfish

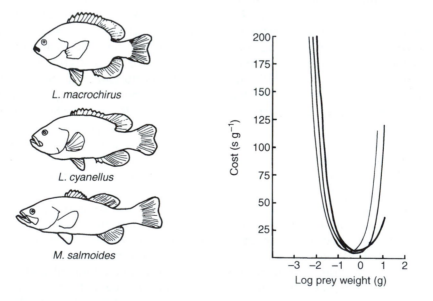

Fig. 9.5 Morphology and cost curves of three centrarchid species which coexist in North American lakes. Thin curve, bluegill sunfish (*Lepomis macrochirus*); medium curve, green sunfish (*L. cyanellus*); thick curve, largemouth bass (*Micropterus salmoides*). Cost curves were calculated for fish weighing 200 g and are terminated at maximum prey size for each species. Redrawn from Werner (1977).

showed no change. The bluegill concentrated on prey from the water column, especially zooplankton, and the pumpkinseed fed mostly on prey from the sediments. All three species grew less when together than when separate, but the reduction in growth was greatest for the bluegill. Associated with the reduction in growth was a reduction in the average size, and hence profitability, of prey and again the bluegill suffered the greatest decline in prey size. When the three species were together in a pond that had distinct areas of vegetation and sediment, all three fed initially in the vegetation. As this favoured area became exploited, first the pumpkinseed switched almost completely to feeding on prey from the sediment, then the bluegill switched partially, but the green sunfish remained in the vegetation. When bluegill and green sunfish were confined together in the vegetation, the growth rate of the bluegill was significantly reduced in the presence of the green sunfish. The green sunfish had a higher survival rate, better growth and a higher weight of stomach contents than the bluegill and did not experience a reduction in growth rate. The green sunfish is morphologically and behaviourally adapted to shallow, inshore, vegetated areas. Its large mouth allows it to feed effectively on large prey associated with the vegetation, it shows strong homing and it is aggressive. In this habitat it is competitively superior to the bluegill and the pumpkinseed. The presence of the green sunfish reduces the profitability of the habitat for less-effective species. In the presence of the green sunfish, other species of sunfish move into different habitats that they can exploit profitably. Their behavioural flexibility allows them to make a niche shift. This flexibility is especially marked In the bluegill. It has a generalized diet and is flexible in its foraging behaviour (Paszkowski, 1986). The morphological specializations of the bluegill in body form and mouth size are complemented by this behavioural plasticity.

However, even in bluegill populations some individuals may be behaviourally and morphologically better adapted for foraging in some areas than in others (Ehlinger and Wilson, 1988; Ehlinger, 1990). This shows a contribution of a between-phenotype component to their niche width (see also page 200).

The bluegill can exploit the water column because it is highly manoeuvrable and can effectively capture zooplankton with its small, protrusible mouth (Chapter 3). The pumpkinseed is morphologically adapted to utilizing the sediments and, because of its ability to crush gastropods, may show relatively little dietary overlap with either the green or the bluegill.

A size-dependent factor complicates these interspecific interactions. The morphological differences between adult fish that allow the species to exploit different types of prey develop during the growth of the fish. For small fish, the morphological differences may be too weakly developed to allow a dietary separation. The ability of the pumpkinseed to crush snails is a function of its size. Fish less than about 45 mm in length cannot process snails, and even fish between 45 and 80 mm can use only small, easily crushed snails (Mittelbach, 1984). In small lakes in Michigan containing both species, bluegill and pumpkinseed larger than 75 mm are segregated both by diet and by habitat. The bluegill feed on zooplankton in open waters and the pumpkinseed on snails associated with vegetation. Smaller fish of both species feed mainly in vegetated areas and have much more similar diets, partly reflecting the inability of the small pumpkinseed to utilize the snails. A comparison of the growth rates in several lakes in Michigan showed that while the growth patterns of small bluegill and pumpkinseed were similar, the growth patterns of the larger fish diverged. The size at which the divergence occurred corresponded to the size at which the diets of the two species diverged (Osenberg *et al.*, 1988). Both laboratory studies and field observations showed that in the vegetated areas, the small bluegill and pumpkinseed are nearly equal competitors. Differences between lakes in the relative abundances of the two species are related to the availability of suitable habitats for the adults. In a lake with a relatively large area of open water in the littoral zone, the bluegill are far more numerous than the pumpkinseed, whereas in a lake with an extensive shallow, vegetated littoral area, the abundances of the two species are about equal (Mittelbach, 1984).

The distinctive ontogenetic niche shift in diet shown by the pumpkinseed has led to the concept of two-stage life histories based on size (Osenberg *et al.*, 1992) and to an analysis of the consequences of inter- and intraspecific competition in the two stages. In a lake in which bluegills had been eliminated by a winter kill, pumpkinseed were abundant. In the absence of bluegill, the juvenile pumpkinseed had good growth. In contrast, the adults had poor growth and the lake had low densities of snails, suggesting intense intraspecific competition between adults. These adults had a weaker development of the pharyngeal apparatus used in crushing molluscs (Chapter 3) than pumpkinseed from lakes holding bluegill, and were less effective in handling thick-shelled snails. Thus, a relaxation of interspecific competition in one stage can lead to an increase in intraspecific competition in the succeeding stage.

The interspecific competition between young bluegill and pumpkinseed sunfish in the vegetation may impose a bottleneck on the two species. A competitive bottleneck is the situation in which resource limitation at juvenile (smaller) stages limits recruitment to adult (larger) stages (Persson and Greenberg, 1990a).

The pond experiments with sunfish have been criticized for using unnaturally high densities of fish and for having inadequate controls and replication (Hanson and Leggett, 1985, 1986). Although these criticisms are partly justified, these experiments

are not ends in themselves but are used to illuminate the distributions of the species in natural lakes. The success of the pond experiments is judged by the extent of that illumination. Interestingly, experiments in which yellow perch, *Perca flavescens*, and pumpkinseed sunfish were held in enclosures in a lake at natural densities provided clear evidence of both intraspecific competition and asymmetric interspecific competition (Hanson and Leggett, 1985, 1986).

Interspecific competition between surfperches

Studies of interspecific competition are not restricted to freshwater systems. The surfperches (Embiotocidae) are coastal marine species. They are viviparous, producing small numbers of large offspring. Two species, the black surfperch, *Embiotoca jacksoni*, and the striped surfperch, *E. lateralis*, live together on shallow rocky reefs on the coast of central California, feeding on benthic invertebrates. Although they live over the same range of depths, striped surfperch are more common in shallow water where foliose algae are abundant. Black surfperch are more abundant in deeper water where the substrate is dominated by a turf of small plants, colonial animals and debris, which contains prey. The foraging behaviour of the two species differs, with striped surfperch more effective when foraging on prey associated with foliose algae, but black surfperch able to exploit prey in the turf. A field experiment studied the effect of the removal of one species on the distribution and abundance of the second species over a 4 year period (Schmitt and Holbrook, 1990). Removal of one species led to an increase of about 40% in the numbers of the other species, but only in the shallower habitats. This long-term change was consistent with the results of a short-term removal experiment (Holbrook and Schmitt, 1989). In shallow water, removal of one species increased the usage by the other species of a microhabitat dominated by the alga *Gelidium*. In deeper water, this alga was rare and there was no change in the use of microhabitats by one species after removal of the other.

Interspecific competition between unrelated species

Unrelated species may compete. European perch are strictly carnivorous while roach are omnivorous. Nevertheless, in some circumstances they may compete. Perch show more pronounced ontogenetic shifts in diet than roach. Before switching to feed on macroinvertebrates 0+ perch feed on zooplankton. Subsequently, if they grow sufficiently large, they become piscivorous. By contrast, the diet of roach changes relatively little as they grow. In a small, fertile, Swedish lake, perch enclosed with roach grew more slowly and switched from feeding on zooplankton to feeding on benthic macroinvertebrates. The presence of the perch had no effect on the roach (Persson, 1987). This experiment provided evidence that asymmetric competition between roach and perch may occur if the perch are forced by a shortage of food to move from the littoral zone of the lake into open water. The roach is less affected because it is more effective at exploiting zooplankton and can also feed on bluegreen algae and detritus. When the roach population in a lake was reduced by 70%, the perch population increased in biomass by 140% (Persson, 1986). At the lower roach densities, the 0+ perch exploited zooplankton for a longer period during the year, before switching to feeding on macroinvertebrates.

The presence of roach can also increase the intercohort competition between 0+ and 1+ perch. This possibility was suggested by experiments in which 0+ and 1+ perch were enclosed with different densities of roach in enclosures in the littoral area of a pond and the open-water zone of a shallow lake (Persson and Greenberg, 1990a,b). As the density of roach present increased, pelagic zooplankton formed a smaller and smaller proportion of the diet of 0+ and 1+ perch, which fed more on benthic prey. For juvenile perch in the presence of roach, the stage of feeding on macroinvertebrates may represent a competitive bottleneck, which restricts recruitment to the piscivorous stage, because intercohort competition decreases growth rates.

When Scandinavian lakes are compared, the abundance of perch first increases as lake productivity increases, but then declines. The abundance of roach and ruffe, *Gymnocephalus cernuus*, increases with lake productivity. The ruffe is a percid that feeds on benthic macroinvertebrates. It can feed at lower light levels than perch. When perch, ruffe and roach were confined in enclosures in a pond, the growth rate of both perch and ruffe decreased as the density of ruffe was increased, but the roach were unaffected. When roach and ruffe are both present, the perch may be at a competitive disadvantage to the roach because the perch is poorer at exploiting zooplankton, and at a disadvantage to the ruffe because the perch is less effective at exploiting benthic prey (Bergman and Greenberg, 1994).

Interspecific competition need not be confined to fish–fish interactions. On coral reefs, the algal mat is grazed by herbivorous fishes such as surgeonfishes (Acanthuridae) and parrotfishes (Scaridae) and by herbivorous invertebrates including the sea urchin *Diadema*. In the Caribbean, a pathogen eliminated 95–98% of the *D. antillarum* in 1983–84. After this epidemic, the feeding rate of herbivorous fishes increased and the biomass of algae removed by them also increased slightly (Carpenter, 1988; Jones, 1991). The algal mat may have been a limiting resource for the fishes and the sea urchin.

At higher trophic levels, there is the possibility of competition between piscivorous fishes and other fish-eating vertebrates including birds such as mergansers, and mammals including otters and seals.

Effect of interspecific competition on niche width

The effect of interspecific competition seems frequently to be a narrowing of the niche width of the competing species along one or more niche dimensions. In some species this may lead to an evolutionary change in niche width, leading to selective segregation. In other species, more phenotypic plasticity is retained so a reduction in niche width occurs if competitors are present. In the absence of the competitor, the full potential of the population can be displayed and an increase in the niche width occurs. This phenomenon is competitive release. When dolly varden and cutthroat trout from a sympatric population were transplanted to two lakes as allopatric populations, the transplanted fish initially retained the depth and feeding preferences that they had when sympatric (Hume and Northcote, 1985). Eight years later, the allopatric cutthroat trout still retained their original depth and diet preferences. However, the allopatric dolly varden had increased their utilization of benthic prey from shallower regions of the lake and their vertical distribution (Hindar *et al.*, 1988). In Scandinavian lakes, charr utilize the littoral zone when brown trout are absent (Svardson, 1976).

If the resource that is the cause of the competition becomes superabundant, then competition through exploitation cannot occur. In such circumstances, the coexisting species may exhibit an increase in niche width and in niche overlap as competition relaxes. In a Panamanian stream, the overlap in diets between the species present was higher in the wet season than in the dry (Zaret and Rand, 1971). Circumstantial evidence suggested that food availability was higher in the wet season than in the dry. In the latter season, the volume of habitat available to the fish was smaller and fish densities were consequently higher.

Evolutionary consequences of competition

Evidence of the evolutionary consequences of competition comes from studies of closely related fishes when living in sympatry (Robinson and Wilson, 1994; Smith and Skulason, 1996). A repeated pattern is that when closely related forms occur in a lake, there is a form (or morph) adapted to feeding on zooplankton in the pelagic (limnetic) habitat and a form adapted to feeding on benthic fauna in the littoral zone. A striking example is the presence in a few lakes on islands in British Columbia of two forms of the three-spine stickleback, a benthic morph and a limnetic morph (McPhail, 1994). The two morphs have not been formally described as species. There is evidence that the two forms are reproductively isolated, although in the laboratory they will hybridize. The two forms differ morphologically. The limnetic morph is slim-bodied, has a narrow mouth and numerous, long gill rakers. The benthic morph is deep-bodied, has a wide mouth, a short, broad snout and a few, short gill rakers. The limnetic morph is more effective when feeding on zooplankton, while the benthic morph is more effective at feeding on benthic invertebrates (Schluter, 1993, 1995). A field experiment demonstrated a trade-off in growth rates between habitats (Schluter, 1995). The benthic morph had a high growth rate in the littoral zone, but a low rate in open water. The limnetic morph had a higher growth rate in the open water than in the littoral zone, although the difference was not as great as for the benthic morph. Most lakes in the stickleback's distribution contain only a single form of the threespine stickleback, which tends to be intermediate in form between the limnetic and benthic forms. The difference between these two morphs is an example of character displacement (Robinson and Wilson, 1994), probably caused by competition between the two. McPhail (1994) suggested that the two morphs in each lake have evolved during the 13 000 years since the last Pleistocene glaciation, as a consequence of a unique sequence of changes in sea level in the area. His model assumed that an initial invasion of a lake by marine sticklebacks led to the evolution of a form that could forage in the littoral zone, the benthic morph. A second invasion, unable to compete in the littoral zone, persisted as planktivores, the limnetic morph. Forms intermediate in morphology would be at a disadvantage to the benthic morph in the littoral zone and to the limnetic morph in the pelagic zone. Unusually for studies on character displacement, an experimental study provided supporting evidence for this hypothesis (Schluter, 1994). In a pond, the presence of the limnetic morph caused a significant reduction in the growth rate of those fish from an experimental population that most closely resembled the limnetic morph in morphology.

Other cases of morphological and habitat differences between closely related forms in lakes including examples of charr (*Salvelinus*) and whitefish (*Coregonus*) have been described (Robinson and Wilson, 1994; Smith and Skulason, 1996). Typically, there are two morphs, a limnetic and a benthic form. In Lake Thingvallavatn (Iceland), four morphs of the Arctic charr coexist (Snorrason *et al.*, 1994). These are small and large benthic forms, a pelagic planktivorous form and a piscivore. The origin of closely related forms within a single lake is still uncertain. Some examples may be a consequence of phenotypic plasticity. Individuals from a common gene pool follow different growth patterns because of different foraging experiences. A second possibility is that the forms have diverged genetically and are partially or wholly reproductively isolated. The reproductive isolation may reflect a history of separate invasions of the lake, as is possibly the case with the sticklebacks. An alternative possibility is that the forms have become reproductively isolated while coexisting in the lake, that is sympatrically. Whatever the

mechanism, the phenomenon seems to occur in lakes in which there are few if any other competing species, and where the most effective ways of exploiting the different habitats require some degree of morphological and behavioural specialization (see also Chapter 3).

Interrelationships between predation and competitive interactions

Predation can reduce the intensity of both intra- and interspecific competition. Pools stocked with monospecific cultures of species such as bluegills frequently develop large populations of small fish (Hackney, 1979). A plausible hypothesis to account for this is that, in the absence of predation, the density of the bluegill population increases, and so the rate of exploitation of the bluegill's prey increases (Mittelbach, 1983, 1986). This increase in predation pressure shifts the distribution of prey sizes towards smaller prey because of the size-selective predation by the bluegill (Chapter 3). As the abundance of large prey declines, there is a decrease in the size of the bluegill at which the net rate of energy income can be maximized, and the net energy incomes of large fish are reduced disproportionately relative to small fish. The fish size at which the net energy income available for growth is zero becomes smaller and smaller. With the reduction in prey abundance, the bluegill population becomes dominated by small individuals that have relatively low absolute rates of consumption. These small fish are able to utilize small prey yet still maintain a positive energy balance. Piscivorous fish such as largemouth bass take the smaller bluegill, and so the introduction of piscivores will reduce the pressure on the food base. This reduction allows some bluegill to grow through the vulnerable size classes. The effect of the piscivory is to relax the intraspecific competition for food by allowing an increase both in the absolute quantity of food and in the availability of prey sizes that are profitable for the larger bluegill.

In contrast, there are other circumstances in which the presence of a population of piscivorous fish may increase the possibility of both intra- and interspecific competition. On the basis of optimal foraging theory and the seasonal changes in the abundance and size distributions of prey in different habitats in a small Michigan lake, Mittelbach (1981a) predicted when the bluegill should switch habitats to maximize their net energy income (Chapter 3). The prediction was reasonably accurate for the large bluegill (101–150 mm), but the medium (51–100 mm) and the small (10–50 mm) failed to make the predicted switch from foraging in the vegetation to foraging in open water early in the summer. Was this because the smaller bluegill were restricted to the less profitable vegetation by the presence of piscivorous largemouth bass, which are more effective predators in the open waters? In small ponds, in the absence of piscivorous largemouth bass, the bluegill of all size classes switched to the more profitable habitat at approximately the time predicted. In the presence of bass, the small bluegill continued to forage predominantly in the less profitable habitat, the vegetated area of the pond (Werner *et al.*, 1983a,b). Only this small size class suffered significant mortality from largemouth bass predation. The small bluegill paid a cost for not moving into the more profitable but riskier foraging habitats. The average growth increment of individual small bluegill was 27% lower in the presence of the predator than in its absence. In contrast, the growth of fish in the larger, invulnerable size classes was better, presumably because the absence of the smallest bluegill from the most profitable habitat made more prey available to them. The presence of the size-selective predator increased intracohort competition but decreased intercohort competition in the bluegill.

Studies of the distribution of fishes in these North American lakes have shown that even when the older and larger fish exhibit habitat or dietary segregation, the younger and smaller fish usually live in vegetation. Piscivorous fish are less effective predators in vegetation (Savino and Stein, 1982) so this habitat forms a refuge for juvenile fish. However, for this refuge they may pay a cost in reduced growth because of intra- and interspecific competition for food. A consequence of this reduced growth would be that they spend longer in the size classes that are most vulnerable to predation.

The juveniles of some species have a higher overlap in their diet than when they are adults, for example the bluegill and pumpkinseed sunfishes (Mittelbach, 1984). The growth rates of bluegill and pumpkinseed juveniles in natural lakes are lower than the rates observed in experimental ponds, which suggests that both species are food limited. Predation could create a competitive bottleneck by constraining the young of several species to a habitat that forms a refuge from predation. Evidence for a bottleneck comes from an experiment in which different densities of small bluegill were confined with two small pumpkinseed sunfish in enclosures in the vegetated littoral margin of a lake (Mittelbach, 1988). The growth rates both of the bluegill and of the pumpkinseed decreased with an increase in the density of the bluegill. The mean size of the invertebrates in the enclosures also decreased with the increase in bluegill density because of a reduction in the abundance of large invertebrates, which are the preferred prey of the sunfishes. But even in the juvenile stage, morphological differences such as mouth size allow some segregation of species by diet (Keast, 1985).

Ontogenetic changes of diet as fish grow mean that even piscivores may experience competitive bottlenecks as juveniles because of the presence of a prey species. In their first summer of life, largemouth bass feed on invertebrates in the littoral zone until the bass reach a size at which they can feed on fish. In this period, they are potentially competitors with, rather than predators, of juvenile bluegill. Thus, the life history of the bass can be interpreted as consisting of two distinct, size-based stages (Olson *et al.*, 1995). These stages will respond differently to changes in bluegill density. In a pond experiment, the growth of juvenile bass was reduced in the presence of juvenile bluegill. The growth of the bluegill was mainly affected by bluegill density. In lakes with a high density of bluegill, the growth of juvenile bass will be inhibited, but that of larger, piscivorous bass enhanced. At low bluegill density, the juvenile bass will suffer less competition and so will become piscivorous sooner, but then will experience an inadequate food supply.

The profitability of different habitats, in terms of the net energy gain to foragers, changes with changes in the size and abundance of fish. The risk of predation is also a function of size (Werner and Gilliam, 1984). A knowledge of changes in size of fish is probably more important for an understanding of the dynamics of habitat use, foraging success and mortality than information on their ages. In some small, dystrophic lakes in Wisconsin, mudminnows, *Umbra limi*, and yellow perch coexist. Experimental evidence suggested that 0+ perch were competitively superior to 0+ mudminnows and inhibited the growth of the latter. This slower growth may make the mudminnows more vulnerable to predation by older perch (Tonn *et al.*, 1986).

These illustrations, taken together with other examples in Chapter 8, highlight the potential importance of size-dependent, ontogenetic changes in habitat and diet for the interactions between fishes. The experimental and observational studies on lacustrine species have led to the concept of stage-structured systems, exemplified by the pumpkinseed sunfish (page 208) and the largemouth bass (page 212). The concept emphasizes the importance of foraging success and its consequence for growth rates, which determine the timing of ontogenetic changes in habitat and diet.

Interelationships between parasitism and competition

The intensity of competitive interactions may also be altered by changes in the intensity of parasitism. Heavily infested fish may be less effective competitors because of the deleterious effects of their parasitic burden (Milinski, 1984b). Yet in the presence of a potential predator, sticklebacks infested with *Schistocephalus solidus* (Chapter 8) are better competitors for food than uninfested fish because the foraging of the parasitized fish is less disturbed by the presence of the predator (Milinski, 1985).

Some circumstantial evidence suggested that parasitism can alter the balance in an interspecific interaction (Burrough and Kennedy, 1979; Burrough *et al.*, 1979). In Slapton Ley, a lake in the south-west of England, an increase in the abundance of the roach population was correlated with a decease in the abundance of rudd, *Scardinius erythrophthalmus*. When the roach population suffered high mortalities because of infestation with the cestode *Ligula intestinalis*, the population of rudd increased. The growth rate of the remaining roach also improved. These observations suggested that both interspecific competition between the roach and rudd, and intraspecific competition within the roach population were alleviated by the reduction in the latter.

9.5 MUTUALISM

Mutualistic interactions have attracted few quantitative or experimental studies, although the natural history of such interactions continues to fascinate biologists.

One example of mutualism is the formation of multispecies shoals (Sale, 1980). The advantages for single species of shoaling as a defence against predation, for the exploitation of food patches or to overwhelm the territorial defence of a competitor may extend to these multispecific groups.

A good example of mutualism is the phenomenon of cleaner fish. These are species that clean the surfaces of client fish, picking off ectoparasites, mucus and scales (Limbaugh, 1961; Feder, 1966). They may also clean wounds (Foster, 1985c). Both marine and freshwater families include species that are cleaners (Losey, 1972). Some species have become specialized cleaners either for a phase during their ontogeny or for much of their life. For example, the bluehead wrasse (Labridae) of the Caribbean is a cleaner as a juvenile. Throughout much of the Indo–Pacific, the labrid genus *Labroides* provides the principal cleaner species, and these act as cleaners even when adult. The mouth and dentition of *Labroides* are extremely modified for picking at small objects. The cleaners occupy specific stations on the coral reef, to which clients come. The tactile stimulation from the cleaning may be important in motivating the clients to visit a cleaning station (Gorlick *et al.*, 1978). Although it is likely that both client and cleaner benefit from the interaction, clear quantitative evidence of this in terms of components of fitness such as survival and fecundity is not available. Suggestions that cleaner fish play a major role in determining the abundance and number of species on reefs do not seem justified (Limbaugh, 1961; Gorlick *et al.*, 1978). Wrasse from the coasts of western Europe are being used in attempts to control infestations of salmon louse on caged Atlantic salmon. The wrasse pick the lice off the skin of the salmon.

The cleaning mutualism is exploited by some species, which mimic the cleaners. The blenny, *Aspidontus taeniatus*, uses its close resemblance to the cleaner *Labroides dimidiatus* to approach other fish, then tears pieces from their fins (Feder, 1966).

The cleaning mutualism is not confined to fish–fish interactions. Several species of shrimps, particularly in the tropical regions, act as fish cleaners, removing ectoparasites from the body of the fish and removing necrotic tissue from wounds (Feder, 1966). Other mutualistic associations with invertebrates also occur. The anemonefish or clownfish, *Amphiprion* spp., lives in association with sea anemones (Allen, 1972; Sale, 1980). The fish probably gain protection from predators because of the nematocysts of their anemone host. These fish only live in association with the anemones, but the anemones occur in the absence of fish. Anemones may benefit because the resident anemonefish protect them against predators, which crop their tentacles. Some anemonefish species are generalists, living in several species of anemone. Other species are specialists with one or two hosts. The protection of fish from the nematocysts of their host anemones develops during the ontogeny of the fish. A series of tests showed that pelagic larvae of anemonefish adhered to the tentacles of anemones, even the tentacles of the species of anemone in which the fish were later to live. Protection from the anemone developed soon after the metamorphosis from the larval to the juvenile form (Elliott and Mariscal, 1996).

Another example of mutualism is provided by goby–burrowing shrimp associations such as that between the goby, *Psilogobius mainlandi*, and shrimps of the genus *Alpheus* (Preston, 1978). The goby and shrimp share a burrow, which is dug by the shrimp. The goby, which sits near the entrance to the burrow, signals the presence of potential danger. This may stimulate the shrimp to retreat into its burrow, followed by the goby. The goby usually reappears from the burrow ahead of the shrimp.

9.6 SUMMARY AND CONCLUSIONS

1. In a competitive interaction, one or more of the participants suffers a net loss of fitness and none shows a net gain. Com-

petition may be through exploitation or contest. In mutualism some of the participants show a net gain and none shows a net loss.

2. The niche defines the environmental ranges within which an organism or population can be reproductively successful. The fundamental niche may be reduced to a realized niche through biotic interactions including interspecific competition.

3. Intraspecific competition may be either intra- or intercohort. Territoriality or dominance relationships may lead to an unequal distribution of resources between individuals in a population. Although intraspecific competition may lead to an increase in the niche width of a population, in some circumstances it may cause a decrease by reducing the number of size classes in a population.

4. Sympatric species may show differences in their exploitation of resources because of selective or interactive segregation. The latter is caused by contemporary interspecific competition. Evidence for both selective and interactive segregation is provided by stream- and lake-dwelling salmonids and by centrarchids.

5. Unrelated species may compete if they utilize the same resource when it is in short supply.

6. The intensity of both intra- and interspecific competition may be modified by predation. If predators remove the competitively successful individuals, competition is relaxed. But competition may be intensified if the predators confine their prey to habitats in which resources for the prey are limited. Both effects are shown in centrarchid populations. Parasitism may also modify the intensity of competition.

7. The concept of mutualism is illustrated by cleaner fishes and species such as the anemonefishes that live in a mutualistic relationship with invertebrate partners.

8. Competition and predation are strongly influenced by the size structure of fish populations. Any interpretation of the interactions within and between populations in the context of niche theory has to incorporate the ontogenetic, size-related changes in diet and habitat that are such a feature of the life histories of most species of fish.

10

DYNAMICS OF POPULATION ABUNDANCE AND PRODUCTION

10.1 INTRODUCTION

Previous chapters describe the effects that environmental factors have on the survival, growth and reproduction of fish. This chapter explores the changes in the abundance and total weight of fish in a population that result from these effects. Fishes have high fecundities (Chapter 7), so the potential lifetime production of offspring is high. Although fish populations vary in abundance (Fig. 10.1) (Cushing, 1982; Rothschild, 1986), increases in abundance are usually several orders of magnitude lower than the potential maximum increase. Even when a population is increasing, the fate of most zygotes is to die before reaching sexual maturity.

The abundance of a population changes as the probabilities of survival and reproductive success of individual fish change. How much information about the fate of individuals is required to predict changes in abundance? Early models of population dynamics tended to treat all individuals as identical and to describe a population in terms of its overall density (N). An obvious example is the logistic model described in all standard textbooks on ecology (Begon *et al.*, 1996). This approach was used in the development of an influential model for fisheries management, the Schaefer surplus-yield model, which applied the principles of the logistic model to an exploited population (Pitcher and Hart, 1982; Hilborn and Walters, 1992). The recognition that the age structure of the population, if known, would be relevant to the dynamics of abundance and exploitation led to the dynamic pool model of Beverton and Holt (1957). This model formed the basis for much fisheries management in the succeeding years. Subsequently, matrix theory has been used to develop age- and size-structured models of population dynamics (Caswell, 1989). The increasing power of computers has now allowed the development of individual-based models of fish population dynamics (DeAngelis and Gross, 1992; Van Winkle *et al.*, 1993). Individual-based models simulate the fate of individuals in a cohort using information such as rate of encounter with prey (Chapter 3) or with predators (Chapter 8) and the bioenergetics of growth (Chapter 6), and so predict changes in abundance and biomass of that cohort. As models increase in complexity, they are increasingly greedy for relevant information. The logistic model requires estimates of two parameters for a population, while an individual-based simulation requires estimates of tens of parameters. An outstanding question is whether changes in the abundance and biomass of a population can be predicted from average rates for cohorts, or whether small subgroups within the age cohorts, with unusual survival or growth rates, drive the changes.

There is a danger of thinking of the population as the fundamental unit of study. A population is a collection of individuals and the fate of these individuals during their life span determines what happens to the population. The adaptations of these individuals

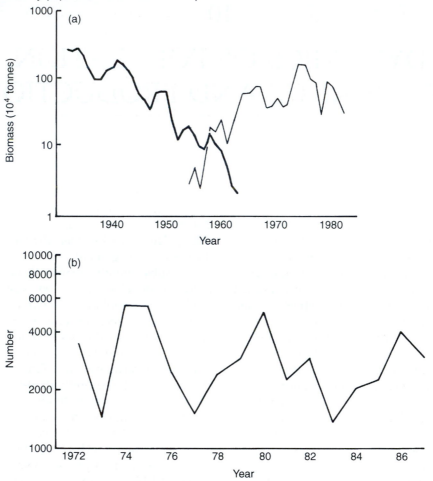

Fig. 10.1 Examples of variability in abundance of fish populations. (a) Spawning biomass of Pacific sardine, *Sardinops sagax caeruleus*, (age 2+) (thick line), and northern anchovy, *Engraulis mordax* (thin line), off coast of California. Redrawn from Rothschild (1986). (b) Abundance of threespine sticklebacks, *Gasterosteus aculeatus*, in October in small backwater of Afon Rheidol in Mid-Wales. note logarithmic scales.

are related to their individual reproductive success, not to the long-term interest of the population (Łomnicki, 1988). These individual adaptations may have effects that seem advantageous at the population level such as minimizing the population's response to environmental change. But it is at the level of the individual that the functional significance of adaptations has to be studied.

10.2 DEFINING THE POPULATION

Population structure

A simple definition of a population is a group of fish of the same species that are alive in a defined area at a given time. The area may be defined arbitrarily for the convenience of the investigators or it may be physically meaningful for the population under study such as a lake or a system of currents in a sea. Fisheries biologists frequently use the term 'stock' to refer to a population that is exploited by a fishery and which may be subject to some form of management.

The most satisfactory situation is when the population (or stock) being studied consists of a collection of individuals that make up a gene pool, which has continuity in time because of the reproductive activities within the population (MacLean and Evans, 1981). The population can be described in terms of the growth, survival, and reproductive rates intrinsic to it. The distinctiveness of gene pools will be reduced by gene flow between populations, caused by individuals moving between populations. A single geographical unit such as a lake may contain several distinct gene pools of the same species.

Some species consist of a relatively small number of self-sustaining populations with wide geographical distributions, for example the Atlantic menhaden. Other species contain a high number of populations, which are largely or completely isolated from each other. The Atlantic salmon, Atlantic herring and American shad show this characteristic. The isolation between populations is maintained because adults have a strong tendency to return to particular spawning localities, often the locality where they were spawned (Chapter 5). Species differ in what Sinclair (1988) has called population richness, that is the number of self-sustaining populations. Differences in population richness between species probably relate to geographical features that promote genetic isolation between populations. Such features can include current systems for marine species and watershed systems for freshwater and anadromous species.

The population richness of a species depends on the pattern of larval and early juvenile recruitment to the sexually active component of the population (Carr and Reed, 1993). In a closed population, the larvae recruit back into the population that spawned them (Fig. 10.2(a)). The frequency of straying between populations is usually too low to prevent genetic differentiation, and the populations are demographically self-sustaining. For exploitation, each popu-

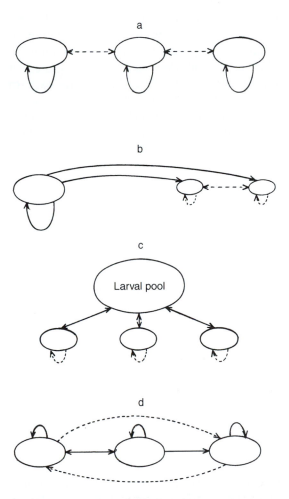

Fig. 10.2 Schematic diagram of population structure: a, closed populations; b, source–sink populations; c, open populations; d, 'stepping stone' (limited distances) populations. Ellipses, geographically separate populations; solid arrows, high recruitment; broken lines, weak recruitment. Modified after Carr and Reed (1993).

lation should be treated as a management unit. In open populations, the constituent populations contribute to and draw their recruits from a common larval pool (Fig. 10.2(c)). This inhibits the development of genetic divergence between the contributing populations. It also decouples the rate at which a constituent population contributes recruits to the larval pool from the rate at which that population recruits from that pool subsequently. The former will depend on the conditions within the population in relation to reproductive success (Chapter 7). The latter will depend on the vagaries of the processes that bring the recruits back to the population from the common pool. Many fishes of coral reef have open populations with pelagic eggs and larvae dispersing greater or lesser distances from the natal reef (Thresher, 1984; Sale, 1991). Management of such open populations should maintain the size of the larval pool sufficient to sustain the contributing populations. Some populations have a source–sink relationship (Fig. 10.2(b)). The source population is fully self-sustaining, but the sink populations depend to a greater or lesser extent on emigrants from that source to maintain their existence. An intermediate pattern between open and closed populations is the situation in which populations that are geographically close exchange recruits, but populations that are geographically remote have little or no direct exchange (Fig. 10.2(d)).

Populations that are linked by recruitment form a metapopulation (Hanski and Gilpin, 1991). Within a metapopulation, individual populations may go extinct but be resurrected by recruits from other populations. McQuinn (1997) interpreted the population structure of Atlantic herring in terms of the concept of the metapopulation. He drew attention to the importance of discrete spawning sites for local populations, but also to the role of migrants between these populations.

Genetic and other population markers

Powerful techniques from molecular biology can now be applied to the problem of identifying distinct gene pools, that is distinct biological populations, using genetic markers (Carvalho and Hauser, 1994; Ferguson *et al.*, 1995). The aim is to detect whether there are statistically significant differences in the frequencies of genetic markers between the populations of interest. Although these molecular techniques are frequently used to detect genetic differences between populations, they can be deployed to investigate evolutionary relationships between closely related species. They are also used for species identification of eggs and larvae, which can be difficult to identify in other ways. The technique of DNA fingerprinting is important for identifying familial relationships within a population, especially parentage.

In this rapidly developing field, there are two types of marker, protein and DNA. If a genetic locus codes for a structural protein or an enzyme, different alleles at that locus will code for slightly different proteins. If this difference is reflected in the electrical charge carried by each protein, then alleles can be distinguished by how far the protein coded by a given allele migrates in the electrical field generated in electrophoresis equipment (Ferguson, 1980). The genotypes of individuals in a population are defined by the alleles thus identified. Use of DNA allows genetic differences between individuals to be identified directly. Two types of DNA are available. Both mother and father of an individual contribute to the nuclear DNA. Only the mother contributes to the DNA found in the mitochondria in the cell cytoplasm. A variety of techniques, including cutting up DNA at specific sites using restriction enzymes, gene cloning, and amplification of fragments of DNA with the polymerase chain reaction technique (PCR), are deployed to allow comparisons of DNA sequences between individuals. For nuclear DNA, the methods are usually targeted at satellite DNA. This does not code for proteins, but consists of repeated copies of segments of DNA. Mini- and microsatellite DNA are of particular interest. They can be used to generate either multilocus (DNA fingerprinting) or single-locus definitions of the genotype of an individual. An advantage of the use of micro- and minisatellites is that only tiny amounts of material are required, for example a few scales or a clipping from a fin, and even dried or preserved material are sometimes suitable.

An example of the application of these molecular techniques is the analysis of the genetic structure of brown trout in Lough Melvin in north-western Ireland. This 21 km^2 lake contains at least three, reproductively isolated populations of trout (Ferguson and Mason, 1981; Ferguson *et al.*, 1995). The trout populations identified by the molecular systematics analyses correspond to forms that had previously been recognized by visual features such as size and colour. Protein allozymes, mitochondrial DNA and single-locus minisatellites all indicated similar levels of genetic divergence between the three forms.

A failure to demonstrate differences between populations using molecular methods does not necessarily mean that the populations can be treated as a single unit in terms of demographic variables such as survival and reproductive rates, or as a single unit for fishery management (Carvalho and Hauser, 1994). Several processes may contribute to a lack of genetic differentiation. Relatively low levels of migration can maintain sufficient gene flow to prevent gene pools becoming differentiated, although the rate of migration may be too low to have a detectable demographic effect. There may be sporadic recruitment to a population from distant areas. Populations that are demographically distinct but occupying similar environments may be genetically similar because of stabilizing selection. The time since the separation of populations may have been too short for divergence of their genetic characteristics. Other types of data should supplement molecular information.

An obvious, additional method for identifying separate populations is on the basis of the morphological characteristics of individuals. These may be metric characters such as length, which are measured on a continuous scale, or meristic characters such as the number of scales along the lateral line, which take discrete values. Many metric and meristic characters or variates can be measured on each fish in a sample. Multivariate statistical techniques, now commonly available in computer statistical packages, allow these many characters to be used to characterize the morphology in a given population and to compare populations. These statistical techniques, which include principal components analysis, discriminant analysis, and multivariate analysis of variance and covariance, provide a variety of related methods for detecting the presence of different populations (Manly, 1994). An example was the use of multivariate analysis of covariance to discriminate between populations of the capelin from around Newfoundland (Misra and Carscadden, 1987). On the basis of seven metric characters, the analysis suggested the presence of at least five populations of capelin in the region. Clearly, the analysis of metric and meristic characters can be combined with molecular systematics to explore the discreteness of the populations. Multivariate morphological analysis and variation in mitochondrial DNA illustrated the divergence of a population of a freshwater clupeid, *Limnothrissa miodon*, in the 34 years after the species had been introduced into Lake Kivu from Lake Tanganyika in East Africa (Hauser *et al.*, 1995).

In some cases, populations may be identified by their characteristic parasite fauna (Chapter 8).

10.3 ESTIMATION OF FISH ABUNDANCE

Once the population has been defined, the next problem is to estimate its abundance. The estimation of fish abundance presents two methodological problems: how can fish of a given age be sampled quantitatively, and how can the absolute abundance of the fish be estimated from the quantitative samples?

Methods of sampling

In a few unusual situations, all or a component of a population can be enumerated directly. Small enclosed volumes of water such as ponds or sections of canals can sometimes be drained and the fish collected. The migratory component of a population may pass a natural or human-made structure that allows direct counts of the number of migrating fish. In a study of the biology of the sockeye salmon of the Skeena River system in British Columbia, a counting fence was installed on the Lower Babine River. This allowed a complete count of all the adult salmon migrating upstream to spawn (Brett, 1983, 1986) (Chapter 5).

More usually, a population must be sampled (Lagler, 1971; Pitcher and Hart, 1982). Fish differ greatly in size at the various stages in their life cycle (Chapter 6). A method that samples eggs quantitatively is unlikely to work for sampling the abundance of the adults that spawned the eggs. As fish grow, and as their behaviour changes, their chance of being captured by a given method of fishing usually changes (Heath, 1992; O'Hara, 1993; Wardle, 1993). The choice of sampling method will depend on the physical characteristics of the water body in which the population lives, the behavioural characteristics of the fish and the selectivity of fishing apparatus used (Lagler, 1971).

Pelagic eggs and larvae are often sampled with modified plankton nets, which are usually hauled vertically upwards (Bagenal

and Braum, 1971; Smith *et al.*, 1985; Heath, 1992). As the young fish become stronger swimmers, their ability to evade a plankton net increases and such fish may be sampled with a small purse seine. In shallow water, the seine may be sunk on the bottom but timed to release and float to the surface, sampling the young fish in the water column as it ascends. Demersal eggs, deposited on the surface of the substratum, may be estimated using quadrats. When eggs are buried in the substratum, as in the redds of the salmonids (Chapter 7), they can be carefully dug out and any drifting eggs caught in downstream nets (Elliott, 1984).

The most frequently used methods for sampling juvenile and adult fish are netting, hook-and-line fishing and electrofishing (Lagler, 1971; Cushing, 1982; Pitcher and Hart, 1982). Netting techniques depend either on active capture such as trawling or seining, or on passive capture including gill netting and trapping. Electrofishing using either AC or DC apparatus is particularly valuable for some species in small bodies of fresh water (Cowx and LaMarque, 1990). In habitats that are topologically complex, including rocky littoral areas and coral reefs, fish may be sampled with poisons such as rotenone.

Methods for estimating abundance

Common methods of estimating the abundance of a fish population are mark–recapture, catch per unit effort and cohort (or virtual population) analysis. In concept, each of these three methods is simple, but their application to natural populations has required the development of a large body of statistical and mathematical models (Seber, 1973; Ricker, 1975; Gulland, 1977).

Mark–recapture

The principle of mark–recapture is that a sample of fish, n_1, is taken from the population at time 1, each fish is given a recognizable mark and then the sample is returned to the population. At time 2, a second sample is taken, n_2, of which m are marked. If it is assumed that the proportion of marked fish in the second sample is the same as the proportion of marked fish in the total population, N, then:

$$n_1 / N = m / n_2 \qquad (10.1)$$

from which an estimate of N, $\hat{N}$, can be calculated as:

$$\hat{N} = n_1 n_2 / m. \qquad (10.2)$$

This is the Petersen estimate, the simplest form of mark–recapture estimation. It requires only two samples and only one type of mark. The precision of the estimate is improved if the number of recaptures is high. The method makes the following assumptions about the population and the marking process (Seber, 1973; Begon, 1979).

1. Marks are not lost in the period between the two samples and are correctly recognized on recaptured fish.
2. Being caught, handled and marked has no effect on the probability that the fish will be recaptured.
3. Being caught, handled and marked has no effect on the probability of the fish dying or emigrating.
4. The population is sampled at random, without regard to the age, sex, physiological condition or size of the fish.
5. All fish, marked or unmarked, have the same probability of dying or emigrating.
6. Either there are no births or immigrations, or there are no deaths or emigrations, or there are none of these, in other words a closed population is assumed.

These assumptions are extremely restrictive. More sophisticated mark–recapture methods have been developed to allow some to be relaxed, particularly the assumption of a closed population. The best-known of these is the Jolly–Seber method. This requires that fish be captured, marked and released on several successive occasions and that on each occasion a mark specific to that sample be applied to the fish. It is known when a marked fish was last captured. This makes it possible to estimate the number of marked fish that are at risk of being captured immediately before a sample is taken, so that population abundance when the ith sample is taken, $\hat{N}$, can be estimated as:

$$\hat{N} = (n_i \hat{M}_i) / m_i \qquad (10.3)$$

where n_i is the number of fish sampled at time i, m_i is the number of marked fish in the sample and $\hat{M}_i$ is the estimated number of marked fish in the population immediately before the ith sample is taken. The method also gives estimates of loss (deaths and emigration) and recruitment (births and immigration) rates. The details of the method are lucidly described by Begon (1979). A comparison of the results of applying the Petersen and Jolly–Seber methods to a small population of threespine sticklebacks is shown in Table 10.1. For this population, the estimates of abundance obtained in autumn using the two methods were sufficiently alike to suggest that the more restrictive assumptions of the Petersen method were not leading to unrealistic estimates. For this population at other times of the year or for other populations, the simple Petersen estimate may be inadequate.

Various methods are used to mark fish (Parker *et al.*, 1990). Fin or spine clipping involves mutilation, but the adipose fin of a salmonid is sometimes clipped to indicate that the fish is also carrying a more informative mark. Tags, which can be attached to fish, come in a range of designs and sizes, but are difficult to use on small fish. Metallic microtags injected into the fish with a specially designed gun and detected magnetically are coming into more common use. PIT tags inserted into the body cavity allow identification of individual fish (page 89). Cold branding, using a distinctive branding iron cooled in liquid nitrogen, can be used to

Table 10.1 Comparison of population abundance in autumn estimates* of threespine sticklebacks, *Gasterosteus aculeatus*†: results suggest that an assumption that the population was closed was reasonable

Year	Petersen method	Jolly–Seber method
1978	2450	2650
1979	2950	2650
1980	5150	5050
1981	2150	2150
1982	2900	2850
1983	1350	1200

*Derived from Petersen (single recapture) and Jolly–Seber (multiple recapture) mark–recapture methods.
†The population lived in a small (~ 200 m²) backwater of the Afon Rheidol in Mid-Wales.

apply a date-specific mark on the flank of the fish. Such a mark can persist for several months. Coloured marks can be applied to the fins or ventral surface by using a 'Panjet' to fire a jet of dye such as alcian blue (Hart and Pitcher, 1969), or by injecting a small blob of colour under the skin with a hypodermic needle. Many small fish can be marked at one time by spraying them with particles of fluorescent dye using a modified paint spray. The particles fluoresce when illuminated with UV light (Moodie and Salfert, 1982). When choosing a method of marking, the effects of the marking procedure and of the mark on the survival and behaviour of the fish have to be considered. The goals are to provide a mark that can be detected reliably but has a minimal effect on the fish.

The use of mark–recapture methods for estimating the size of fish populations is complicated by two factors. The first is that many methods of sampling fish are size-selective. In these cases, the sample taken does not represent a random sample of the whole population; some size classes are over-represented and some underrepresented. The second is that the survival of fish between successive samples may be a function of their size (Chapter 8): often it may be necessary to make estimates of abundance, loss and recruitment separately for the different size classes present in a population (Pollock and Mann, 1983).

Catch-per-unit-effort methods

An estimate of population size that avoids the need to mark the fish can be obtained from changes in the catch per unit of fishing effort (CPUE). As the abundance of a cohort declines through fishing and natural mortality, the number of fish caught per unit of fishing effort will usually decline. Consider a population from which two samples of abundance n_1 and n_2 are taken with the same fishing effort over a short time period so that natural mortality and recruitment are negligible. If the initial population abundance is N, then the probability of capture in the first sample is n_1 / N. The probability of capture in the second sample is $n_2 / (N - n_1)$. Assume that the probability of capture is constant. Then:

$$n_1 / N = n_2 / (N - n_1) \qquad (10.4)$$

so that an estimate of the initial abundance, $\hat{N}$, is given by:

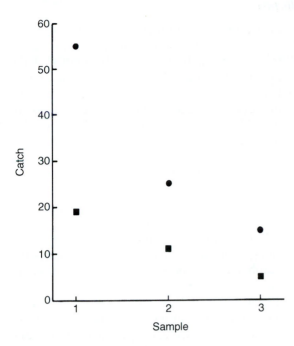

Fig. 10.3 Decline in catch per unit effort with successive samples illustrated by decline in catch of brown trout, *Salmo trutta*, from a 50 m stretch of a small stream in Mid-Wales, electrofished three times in succession over a 3 h period in (●) March 1981 and (■) March 1988.

$$\hat{N} = n_1^2 / (n_1 - n_2) \qquad (10.5)$$

(Seber and LeCren, 1967). If more than two samples are taken, a graph of catch per sample against sample number (Fig. 10.3) can be plotted. The problem then is to estimate the initial population size that would have generated the observed decline in catch. Zippin (1956, 1958) described a maximum likelihood method of estimation and Higgins (1985) provided a simple BASIC program to ease the calculation. An alternative technique is to calculate the linear regression of catch per sample against cumulative catch, from which an estimate of the total population is given by the intercept on the *x*-axis, that is the cumulative catch axis (Ricker, 1975; Cowx, 1983). The method has to be modified

if the population is experiencing significant natural mortality, immigration or emigration while the successive catches are being made (Ricker, 1975).

A Petersen mark–recapture study can be combined with a removal method of estimating population size (Gatz and Loar, 1988). These authors emphasize the need to test the assumptions of both methods and if necessary to adjust the estimates of population size when violations are identified.

Catch-per-unit-effort methods typically assume that the probability of capture is constant, so that as the density of a population declines, the CPUE declines proportionately. However, the technology used in modern commercial fisheries for detecting and capturing fish is extremely effective. Even when the overall abundance of a stock is low, local concentrations of fish can be found. These concentrations can be exploited. This means that CPUE does not reliably reflect abundance, and its use tends to lead to overestimates of abundance at low stock abundance. This increases the danger of the excessive exploitation of a stock when it is most vulnerable to overexploitation.

Virtual population analysis (VPA, cohort analysis)

Consider a population that consists of several age classes (cohorts) and is being fished. The total number of fish from a given cohort that is caught by the fishery gives an estimate of the minimum number of fish that were present in that cohort at the age of recruitment to the fishery. An improved estimate of numbers in a given cohort can be obtained if information is available about the natural mortality rate prevailing in the population. From information on the numbers caught at each age and estimates of fishing and natural mortality, VPA allows a reconstruction of the successive abundances of a cohort once it has recruited to the fished portion of the population (Gulland, 1977; Hilborn and Walters,

1992). A clear description of the successive steps involved in a simple VPA was provided by Pitcher and Hart (1982). The technique has been extended to allow simultaneous estimates of the abundances of several species, taking into account both fishing mortality and losses to other predators (Daan, 1987; Daan and Sissenwine, 1991).

Acoustic methods

The development of echosounding techniques to detect submarines led to the use of similar techniques to detect fishes. A train of sound pulses is emitted from a transmitter and reflections from objects in the train of the pulses are detected as echoes. In echosounding the pulse is directed vertically downwards, while in sonar the pulse is directed at an angle to the vertical. These techniques are used by fishing boats to detect the presence of concentrations of fish suitable for capture. They are also used to obtain indications of the abundance of fish stocks (MacLennan and Simmonds, 1992; Misund, 1997). The application of acoustic methods presents two methodological problems. The first is the translation of the information in the echoes into an estimate of the abundance of the fish responsible for the echoes. The second is the identification of the species of fish responsible. The effectiveness of modern acoustic methods in detecting and indicating the abundance of fish has probably contributed to the overfishing which now threatens many exploited stocks.

Other methods

An indirect method of estimating the abundance of the spawning component of a population is to sample the spawned eggs quantitatively. Then, if the batch fecundity and the rate of spawning are known, the abundance of spawning females can be estimated. This technique has been particularly well developed for estimating the abundance

of spawning northern anchovy (Lasker, 1985). In a study of migratory European trout in a small stream in the English Lake District, the annual abundance of spawning females was estimated by counting the number of redds (Elliott, 1984). Males of mouthbrooding cichlids typically dig distinctive nest pits in which spawning takes place (Fryer and Iles, 1972). An estimate of the number of sexually active males can be obtained from the density of the nest pits, which is sometimes so high that the pits abut each other (Payne, 1986).

Line transects, in which an observer swims along a pre-determined route and counts the number of fish observed, have been used to estimate abundances on coral reefs, in the littoral zone of lakes and in streams. Given the light-transmitting properties of water (Chapter 2), it is usually necessary to take into account the decreased probability that the observer will see a fish with increasing distance between fish and observer (Ensign *et al.*, 1995).

10.4 MEASURES OF THE RATE OF POPULATION CHANGE

Changes in the abundance of a population are caused by changes in the rates at which individuals are added through births or immigration, and lost through death or emigration (Fig. 10.4). The basic tool for analysing the dynamics of a population is the cohort life table. This itemizes the age-specific survival and reproduction of a cohort. In this way the fates of the individuals spawned (or born) at approximately the same time are summarized. The construction of such life tables is described in general ecological texts (Krebs, 1994; Begon *et al.*, 1996). An example for a cohort of brook trout from a small stream in Michigan is shown in Table 10.2. The important features are the proportion of the original cohort that survives to reach age x, l_x, and the number of female eggs produced per female age x, m_x. (Life tables for

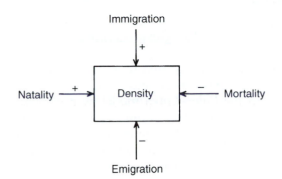

Fig. 10.4 Population density as a function of birth, death, immigration and emigration rates. Modified after Krebs (1994).

Table 10.2 Life table* for the cohort of 1952 in a population of brook trout, *Salvelinus fontinalis*, living in Hunt Creek, Michigan[†]

x (years)	l_x	m_x	$l_x m_x$
0	1.00000	0	0
1	0.05283	0	0
2	0.02063	33.7	0.69523
3	0.00390	125.6	0.48984
4	0.00051	326.9	0.16672

$\Sigma m_x = 486.2$ $\Sigma l_x m_x = 1.352$

*Entries: x, age (years); l_x, probability of surviving to age x; m_x, mean number of daughters produced by a female aged x.
[†]Source: McFadden *et al.* (1967).

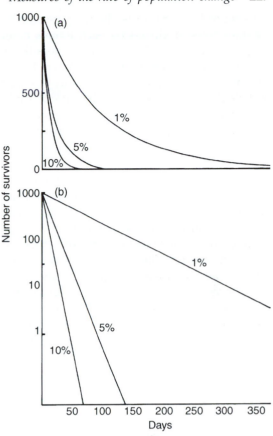

Fig. 10.5 Effects of constant mortality rates of 1%, 5% and 10% per day on number of survivors from a cohort with an initial abundance of 1000. (a) Arithmetic survivorship curve; (b) logarithmic survivorship curve.

males can also be constructed but this is usually more difficult because of problems of estimating the number of eggs fertilized on average by a male of age x.) A plot of l_x, against x gives the survivorship curve for the cohort (Fig. 10.5(a)). It is usual to plot the logarithm of l_x against x, because this yields a straight line if the per capita mortality rate is constant (Fig. 10.5b). A constant mortality rate indicates that the risk of an individual dying in a given time interval does not change with the age of the individual.

The gross reproductive rate (*GRR*) is the average number of daughters produced by a female that lives to the maximum age reached in that cohort:

$$GRR = \sum m_x. \tag{10.6}$$

Most females do not survive to that maximum age, indeed most die before reaching sexual maturity, so a more useful quantity is the net reproductive rate, R_0:

$$R_0 = \sum l_x m_x. \tag{10.7}$$

Net reproductive rate is the average number of daughters produced by each female entering the population. If $R_0 > 1$, the population is increasing in abundance, if $R_0 = 1$, it is stable and if $R_0 < 1$, it is declining. An alternative description of R_0 is the rate of multiplication of the population per generation. For the brook trout cohort shown in Table 10.2, the GRR was 486.2, but R_0 was only 1.352, which illustrates the difference between the potential and achieved rate of increase. The trout life table also shows that the age class contributing most to the R_0 was the two-year-old females. Older and larger females were more fecund, but the chance of a female reaching these older ages was low.

R_0 is an important indicator of the status of a population. It has the disadvantage that it is the rate of change of a population per generation and so depends on the generation length for that population, a biological time unit, rather than an absolute time unit. It is not suitable for use in comparisons of the rates of change of populations per absolute time unit, for example per year.

The life table can also be used to obtain a measure of the rate of increase of the population in absolute time units. This measure is the intrinsic rate of increase of the population, r_m. It is the instantaneous per capita rate of increase of a population with a given l_x and m_x schedule when that population has achieved a stable age distribution and l_x and m_x are not functions of population density (page 238). A stable age distribution is achieved when the relative abundances of fish in each age class in the population do not change even if the absolute abundance of the population changes. Thus if n is the number of females in the population

$$r_m = dn \,/\, ndt \qquad (10.8)$$

for a population with a stable age distribution. It is measured in units of females per female per time unit. The value of r_m for a given population is defined by the Euler–Lotka equation:

$$1 = \sum_{x=\alpha}^{\infty} \exp(-r_m x) l_x m_x \qquad (10.9)$$

where α is age at first reproduction (Charlesworth, 1994). Estimates of r_m are obtained from the Euler–Lotka model by a process of iteration, in which values of r_m are entered successively until summation of the right-hand side of the equation approaches unity to within a predetermined value. An approximate value of r_m, which can act as a useful starting value for the iteration process, can be obtained from the expression:

$$r_m = (\log_e R_0) \,/\, T \qquad (10.10)$$

where T is the generation length of the population in absolute time units and can be approximated as:

$$T = \left(\sum l_x m_x x\right) / \left(\sum l_x m_x\right). \qquad (10.11)$$

For a population that is growing in abundance, r_m is greater than zero; if the population is numerically stable, r_m is zero; and if the population is declining, r_m is negative.

The finite rate of increase of the population (λ), is defined as:

$$\lambda = (N_{t+1} \,/\, N_t) \qquad (10.12)$$

where N_{t+1} and N_t are the population abundances at times $t + 1$ and t. For a population at a stable age distribution it is simply related to r_m:

$$\log_e \lambda = r_m \qquad (10.13)$$

For the cohort of brook trout (life table, Table 10.2), the values were: $R_0 = 1.352$; $r_m = 0.117$ females per female per year; $\lambda = 1.124$; $T = 2.61$ years. These values can then be used to predict the change in abundance of the population on the assumption that the age-specific survival and fecundity schedules for the population do not change (Fig. 10.6).

A disadvantage of these methods is that they use average values for each cohort and so do not indicate whether the values observed are representative of the whole cohort or are generated by subcohorts within

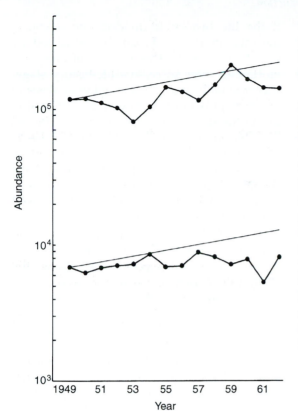

Fig. 10.6 Predicted (thin line) and observed (thick line) abundance of brook trout, *Salvelinus fontinalis*, eggs (upper) and fish (lower) in a Michigan stream. Prediction assumes a net reproductive rate of 1.352. Based on data in McFadden *et al.* (1967).

events (Ruzzante *et al.*, 1996). However, a subcohort of the cod larvae identified on the basis of age at length probably did originate from a single spawning event.

A further problem in the analysis of demographic patterns in fish is that in many populations, a stable age distribution is not attained. Instead, the phenomenon of dominant age classes is observed. At irregular intervals, a cohort achieves an unusually high level of abundance and this cohort is then prominent for several years in the population (Fig. 10.7). Such dominant year classes are important because of the contribution that they frequently make to the yield provided by the population to a fishery. An understanding of the factors that lead to dominant year classes is an important area of research for fisheries biologists (Cushing, 1975, 1982, 1995; Rothschild, 1986).

the cohort. In practice, a cohort may be made up of several groups, each of which has different schedules of growth, survival and fecundity and so different rates of increase. The consequences of an environmental change for these schedules may be different. Molecular genetics now offer the potential for identifying such subcohorts. An analysis of six microsatellite loci (page 221) of cod larvae collected over a three-week period from an aggregation off the Atlantic coast of Canada demonstrated that the aggregation originated from several distinct spawning

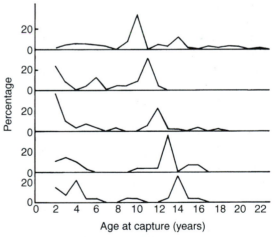

Fig. 10.7 Dominant age class illustrated by the 1959 year class in a population of chub, *Leuciscus (Squalius) cephalus*, in the River Stour, southern England. The prominent peak representing the 1959 year class is clearly visible in catches from (top) 1969–70 (67 fish), 1970–71 (25 fish), 1971–72 (59 fish), 1972–73 (28 fish) and (bottom) 1973–74 (25 fish). Redrawn from Mann (1976b).

10.5 MEASUREMENT OF MORTALITY RATES

The measurement of growth rates (Chapter 6) and age-specific fecundity (Chapter 7) are described earlier. The third important life-history trait is age-specific mortality. Mortality rates in a population of fish can be described by two closely related concepts. The instantaneous per capita mortality rate (Z) is defined by:

$$dN / Ndt = -Z \qquad (10.14)$$

where N is the abundance of the cohort (compare this with the specific growth rate defined in Chapter 6). For a given time interval, t_1 to t_2, Z can be calculated as:

$$Z = (\log_e N_2 - \log_e N_1 / t_2 - t_1) \quad (10.15)$$

where N_2 and N_1 are the abundances of a cohort at times t_2 and t_1. Survivorship, S, is defined as N_2/N_1. So for a unit time interval t to $t + 1$:

$$\log_e S = -Z. \qquad (10.16)$$

The use of Z implies that the decline in abundance of a cohort over the defined time interval is exponential, although the true pattern of mortality within the time interval is unknown. If the number in a cohort at time t_1 is N_1, then the number in that cohort at time t_2 is given by:

$$N_2 = N_1 \exp[-Z(t_2 - t_1)]. \qquad (10.17)$$

Mortality rates measured as Z are in absolute time units. A cohort that declines in abundance from 1000 fish to 100 fish in a year has a Z value of: $(\log_e 1000 - \log_e 100)/1$ per year, i.e. $2.303 \ \text{year}^{-1}$, a survival rate of $e^{-2.303}$ or 10%.

An alternative way of quantifying mortality is to consider the decline in numbers in a cohort over a defined life-history stage, for example the mortality of eggs from fertilization to hatching. A measure of the mortality over this stage, k_{eggs}, is given by:

$$k_{\text{eggs}} = \log_e \text{(number of eggs fertilized)}$$
$$- \log_e \text{(number of eggs hatched)}. \ (10.18)$$

If the life history is divided into n stages, the total mortality, K, can be calculated as: $k_1 + k_2 + \ldots + k_n$. This method of analysing mortalities over successive life-history stages was originally introduced for the analysis of insect populations (Varley *et al.*, 1973), but can be used for fish populations if sufficient life-table information is available. It helps to identify those life-history stages in which changes in the mortality rates cause large changes in population abundance, and those stages in which mortality is related to cohort density. An example of its use is given on page 244. The calculations of the k values require information from the individual cohorts within the population (Dempster, 1975) and so depend on the accurate ageing of the fish (Chapter 6).

10.6 PATTERNS OF MORTALITY IN FISH POPULATIONS

Fish die from intrinsic or extrinsic causes. Intrinsic causes include genetic deaths (presence of lethal alleles in the genotype), physiological failures and diseases such as cancers. Extrinsic causes include the lethal effects of abiotic factors such as temperature, salinity and turbulence, and biotic factors, which include predation (including fishing), parasitism, infectious diseases and malnutrition (Fig. 10.8). There may be interactions between these factors. An inadequate food supply that causes malnutrition may make the fish more susceptible to the effects of parasitism. The intensity and relative importance of the effects caused by these factors will change as the fish get older and bigger. In general, the pattern in fish is for the young stages to suffer much higher rates of mortality than the larger juveniles and adults. This pattern of mortality is described by a Type III survivorship curve (Fig. 10.9). An obvious modification of this pattern is the total mortality of adults associated with reproduction in semelparous populations. Senescent adults also have high mortality

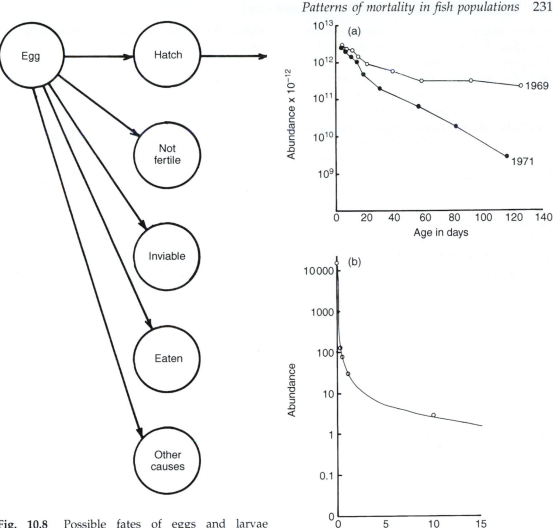

Fig. 10.8 Possible fates of eggs and larvae showing major causes of death in the early life-history stages of fishes and the link between stock (expressed as abundance of eggs) and recruitment. Redrawn from Rothschild (1986).

Fig. 10.9 Survivorship curves illustrated by North Sea plaice, *Pleuronectes platessa*. (a) Survivorship curve for eggs and larvae. Redrawn from Bannister *et al.* (1974). (b) Survivorship curve for juveniles and adults. Redrawn fron Cushing (1975).

rates (Craig, 1985). For large juveniles and adults, it is often assumed that the mortality rate is constant, a Type II survivorship curve. (A Type I survivorship curve shows low rates of mortality at all ages until senescence is indicated by high rates of mortality.)

Mortality of eggs and yolk-sac larvae

Eggs and yolk-sac larvae do not depend on the environment for their food supply, which has been provided by the female. They have little or no capacity to evade predators or malign abiotic factors, unless protection is provided by some form of parental care (Chapter 7). In the absence of parental care, the mortality rates of eggs and larvae can be extremely high. For marine pelagic and demersal eggs, the instantaneous daily mortality rate (Z) has a range of the order of 1.0 to 0.04 per day, that is a survival rate of 37% to 96% per day (Fig. 10.10) (McGurk, 1986). A careful study, in which the survival of a cohort of yolk-sac embryos of the capelin in the St Lawrence Estuary was followed for 48 hours, provided an estimate of Z of 0.8 day^{-1}, a 44% daily survival rate (Fortier and Leggett, 1985).

In marine organisms, there is an inverse relationship between Z and body weight (W), which takes the form:

$$Z = aW^{-b} \qquad (10.19)$$

or in the linear form:

$$\log_e Z = \log_e a - b \log_e W \qquad (10.20)$$

(Peterson and Wroblewski, 1984). Thus, other things being equal, larger, heavier eggs suffer a lower instantaneous mortality rate than smaller eggs. However, the incubation period for larger eggs tends to be longer, so they spend a longer period in this highly vulnerable stage of their life history (Ware, 1975b).

Z for pelagic fish eggs and larvae is higher

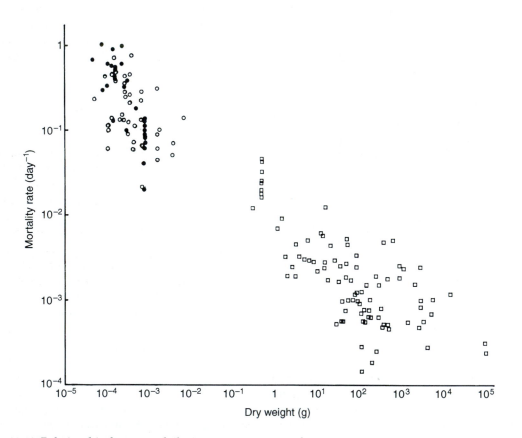

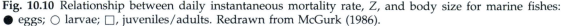

Fig. 10.10 Relationship between daily instantaneous mortality rate, Z, and body size for marine fishes: ● eggs; ○ larvae; □, juveniles/adults. Redrawn from McGurk (1986).

than would be expected from the trend of mortality rate with body weight (McGurk, 1986). This higher mortality rate may be due to the feeding of predators on patches of eggs and larvae. Such patches would be a consequence of the spawning aggregations of the adult fish and the hydrodynamics of the water body in which the eggs are spawned. Schooling, pelagic fishes, both juvenile and adult, can be important predators of pelagic fish eggs and larvae (Hunter, 1981). Predation by northern anchovy adults has been estimated to account for 32% of the daily mortality of their eggs, and there is evidence that the adults feed more intensively on egg patches (Hunter, 1981). Possible invertebrate predators of pelagic fish eggs include ctenophores, cnidarians, chaetognaths and predatory zooplanktonic crustaceams (Bailey and Houde, 1989; Heath, 1992).

The quantitative effects of disease and parasitism on the mortality of eggs and yolk-sac larvae in natural populations have received little study.

Unusual water turbulence may scatter and damage eggs. In a small stream in the English Lake District, which is subject to spates, the number of brown trout eggs dislodged from the redds and drifting downstream increased with an increase in water velocity (Elliott, 1987). The eggs of species that spawn near water margins or in flood waters may become desiccated if water levels fall (Welcomme, 1985).

Parental care does not fully protect developing eggs. In Lake Malawi, cichlids of the genus *Cyrtocara* feed on the developing eggs and larvae of mouthbrooding cichlids. These paedophagous cichlids ram brooding females, causing the latter to disgorge their eggs, which are then consumed (Yamaoka, 1991). In some three-spine stickleback populations, shoals of females raid the nests of parental males, devouring the eggs (Whoriskey and FitzGerald, 1994).

Mortality of larvae and young juveniles

Initially, the mean mortality rates of larval fish that have commenced feeding are similar to those of the eggs (McGurk, 1986). Three factors are probably most important in determining the mortality rate: the availability of suitable prey, the abundance of predators and the spatial distribution of the larvae in relation to patches of suitable food or aggregations of potential predators. The relative importance of these factors in setting the observed mortality rates is not yet clearly understood. The heterogeneous distribution of food and natural enemies, often over relatively small distances, means that different cohorts of larvae, or even different groups of fish belonging to a single cohort, may encounter food and predators at quite different rates. The effects of mortality at this life-history stage on the abundance of the population will depend not only on the mean mortality rate but also on its variance.

Effect of food availability

With the absorption of their yolk sac, the larval fish begin to feed exogenously, and so become dependent on the abundance of food provided by the environment. The transition from endogenous feeding on yolk to exogenous feeding on prey could be a critical period in the life of the larval fish, because the larvae will starve if they do not encounter suitable food soon after absorbing their yolk. A critical period is a time when the rate of mortality is unusually high. Variations in this high mortality rate could be important in determining the subsequent numerical abundance of the cohort, that is its year-class strength (May, 1974; Heath, 1992). Because of the difficulties of quantitatively sampling pelagic larvae from an identifiable cohort, a period of unusually high mortality around the time of yolk-sac absorption has been difficult to demonstrate. In their study of the

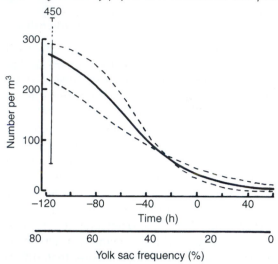

Fig. 10.11 Survivorship in a cohort of capelin, *Mallotus villosus*, in the estuary of the St Lawrence River (Canada), modelled by a sigmoidal curve suggesting higher mortality rates at the time of yolk absorption. Model used to extrapolate survival curve from emergence to final yolk-sac exhaustion; emergence of larvae considered to have occurred at yolk-sac frequency of 75.8% at −120 h. Broken lines give 95% confidence limits for predicted abundance; vertical line at −120 h illustrates the range of abundance at peak emergence. Redrawn from Fortier and Leggett (1985).

survival of capelin larvae in the St Lawrence, Fortier and Leggett (1985) found that a sigmoidal survival curve did fit their data (Fig. 10.11). This curve suggested high rates of mortality about the time of yolk-sac absorption, but the change in mortality was not dramatic (Leggett, 1986). As yet there is no unambiguous evidence that year-class strengths are determined by a critical period at the time of yolk absorption.

Most larval fish are less than 10 mm in total length when they start to feed independently. At this stage their swimming speed (a function of body length – Chapter 4) is low and the distance at which they can perceive prey (the reactive distance – Chapter 3) is short (Miller *et al.*, 1988). Consequently, the volume searched by the larvae is low. Northern anchovy larvae 6–10 mm long may search 0.1–1.0 l per hour (Hunter 1972). The small gape size of the larval fish restricts the size range of prey that can be taken (Chapter 3). The young larvae also lack experience of hunting and seizing prey, and at the onset of feeding their capture success can be low. In the northern anchovy, the initial capture success was 10%, rising to 90% in about three weeks (Hunter, 1972).

Prey suitable for larval fish are not distributed uniformly. In pelagic systems, phytoplankton and zooplankton have patchy distributions. In river systems, prey of larval fish are frequently found in shallow backwaters and regions of slow current (Mills and Mann, 1985). It is only in zones of dead water in rivers that plankton populations can build up high densities. The survival of larvae depends on their encountering a supply of prey of sufficient density, nutritional value and size to allow them to grow. Because mortality rates are inversely related to size, the faster the larvae grow, the less time they will spend in the size classes with the high mortality rates. A significant portion of this size-dependent mortality is probably caused by predation, so food availability and predation rates may interact to determine the overall mortality rate of the larvae (Shepherd and Cushing, 1980).

Temperature may also be an important abiotic factor if higher temperatures increase the supply of food and permit faster growth rates for the larvae (Chapter 6). Mills and Mann (1985) found that year-class strength in cyprinid populations in a river in southern England was often greater when temperatures were high in the first summer of life of the year class. The abundance of two-year-old perch in Windermere was correlated with the temperatures experienced by the fish in their first summer of life (Craig, 1987; LeCren, 1987).

Laboratory studies on the growth and

feeding rates of larval fish have been used to define the concentrations of prey that are required for the survival and growth of the larvae (Hunter, 1981). Northern anchovy larvae stocked at an initial rate of 10 eggs l^{-1} had a survival rate of 51% over a 12 day period at a prey density of 4000 nauplii l^{-1}. At a concentration of 900 nauplii l^{-1}, only 12% survived the 12 days. Average densities of microcopepods in the open sea are much lower, of the order of 50 l^{-1}, off the coast of California, and perhaps an order of magnitude higher in partly enclosed bodies (Hunter, 1981). Studies on the rates of starvation of larval fish using measures of condition (Chapter 6) or histological criteria of starvation (O'Connell, 1980) provided some information on malnutrition of larvae. In the Los Angeles Bight in March, about 8% of the northern anchovy larvae were close to starvation (O'Connell, 1981). This represented about 40% of the average daily mortality.

The apparent discrepancy between the prey densities required by the larvae and the densities in the environment led to the suggestion that the larvae depend on small-scale patchiness in the distribution of the plankton. In these patches the concentrations of prey are sufficient for the larvae, but elsewhere the larvae would suffer much higher rates of mortality and slower growth (Hunter, 1981). The high variability in mortality rates between samples of anchovy larvae in the Los Angeles Bight may indicate the importance of the patchiness of the prey. Lasker (1975) suggested that a stable layer of a high concentration of plankton is important for the survival of anchovy larvae. If this layer is disturbed by storms, the mortality will be higher. Observations on other species, including the menhaden, also suggested that year classes are stronger when the prevailing weather conditions lead to an association between larvae and concentrations of their potential prey (Leggett, 1986).

The results obtained for the northern anchovy may not generalize to other species.

Herring larvae initiated feeding even at densities of nauplii as low as 7.5 l^{-1} (Heath, 1992). Other laboratory studies, experiments that used enclosures in the sea (mesocosms), and estimates of feeding rates of free-living larvae suggested that larvae can feed at or close to satiation levels at the concentrations of prey organisms recorded in the sea (Leggett, 1986; Leggett and DeBlois, 1994). A factor that may increase the rate of encounter between larvae and their prey, especially at low prey densities, is small-scale turbulence (Rothschild and Osborn, 1988; Dower *et al.*, 1997). As turbulence increases, the rate of encounter increases but the rate of successful pursuit tends to decrease. Consequently, there is likely to be an intermediate level of turbulence at which feeding rate is maximized.

The earlier studies on the feeding of pelagic larvae may also have underestimated the importance of some phytoplankton species as food soon after hatching.

Under laboratory conditions, the survival rates of groups of larvae at relatively low prey densities can be variable (Houde, 1978; Pitcher and Hart, 1982). At naturally low prey densities, some larvae may have relatively high survival rates and these would determine the abundance of the year class. The high fecundity of fishes means that it is the exception, rather than the rule, that an individual survives its first few weeks of life.

The importance of the prey size spectrum to larval fish was illustrated by a field experiment on the larvae of the capelin (Frank and Leggett, 1986). The larvae were maintained in large enclosures in Conception Bay, Newfoundland. Predators were excluded from the enclosures and the larvae were fed with zooplankton that varied both in concentration and in size spectrum. The daily growth rate of the larvae was strongly correlated with the density of plankton in the 40–51 µm size class (assuming the particles to be spherical), but not with the total plankton concentration. The daily mortality was inversely correlated with the concentration of prey particles in the 28–80 μm classes, but positively correlated with the concentration of particles in size classes above 81 μm. Along the Newfoundland coast, the size spectrum of the plankton depends on the direction of the prevailing winds. The water mass associated with onshore winds is richer in zooplankton in the small size classes, which support the growth and survival of the larvae. In contrast, offshore winds result in the upwelling of a water mass richer in larger plankton and so less suitable for the larvae. Thus, for the capelin, the survival and growth rate of the larvae partly depend on an interaction between a climatic factor, wind direction, and a biotic factor, the composition of the zooplankton.

As a general hypothesis, Cushing (1975, 1982, 1990) suggested that much of the variability in the strength of age classes in temperate marine fishes is related to the effect of climatic conditions on the timing of the production cycle in the sea. He argued that the timing of the breeding season of the fish shows relatively little temporal variation, but the timing of the production cycle, hence the appearance of suitable prey for the larval fish, varies from year to year because of climatic variation. In some years, the timing of the appearance of the larvae may be a better match to the availability of food than in other years: the match–mismatch hypothesis. Those years in which the match is good should be characterized by strong year classes. Those years in which the appearance of larvae is mismatched to the availability of prey would generate weak year classes. This hypothesis assumes that the important source of variability in larval survival and growth is their food supply. A critique of the match–mismatch hypothesis suggested that if there is a mismatch between the appearance of the larval cohorts and their food supply, a weak year class is likely. However, if the match is good, both strong and weak year classes can be the result (Leggett and DeBlois, 1994). This suggests that there is a threshold level of food. If the level is below the threshold, survival of larvae is poor. If the level is above the threshold, other factors will determine whether survival of larvae is poor or good. An obvious factor is predation.

Effects of predation

The quantitative effects of predators on larval mortality rates are still poorly known (Bailey and Houde, 1989; Heath, 1992). The rates for capelin larvae in enclosures were significantly lower than rates for unenclosed larvae (Frank and Leggett, 1986). In the enclosures, the rate was about 10% per day, but rates of 55–65% day^{-1} were estimated for the free-living larvae. Predators were excluded from the enclosures so the significant difference in mortality rates may reflect the important of predation. Similar results have been obtained for herring, cod and flounder, suggesting that predation is a major cause of mortality of larvae (Bailey and Houde, 1989).

Effects of predation may not be independent of larval food abundance. Poorly fed larvae will grow slowly and so may remain vulnerable to predators for longer than fast-growing larvae. Smaller larvae have poorer sensory (Chapter 2) and swimming capacities (Chapter 4) and so are less likely to escape from a predator if detected. Their small size also makes them vulnerable to a wider range of predators, both size- and gape-limited (Chapter 8). A disadvantage of large size is that the reaction distance of predators will be greater (Chapter 3), and larger larvae may be more profitable for the predators (Chapter 3). Although it is often assumed that the risk of being eaten by a predator declines as larvae increase in size, the relationship between mortality and size will reflect the balance between the advantages that accrue with size (Miller *et al.*, 1988) and the disadvantages (Leggett and DeBlois, 1994) (Chapter 8).

Effect of spatial processes

Eggs and larvae may be lost from a population, not through predation or starvation, but because water movements take them out of the usual localities of that population (Sinclair, 1988). Consequently, even if they survive, these vagrants will not form part of the self-sustaining population that produced them. They may enter other populations as migrants or simply eventually die. Sinclair (1988) drew a distinction between losses from a population caused by these spatial processes and losses through energetic (food-chain) processes such as starvation and predation. He argued that spatial processes can be more important in determining the abundance of a population than energetic processes. The flux of fish larvae into and out of Conception Bay, Newfoundland, was estimated in midsummer (Pepin *et al.*, 1995). Water movement accounted for 12–75% of the variation in abundance of the larvae (Pepin *et al.*, 1995). In assessing the importance of starvation and predation, the effects of physical processes on eggs and larvae have to be taken into account. This requires careful design of sampling programmes for both the abundance of larvae and physical processes, including current speed and direction. The sampling programme has to select the appropriate spatial and temporal scales to reveal the effects of spatial and food chain processes (Pepin *et al.*, 1995).

Consequence of early mortality

Because of the high fecundities of fishes, slight differences in the mortality rates of the eggs and larvae can produce major differences in the number of juveniles produced. A simple calculation illustrates this. Assume that two cohorts of spawners each produce 10^7 eggs. The eggs and larvae of one cohort experience a daily instantaneous mortality rate, Z, of 0.1 day^{-1}. This is a survival rate of 90.5%. After 60 days there are 24 787 survivors. The eggs and larvae from the other cohort have a daily mortality rate of 0.05, a survival rate of 95.1%. After 60 days, there are 497 871 survivors. In practice it might be difficult to detect the difference between the two survival rates, yet the absolute number of recruits produced differs by over 400 000. This illustrates the potential that variations in egg and larval mortality rates have for generating variations in the numbers of juvenile and adult fish. Unfortunately, the estimation of the abundance and mortality rates of these early life-history stages to a useful level of accuracy and precision is frequently far more difficult than the estimation of similar population variables for larger fish. The problem becomes even more difficult if the mortality rates vary spatially and temporally because of the patchiness of food or of predators. Furthermore, the critical events may be taking place over a short time scale, a matter of a few hours or days, and in small volumes of water (Leggett, 1986). These sampling difficulties make experimental studies important. Well-designed experiments, both in the laboratory and in the field, permit a quantitative analysis of the effects of factors such as prey density and size spectrum, predators and abiotic factors. From such studies, predictive models can be developed which can then be tested under natural condition (Laurence, 1981).

Mortality of older juveniles and adults

The instantaneous mortality rates of juvenile and adult fish are much lower than for eggs and larvae (Fig. 10.10). McGurk (1986) lists the daily Z values for juveniles and adult marine fish. These vary from 1.9×10^{-2} (survival: 98.1%) for juvenile plaice weighing 0.5 g, to 9.9×10^{-4} (survival: 99.9%) for *Trichiurus lepturus* weighing 61 g.

The causes of death of juveniles and adults probably include predation (including fishing), disease and parasitism (Chapter 8) and, more rarely under natural conditions, exposure to lethal abiotic conditions (Chapter 4) or senility (Craig, 1985). The plasticity of growth of fish, their ability to lay down reserves of lipids and their capacity to survive long periods without food may reduce the importance of starvation as a cause of death.

There is evidence of size-selective, over-wintering mortality in several species living in mid to high latitudes, particularly in young-of-the-year juveniles experiencing their first winter. Examples include coastal marine species such as the Atlantic silverside (Conover, 1992) and freshwater species such as yellow perch (Shuter and Post, 1990). Smaller fish that enter the winter period with inadequate reserves have poorer survival. Insufficient energy stores may cause die-offs of the alewife in Lake Michigan during winter (Flath and Diana, 1985).

Once the fish have reached sexual maturity, there may be an increased rate of mortality associated with reproduction (Chapter 11).

With the exception of a few unusual circumstances, mortality rates of the older juveniles and adult fishes show too little variation to account for the variability in recruitment that is a characteristic of so many fish populations (Rothschild, 1986). Year-class abundance is usually assumed to be determined by variations in the fecundity of the spawning females and variations in the mortality rates experienced by cohorts in their first year of life, for many species perhaps early in that first year (Cushing, 1982). A simple regression model developed for perch in Windermere accounted for a high proportion of the variance in the abundance of two-year-olds (Craig, 1987; LeCren, 1987). The regression predicted that abundance increased with higher temperatures in the year of hatch, but decreased with an increase in the biomass of adult perch in the year of hatch and in the abundance of the pike cohort in the same year. These last two terms probably reflect the effect of intra- and interspecific predation on the young perch.

10.7 REGULATION OF FISH POPULATIONS AND THE STOCK–RECRUITMENT RELATIONSHIP

Although fish populations vary to a greater or lesser extent in abundance, archaeological and historical evidence suggests that populations may persist in a geographical area for long periods of time. This persistence in time raises the possibility that there are effects which tend to produce an increase in population density at low levels of density and a decrease at high levels – the possibility of population regulation. The high fecundity that characterizes most teleosts could result in far higher levels of variation in population density than is usually observed. One female cod, if all her eggs survived, would replace herself by several million offspring, but if no eggs survived by none, a potential variation of about seven orders of magnitude. The abundance of a population will be defined by mean density (numbers per unit area) and the area over which the population is distributed. Abundance may be determined by the size of the spawning locality used by that population (Sinclair, 1988).

Density dependence and regulation of population abundance

A continuing debate in population ecology is the extent to which the density of a population has an effect on its age-specific mortality and fecundity schedules. The existence of such density-dependent effects has important implications for the dynamics of the population, especially for its response to environmental changes. Fuller discussions of this debate, with examples from a range of organisms, can be found in recent textbooks on ecology (Krebs, 1994; Begon *et al.*, 1996). The crux of the discussion is the concept of regulation. If per capita mortality and fecundity rates are independent of density, then the density of a population will fluctuate through time. These fluctuations are driven directly by those factors that affect death and birth rates, unmodified by any effects of population density. Except by chance, reduced mortality and increased fecundity will not ameliorate low densities, nor increased mortality and reduced fecundity reduce high densities. There is no sense in which the population can be thought of as regulated.

If the per capita death rate tends to increase with an increase in population density (density-dependent mortality), or if the birth rate tends to decrease with an increase in population density (density-dependent natality), then there may be an equilibrium population density at which the birth and death rates balance each other. If the density drops below this equilibrium, the death rate will decrease and/or the birth rate will increase and so the population will return to its equilibrium density. An increase in population density above the equilibrium has the reverse effect on birth and death rates and so again the population tends to return to equilibrium (Fig. 10.12). The value of the equilibrium population density depends on the overall, prevailing birth and death rates (Fig. 10.12). Emigration and immigration rates may also be functions of population density, helping to keep the density in a locality at equilibrium (Chapter 5). Such density-dependent effects tend to buffer the direct effects of the factors that cause changes in mortality or fecundity. For fishes and similar organisms in which fecundity (Chapter 7) and the risk of predation are functions of body size (Chapter 8), density-dependent growth will also be an important demographic process.

This debate has implications for the management of fisheries. If density-dependent mortality is important in a fish population, then fishing will substitute one form of mortality for another. Density-dependent growth and fecundity may also compensate for the reduction in population density caused by fishing (Rijnsdorp, 1994). If density-depen-

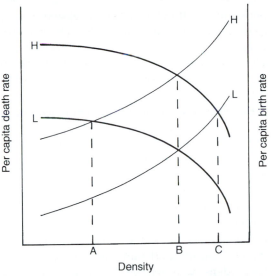

Fig. 10.12 Density-dependent birth (thick lines) and death (thin lines) rates and the concept of an equilibrium population density. H and L represent environments in which the rates, independent of density, are high and low. In the two environments, the effect of a unit increase in density is the same and it is assumed that the relationship between rates and density is deterministic. Equilibrium densities A, B and C are defined by the densities at which birth and death rates are equal: A, high death rate, low birth rate; B, high birth rate and high death rate or low birth rate and low death rate; C, low death rate and high birth rate.

dent effects are absent in the population, the fishing will be an additional burden of mortality imposed on the population. The reproductive rate of the population must be sufficiently high to cover this extra burden or the average net reproductive rate of the population will drop below unity and the population will decline. The crucial questions then are empirical. In a population, can density-dependent effects be detected? If they are present, what are the consequences for the dynamics of the population? A density-dependent effect may be present, but may be too weak to result in any significant regulation of the population. A strong density-dependent effect can even destabilize a population by increasing the chance that very low densities are reached (Boer, 1987). The strength of density-dependent effects may vary both spatially and temporally (Bailey, 1994).

In principle, density dependence in a natural population is detected by plotting the per capita mortality or fecundity against density and testing for a significant relationship. A common method is to test for density-dependent effects operating during a given life-history stage by plotting the per capita mortality within that stage, the k value, against the initial density of that stage (or its logarithm) (Dempster, 1975). In practice, density-dependent effects may be obscured by the variability of the data, by curvilinear relationships between mortality (or fecundity) and density, by too few data points, or by the data covering an insufficiently wide range of densities. Even after the detection of apparently density-dependent effects in a natural population, it may be difficult to determine the relevant causal relationships. A further problem is that the effect of a given population density on the rates of mortality, fecundity or growth may vary depending on other environmental factors and the physiological condition of the individuals. There is the danger of treating the population as the unit experiencing the density-dependent effects. Any effects are generated because of the response of individual fish to some factors that are affected by the presence of fish of the same species.

A fastidious analysis of mortality in the early life-history stages of plaice in European seas illustrated the potential of density-dependent mortality to stabilize population density (Iles and Beverton, 1991; Beverton and Iles, 1992a,b). The process of regulation was quantified by calculating the ratio of the highest to the lowest density at a given age over a series of year classes. In the analysis, data from 1970 to 1985 were used. The eggs of plaice hatch into pelagic larvae, which are carried by currents to coastal nursery areas. On the nursery grounds, the larvae complete the metamorphosis that transforms them into the bottom-living, laterally compressed flatfish. The analysis estimated that the egg stage was dominated by density-independent mortality that was remarkably constant from year to year. However, during the pelagic larval phase, density-independent mortality showed great variability. Consequently, at the time of arrival on the nursery grounds (June), the ratio of maximum to minimum densities was in the region of 190–300. The highest density of larvae observed in June was about 200 times greater than the lowest observed. Within a month on the nursery grounds, the ratio had reduced to 29–38. By October, the ratio was 10–13. By October of the following year, when the juvenile fish were over one year old, the ratio was down to 4.0.–4.5. This analysis helps to explain the stability of plaice populations in areas such as the North Sea (Fig. 10.14 below). Density-dependent mortality over the first two years of life has been detected in several other demersal, marine species (Myers and Cadigan, 1993).

Stock–recruitment relationship

The relationship between the number of offspring and the number in the parental cohort is called the stock–recruitment relationship. These terms reflect the interest of fisheries biologists concerned with the management of a stock of fish, which is exploited by a fishery. That stock, through its reproduction, must also supply the recruits to the fishery. In the context of fisheries management, the age at recruitment refers to the age at which the fish in a population first become available to the fishery.

In the absence of any density dependence, the number of offspring of a defined age will be an increasing function of the abundance of the parental cohort (Fig. 10.13(a)). If density-dependent effects on mortality, fecundity and growth are important, then this should be reflected in the stock–recruitment relationship. Two types of stock–recruitment curve incorporating density dependence have been the basis for much study. A stock–recruitment curve proposed by Beverton and Holt (1957) assumes that there is an asymptote which the curve approaches at high stock densities (Fig. 10.13(b)). The curve is described by:

$$R = 1 / [a + (b / P)] \qquad (10.21)$$

where R is the number of recruits (offspring), P is the size of the parental stock, which can be measured in numbers of adults or egg production, with a and b the parameters of the curve. Ricker (1975) suggested a dome-shaped curve (Fig. 10.13(c)), which can be described by the relationship:

$$R = aP \exp (-bP) \qquad (10.22)$$

in which a is the slope of the curve when P is very small and so is an index of density-independent effects, and b is an index of density dependence. Ricker (1975) suggested that his curve is more appropriate when the cause of the density dependence is cannibalism of the young by adults, or there is an increase in the time it takes for the young to grow through a vulnerable size range, or when there is a time lag in the response of a predator or parasite to the abundance of the fish being attacked. The Beverton–Holt curve is more appropriate if there is a maximum abundance imposed by food availability or space, or if the predator can adjust its predatory activity immediately to changes in prey abundance. Stock–recruitment relationships derived from bioenergetics principles are described in Chapter 11.

Much ingenuity has been spent in fitting these curves to data sets and to elaborating the basic models. All this effort has largely foundered in the face of the variability in the relationships between stock and recruitment shown by most natural populations. The curves can be fitted, but it takes an act of

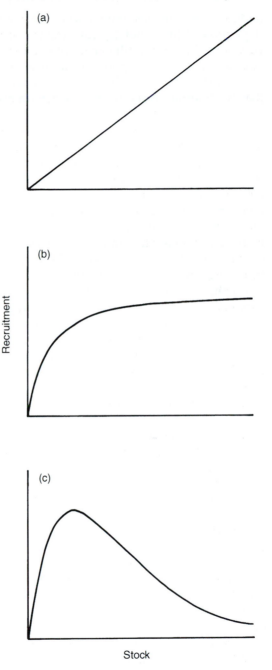

Fig. 10.13 Stock–recruitment relationships: (a) density-independent relationship; (b) Beverton and Holt curve; (c) Ricker curve.

faith to take the resulting curves seriously (Fig. 10.14). This is not to deny that there is an underlying relationship between spawner abundance and subsequent recruitment, but to question whether it is useful to describe that relationship in the form of a deterministic curve (cf. Iles, 1994).

An analysis of over 300 time series describing stock and recruitment in marine, freshwater and diadromous species revealed some basic characteristics of the relationship (Myers and Barrowman, 1996). Highest recruitment tended to occur when spawner abundance was high, and the lowest recruitment when spawner abundance was low. Mean recruitment was usually higher when spawner abundance was higher rather than lower than the median abundance. However, fishes typically have high fecundities (Chapter 7). It requires only moderate variability in the mortality rate experienced by pre-recruits to prevent the definition of a simple stock–recruitment relationship (Koslow, 1992).

A further factor makes a simple, well-defined relationship unlikely. It is unrealistic to expect a population-specific value for a parameter that describes density-dependence such as the b in the Ricker equation. Density dependence results from a response of individuals to the consequences of population density. This response will depend on the characteristics of the individuals. The same density of spawners may generate different numbers of recruits because of the individual qualities of both spawners and potential recruits. In describing the stock–recruitment relationship for a given population, the problem is to define the likely range of recruits from a given spawner density. In fisheries management, the risk of recruitment falling below the level required to sustain an exploited population has to be assessed on the basis of the variability shown in the stock–recruitment relationship.

At the core of the problem is the failure of simple models to incorporate sufficient biolo-

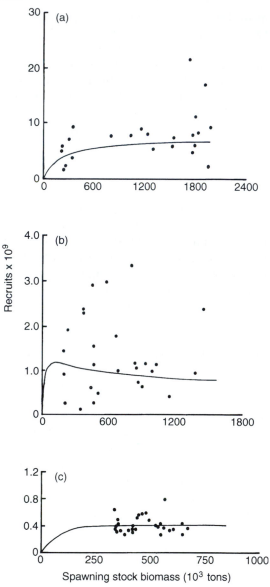

Fig. 10.14 Examples of variability in relationship between recruitment and stock. Curves are stock–recruitment function of Shepherd (1982). (a) North Sea herring, *Clupea harengus*; (b) Arcto-Norwegian cod, *Gadus morhua*; (c) North Sea plaice, *Pleuronectes platessa*. Redrawn from Rothschild (1986) and based on figures in Garrod (1982).

gical and environmental detail to provide useful summaries of the biological data (Rothschild, 1986). An improvement in the understanding of stock–recruitment relationships will only come with improved understanding of the biological processes that underlie the relationship. Such an understanding requires a combination of detailed, long-term demographic studies of species of interest, combined with experimental studies on the quantitative effects of biotic and abiotic factors on mortality, fecundity and growth which provide an insight into the causal factors. The stock–recruitment relationship is an effect of what is happening to individual fish, not a fixed property of the population to which individuals conform.

There is an important message for fisheries managers. Although the stock–recruitment relationship may be diffuse, low spawner densities tend to generate low recruitment. There is a danger that if excessive fishing reduces the density of spawners too low, the number of recruits will be, in the long term, too few to sustain the fishery – recruitment overfishing.

For open populations (page 220), the decoupling between the size of the parental population at a locality and recruitment at that locality precludes the definition of a true stock–recruitment relationship, except at the metapopulation level. For such populations, the abundance of spawners at a locality will depend on factors that determine the rate of recruitment to that locality from the larval pool and factors intrinsic to that locality that determine post-recruitment survival and growth (Pfister, 1996; Steele, 1997) (Chapter 12).

Use of otoliths and genetic markers in recruitment studies

Two technical advances promise to provide an increased understanding of the factors that influence the stock–recruitment relationship. The daily increments laid down in the otoliths can be used to calculate an approximate date of hatching for juvenile fish caught months after hatching (Chapter 6). An analysis of the otoliths of one-year-old bluegill sunfish in Lake Opinicon (Canada) revealed the effect of differences in time of hatching in 1993 for survival into 1994 (Cargnelli and Gross, 1996). Although the fry produced early in the 1993 breeding season formed only a small proportion of the total fry hatched in 1993, they formed a relatively high proportion of the survivors in 1994. Fry hatched early in the breeding season had the advantage of a longer growing season before the onset of winter. The use of otoliths will answer the question of whether there are particular, favourable times in a breeding season during which a high proportion of the future recruits is produced.

Family-specific DNA fingerprints were used to identify male smallmouth bass, *Micropterus dolomieu*, that had fathered young-of-the-year juveniles collected in September (Gross and Kapuscinski, 1997). Only 5.4% of spawning males produced nearly 55% of the fry collected, although the study was not able to identify the factors responsible for this differential success. The use of these molecular methods will answer the question of whether recruits are drawn from a restricted subset of the parental generation.

The intensive application of these two methods to a range of species will help to identify the extent to which each recruitment episode is unique and the extent to which generalizations on the stock–recruitment relationships can be generated.

10.8 EXAMPLES OF FISH POPULATION DYNAMICS AND THE STOCK–RECRUITMENT RELATIONSHIP

Salmonid populations

Salmonid populations in small streams have provided some of the best examples of long-term population studies. A small stream facilitates the quantitative sampling required to obtain life-table information.

A pioneering study of a non-migratory population of brook trout in a stable Michigan stream obtained life tables for eleven successive annual cohorts for the period 1949 to 1959 (McFadden *et al.*, 1967). The mean net reproductive rate was 1.14, indicating that the population tended to increase slightly. This was because of a slight improvement in the survival of adult fish. The range of densities recorded during the study was small, which made the detection of density-dependent effects difficult. The most striking result was that the absolute number of fish that survived to reach an age of 12 months varied little regardless of the parental egg stock, which ranged from 8×10^4 to 21.2×10^4.

Elliott has conducted an outstanding long-term study, starting in 1967, of the population dynamics of a migratory trout, *Salmo trutta*, in a stable stream, Black Brows Beck, in the English Lake District (review, Elliott, 1994). Density-dependent mortality in the first year of life was also a feature of this population (Elliott, 1984, 1985a,b,c). The trout migrate from the stream to the estuary and the sea, usually at the start of their third year of life. Mature fish return to spawn after spending a summer or more than a year at sea. Sufficient quantitative data were obtained on this population to allow an analysis of the population dynamics using the *k* values for successive life-cycle stages (Fig. 10.15).

The *k*-factor analysis showed that the number of eggs or alevins (newly hatched young) in a cohort was the main factor affecting the number of survivors in that cohort over its life span. The eggs buried in redds could have a survival rate of over 90% for a three-month period, and the buried alevins could have similarly high rates of survival (Elliott, 1984). Losses in the first weeks after the emergence of the alevins from the gravel in spring were strongly density dependent. During the first summer

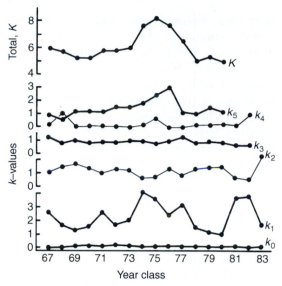

Fig. 10.15 A *k*-factor analysis of cohorts of migratory trout, *Salmo trutta*, population of Black Brows Beck, NW England. Loss rates: k_0, alevins; k_1, first spring; k_2, first summer; k_3, first winter; k_4, second summer; k_5, second winter and post-migration; K, total loss, i.e. $k_0 + k_1 + \ldots + k_5$. Redrawn from Elliott (1985a).

of life, losses were weakly density dependent. The relationship between egg stock and number of survivors one month after emergence was strongly dome-shaped and was described well by the Ricker stock–recruitment curve. Although the effects of this density dependence were detectable in the relationship between parental egg stock and progeny egg recruitment, the data points were widely scattered around the fitted Ricker curve (Fig. 10.16).

The growth in body size of trout in a cohort was not related to density, but the coefficient of variation in size at a given age was. Variation in the weights of fish in a cohort when they had reached a given age was greater the smaller the number of eggs (or alevins) that had generated that cohort. This effect on growth was established in the first weeks after emergence of the fry from

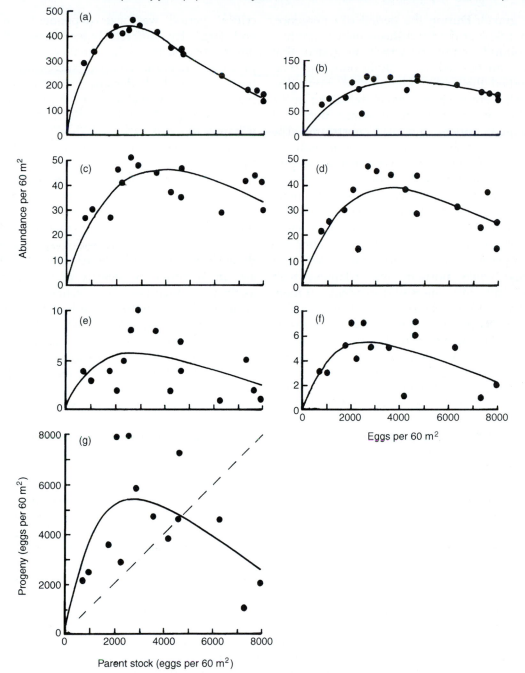

Fig. 10.16 Relationships between mean number of eggs and number of survivors for 14 cohorts of the migratory trout, *Salmo trutta*, population of Black Brows Beck. Lines are the fit of Ricker's stock–recruitment relationship. Number of survivors at (a) 0+, May/June; (b) 0+, August/September; (c) 1+, May/June; (d) 1+, August/September; (e) 2+, May/June; (f) 3+, November/December; (g) total eggs produced by cohort. Diagonal shows line of exact replacement. Redrawn from Elliott (1985b).

the gravel. During the period of emergence, about 80% of the fry drifted downstream in a moribund condition, which suggested that they had not fed successfully once they had resorbed their yolk (Elliott, 1986). Some fry, which were feeding actively, remained in the region of the redd from which they had emerged; others migrated upstream. The proportion that migrated increased with the number of eggs in the redd, that is the proportion was density dependent.

Clearly, the events occurring in these early weeks were critical to the demography of this population. Detailed studies of eight cohorts identified a critical period after the emergence of the fry from the gravel as a period when mortality was relatively high and strongly density-dependent (Elliott, 1989a,b, 1994). Subsequent to this critical period, mortality was not related to density. The length of this critical period was inversely related to density, with the period lasting longer at lower densities. The end of the critical period was not defined by when the fish reached a given size, rather the size of the fish at the end of the period was greater the longer the period (Elliott, 1990a).

Observations on the behaviour of fry of four cohorts in the first weeks after emergence suggested an interpretation of these observations (Elliott, 1990b, 1994). After emergence, the fry fell into two groups, fish defending feeding territories and non-territorial fry. The territorial fry formed groups structured by a dominance hierarchy (Chapter 5). Within a group, territory size was positively related to fish size (Chapter 5). Only territorial fish survived to the end of the critical period. During this period, territorial fish directed high levels of aggression at non-territorial fry. By the end of the critical period, the number of groups had stabilized and the variation in the number of fry within each group had decreased. At low densities, the variation in the size of the fish at the end of the critical period was high. At high densities, mortality during the

critical period was size dependent. Small and large fry suffered higher losses than intermediate-sized fish. Smaller fry were probably unable to compete successfully for territories. At high densities, the largest fry may have been unable to meet the energy costs associated with territorial defence (Chapter 4).

Thus, the territorial behaviour of the fry in the first weeks of life plays an important role in population regulation in this population of trout. A survey of several salmonid species also provided evidence that territory size can limit maximum density of juvenile salmonids in shallow streams (Grant and Kramer, 1990). Because territory size increases with body size in salmonid fry, density of fry declines as their size increases. This decline in density with an increase in body size had been interpreted as an example of self-thinning, in which the rate of decline is related to the increase in metabolic demand with an increase in body size (Chapter 4) (Elliott, 1993). A more detailed analysis of the pattern of decline in density with body size for the Black Brows Beck population suggested that there was not a simple and consistent self-thinning rule applicable after the critical period (Armstrong, 1997).

Years that were unusually dry had an adverse effect on the Black Brows Beck trout populations, particularly on the survival of trout in their second year of life. Despite the strong density-dependent effects early in the life cycle of the fish, the effect of unfavourable abiotic factors such as drought meant that the abundance of the population did not come to an equilibrium value, but fluctuated in time. Surprisingly, these fluctuations in abundance were not smaller relative to total abundance than the fluctuations in a nearby population of non-migratory trout in a more unstable stream, Wilfin Beck (Elliott, 1987). The density of trout in Wilfin Beck was lower than in Black Brows Beck, and there was no evidence of density-dependent losses at any stage in the life history. Recruitment tended to be directly dependent on parental stock. This is probably because the effects of drought and spates in Wilfin Beck kept the density of the population below that at which density-dependent effects became important. The demography of this trout population was dominated by the density-independent effects of abiotic factors. This comparison of two geographically close trout populations gives support to the hypothesis that the importance of density-dependent effects on life-history traits increases in benign environments (Chapter 11). In hostile environments, the effects of abiotic factors, which are independent of density, mean that populations rarely, if ever, reach abundances at which density-dependent effects can become manifest (Krebs, 1994).

The long-term study of the Black Brows Beck trout has provided a unique picture of the inter-relationships between behaviour, growth and demography in a fish population. It is one of the few studies in which a critical period of unusually high mortality early in life has been clearly identified. Unfortunately, the relatively unusual characteristics of salmonids – anadromy, large egg size, large size of fry on emergence and relatively low fecundity – makes it difficult to generalize its results to other, non-salmonid species.

Population dynamics of clupeid fishes

Clupeids (Clupeidae) are typically pelagic, shoaling fishes. They usually have a short or medium life span, reaching sexual maturity at the age of 1 year or 2 years, so the populations have only a small number of adult age classes. In many parts of the world, they are or have been the prey of major commercial fisheries. Modern fisheries rely on sonar to detect shoals, which are then fished with purse seines. These techniques are effective and allow shoals to be exploited even when the overall population abundance is low. Exploited clupeid populations have a dis-

concerting tendency to crash, with adverse effects on the fisheries (Murphy, 1977; Beverton, 1990). Two classic examples will be described.

The Peruvian anchoveta, *Engraulis ringens*, is the basis of a fishery off the coast of Peru. In most years cool, nutrient-rich water wells up along this coast and supports the productive plankton community which provides the food of both the larval and adult anchoveta. In turn the anchoveta is the prey of piscivorous birds, including cormorants, boobies and pelicans, which establish dense breeding colonies along the coast of Peru. The droppings of these birds form deposits of guano which are mined for fertilizer.

From a small base in the 1950s, the anchoveta fishery grew until it was the largest fishery in the world in the late 1960s (Cushing, 1982). In 1972 (and again in 1983), the exploited stock collapsed, with the catch dropping to one-third of peak levels (Fig. 10.17). The collapse was associated with an El Niño event. El Niño is a current which periodically floods the coastal Peruvian waters with warm, nutrient-poor waters and

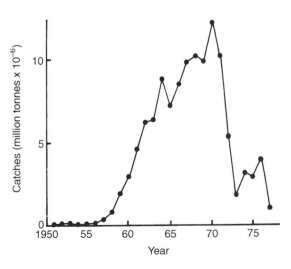

Fig. 10.17 Growth and collapse of fishery for the Peruvian anchoveta, *Engraulis ringens*. Redrawn from Cushing (1982).

persists with some fluctuations for up to 18 months. During El Niño, the production of plankton is reduced. The current is caused by atmospheric and oceanic events, which encompass most of the Pacific at low latitudes (El Niño–Southern Oscillation, ENSO). During El Niño, part of the stock of anchoveta shifts southwards into more fertile waters until El Niño recedes. If El Niño is particularly intense, as it was in 1972 and again in 1982–83, recruitment may also decline. The guano birds also suffer a decline in abundance during an intense El Niño, with the adults abandoning their breeding attempt as a response to the decline in their prey. In the absence of a fishery, the anchoveta stocks can recover quickly as normal upwelling conditions resume. The collapse in 1972–73 was probably caused by the intensity of El Niño combined with a continued fishing pressure on a reduced stock, which prevented a full recovery once El Niño had receded (Cushing, 1982).

Off the coast of California, populations of the Californian sardine, *Sardinops sagax caeruleus*, supported a fishery immortalized by the novelist John Steinbeck in *Cannery Row*. The fishery collapsed in 1949–50. Subsequently the northern anchovy has become the dominant clupeid in the area (Fig. 10.1). The causes are not fully understood but the collapse was probably a result of overfishing exacerbated by a climatic change (Murphy, 1977; Cushing, 1982). Evidence that long-term climatic changes may be a causal factor come from analyses of fish scales that are preserved in anoxic sediments off the Californian coast (Soutar and Isaacs, 1974). Their analysis suggested that in that locality from about 1785 onwards, the sardine and anchovy have alternated in dominance. Sardines were abundant from about 1820 to 1865 and from 1890 to 1945. Other examples of long-term fluctuations in clupeid populations, which seem to be correlated with climatic changes, are given in Cushing (1982). The causal link between climate and fish

abundance is poorly understood (Shepherd *et al.*, 1984).

The northern anchovy populations have been the subject of an investigation which has provided new insights into the dynamics of clupeid populations (Lasker, 1985; Peterman *et al.*, 1988). Cohort analysis (page 225) provided estimates of year-class strength over a decade (Table 10.3) (Peterman *et al.*, 1988). Quantitative analyses of fecundity and the survival rates of eggs, yolk-sac larvae and feeding larvae up to 19 days old were obtained by careful sampling programmes (Lasker, 1985; Lo. 1986). The analyses suggested that survival up to the age of 19 days was correlated with stable weather conditions. These conditions allowed the development of concentrations of plankton on which the larvae feed. But there was no correlation between the number of larvae surviving to 19 days and the subsequent number reaching 1 year of age (Fig. 10.18). There was no evidence for a critical period in the first few days of life, which would determine year-class strength (Peterman *et al.*, 1988). Critical periods may be important in other species, but the study of the northern anchovy did show the importance of information about the fate of fishes throughout the first year of their life for an understanding of population dynamics.

10.9 MODELS OF POPULATION GROWTH

A central aim of demographic studies must be the development of predictive models for populations of interest. The stock–recruitment models of Ricker and other workers represented a step in the development of such models, but their lack of biological detail makes them caricatures of reality. A similar lack of biological detail renders models such as the influential logistic model of population growth and the Lotka–Volterra model of predator–prey dynamics inadequate except as teaching tools used to

Table 10.3 Abundance* of central population of northern anchovy, *Engraulis mordax*, estimated by virtual population (cohort) analysis[†] (see also Fig. 10.18)

Year	Age 1	Age 2	Age 3	Age 4	Age 5+
1965	10 794	5 744	1 965	1 339	635
1966	5 494	5 695	2 917	959	924
1969	14 241	7 779	1 707	473	528
1972	56 233	24 496	5 151	1 013	375
1975	17 806	21 822	17 177	6 118	2 547
1978	9 390	13 462	1 211	754	1 099
1979	76 561	3 457	4 276	355	500
1980	58 144	24 737	885	966	170
1981	45 273	18 913	6 634	208	233
1982	15 537	11 063	3 319	908	47
1983	51 305	3 667	1 728	392	84
1984	24 174	17 641	1 046	438	107
1985	54 693	8 980	5 718	313	150

*Expressed as millions of fish in each age class at time of spawning.
[†]Source: Peterman *et al.* (1988).

highlight some basic principles of population dynamics. Promising techniques for modelling the dynamics of fish populations require a richness of detail that is provided by the information such as that found in life tables.

Life-table methods

An analysis of the losses over successive stages in the life cycle using the k values can provide the basis for a predictive model (Elliott, 1985a, 1994). From a given starting abundance of eggs, the k values for each succeeding stage can be used to predict the number of survivors at the end of the stage, and so the number of fish that become sexually mature adults, and hence the number of eggs that will be produced by that cohort.

The Leslie matrix provides an alternative method of using life table data. This matrix, **A**, defines the age-specific survival rates and fecundities (Caswell, 1989; Stearns, 1992).

$$\mathbf{A} = \begin{bmatrix} F(0) & F(1) & F(2) & \cdots & F(l-1) & F(l) \\ p(0) & 0 & 0 & \cdots & 0 & 0 \\ 0 & p(1) & 0 & \cdots & 0 & 0 \\ 0 & 0 & p(2) & \cdots & 0 & 0 \\ 0 & 0 & 0 & \cdots & p(l-1) & 0 \end{bmatrix}$$

where $F(i)$ is the age-specific fecundity of females in age class i weighted by the probability that they survived from age $i-1$ to i, $p(i)$ is the probability of survival from age class i to $i+1$, and l is the maximum number of age classes in the population (it is assumed that numbers have been counted once a year at the end of the breeding season). If the abundances of fish in each age class at time t are given by the column vector $\mathbf{n}_t$:

$$\mathbf{n}_t = \begin{bmatrix} n(0) \\ n(1) \\ n(2) \\ \cdots \\ n(l-1) \\ n(l) \end{bmatrix}$$

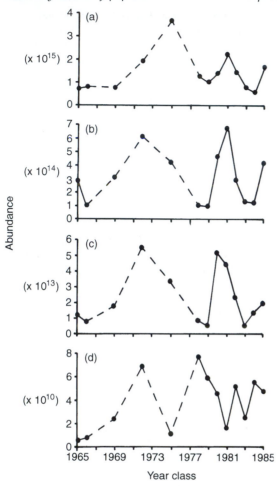

Fig. 10.18 Estimates of the abundance of northern anchovy, *Engraulis mordax*, in first year of life, showing relationship between stock (number of eggs) and recruits (number of one-year-olds). (a) Eggs; (b) yolk-sac larvae; (c) pre-recruits; (d) recruits. Broken line used to link points not at yearly intervals. Redrawn from Peterman *et al.* (1988).

where n_t, is the total abundance at time t and $n(i)$ is the abundance in age class i, then the abundances at time $t + 1$ are given by:

$$n_{t+1} = \mathbf{A} . n_t. \qquad (10.23)$$

Because computers handle such matrix multiplications easily, this model potentially pro-

vides a powerful method of predicting the population dynamics of fish populations that contain several age classes. Although not shown explicitly in this simple form of the model, the age-specific fecundities and survivorships can be defined as functions both of abiotic and of biotic factors such as temperature, food abundance and population density. Matrix models have been used to predict the potential impact of pollution on population abundance. Sadler (1983) used the method to evaluate the possible effects of the acidification of Norwegian lakes on the brown trout populations. Schaaf *et al.* (1987) simulated the effect of pollution on estuary-dependent species including the menhaden and striped bass. They assumed that pollution has its main effect on the survival of fish in their first year of life, that is on the value of $p(0)$.

Individual-based models

These models exploit the computational power of computers to simulate the fate of individuals that are representative of a cohort (DeAngelis and Gross, 1992; Van Winkle *et al.*, 1993). An interesting example simulated the population dynamics of a flatfish, the winter flounder, in the Niantic River on the north-east coast of the USA (Rose *et al.*, 1996). The full model had two components. An individual-based model (IBM) simulated the growth and mortality of cohorts in their first year of life. The numbers that survived to reach one year old were then entered into an age-structured matrix model based on yearly intervals. This matrix model predicted the dynamics of fish older than one year and generated the numbers of mature females that spawn the next cohort of eggs. The IBM at its starting point predicted the day of spawning, number of eggs spawned and the mean weight of the eggs for each mature female. The model then predicted the time to hatching and first feeding in relation to water temperature. The size of the larvae at first feeding was predicted from the mean egg weight. The proportion of eggs and yolk-sac larvae dying per day was calculated for a given water temperature and a baseline mortality rate. The fate of individual larvae was then followed from day of first feeding. Growth was estimated from a bioenergetics model (Chapter 6). Food consumption was simulated using a model for prey encounter linked with an optimal foraging model for prey selection (Chapter 3). The rates of mortality caused by starvation and predation were predicted for larvae and juveniles. Thus, the number and size of survivors from the original cohort that reached one year old was calculated. The predictions of the model for the temporal pattern of abundance over a year, duration of particular life stages, growth rate and mortality rate were comparable to field estimates. For the model, recruitment (defined as the number surviving to age 1) was correlated with the number of larvae metamorphosing, but not with abundances of earlier stages in the life history.

The advantage of such models is that the effects of varying factors such as density of food, temperature regimes and hours of daylight on recruitment can be simulated. Their disadvantage is the level of detail that has to be known about the species being simulated. The models will help to indicate how much of that detail is required to generate realistic predictions.

Size-based models

Both survival and fecundity are size dependent rather than age dependent (Chapters 7, 8 and 9). While it will usually be advantageous for a fish to grow quickly through the life-history stages that suffer the highest mortality rates (Fig. 10.10), it must also avoid foraging in areas that are risky because of the presence of predators (Chapter 8). The development of predictive models for fish population dynamics will have to incorporate these size-dependent relationships between foraging, growth, survival and fecundity (Werner and Gilliam, 1984). Beyer (1989) developed a size-based model that predicted the stock–recruitment relationships.

10.10 POPULATION PRODUCTION

Concept of production

As the fish in a cohort grow and die, biomass, the total weight of fish present in that cohort, changes (Fig. 10.19). This trajectory of changing biomass integrates the effects of the growth and deaths of individuals in the population. Production, P, is the total growth in weight of fish in the cohort over a defined time interval and includes that part of the cohort that dies

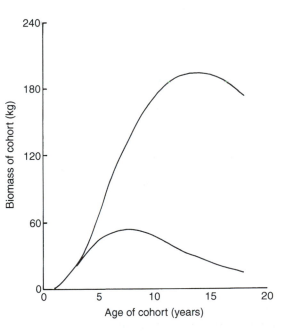

Fig. 10.19 Elaboration of biomass by a cohort of North Sea plaice, *Pleuronectes platessa*, showing dimorphism in growth. Upper curve, females; lower curve, males. Redrawn from Rothschild (1986).

during that interval (Ricker, 1975). It is a function of the mean growth rate of fish in the cohort, measured by the specific growth rate, g, and the rate of mortality measured by Z. At a given time, t, the total weight of fish present in a cohort is the biomass, B_t, which is estimated from the abundance of fish in the cohort, N_t, and the mean weight of fish, w_t, so that:

$$B_t = N_t w_t. \tag{10.24}$$

Production over a time interval t_1 to t_2 is then:

$$P_t = g\overline{B} \tag{10.25}$$

where $\overline{B}$ is mean biomass of the cohort during that interval.

Production is of interest because it is a measure of the fish flesh generated by a cohort potentially available to predators including fisheries. Yield is the portion of production that is taken by a defined predator or group of predators. However, it is important to recognize that the basis of a cohort's production is the growth of individual fish in that cohort. The pattern of growth of an individual is that which will tend to maximize the number of offspring produced by that individual in its lifetime. There is no selection for the maximization of production by a cohort, though that might be a coincidental by-product of selection processes at the individual level. In other words, it is not the function of a cohort to generate production, rather production is an effect of the bioenergetic processes by which individual fish seek to maximize their individual reproductive success. An understanding of the factors that determine the magnitude of production by a cohort will depend on an analysis of the effects of the relevant factors on the growth, reproduction and survival of individual fish.

The relationship between the total production by all the fish species coexisting in a water body and the environmental conditions will be considered in Chapter 12.

Measurement of production

The definitions of production given above refer only to production in the form of somatic growth. The total production by a cohort will be the sum of somatic production and the biomass of gametes produced during the interval over which production is measured. The total production by a population for a given period is the sum of the production by each cohort. There are two main methods used to estimate somatic production by a cohort. The first is Allen's graphical method. The second assumes exponential rates of growth and mortality over the time period in question.

Biomass can be measured in units of weight or of energy and is usually expressed per unit area, while production has the units of weight (or energy) per unit area per unit time, for example g wet wt m^{-2} year^{-1}. The ratio of production to biomass, that is P / B, is the turnover ratio, and has the units of time^{-1}. It provides an index of the rate of production per unit of biomass in the population.

Allen's graphical method of estimating production

An Allen curve is a plot of N_t, the number of survivors in a cohort at time t, against w_t, the mean weight of fish in that cohort at time t (Fig. 10.20(a)). The area under the curve between w_1 and w_2 gives the production over the interval t_1 to t_2, where w_1 and w_2 are the mean weights at times t_1 and t_2 (Pitcher and Hart, 1982). The total production by the cohort is the total area under the curve from the time that the fish start to feed exogenously until there are no more survivors. An example of an Allen curve for a population of bullhead, *Cottus gobio*, in Bere Stream in southern England (Fig. 10.20(b)) showed that during the spawning period, there was a period of negative production which reflects the loss of gametes (Mann, 1971).

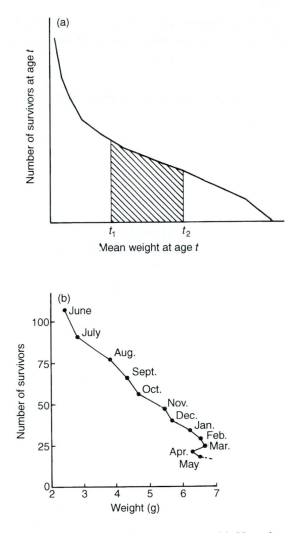

Fig. 10.20 Allen production curves. (a) Hypothetical Allen curve, with shading showing production by cohort over period t_1 to t_2. (b) Allen curve for a population of *Cottus gobio* from a stream in southern England and showing a period of negative production during the spawning period. Redrawn from Mann (1971).

Ricker's production equation

The Allen curve makes no assumptions about the pattern of survival and growth for the cohort, but requires that abundance and mean weight for the cohort are estimated at short intervals of time. An alternative approach is to assume that over the interval for which production is to be estimated, specific growth rate, g, and instantaneous mortality rate, Z, are constant. Then production for that interval is given by the expression (Ricker, 1975):

$$P = gB_0[\exp(g - Z) - 1] \, / \, (g - Z) \quad (10.26)$$

where B_0 is the biomass present at the start of the interval.

Bioenergetic basis of production

The bioenergetics equation for an individual fish (Chapter 4):

$$C = F + U + R + P \qquad (10.27)$$

clearly identifies the origin of production: it is that portion of the energy income that is not lost in the form of faecal and excretory losses or through respiration. Production by a cohort is simply the summation of the individual P values, taking into account values for fish that die during the interval over which production is measured. This principle was illustrated by a study of production by a population of American plaice off the coast of Nova Scotia, Canada (MacKinnon, 1973). From a simple bioenergetic model, together with estimates of the annual mortality rate, he developed a model for the energy allocation of individual fish and for each age class (Fig. 10.21). During the year, there were periods when production was negative because the energy provided by the food was not sufficient to cover the outputs. The analysis of the bioenergetics of individual fish helps to reveal the mechanisms underlying the changes in production in the cohort or population.

The bioenergetics basis of production is also reflected in the strong relationship between the respiratory energy losses by a population per unit time, R, and production (Humphreys, 1979). From studies of the energy budgets of nine species the following relationship was calculated:

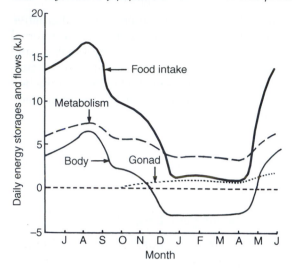

Fig. 10.21 Mean daily energy budget for mature (age 11) American plaice, *Hippoglossoides platessoides*, showing negative production between November and May. Redrawn from MacKinnon (1973).

$$\log_{10} P = 0.834 \log_{10} R - 0.249 \quad (10.28)$$

where P and R were expressed as cal m^{-2} year^{-1} and refer to the total population. The value of the slope, 0.834, was not significantly different from 1.0, indicating that production was linearly related to respiration.

Magnitude of population production

Estimates of population production in a variety of temperate fresh waters for a range of species covered more than three orders of magnitude from about 0.1 g m^{-2} year^{-1} to nearly 200 g m^{-2} year^{-1} (Chapman, 1978; Mann and Penczak, 1986). This range reflected the differences between the species in their abundance, growth rates and mortality. A clearer picture of the importance of these factors can be obtained by examining the production by the successive cohorts in a population over a series of years. This was illustrated by the production in the migra-

tory trout population the dynamics of which are described on page 243 and the perch population in Windermere.

Production in a trout population

The annual production of migratory trout in Black Brows Beck for the period between emergence from the gravel as alevins and migration to the sea ranged from 8.86 to 33.9 g m^{-2} year^{-1} between 1967 and 1990 (Elliott, 1985d, 1993, 1994). The lowest values were in 1983 and 1984, drought years. However, production in other drought years, 1976 and 1989, was not unusually low. Production for individual cohorts varied from 7.59 g m^{-2} for the 1983 cohort to 33.88 g m^{-2} for the 1970 cohort. Because of the detail in which this population was studied, it was possible to estimate that most of the variation in the production by cohorts could be accounted for by density-dependent survival (57% of the variation) and density-independent growth (17.5%). Temperature was the chief factor that determined the growth rate in the Beck. Although growth was density independent, the relative variation in size at a given age varied greatly between cohorts, and this relative variation was density dependent (page 244). Cohorts that were numerically small at the start of their life showed more variability in individual body size relative to that cohort's mean body size than did abundant cohorts. The use only of mean sizes and mean growth rates would have obscured this source of variation in the production process, which is generated by the effects of density on the individual fish (Elliott, 1985d, 1994).

Windermere perch

Wide variations in annual production were also shown by the perch population of Windermere (Craig, 1980). For example in the southern basin of the lake, in the period from 1961 to 1972, annual perch production ranged from 23 kJ m^{-2} year^{-1} in 1966 to 158.2

kJ m^{-2} year^{-1} in 1972. Such intrapopulation variation almost equals the interpopulation and interspecific variation in production, and prevents the development of any generalizations about population production. Because of their effect on growth, survival and reproduction of fish in a cohort, local environmental conditions are important in determining production.

Contribution of young age classes to production

A feature of production in many populations is the high proportion contributed by the youngest age classes in a cohort. From his pioneering study of fish production in the River Thames near Reading (UK), Mann (1965; Mann *et al.*, 1972) estimated that populations of roach and bleak, two cyprinids, together had a production of 573 kJ m^{-2} year^{-1}. Fish in their first year of life generated 69% of this. For the perch in Windermere, the production in the first 2 years of life accounted for between 60% and 80% of the total production by a cohort (Craig, 1980). The juvenile perch generated a higher proportion of the total production in cohorts in which growth was faster, than in cohorts that had slower growth rates. This high contribution to production by young fish is a reflection of the high growth rates in the early life-history stages, which are characterized by higher P / B ratios than larger and older fish. Predators exploiting the younger age classes can potentially take a higher yield than predators dependent on older and slower-growing fish.

Contribution of reproductive products

A portion of the production by a cohort will consist of eggs and sperm. Windermere perch showed large changes in this proportion over a 40 year period (Craig, 1980). Before 1964, the proportion of adult production formed by gonadal products averaged 20.8% for males and 87.4% for females; after 1964, the comparable proportions were 11.1% and 36.0% (Craig, 1980). Before 1964, the perch population was dominated by a few strong cohorts, whereas from 1966 onwards, the population consisted of cohorts that grew rapidly and had high rates of production before reaching sexual maturity. These observations again illustrate the sensitivity of production to changes in the more fundamental life-history traits of growth, survival and fecundity. An understanding of the factors that determine the production by a population can be gained by analysing the factors that affect the fundamental life-history traits. Little insight into the life-history traits can be gained merely by a study of the production of a population.

Lifetime bioenergetics and production of Skeena sockeye salmon

An ambitious attempt to synthesize life-table information and the bioenergetics of individual fish to produce a picture of cohort production was the analysis by Brett (1983, 1986) of the Babine Lake (Skeena River) sockeye salmon. The energetics of an individual male and female sockeye that return to fresh water to spawn in their fourth year are described in Chapter 4. This information was combined with values for age-specific mortality which were obtained partly from observation and partly by assuming that mortality in the sea was size dependent (Fig. 10.10). The change in biomass of a cohort was then calculated. This analysis quantified the importance of the marine phase of the life history for cohort production. During the freshwater phase in Babine Lake, production by juvenile sockeye was about 128 tonnes, but the accumulated weight of adult sockeye returning to the coastal area was 3400 t. After their downstream migration from Babine Lake as smolts, the young sockeye spent about five months in coastal waters. During this time they grew from about 5 g to 50–60 g, and mortality accounted for about 90% of the smolt run. In the five month period, only 4% of a cohort's marine production was accumulated. The balance was accumulated by the remaining 10% of the population that entered the offshore Alaskan Gyre, and most of the marine production was elaborated in the final year at sea, before the spawning migration into fresh water. The biomass of mature sockeye returning to the Skeena was 3382 t, while the total production during the lifetime of the cohort was 3894 t. In energetic terms, the biomass of returning sockeye represented about 23% of the biomass of food consumed by the cohort. From his analysis, Brett (1986) drew a picture of the sockeye salmon as a fish that saturates its predators during its early sea life, because large numbers of small fish are present for a restricted period of time. The rapidly growing young fish pass quickly out of the most vulnerable prey size and the sockeye itself becomes a dominant offshore predator. High levels of growth are maintained during the marine phase, when the sockeye remains a physiologically young fish. The sexually adult phase, which is often characterized by reduced growth rates in fish, is curtailed in the sockeye: no adults survive their first breeding attempt. In Brett's words:

> As a fishery, salmon are ideal. They comb the ocean for its abundant food, convert this to delectable flesh, and return regularly in hordes to funnel through a limited number of river mouths exposing themselves to the simplest of capture – a gill net, or seine net.

The circumstances that might favour this life-history pattern are discussed in Chapter 11.

10.11 SUMMARY AND CONCLUSIONS

1. A biological population constitutes a gene pool that has continuity through time because of the reproductive activ-ities of the individuals in the population. Population abundance changes as the probabilities of survival and fecundities of individuals in the population change. In closed populations, recruitment is primarily from progeny produced by that population. In open populations, the progeny disperse into a common larval pool from which recruits are subsequently drawn.

2. Populations can be identified as distinct gene pools by studies of genetic markers including allozymes and DNA sequences, and indirectly from morphological characteristics of fish in the populations.

3. Estimates of population abundance can be obtained by indirect methods. These include mark–recapture, change in catch per unit effort, cohort (virtual population) analysis, acoustic techniques, and a variety of other methods. When estimating the abundance of a population, care has to be taken that samples are representative of the population as a whole.

4. Measures of the rate of change of abundance include net reproductive rate, the intrinsic rate of natural increase and the finite rate of increase. Estimates of these rates can be obtained from cohort life tables, which itemize the age-specific mortality and fecundity rates in the cohort.

5. Mortality rates in fish populations are size-related, with the rate declining with an increase in body size. The relative importance of starvation and predation as causes of mortality, especially in the early life-history stages (larvae and young juveniles), is still poorly understood. In some species there may be a critical period of high mortality. In larval fish, this may occur when the yolk is finally absorbed. This critical period could be important in generating the variation in age-class strength seen in many fish populations.

6. Mortality rates, growth and fecundity

may be functions of the density of the population (density dependence). Such density dependence could lead to a population fluctuating about a long-term mean level of abundance (equilibrium density). But density dependence may be too weak, too variable or too strong to stabilize population abundance.

7. The relationship between the abundance of the sexually mature portion of the population and recruitment to a defined age (or size) class is the stock–recruitment relationship. It is typically highly variable, obscuring any underlying relationships including density dependence. The use of otoliths and molecular methods to identify the time of hatching and the familial relationships of recruits will help to identify the causes of the variation.

8. Models that seek to predict the dynamics of the abundance of fish populations must include sufficient biological detail if they are to generate realistic predictions. Promising models include age-structured matrix models and individual-based models. The effects of fish size on mortality, growth and fecundity are particularly important.

9. Production measures the amount of fish flesh generated by a cohort as the fish grow and die. It can be estimated using an Allen curve or from estimates of specific growth rates and instantaneous mortality rates.

10. The magnitude of production depends on the bioenergetics of the individual fish in a population. Because of their high growth rates, the young age classes frequently contribute a high proportion of the total production of a cohort. Gametes may also represent an important component of cohort production.

11. The analysis of the dynamics of population abundance and production is complicated by the uniqueness of the individuals that make up populations. In contrast to physical and chemical systems, biological populations cannot realistically be considered as aggregates of identical units. Models of biological populations have to incorporate the consequences of individual variation.

11

LIFE-HISTORY STRATEGIES

11.1 INTRODUCTION

A life table encapsulates in a quantitative form the life-history pattern of a population. Fishes have evolved a diversity of life-history patterns (Breder and Rosen, 1966). In some species, sexual maturity is reached within a few weeks of hatching, in others only after several years. Some species are semelparous, others iteroparous. Some have short life spans, others may live for many decades. Even within a species, there may be major variations in the life-history patterns shown by different populations, including variations between populations living at different latitudes (Jonsson and L'Abee-Lund, 1993; Baker, 1994). Intraspecific differences in migration, growth, age at first reproduction, life span and fecundity are described in earlier chapters. What are the environmental factors that favour the evolution of a particular life-history pattern? How will a life-history pattern change as environmental conditions change? This second question is relevant to the effects of fishing and pollution, which impose new regimes of mortality or other adverse effects on a population, and to the consequences of long-term changes such as global warming.

As the environmental conditions experienced by a population change, adjustments in the age-specific patterns of mortality, growth and fecundity can occur by two processes. Firstly, the individual fish already present in the population may display phenotypic plasticity. This allows them to respond adaptively to the environmental change by changes in their physiology and behaviour, with consequences for their growth rate, reproduction or survival that mitigate the effects of the environmental change (Chapter 1). Such phenotypic adaptations will not result in changes in the genotypic frequencies in the population. Secondly, some genotypes may be better adapted to the changed environment than others, so their offspring form a higher proportion of the population in succeeding generations. This will cause a change in the allelic and genotypic frequencies in the population, which is an evolutionary response. These two mechanisms of adaptation, phenotypic plasticity and genetic selection, are not incompatible, but their time courses are different. Phenotypic adaptation takes place within a generation, but genetic adaptation occurs between generations. The former can be regarded as a tactical response by individuals to their encounter with the environment: the latter is a strategic response observable at the level of the population (gene pool), but driven by adaptive differences between individuals (Wootton, 1984b).

11.2 EVOLUTION OF LIFE-HISTORY PATTERNS

There are three main approaches to understanding the relationship between life-history patterns and environmental conditions. The first has developed a body of theory based on the life table, comprehensively reviewed in Roff (1992) and Stearns (1992). The second identifies regularities in the relationships between life-history traits such as age at maturity and mortality rate or length at maturity and asymptotic length (Charnov,

1993). The third attempts to relate syndromes of life-history traits to syndromes of habitat characteristics (Southwood, 1988). Each of these focuses attention on different facets of life-history evolution. A definitive synthesis of the three approaches has still to be developed.

Life-history theory based on demography

A body of theory has been developed which seeks to predict the life-history patterns that will evolve in environments that have defined effects on age-specific schedules of mortality, growth and fecundity. The evolution of the life-history pattern is driven by the effects of environmental conditions on these demographic variables. These conditions may include biotic interactions such as predation and intra- and interspecific competition (Chapters 8 and 9). The theory assumes that the process of selection favours those genotypes that have age-specific schedules of growth, fecundity and mortality that generate the highest per capita rates of increase relative to other genotypes in the population. Selection tends to maximize the quantity r_m as defined by the Euler–Lotka equation:

$$1 = \sum_{x=\alpha}^{\infty} \exp(-r_m x) l_x m_x \qquad (11.1)$$

Chapter 10) (Roff, 1992; Stearns, 1992; Charlesworth, 1994). Further analysis shows that r_m will be maximixed if at each age, an individual maximixes the value of:

$$m_i + (p_i v_{i+1} / v_0)$$

where m_i is the fecundity of a fish aged i, p_i is the probability of a fish surviving from age i to $i+1$ and v_{i+1}/v_0 is the reproductive value of a fish aged $i+1$. In words, the per capita rate of increase is maximized if at each age i, the fish maximizes the sum of its present fecundity (m_i) and its future expected fecundity ($p_i v_{i+1}/v_0$) (Williams, 1966). The reproductive value of a fish aged i is defined by:

$$v_i / v_0 = [\exp(r_m i) / l_i] \sum_{i=y}^{\infty} [\exp(-r_m y) l_y m_y] \qquad (11.2)$$

and is the average number of young that a female aged i can expect to have over the remainder of her life, discounted back to the present, and expressed relative to the reproductive value of a female at birth (Stearns, 1992). (The discounting is done to reflect the principle that early births contribute more to future population growth than late births – compare the consequences of investing a sum of money in an interest-bearing account in a bank now or 10 years hence on the amount 20 years from now.) Reproductive value weights the contributions of individuals of different ages to population growth (Stearns, 1992).

A limitation of this theory is that it is based on the Euler–Lotka stable-age-distribution equation. As Chapter 10 describes, many fish populations do not achieve a stable age structure but are characterized by strong and weak age classes as recruitment varies in time. The consequences for the theory of such variation in age-class strength have still to be explored.

A second disadvantage of the theory is that it assumes that fecundity, survival and growth rates are functions of age. In reality, the performance of an individual of a given age is likely to depend on the physiological and maturational state of the individual rather than just its age. This state is sometimes referred to as condition. Some attempts have been made to develop models that take account of the state rather than the age of individuals (Caswell, 1989; McNamara and Houston, 1996). Such developments may prove to be important for understanding the evolution of fish life histories because of size-related traits such as growth rate (Chapter 6), fecundity (Chapter 7) and mortality (Chapter 10).

Further developments of life-history theory will explore situations in which r_m may not be the appropriate measure of fitness. The consequences of density dependence (Chapter 10), frequency dependence (Chapter 7) and environmental stochasticity are of particular importance (Stearns, 1992; Charlesworth, 1994).

Dimensionless numbers in life histories

Estimates of natural mortality, M, are important in the application of fisheries management models (Chapter 10), but it is often difficult to obtain reliable quantitative estimates for natural populations. This led fisheries biologists to seek other variables that were highly correlated with M but were easier to measure in natural populations (Beverton, 1963). An empirical study of 175 different populations of fish representing 84 species from freshwater and marine environments found a significant inverse correlation between the natural instantaneous mortality rate and asymptotic size estimated from the von Bertalanffy growth model (Chapter 6), but a positive correlation between mortality and the growth rate coefficient from the same model (Pauly, 1980). The calculated multiple regression took the form:

$$\log_{10} M = -0.0066 - 0.279 \log_{10} L_\infty + 0.654 \log_{10} K + 0.4634 \log_{10} T \quad (11.3)$$

where M is annual, natural mortality rate, L_∞ is asymptotic length in cm, K is the growth rate coefficient and T is average annual temperature in °C. In general, fish that grow faster and have a relatively small asymptotic length have a higher mortality rate than slow-growing fish with a high asymptotic length.

Within limited taxonomic boundaries, such as within the Clupeidae, quantities including M / K, L_α / L_∞, and αM are approximately constant (Fig. 11.1). M is the instantaneous per capita mortality rate (dimension, time^{-1}); α is the age at maturity (dimension, time); L_∞ and L_α are the asymptotic length and length at maturity respectively (dimension,

length); and K is the growth rate from the von Bertallanfy growth model (dimension, time^{-1}) (Chapter 6) (Charnov, 1993). A characteristic of quantities such as αM is that they are dimensionless. Charnov (1993, 1997) argued that the constancy of these dimensionless variables for some taxonomic groupings reflects a constraint on the evolution of life-history traits imposed by some general principles of life-history trade-offs. The concept of life-history trade-offs is central to life-history theory and is discussed below.

Habitat as a templet for life histories

A seductive metaphor has likened the habitat to a templet on which natural selection moulds characteristic combinations of life-history traits (Southwood, 1977, 1988). This approach builds on the concept of r- and K-selection introduced by MacArthur and Wilson (1967). This influential model suggested that high mortality caused by density-independent effects selects for early reproduction, high fecundity and short life expectancy. High mortality caused by density-dependent effects selects for delayed reproduction, low fecundity and a long life expectancy. High density-independent mortality would be expected in harsh or unpredictable environments, whereas high density-dependent mortality would be expected in abiotically benign environments in which populations are at or close to their equilibrium population densities (Fig. 10.12).

One habitat dimension, the strength of density dependence, has proved inadequate to explain the diversity of life-history patterns observed. Southwood (1977) argued that there is a need to distinguish between the temporal heterogeneity and the spatial heterogeneity of habitats. Temporal heterogeneity relates to the rate and pattern of change over time of conditions in a habitat, whereas spatial heterogeneity relates to the size and spatial pattern of patches of different quality in a habitat.

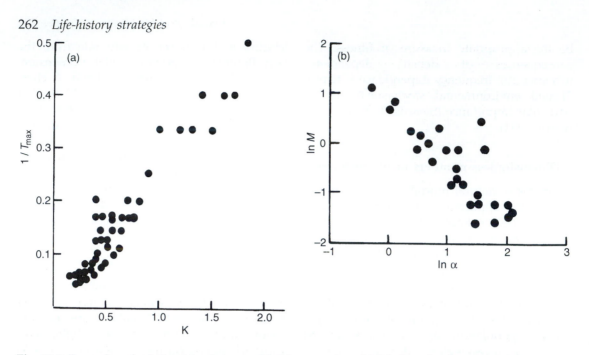

Fig. 11.1 Examples of relationships between life-history traits which led to development of the concept of dimensionless numbers in life-history theory. (a) Relationship between von Bertalanffy growth coefficient, K, and inverse of maximum life span, $1/T_{max}$ (T_{max} is inversely proportional to instantaneous mortality rate, M). (b) Relationship between M and age at first reproduction, α. Redrawn from Charnov and Berrigan (1991).

In a synthesis of studies that have identified habitat dimensions, Southwood (1988) identified three axes and suggested that particular combinations of life-history traits can be sited at different points with reference to these three axes. The first dimension describes a 'disturbance' axis, which relates to the temporal stability of the habitat. Compare the temporal stability of abiotic conditions at abyssal depths in an ocean with those of a small pool on the floodplain of a river. The second dimension describes an 'adversity' axis, which relates to the adversity of the habitat in the context of the conditions required for life. Habitats with temperature regimes close to the lower or upper thresholds at which life can exist are at one extreme of an adversity axis (Chapter 4). The third dimension is a 'biotic interactions' axis, which describes the strength of interactions such as predation, competition and parasitism (Chapters 8 and 9).

Can the recognizable combinations of life-history traits required by the habitat templet model be identified? Winemiller and Rose (1992) used 16 life-history traits for 216 North American freshwater and marine species of fishes in the multivariate statistical technique of principal components analysis (PCA) to explore associations between the traits. They recognized two primary gradients of variation. There was an association of larger adult body size with delayed maturation, longer life span, larger clutches, smaller eggs and fewer spawning bouts per year. A second association was between parental care, larger eggs, longer spawning season and multiple spawning bouts within a year. From this analysis three syndromes of life-history traits were labelled 'periodic',

Table 11.1 Triangular life-history continuum model of Winemiller and Rose (1992).

Life history strategy	Typical life history traits	Habitat qualities
'Opportunistic'	Short generation length, T Small age-specific fecundity, m_x Low age-specific survival, l_x	Change frequently or stochastically on small temporal and spatial scales
'Equilibrium'	Long generation length, T Small age-specific fecundity, m_x High age-specific survival, l_x	Low variation in quality Strong direct and indirect biotic interactions
'Periodic'	Long generation length, T High age-specific fecundity, m_x Low age-specific survival, l_x	Large-scale cyclic or spatial variation

'opportunistic' and 'equilibrium' (Table 11.1). The opportunistic syndrome is characterized by early sexual maturity, small clutches and low survivorship. Many cyprinodonts such as the killifishes provide examples. The equilibrium syndrome is defined by late maturity, small clutches and high survivorship, traits characteristic of many cichlids. The periodic syndrome is shown by many fishes. It could almost be regarded as the 'typical' teleost pattern of late maturity, large clutches, and low survivorship through the early life-history stages (Chapters 7 and 10). A multivariate classification of the fish fauna of the Rhone River on the basis of life-history traits yielded comparable, though not identical, associations (Persat *et al.*, 1994).

Constraints on life-history traits

The ideal fish would have a long life, reach maturity at an early age, then reproduce often and produce numerous young at each breeding attempt. But there are constraints that limit the capacity of fish to achieve these desirable demographic characteristics simultaneously (Roff, 1992; Wootton, 1992b). A concept central to the development of life-history theory is that of trade-off constraints. Fish live in a finite world. An increase in the time or resources invested in one activity

may be traded-off against a decrease in the time or resources invested in others. Any reductions in survival, growth or future fecundity that result from a current breeding attempt can be regarded as the cost of reproduction (Williams, 1966). The evidence that such trade-offs exist and that reproduction does exert a cost on other life-history traits is discussed in detail on page 269. Stearns (1992) identifies at least 45 possible trade-offs between life-history traits, including those between present reproduction and survival and between present reproduction and future reproduction. Some examples are given in Table 11.2. In fish, a potentially important trade-off is between present reproduction and growth because female fecundity is a function of body size.

A second type of constraint is imposed because the demographic traits have to be compatible with other adaptations of the fish, such as those related to locomotion. The per capita rate of increase, r, can be raised by an increase in fecundity. This could require an increase in the volume of the abdominal cavity to accommodate the extra eggs and would cause a change in body shape. But physical factors impose design constraints on body shape in relation to locomotion (Chapter 2). Comparisons of the relationship between the total volume of eggs and body

Table 11.2 Pairwise trade-offs between life-history traits

Trade-off[†]	Interpretation
n vs. s_2	Risky reproduction
n vs. s_1	Parental feeding critical for offspring survival (rare in teleosts)
s_2 vs. s_1	Parent(s) guard offspring
s_2 vs. t_1	Parents feed offspring (rare in teleosts)
s_1 vs. t_1	Risky feeding for juveniles
s_2 vs. t_2	Risky feeding for adults
n vs. t_1	More, smaller vs. fewer, larger offspring
n vs. t_2	Adults with more offspring take longer to recover
t_2 vs. t_1	Parental investment speeds offspring development
t_2 vs. s_1	Parental investment protects offspring

Source: Sibly and Calow (1983).
[†] Life-history traits: n, number of female offspring produced per female per breeding attempt; s_1, survivorship from birth until first breeding; s_2, survivorship of adults between successive breedings; t_1, age at first breeding; t_2, interval between successive breedings.

size indicate less interspecific variation than do comparisons of the relationship between fecundity and body size (Roff, 1982; Wootton, 1984b) (Fig. 7.4). The uptake of nutrients and oxygen will depend on the surface area available for digestion and respiration. However, the rate of use of nutrients and oxygen will depend on the volume of the organism. Consequently, many physiological processes have an allometric relationship to body mass (Chapter 4). Demographic traits will be constrained by such allometric relationships (Wootton, 1992b).

A reduction in the age of first reproduction or in the time between successive breedings can be potent in increasing r_m. An evolutionary tendency to reduce these times may meet a limit that is set by the rate at which the physiological processes of growth and gametogenesis can take place. Roff (1982)

suggested that the tendency for some species of flatfish to increase their size at first reproduction without showing a decrease in the age at first reproduction reflects such a physiological constraint.

Selection of life-history traits can occur only if there is genetic variation in the population for those traits. Even if an environmental change favoured the selection of an increase in fecundity, this can occur only if some of the interindividual variation in fecundity in the population reflected additive genetic variance for fecundity (Falconer, 1989).

The evolutionary history of a population may create a phylogenetic constraint. Groups of related species become trapped in their life-history patterns by their previous evolutionary histories. An analysis of freshwater fishes in Canada which clustered species together on the basis of their life-history patterns linked together the salmonids in a closely related group and the centrarchids in another group (Wootton, 1984b). The classification based on life-history patterns partly reflected the phylogenetic classification. The importance of using phylogenetic information when interpreting patterns in life-history traits and the methods available for using such information are discussed in Harvey and Pagel (1991). However, the power of phylogenetic constraints should not be overemphasized. The male threespine stickleback is a textbook example of diligent male parental care in fish (Chapter 7). On the east coast of Canada, there lives a form of the threespine stickleback, the 'white' stickleback, in which the male scatters the fertilized eggs amongst filamentous algae and shows no further parental care (MacDonald *et al.*, 1995).

Reproductive effort and life-history theory

An important step in the development of life-history theory came when the theory (page 260) was extended by the incorporation of the concept of an allocation or trade-off constraint (Schaffer, 1974a,b, 1979; Schaffer and Rosenzweig, 1977). This extended theory assumed that at each age, an individual fish can allocate a fraction of its energy resources to reproduction. The fraction allocated is called the reproductive effort, E, so that E_i is the reproductive effort of a fish aged i. From the principle of an allocation constraint, it is assumed that an increase in E_i produces an increase in m_i (fecundity) but a decrease in p_i (survival). Consequently, there is a trade-off between m_i and p_i (Fig. 11.2). (Schaffer's model also takes into account a trade-off between fecundity and post-reproductive growth.) A set of combinations of p_i and m_i can be defined which generate the same value for r_m. A real animal is physiologically capable of only certain combinations of p_i and m_i. The model assumes that at each age i, there is selection for that achievable combination of p_i and m_i that maximises r_m. This optimal combination depends on the shape of the trade-off curve. If the curve is concave, the optimal solution is either no reproduction, that is maximize p_i, or no post-breeding survival, that is maximize m_i, adopting a semelparous life history (Fig. 11.2(b)). For a convex trade-off curve, the optimal solution is a positive p_i and m_i – an iteroparous life history (Fig. 11.2(a)). If the trade-off curve is more complex, more than one optimal combination of p_i and m_i is possible: for example one may represent a semelparous solution, another an iteroparous solution. This recalls the life histories of the Pacific salmon and trout, *Oncorhynchus* spp. The two coexist in the same rivers and both may undertake an anadromous migration, but Pacific salmon are semelparous while the trout are iteroparous (Schaffer, 1979).

Schaffer's model suggested that if environmental conditions cause a reduction in juvenile survival or in age-specific fecundity per unit of reproductive expenditure, selection will favour an increase in the age of first reproduction and a reduction in the repro-

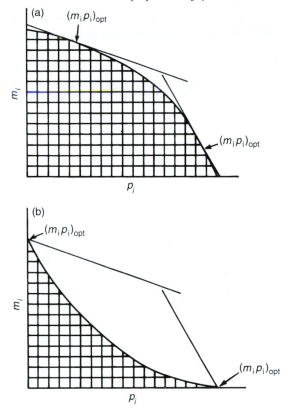

Fig. 11.2 Trade-off curves between fecundity at age i, m_i, and probability of surviving from age i to $i+1$, Pi. (a) Convex trade-off curve for which optimal combinations are intermediate values of m_i and p_i (iteroparity). Two examples are shown, one with high fecundity and poor post-breeding survival and one with low fecundity and high post-breeding survival. (b) Convex trade-off curve for which optimal combinations are either no reproduction at age i, $m_i = 0$, or no post-breeding survival, p_i (semelparity). Straight lines define combinations of m_i and p_i, which yield the same intrinsic rate of increase, and shaded areas the combinations of m_i and p_i that are physiologically possible.

ductive effort at each age. An environmental change that decreases the survival per unit of reproductive expenditure between breeding seasons selects for an increase in reproductive effort and a decrease in the age of first breeding. A reduction in rate at which fecundity increases with age has the same effect.

Life-history theory based on demographic theory made the following predictions. High, variable or unpredictable adult mortality rates select for increased reproduction early in life. High, variable or unpredictable juvenile mortality rates select for decreased reproduction and longer adult life (Schaffer, 1979; Sibly and Calow, 1983; Stearns, 1983a). Any changes that reduce the value of juveniles and increase the value of adults will favour iteroparity (Stearns, 1992).

If the survival rates of the immature stages vary from one reproductive event to the next, selection will favour a longer reproductive life span (Charlesworth, 1994). Iteroparity over an extended reproductive life span will reduce the risk that all the offspring produced by an individual in its lifetime are lost because of unfavourable conditions. This danger attends life histories that are semelparous or have restricted iteroparity. The strategy of spreading the risk over several breeding attempts is sometimes called 'bet hedging' (Stearns, 1992). An early demonstration of bet hedging came from a simulation model based on the life-history patterns of clupeid fishes (Murphy, 1967, 1968). The simulations suggested that in an environment in which there is considerable variation in reproductive success, populations in which the fish have a later age at maturity, a lower age-specific fecundity, but a longer life span, reach higher levels of abundance than populations in which the fish have an early age at maturity, higher fecundity and a shorter life span. Other methods by which the risk of leaving no offspring may be reduced include the variation in age of maturation seen in some semelparous salmonid species and the variable lengths of diapause found in the eggs of annual fishes (Chapter 7).

These predictions emphasize that life-history theory assumes that environmental factors will drive evolutionary changes in life-history patterns through their effects on the age-specific mortality and fecundity rates. The evolutionary response is in the form of changes in the age-specific schedules of mortality, growth and fecundity (Roff, 1992; Stearns, 1992).

11.3 EVALUATION OF THEORY

Evaluation of demographic life history theory

Life-history evolution in Trinidadian guppies

The most compelling evidence of an environmental factor driving the evolution of life-history traits comes from studies of guppy populations in Trinidadian streams. The relevant environmental factor is predation on the guppy by piscivorous fishes (Chapter 8) (Reznick and Endler, 1982; Reznick *et al.*, 1990). Sites in the streams are classified by the dominant fish predator. *Crenicichla* sites are characterized as sites of high predation intensity; *Rivulus* sites have moderate predation intensity, predominantly on juveniles; *Aequidens* sites have low predation on all size classes of guppies. Guppies from *Crenicichla* sites mature at a smaller size, have a higher reproductive allotment (dry weight of embryos / total dry weight of female), produce more and smaller offspring and reproduce more frequently than guppies from the *Rivulus* and *Aequidens* sites. These differences are all in the direction predicted by life-history theory if it is assumed that at the *Crenicichla* sites, larger (and older) guppies are preferentially predated, while at the *Rivulus* sites, the smaller (and younger) fish are more at risk.

In a long-term field experiment, guppies from a *Crenicichla* site were transferred to a *Rivulus* site from which guppies were naturally absent (Reznick *et al.*, 1990). Over an 11 year period, representing about 30 to 60 guppy generations, the transferred population showed shifts in its life-history traits towards those characteristic of guppies found naturally at *Rivulus* sites. The shifts included an increase in offspring size and an increase in the body size at which reproduction started. In samples taken in the dry season, the females from the transferred population had smaller broods and a lower reproductive allocation than females from the original population. In the wet season, the differences were not significant. This illustrates that other environmental factors probably interact with predation to influence the traits. Further evidence for an effect of other factors came from a mark–recapture study of guppies. The marks allowed estimates of survival rates of guppies of different sizes (Reznick *et al.*, 1996b). Guppies at *Crenicichla* sites did have poorer survival than at *Rivulus* sites, but the higher mortality was approximately the same for all the size classes. The differences in life-history traits between guppies from high- and low-predation sites are those expected if, at the high-predation sites, the larger guppies suffered a higher rate of predation.

Breeding experiments under controlled conditions in the laboratory over two generations showed that the changes in the transferred population were heritable. The number of females originally transferred in the field experiment and the size of the transferred population thereafter suggest that the changes were not the consequence of a 'founder effect' nor genetic drift, but represented an adaptive change.

These studies had concentrated on guppy populations in the watersheds of the southern slopes of Trinidad's Northern Range Mountains. On the northern slopes, guppies are exposed to a largely different assemblage of fish predators, notably mullets and gobies derived from marine stocks. Comparisons of the life-history traits of guppy populations from high- and low-predation sites on these northern slopes showed differences between the populations that parallel those found on

the southern slopes (Reznick and Brygla, 1996; Reznick *et al.*, 1996c). This suggests that these north slope populations have responded to a similar environmental factor, predation, by similar adaptive changes in their life-history traits. They show parallel evolution.

The studies of the Trinidadian guppy populations are examples of original and imaginative use of comparative field and laboratory studies to analyse the effect of natural selection on behavioural and life-history traits. They have shown that rates of change in life-history traits under natural selection can be of the same magnitude as changes achieved under artificial selection (Reznick *et al.*, 1997).

Life-history evolution in brook trout

Life-history theory provided the framework for an analysis of the differences in life history traits between populations of the brook trout isolated in small rivers on Cape Race, Newfoundland (Hutchings, 1993b, 1994). The rivers, although close geographically, differ in the biomass of invertebrates and in the availability of overwintering refuges. As a consequence, growth rates and survivorship differ between the trout populations. As predicted by life-history theory, the population with the highest adult survival rate relative to juveniles had the latest age of maturation, expended the least effort on reproduction and experienced the lowest survival cost of reproduction. High juvenile growth rate relative to adult rate was correlated with early reproduction, a high investment in reproduction but a high survival cost to reproduction. Importantly, further modelling studies suggested that the observed age at first reproduction and age-specific rates of maturity in the brook trout populations were close to those that maximized individual fitness. The differences between populations are adaptive and do not represent just random variations. This study also illustrated the importance of growth rates in the evolution of life histories in fishes.

Other examples

Along the eastern seaboard of North America, there is a north–south cline in the life-history characteristics of the American shad (Leggett and Carscadden, 1978). The shad is anadromous, and does not feed during its freshwater spawning migration. Spawning fish home to their natal stream, allowing the evolution of adaptations to the conditions of the home river. The percentage of fish in a spawning run that are repeat spawners increases northwards with the latitude of the home river (Fig. 11.3). In Florida, the shad are semelparous, whereas in New Brunswick, the populations are strongly iteroparous, with repeat spawners accounting for a high proportion of the spawning run. Fecundity per spawning at a given weight increases from north to south, while mean age at maturity tends to decrease. Differences in life-history characteristics are unlikely to be generated during the marine phase in the life cycle. Temperatures in the rivers during

the spawning season are more predictable in the south than in the north (Glebe and Leggett, 1981b). Water temperatures during egg and larval development have a significant effect on spawning success, year-class strength and recruitment. Egg, larval and juvenile survival may be less variable in southern rivers. Life-history theory suggests that conditions in southern rivers will favour the evolution of reduced age at first reproduction and increased expenditure on reproduction. The interpopulation differences found in the American shad may depend on genetic differences and may have arisen by selection, but direct evidence of this is lacking. The possibility that the differences are simply a manifestation of the phenotypic plasticity of these species cannot be discounted.

In 1905, 150 mosquito fish, *Gambusia affinis*, were introduced into Hawaii for mosquito control. They and their descendants were distributed among reservoirs on several of the Hawaiian Islands. Seventy years and about 140 generations later, the life-history characteristics of populations from several reservoirs were studied using fish collected in the field and fish held under controlled laboratory conditions from birth (Stearns, 1983a,b,c). Evidence for both phenotypic plasticity and genetic control of life-history traits was obtained. The problem of phenotypic plasticity is discussed in Section 11.6. The laboratory studies, which probably reveal genetically controlled differences among the populations, suggested that strong and frequent fluctuations in the water level of a reservoir had selected for fish that mature at small sizes but have high fecundities. Field observations indicated that rapid fluctuations affected large fish more adversely than small fish. There was a positive correlation between size of progeny and reservoir volume. This correlation was compatible with the observations that larger reservoirs cannot fluctuate rapidly and that long periods of low water select against

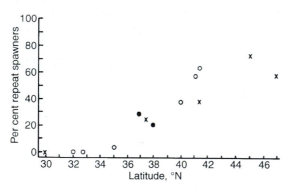

Fig. 11.3 Relationship between per cent repeat spawners (degree of iteroparity) and latitude for populations of American shad, *Alosa sapidissima*, along the eastern seaboard of North America. Symbols: ×, their study; ●, one-year sample; ○, multiple-year sample. Redrawn from Leggett and Carscadden (1978).

small fish. The study of mosquito fish emphasizes the importance of understanding the nature of size- (or age-related) mortalities for interpreting the evolution of life-history traits.

Also compatible with life-history theory is the general observation that in marine and pelagic fishes, the age at first maturity and reproductive life span tend to increase as the harshness and unpredictability of the environment increase (Cushing, 1975, 1982; Mann and Mills, 1979). Cod in the Celtic Sea, between Britain and Ireland, mature at the age of 3 years, at a length of 700 mm, and live for about 8 years. In the Barents Sea, within the Arctic Circle, the fish start to mature at the age of 7 years, but may live for more than 20 years. If the harshness and unpredictability have their major effects on the early life-history stages, then the theory suggests the trends that are observed.

Evaluation of importance of dimensionless numbers in life histories

Although studies on dimensionless numbers in fish populations started at about the same time as the early developments in life-history theory (Beverton, 1963), there has been far less development of these ideas. Not all studies have found evidence of the constancy of dimensionless numbers such as αM or L_α / L_∞ discussed by Charnov (1993) (see page 261). Such exceptions include walleye populations in North America (Beverton, 1987) and brown trout populations in Norway (Vollestad *et al.*, 1993). Further empirical studies are required to evaluate how often, and under what circumstances, fish taxa show invariance in these dimensionless numbers.

Evaluation of the habitat templet model

The idea that there should be a match between the spatial and temporal characteristics of habitats and the life-history traits of populations utilizing the habitats seems so reasonable. Consequently, it comes as a surprise to find that the empirical evidence supporting the habitat templet model for fishes is still weak. An attempt to validate the templet model makes use of an ambitious study of the habitats and fauna of the upper Rhone River in France (Statzner *et al.*, 1994). Multivariate statistical techniques were used to describe the riverine habitats and the life-history syndromes of the fish found in the upper Rhone. However, there was no good correspondence between habitat characteristics and life-history syndromes (Persat *et al.*, 1994). The investigators were reluctant to abandon the templet model. They suggested that the spatial scales chosen for the sampling programme, the paucity of fish species in the Rhone and historical changes in the river may have obscured the predicted match between habitats and life histories. A problem for the templet model is that species with different life-history syndromes often coexist in the same habitats (Winemiller and Rose, 1992). Perhaps it is better to ask what habitats cannot be occupied by a species with a given syndrome of life-history traits?

A danger of the habitat templet model is that a given syndrome of life-history traits is assumed to identify the selective processes acting in a habitat without those processes being identified empirically. The model is assumed rather than tested (Roff, 1992; Charlesworth, 1994).

11.4 COST OF REPRODUCTION AND THE CONSEQUENCES

Nature of the cost of reproduction

Life-history theory has at its centre the assumption that there is a trade-off between current reproduction and future expected reproductive output. Without such a trade-off, only physiological, allometric or phylogenetic constraints could account for the delayed maturation found in many popula-

tions and species. The cost of current reproduction could consist of reduced survival to
the next breeding season, reduced growth
with its correlate of reduced fecundity, or an
increase in the time before the next breeding
attempt. Evidence of a reproductive cost
would be a negative correlation between
current reproductive expenditure and some
other component of fitness such as survivorship, future fecundity or survivorship of offspring (Bell and Koufopanou, 1986).

Costs can be classified as physiological or
ecological. A physiological cost is when the
allocation of resources to reproduction
detracts from allocation to physiological processes required to maintain the long-term
condition of the individual (Chapter 7). An
ecological cost is when an activity associated
with reproduction puts the individual at risk
from predation, infection by disease, or some
other deleterious consequence (Chapter 8).

The search for negative correlations is
complicated by any variation in the physiological condition of the fish. A fish in poor
condition may have both low current fecundity and poor post-breeding survival because
both are a consequence of its poor condition,
whereas a fish in good condition might have
both high current fecundity and good survivorship. Fish can differ in condition, both
within populations and between populations.
Female bluehead wrasse living on coral
patch reefs in Panama showed no inverse
correlation between fecundity and growth
rate. All the significant correlations were
positive, the opposite direction to that predicted by a cost-of-reproduction hypothesis
(Schultz and Warner, 1989, 1991). This
example illustrated the difficulty in detecting
a cost if there is a confounding effect of differences in condition. When populations are
compared, the interactions between the condition of individuals and habitat quality can
generate both positive and negative correlations between pairs of life-history traits.

A second and fundamental problem in the
search is to identify the source of an
observed negative correlation (Reznick,
1985). Life-history theory attempts to predict
the direction of evolutionary change in life-
history traits in response to an environmental
change that has defined effects on age-specific schedules of mortality and/or fecundity.
The theory assumes that selection for higher
fecundity at age i has, as its consequence, a
reduction in future expected reproductive
output beyond age i. This assumption
implies that the cost of reproduction arises
from negative genetic correlations between
fecundity and some other components of
fitness. A negative genetic correlation means
that selection to increase one trait such as
fecundity results concomitantly in a decrease
in another trait such as viability. Negative
genetic correlations may be detected by
selection experiments or by the resemblances
between parents and progeny for the traits
under consideration. Even in the absence of a
negative genetic correlation between some
life-history traits, fish may still show a phenotypic cost of reproduction. Thus, if two
fish, with essentially the same genotypes and
in the same condition, make different reproductive expenditures at age i, they will probably also differ in some component of
postbreeding survival or growth. But in the
absence of genetic variation for the life-
history traits, there will be no evolutionary
change in the population. Selection can only
occur if the phenotypic variations reflect an
underlying genotypic variation. Interestingly,
Maynard Smith (1991) has speculated that
the only genetic correlations likely to be sufficiently stable to be interesting are likely to
arise from physiological interactions detectable by direct experimental manipulation
rather than long-term selection experiments.

Evidence for a cost of reproduction

A cost of reproduction may be expressed as
an increase in mortality, a decrease in
growth, a decrease in future fecundity or an
increase in the time between successive

breedings. In semelparous species such as the Pacific salmon, the cost of reproduction is death. Even in some iteroparous species, the breeding and immediate postbreeding period is a time of high mortality. A detailed life-table study of European minnows in Sea-court Stream in England identified heavy spawning mortality (Pitcher and Hart, 1982). Aspects of reproductive biology of species like threespine stickleback put breeding fish at a greater risk of predation than non-breeding sticklebacks (Wootton, 1984a). There is evidence that the bright nuptial coloration of male sticklebacks increases their risk of being eaten by predators such as rainbow and cut-throat trout (Semler, 1971). During courtship, a male stickleback makes abrupt, con-spicuous movements. Finally, during the par-ental phase, when guarding the eggs in his nest, the male shows a high level of boldness to potential predators (Pressley, 1981).

Other evidence for a mortality cost of reproduction comes from comparisons between closely related species. In popula-tions in which the mean age of first repro-duction is low, the mean life span also tends to be short (Roff, 1981).

Hirshfield's (1980) experimental study of the medaka (Chapter 7) provided direct evi-dence of a mortality cost to reproduction. There was a strong positive correlation between the number of sick and dead fish and average reproductive effort. The latter was measured as the ratio of the total energy content of the eggs spawned to the total energy of the food consumed by the fish. These experiments also provided evidence for a growth cost to reproduction because growth was negatively correlated with repro-ductive effort.

An experiment with guppies provided a comparison of the growth of reproducing and non-reproducing females fed identical rations (Reznick, 1983). Although the non-reproducing females grew more than the reproductively active females, most of the growth was in the form of fat reserves and not somatic protein growth. Energy budgets for the two types of female showed that at a given ration, the total production, that is somatic production plus production of progeny, expressed in energy units, was greater for the reproductively active females than for the non-reproductive fish. The fate of the energy lost by the latter females was unknown. This experiment suggested that guppies had only a limited capacity to redir-ect energy not invested in reproduction into somatic growth. In contrast to this result, when the allocation of energy between growth and reproduction from different populations of guppies was compared, there was a complementary relationship between the two. Fish from populations that invested more in reproduction showed less somatic growth. Thus, in the guppies studied, within a population there was only a slight growth cost to reproduction, but a comparison among populations revealed a cost. This sug-gests that selection for either higher repro-ductive expenditure or faster somatic growth is at the expense of a decrease in the other trait, but that there is little phenotypic plasti-city in the pattern of allocation evolved within a population.

The difference in the growth rate at a given ration between a reproductively active female threespine stickleback and the same fish after the breeding season is shown in Fig. 11.4. At a given ration, reproductively active female sticklebacks had smaller livers and a higher water content in their tissues than females that had ceased spawning (Wootton, 1977). This was evidence that the spawning females were in poorer somatic condition.

Studies on populations of salmonids at high latitudes suggested that it can take 2 years or more for a fish to recover suffi-ciently from spawning to spawn again (Dutil, 1986). The populations are in harsh environ-ments in which juvenile mortality is prob-ably high and variable. These populations are usually iteroparous as predicted by life-

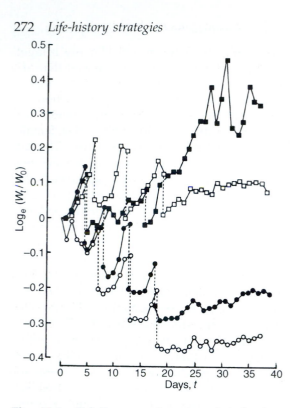

Fig. 11.4 Relative growth rates of individual female sticklebacks, *Gasterosteus aculeatus*, during and after spawning. Females were fed rations of 2% (○); 4% (●); 8% (□) or 16% (■) of initial body weight per day. Vertical broken line indicates spawning. W_t, fresh weight at time t; W_o, fresh weight on day 0.

history theory. But the expenditure on reproduction is such that annual breeding may not be achieved. Dutil (1986) compared the energy reserves of a population of Arctic charr in northern Canada. The charr overwintered in fresh water, a 10 month period during which little feeding occurred. In summer, the non-breeding fish migrated to the sea to feed, but the breeding fish remained in fresh water to spawn. Over winter, non-breeding fish lost 30% of their energy reserves, but this was restored during the two months spent in the sea. Postbreeding fish that migrated to the sea in the spring had energy reserves 46% lower than non-

breeding fish migrating at the same time. Even after the period in the sea, the post-spawners still had lower energy reserves than the non-reproductives, and this difference was greater for bigger fish. The data suggested that it took at least 2 years for the postspawners to restore their energy reserves to the non-reproductive level. There was some evidence that the largest fish were unable to restore the reserves to a level at which reproduction could occur and entered a postreproductive, senescent state (compare with the 'jellied' state described for large flatfish on page 274).

Consequences of a cost of reproduction

Roff (1984, 1992) used stable-age demographic theory (see Chapter 10 and above) together with the assumption that there is a significant cost of reproduction, to model the relationships between life-history traits in teleosts. This cost was modelled by first assuming an ideal age-specific schedule for the product $l_x m_x$ (where l_x and m_x, are the age-specific survival and fecundity respectively – Chapter 10) in which no cost was met, then modifying this ideal schedule to incorporate a cost. The model was then developed to predict the age of first reproduction from the instantaneous mortality rate for postlarval stages (Z), the probability of surviving the larval stage (p), the parameters of the von Bertalanffy growth equation (K and L_∞) and the relationship between fecundity and length. The model accounted for 60% of the variance in the age of first reproduction for 31 populations representing 23 species (Fig. 11.5). By assuming explicit forms for the cost of reproduction, Roff (1984) also predicted the optimal age at maturity for fish from given values for K, L_∞ and p and predicted the maximum rates of mortality that a population could sustain. In general, the relationships between variables predicted by the models were consistent with the relationships observed for real populations. These models

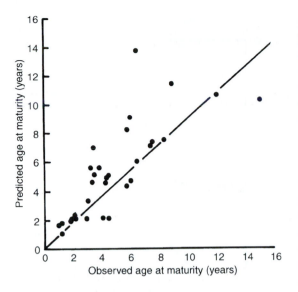

Fig. 11.5 Relationship between observed and predicted age at maturity, where predicted age is obtained from a demographic model incorporating a cost of reproduction. Diagonal line, predicted age equals observed age. Redrawn from Roff (1984).

suggested that one consequence of a cost of reproduction is the empirical relationships observed by Pauly (1980) (page 261). Charnov (1993) has also interpreted regularities in the relationships between traits such as age at reproduction and mortality as the consequence of trade-offs (page 261).

The concept of a cost of reproduction has been reviewed critically (Tuomi *et al.*, 1983; Bell and Koufopanou, 1986; Roff, 1992; Stearns, 1992). In no organism has the shape of a trade-off curve been unambiguously determined. However, as Roff's analysis shows, the incorporation of the concept into demographic theory offers a valuable approach to the analysis of life-history patterns.

11.5 BIOENERGETICS OF LIFE-HISTORY PATTERNS

Although one component of the cost of reproduction may be an increased suscept-

ibility to predation or disease during the breeding season, most studies have considered the trade-off between energy allocated to reproduction and that to maintenance and growth. Both theoretical and experimental studies have examined this allocation problem.

Surplus energy and life-history traits

The energy assimilated by a fish can be allocated to maintenance, somatic growth (including lipid reserves) or gametes (Chapter 4). If it is assumed that the energy expenditure on maintenance including expenditure on swimming has priority, then any energy surplus to the maintenance requirements can be channelled to growth or reproduction (Ware, 1980; Roff, 1983). Ware (1982), noting that power has the dimensions of energy per unit time, has defined surplus power as the difference between the rate of energy income and the rate of expenditure on maintenance per unit time. In this context, the term 'surplus' is slightly misleading because growth and reproduction cannot be regarded as surplus activities. Furthermore, empirical studies have shown that maintenance does not always have priority (Hirshfield, 1980), so the allocation problem is more complex. Nevertheless, both Roff and Ware have used the concept of surplus energy to provide insights into fish life histories. The use of the power allocated to reproduction as a measure of fitness is discussed by Brown *et al.* (1993). Kozlowski (1991) argued that models of the optimal allocation of energy between growth, reserves and reproduction are to be preferred, because of their predictive power, to models based on the concepts of 'reproductive effort' or 'cost of reproduction'.

Roff (1983) used the surplus-energy model to explore the evolution of the life-history pattern of the American plaice. His starting point was the allocation equation:

$$W_{t+1} = W_t + S_t - G_{t+1} \qquad (11.4)$$

where W_{t+1}, and W_t are the somatic weights at ages $t+1$ and t, S_t is the maximum potential increase in somatic weight (that is maximum surplus energy) at age t, and G_{t+1}, is the weight of gonads at $t+1$. In other words, somatic growth over the period t to $t+1$ is equal to the maximum potential increase in somatic weight less any allocation to the gonads. There is an explicit trade-off between somatic and gonadal growth. From this expression, Roff derived the relationship:

$$W_{t+1} = (W_t + S_t) / (1 + \gamma_{t+1}) \qquad (11.5)$$

where γ_{t+1} is the gonadosomatic index of a fish aged $t+1$. Roff obtained a realistic growth curve for a population of the plaice by using observed values for the growth rate of sexually immature fish. The decline in growth with the onset of maturity indicated the growth cost of reproduction. Further analysis showed that the median age of maturity observed in the population was considerably higher than the optimal age predicted by the model from the observed gonadosomatic index. The trade-off between reproduction and growth cannot account for the life-history pattern of the plaice, but the model did not include a trade-off between reproduction and survival.

The allocation model did cast some light on a condition found in mature female American plaice called jellied flesh, in which the muscle has a low protein but a high water content. Roff's model predicted that there is a size of fish, $L_{\max}$, at which all the surplus energy is allocated to reproduction. This size is an inverse function of the gonadosomatic index, so if a fish at $L_{\max}$ then increases its gonadosomatic index, the allocation to reproduction exceeds the surplus energy available. The difference would have to be met by metabolizing somatic tissue. In the plaice population, the gonadosomatic index does increase with length, and the percentage of jellied females also increases with

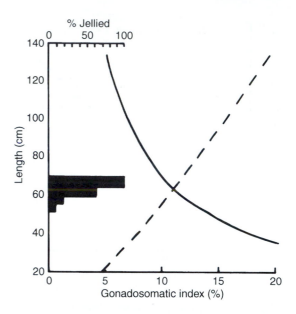

Fig. 11.6 Gonadosomatic index, body size and jellied condition in American plaice, *Hippoglossoides platessoides*, from St Mary's Bay, Newfoundland. Histogram, frequency of jellied condition in relation to length; broken curve, relationship between gonadosomatric index and length; solid curve, relationship between maximum length, $L_{\max}$, and gonadosomatic index predicted by Roff's (1983) growth model. Redrawn from Roff (1983).

length, suggesting that for these females, $L_{\max}$ is exceeded (Fig. 11.6).

Ware (1980, 1982, 1984) developed his concept of surplus power to provide some insight into the stock–recruitment relationship in marine species, thus linking a major demographic problem (Chapter 10) with the energy-allocation problem. One of his aims was to show that, in principle, the shape of a stock–recruitment curve (Fig. 10.13) can be determined by the allocation pattern and the degree to which the density of a stock affects its acquisition of surplus power.

His starting point was the balanced budget (see Chapter 4):

$$pC(t) = R(t) + S(t) = R(t) + W(t) + F(t) \qquad (11.6)$$

where p is the assimilation efficiency, $C(t)$ is the food energy per year in year t, $R(t)$ is the energy expenditure on maintenance, including swimming and food processing per year, $S(t)$ is surplus power per year, $W(t)$ is somatic growth per year and $F(t)$ is energy invested in reproduction per year (reproductive effort). (Note that the symbols used here do not correspond to those in Ware's original publications.) A demographic assumption was that the instantaneous mortality rate of larval and juvenile fish was density dependent (Chapter 10). $F(t)$ was assumed to be a function of $W(t)$ of the general form:

$$F(t) = a\ W^b \qquad (11.7)$$

(Chapter 7). Growth rates and surplus power

were also assumed to be allometrically related to body weight (Chapters 6 and 4).

The analysis showed that when $F(t)$ is strongly density dependent, because surplus power is density dependent, the stock–recruitment curve is dome-shaped (Fig. 11.7; compare Fig 10.13(c)). When $F(t)$ is weakly density dependent or is density independent, the curve is asymptotic (Fig. 11.7; compare Fig. 10.13(b)). Surplus power is density-dependent when the food available to individual fish decreases as their density increases. The model suggested that a dome-shaped recruitment curve can arise when a decrease in the fecundity of individual fish is caused by a density-dependent reduction in the surplus power that can be allocated to repro-

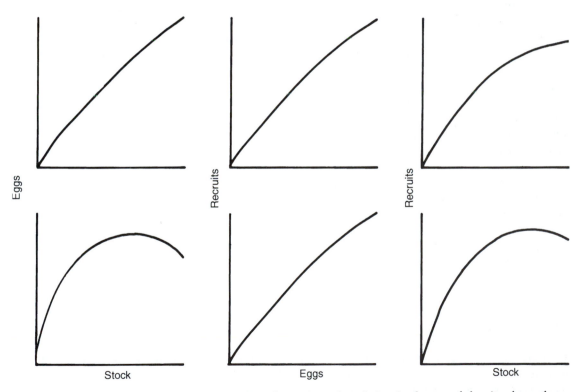

Fig. 11.7 Relationship between recruitment–stock curves and variation in degree of density dependence in egg production and pre-recruit stages. Recruitment–stock curve is asymptotic when density dependence of egg production is weak (upper row), and dome-shaped when density dependence is stronger (lower row). Redrawn from Ware (1982).

duction. The tendency to a dome-shaped recruitment curve is enhanced by density-dependent growth. The model also suggested that those species that divert proportionately more surplus power each year to reproduction have larger ovaries, a smaller maximum body size and an asymptotic or weakly-domed recruitment curve, compared with species in which the surplus power – and so fecundity – is strongly density dependent. Ware (1980) cited herring as an example of the first type of species and cod or haddock as the second type (see also Fig. 10.14). For North Sea plaice, Rijnsdorp (1994) suggested that at a given fish size, there is a threshold value for surplus power. Above this threshold, surplus power is invested in somatic growth rather than in increasing size-specific reproductive output. Below the threshold, reproductive investment decreases.

In its first form, Ware's model did not consider the age structure of the population, but a subsequent version did (Ware, 1984). Using this age-structure model, the consequences of two reproductive strategies were compared. In the first, strategy A, fecundity at a given age increased with an increase in surplus energy. In strategy B, fecundity at a given age was a function only of the size of the fish at that age. Numerical simulation suggested that strategy A yielded a higher fitness for an iteroparous species that experienced a low level of natural mortality after maturation. Strategy B yielded a higher fitness where adult mortality was high. In the simulations it was assumed that there was a mortality cost to reproduction. The results suggested that if a fish has a low chance of survival as an adult, it should not adjust its fecundity to the supply of food but should produce the number of gametes that its body size permits, even if this is at the cost of metabolizing somatic tissues. This prediction that high adult mortality selects for increased reductive effort at a given age, even though this reduces postbreeding survival, recurs in theoretical studies of life-history evolution. Strategy A also tended to generate a higher fitness when there was a size-dependent component to adult mortality. Fish showing this strategy can devote surplus energy to somatic growth, enabling them to move quickly through the size classes that are most at risk.

The importance of the models of Roff (1984) and Ware (1980, 1982, 1984) derives less from their detailed predictions than from the way in which they combine aspects of demographic, life-history and bioenergetics theory in a synthesis. This gives insights into the possible adaptive significance of life-history patterns, the demographic consequences of these patterns and the likely direction of change of life-history traits with a change in environmental conditions. The application of the mathematical techniques of optimal control theory to the problem of energy allocation during ontogeny promises to give further impetus to this approach to life history theory, but at the cost of an increase in the mathematical difficulty of the models (Kozlowski, 1991).

Experimental analysis of energy allocation

Insight may also come from experimental and observational studies on the allocation of energy among maintenance, growth and reproduction (Wootton, 1985). The studies consider two problems: the cost of reproduction (see above) and the change in the pattern of allocation as environmental factors, particularly food availability, change.

Experiments on the effects of ration on the reproductive output of female sticklebacks (Wootton, 1973, 1977; Wootton and Evans, 1976; Fletcher and Wootton, 1995) and the medaka (Hirshfield, 1980) are described in Chapter 7. These species are both small fish that reach sexual maturity at an early age. The experiments showed that both species increase the proportion of the energy income devoted to reproduction as the ration declines. Both species subsidize egg production by the depletion of somatic tissue. On the basis of his experiments with the female medaka, Hirshfield (1980) argued that the cost of maintenance cannot be regarded as a constant that has to be met before energy can be allocated to reproduction or growth (cf. Ware, 1980, 1982). In some circumstances it may be adaptive for a fish to give priority to reproduction even if a lower allocation of energy to maintenance leads to an increase in mortality (Hirshfield, 1980). Compared with the stickleback and the medaka, the winter flounder is a large, long-lived species with potentially many breeding seasons. On low rations this species allocated less energy to the ovaries and tended to maintain somatic condition (Tyler and Dunn, 1976). The winter flounder shows no parental care and its egg and larval mortality are high and variable (Chapter 10). A strategy that gives somatic condition priority over ovarian growth when feeding conditions are poor should result in the high adult survival, with iteroparity, expected by life-history theory. The stickleback has poor adult survival, but there is well-developed parental care of eggs and larvae by the male. The priority given to the ovaries in the allocation of energy is also compatible with the expectations of life-history theory.

In their analysis of the energetics of iteroparous and semelparous populations of American shad, Glebe and Leggett (1981b) made an important point. The pattern of energy allocation is the mechanism by which the fish can achieve an adaptive life history in given environmental circumstances rather than the observed life history simply reflecting a constraint imposed by energy availability. The evolution of life-history patterns involves the evolution of patterns of energy regulation and allocation (Wootton, 1985).

11.6 PHENOTYPIC PLASTICITY OF LIFE-HISTORY TRAITS

The experiments on the threespine stickleback, the medaka and the guppy revealed the phenotypic responses of fish to environmental conditions. They gave some indication of the plasticity of the life-history traits of fecundity, growth and survival. Some changes in the life-history traits in fish populations exposed to fishing pressure (Chapter 7) probably also illustrate such plasticity, because the changes take place over a time scale that is too short for them to arise by selection.

In his study of the mosquito fish populations in reservoirs on the Hawaiian Islands, Stearns (1983a) found that much of the variation in life-history traits among populations was caused by plastic responses to short-term environmental changes such as fluctuations in the water level. A comparison of three species of freshwater fish in western Europe, the bullhead, *Cottus gobio*, the stone loach, *Barbatula (Noemacheilus) barbatulus*, and the gudgeon, *Gobio gobio*, suggested a general trend in life-history traits in relation to the productivity of their habitats (Mann *et al.*, 1984). In low-productivity streams, usually in the north, the tendency is for a single spawning in a breeding season, delayed maturity and long life spans. In high-productivity streams, especially in the south, there tend to be several spawnings in a season, early age at maturity and a short life span. The reciprocal transplants of *C. gobio* between northern and southern sites described in Chapter 7 suggested that most of the differences in the life-history traits between fish from the two sites did not reflect genetic differences, but revealed aspects of the phenotypic plasticity of *C. gobio* (Table 11.3).

The phenotypic plasticity shown in response to environmental change is expected to evolve to minimize the cost to the fish of the change (Chapter 1). The plasticity is adaptive if it forms part of the homeostatic capacity of the fish (Caswell, 1983; Roff, 1992). Both experimental (Alm, 1959) and observational studies (Pitt, 1975) have shown that the age and/or size at maturity can change in response to environmental

Table 11.3. Effect on reproductive biology of *Cottus gobio* of reciprocal transplants between northern and southern sites*[*][†]

Experimental site:	Hury		Waterston	
Source of *Cottus gobio*:	Scur	Bere	Scur	Bere
No. *C. gobio*	0	8	5	19
Spawning period	–	18/4–27/5	26/3–23/5	11/3–23/5
No. eggs per female	–	91	712	1110
No. batches per female	–	0.75	2.60	3.42
No. eggs per batch	–	121	297	357
No. eggs per g total weight	–	33	127	192
Mean dry weight per egg (g)	–	–	0.0013	0.0014

*Source: Mann *et al.* (1984).
[†]Northern sites, Scur Beck and Hury Reservoir; southern sites, Bere Stream and Waterston cress ponds; all fish were taken from Scur or Bere. All Scur Beck fish at Hury that survived were male.

changes. Stearns and his co-workers have developed a model to analyse the effect on the age and size at maturity of an environmental change that causes a reduction in growth rate (Stearns and Crandall, 1984; Stearns and Koella, 1986). The model assumed that the phenotypic change will maximize the per capita rate of increase under the new conditions. Their analysis was based on the Euler–Lotka stable-age equation. The striking result of the analysis was that the model generated a set of age–size maturation trajectories, the shapes of which depended on assumptions on the relationship between growth rate and juvenile or adult mortality (Fig. 11.8). For example, when neither juvenile nor adult mortality depended on growth rate, the predicted optimal age–size trajectory had the following properties.

1. When growth is rapid, changes in growth produce large changes in age at maturity with small changes in size at maturity: the fish seem to have a fixed size at which they mature.

2. When growth rates are intermediate, changes in growth produce large changes in size at maturity but small changes in age at maturity: the fish seem to have a fixed age at which they mature.

3. When growth is slow, changes in growth produce large changes in age at maturity but small changes in size at maturity.

It is the shape of the age–size maturation trajectory that is adaptive and responds to selection, rather than either age or size as a single trait.

The model predicted the age at maturity of 19 populations of fish for which sufficient information on growth, fecundity and mortality rates were available. The model accounted for 82% of the variance in the age of maturity. Its predictions were also compatible with differences in the shifts in age at maturity and size at maturity of populations of mosquito fish from Hawaii reared under controlled laboratory conditions but experiencing different levels of crowding in the first 10 days of life.

A clear picture of the difference between

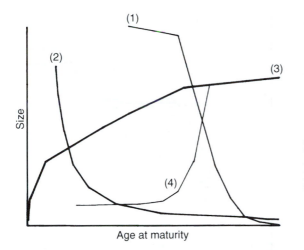

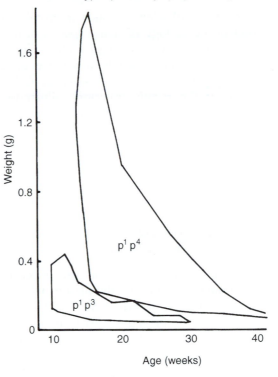

Fig. 11.8 Relationships between age at maturity and size at maturity predicted by Stearns and Crandall's (1984) model when an environmental stress causes change in growth rates (growth decreases from upper left to lower right). Trajectories predicted when: 1, neither juvenile nor adult mortality correlated with growth rates; 2, juvenile mortality increases slowly as growth rates decrease; 3, juvenile mortality increases rapidly as growth rates decrease; 4, adult mortality increases rapidly as growth rates decrease. In 1–3, maturity is delayed as growth rate decreases; in 4, maturity is advanced when growth rate decreases; in 1 and 2, later maturity occurs at smaller size; in 3 and 4, later maturity occurs at larger sizes. Redrawn from Stearns and Crandall (1984).

Fig. 11.9 Relationship between size (weight) at maturity and age at maturity in two genotypes of *Xiphophorus maculatus* reared over a range of rations. Envelopes enclose range of values obtained from each genotype. Redrawn from McKenzie *et al.* (1983).

phenotypic plasticity and genetic variation in the age–size maturation trajectory emerged from a study of two genotypes of the platyfish, *Xiphophorus maculatus*, which are known to differ in their age at maturation (McKenzie *et al.*, 1983; Policansky, 1983). Fish of known genotype, reared in isolation, were maintained on different rations and so showed different growth rates. Within a genotype, there was considerable phenotypic variation in age and weight at the initiation of maturity, which can be approximately described by an L-shaped trajectory (Fig. 11.9). But there was little overlap between the two genotypes. In terms of the Stearns and Crandall model, the difference in the position of the two trajectories could be predicted if the genotypes differed in their adult mortality rates (Stearns and Crandall, 1984).

The precocial maturation of male Atlantic salmon probably provides another example of a life-history trait, age at first reproduction, that illustrates phenotypic plasticity and genetic variation (Hutchings and Myers, 1994) (Chapter 7). High juvenile growth rates promote precocial maturation, but different genotypes have different thresholds for the initiation of maturation.

Phenotypic plasticity is such a prominent aspect of the biology of many species of fish that the development of a predictive theory of the phenomenon must be a major goal of fish ecologists. An important advance is the concept of the plastic trajectory, illustrated by the age–size at maturity trajectory (Fig. 11.9), as a trait under selection. Such theoretical developments are important additions to the study of life-history patterns when linked with laboratory and field experiments on the shape of the relevant trajectories.

11.7 LIFE-HISTORY PATTERNS AND EXPLOITATION

An enormous experiment on the effect of environmental change on the life-history patterns of fish is conducted, albeit inadvertently, by the fishing industry throughout the world. Fishing imposes a mortality on a population that is usually strongly size selective. This new form of mortality may result just in phenotypic changes as the fish respond to the new environmental circumstances, but it may impose selection on the population. The studies of the effects of fish predation on life history and other traits of guppies in Trinidadian streams have illustrated the size and the speed of the evolutionary changes that differential predation can drive. What is the relationship between the response of a population to fishing and the life-history pattern of the population that had evolved in the absence of fishing? The capacity of a population to support a fishery will depend on its ability to meet the losses caused by the fishery by compensatory changes in survivorship, growth and fecundity. Such compensatory changes can include decreases in the age at maturity, increases in the size at maturity with the correlated increase in fecundity, or increases in the size-specific fecundity (Stearns and Crandall, 1984). The relative importance of phenotypic plasticity and genetic change in these responses is not known, but in some cases the

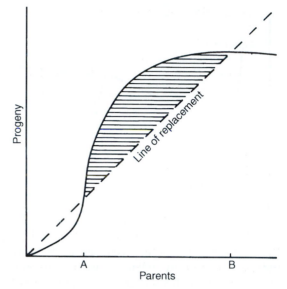

Fig. 11.10 Idealized stock–recruitment curve showing concept of surplus production (shaded area). A, lower critical density – a population reduced to abundance lower than A declines to extinction because recruitment falls below line of replacement; B, upper critical density – a population with abundance in excess of B declines and the extent of the decline depends on the strength of density dependence.

changes have occurred over a period sufficiently short to implicate phenotypic plasticity (Pitt, 1975).

The basic principle of compensation in response to fishing can be illustrated by a simple stock–recruitment curve (Fig. 11.10). The three features of importance are as follows.

1. The line of equal replacement: the stock just replaces itself.
2. The minimum critical density: at stock levels below this density, the stock cannot replace itself. If fishing mortality combined with natural mortality reduces the population to below this level, the population will decrease to extinction even if fishing stops.

3. The maximum critical density: at stock densities above this point, recruitment is insufficient to replace the adult stock. In an unexploited population, this is the equilibrium stock density; the population will return to this density after a disturbance.

Over the range of stock densities between the minimum and maximum critical densities, the population has the capacity to produce recruits in excess of the number required to replace the stock at a given density. This excess production is, in principle, available to the fishery. Unfortunately, this stock–recruitment curve is close to a fiction (Chapter 10). The variability caused by environmental fluctuations and the different responses of individual fish to the environment mean that the relationship between stock and recruitment has to be expressed probabilistically (Chapter 10).

Nevertheless, there does seem to be some relationship between life-history pattern and the response to fishing (Beverton, 1984). Garrod and Knights (1979) contrasted the response of pelagic schooling species such as the Peruvian anchoveta and the Californian sardine exploiting phyto- or zooplankton with that of demersal fishes such as cod or haddock feeding on benthic invertebrates or small fish. The pelagic species are short-lived with a short generation time (cf. opportunistic life-history syndrome, page 263). They mature at a small size, but augment their fecundity by batch spawning (Chapter 7). This life-history pattern means that the population abundance tracks environmental changes closely, because the time lag between the environmental change and changes in fecundity and mortality rates is short. Both the anchoveta and the sardine have supported large fisheries. Both have suffered massive population collapses, probably caused by excessive fishing at times of unfavourable environmental conditions (Chapter 10) (Cushing, 1982). The strong

schooling and pelagic mode of life of these species makes them vulnerable to modern fishing techniques such as purse seining. During unfavourable environmental conditions, they may retreat to localized areas where the conditions are still relatively favourable (MacCall, 1990). There, they maintain locally high densities open to exploitation. There is a small number of age classes of mature fish on which the exploitation concentrates. The life-history characteristics that allow the species to respond rapidly to favourable environmental conditions also make them vulnerable to overexploitation. In contrast, the demersal species tend to be long-lived with many age classes that contribute to recruitment (cf. periodic life-history syndrome, page 263). The fishing pressure is thus spread across several reproductively active age classes even when much of the yield is being taken from a dominant year class. The large number of age classes provides more opportunities for compensatory changes in survival, growth and fecundity. If the fishing pressure is such that the number of age classes is reduced, then the scope for compensatory changes will be reduced and the risk of a collapse in the stock increased. Cod populations typically contain several age classes, but the Canadian cod fishery collapsed in the early 1990s, probably because of overfishing (Myers *et al.*, 1997).

Adams (1980) argued that species that had delayed maturity, reduced growth rates, low rates of natural mortality, relatively long life spans and large body size (cf. equilibrium life history syndrome, page 263), would be particularly sensitive to overfishing. In many fisheries, it is the larger-bodied species that disappear first from the fishery as exploitation increases (Pauly, 1994).

Clearly, it would be valuable to fisheries managers if it can be shown that particular life-history syndromes respond in consistent ways to fisheries exploitation. This is especially true for the multispecies fisheries typical of many tropical marine and freshwater systems. In such fisheries, the number of species tends to preclude regimes of fishery management that require detailed information on growth and mortality of individual species.

In analysing the consequences that fishing has on the life-history patterns of fish, care has to be taken to define the likely effects of fishing on the reproductive success of individual fish. The life-history traits of fish are not selected to maximize the yield to a fishery, nor to ensure that the population remains stable in the face of environmental fluctuations. The goal of individual fish is to increase their contribution of offspring to the next generation relative to that of other individuals in the population. Consequences at the population level are the effects of the responses of the individual fish. Coincidentally, these effects may tend to stabilize the population or make it resilient to fishing pressure. The goal of rational management should be to control the fishery so that the changes in age-specific survival, growth and fecundity shown in the population in response to the exploitation have the effect of maximizing the economic returns to the fishery while minimizing the risk of a recruitment failure (Law and Grey, 1989).

11.8 SUMMARY AND CONCLUSIONS

1. Life-history theory has been developed to analyse the relationship between the characteristics of an environment and the life-history patterns of populations experiencing that environment. It assumes that age-specific schedules of growth, survival and fecundity have evolved through the tendency to maximize the per capita rate of increase (or an equivalent measure of fitness).

2. Some fish taxa, such as the Clupeidae (herrings), are characterized by relatively constant values for dimensionless numbers such as the product of age at maturity, α, and instantaneous mortality, M.

3. Multivariate statistical methods can be used to define clusters (syndromes) of life-history traits, which may then be related to habitat characteristics (habitat templet model).

4. The evolution of life-history traits is constrained by trade-offs between the traits, their compatibility with other traits, the amount of genetic variation in the population and phylogenetic inertia. Trade-offs may be the explanation for the observed constancy of dimensionless numbers associated with life histories.

5. Models that incorporate the trade-off constraint suggest that in environments in which juvenile mortality is high or variable, delayed reproduction and iteroparity are favoured. In environments in which adult mortality is high or variable, early reproduction and restricted iteroparity or semelparity are favoured. These predictions are generally supported by empirical evidence. Empirical evidence for the habitat templet model is, as yet, much weaker.

6. The trade-off constraint implies that there is a cost to reproduction. Evidence for such a cost in terms of reduced postbreeding growth or increased mortality comes from observations on natural populations and from experimental studies.

7. The energy costs of reproduction have been used to develop models of growth and the stock–recruitment relationship in fishes. These models combine demographic and bioenergetic principles in a promising synthesis. Experimental studies also show that energy expenditures on reproduction are related to life-history patterns.

8. Life-history traits may show considerable phenotypic plasticity. Such plasticity may be an important adaptive trait for individuals, allowing them to respond to environmental changes during their lifetime.

9. The impacts of fishing and other disturbances caused by the activities of humans are related to the life-history characteristics of populations. These characteristics make some populations sensitive to overexploitation because of an inability to compensate for a decrease in the abundance of sexually mature fish.

12

FISH ASSEMBLAGES

12.1 INTRODUCTION

An individual fish lives out its life within a complex of interactions and processes, which can both affect and be affected by the individual. The interactions include predator–prey relations (Chapters 3 and 8), competition (Chapter 9) and reproductive activities (Chapter 7). The processes include the flow of energy and nutrients through a trophic web (Chapters 3 and 4). All those organisms in a defined area or habitat, irrespective of taxonomic identity, that interact either directly or indirectly, form a community. Those organisms of the same taxonomic identity, for example all the fishes in a community, form a subcommunity (Giller and Gee, 1987). A guild describes a group of species in a community, which exploit the same class of resources in a similar way (Root, 1967). The term 'assemblage' will describe all the fish species in a defined area irrespective of whether they interact or not. This chapter discusses the factors that determine the species richness of fish assemblages. It also briefly considers the effects that the fishes have on the community of which they are a part.

The pattern of interactions of individuals defines the organization or structure of a community or subcommunity. Such organization is contingent. It is likely to change as conditions alter and drive changes at the phenotypic and genotypic level (Chapter 1). The frequency and nature of the interactions experienced by an individual will depend on the species diversity of the community.

Measurement of species diversity

Ideally, a measure of the species diversity should indicate the probability with which individuals encounter individuals of other species, weighted by the nature of the encounter. At present only cruder measures are available. The simplest measure is species richness, the number of species present. A more comprehensive measure will also include information on the relative abundances of the species present. If a few species in a community have high abundances, then the probability of encounter with a numerically rare species is much lower than if all the species in the community have roughly equal abundances. A frequently used measure of species diversity that includes the concept of relative abundance is Simpson's diversity index, D:

$$D = 1 \ / \ \sum_{i=1}^{S} p_1{}^2 \qquad (12.1)$$

where p_i is the proportion that the ith species contributes to the total abundance (or biomass) of the sample and S is the number of species in the sample. Another common measure of species diversity is the Shannon diversity index, H:

$$H = - \sum_{i=1}^{S} p_i \log_2 p_i. \qquad (12.2)$$

For any sample, a maximum diversity, D_{max}, or H_{max}, can be calculated by assuming that individuals in the sample are evenly distributed among the species in the sample. A measure, equitability, is then defined as:

$$E_D = D_{obs} / D_{max}; E_H = H_{obs} / H_{max}$$
$$(12.3)$$

where D_{obs} or H_{obs}, are the observed diversities.

The interrelationships between diversity measures were discussed by Hill (1973) and the concept was considered in detail by Magurran (1988). A weakness of diversity measurements is that in themselves they contain no biological information: they simply provide numerical summaries of samples and cannot be used to deduce the biological interrelationships among the individuals that make up the samples. Used with care, they and related measures may provide useful preliminary information on assemblages. For example, the effect of pollution on an assemblage can be described by changes in diversity and equitability indices, although the processes that cause the changes will not be revealed (Cornell *et al.*, 1976). For many purposes, species richness will be adequate as a first description of an assemblage.

The total species diversity of a system such as a river basin, its gamma diversity (γ), has two components (Whittaker, 1960). The first, alpha diversity (α), is the species diversity within a distinct habitat. The second, beta diversity (β) is the degree to which different habitats in the system differ in their species composition.

12.2 GENERAL PATTERNS OF SPECIES RICHNESS

Global patterns of species richness

There are about 25 000 recognized species of teleost fishes, of which some 40% are found in fresh waters, although fresh waters form only 0.0093% of the total water on earth (Nelson, 1994; Bone *et al.*, 1995). In both marine and fresh waters, the largest number of species is found in the tropics with a progressive reduction towards the polar regions.

In the sea, the greatest diversity is associated with warm waters fringing the land, especially where corals are highly productive. For coral reef assemblages, the areas with the highest species richness are in the central Indo–Pacific, with over 2000 species recorded from the Philippines. On a global scale, the number declines the greater the distance of the reef from the Philippines (Fig. 12.1) (Sale, 1980). In addition to the reduction in the number of shallow-water marine species with an increase in latitude, there are also changes in species richness with depth (Parin, 1984). There are about 3000 species of the open ocean, defined as water beyond the fringe of land and deeper than 200 m. Of the oceanic waters, the richest in species are the mesopelagic and mesobenthic zones, that is depths between 200 and 1000 m. Below 1000 m, the number of species falls off sharply. Even for these open-ocean faunas, the areas richest in species are under the tropical and warm temperate climatic zones.

Within a latitudinal zone, there are longitudinal differences. The Pacific coastline of North America is more species-rich than the equivalent zone in the eastern Atlantic (Nelson, 1994; Bone *et al.*, 1995).

In fresh water, the richest areas are the tropical zones of South East Asia, Central Africa and South and Central America (Lowe-McConnell, 1975, 1987; Welcomme, 1985; Nelson, 1994). In South America there are over 2400 neotropical freshwater fishes. This fauna is dominated by characins (Characiformes) and siluroid catfish (Siluriformes), both ostariophysans. These two groups make up about 80% of the Amazonian fish fauna of some 1300 species. The neotropical characins represent 'one of the most extreme cases of evolutionary radiation and adaptation amongst living vertebrates' (Goulding, 1980). The African fish fauna of over 2500 species is also dominated by Ostariophysi, though not to the same extent as the neotropical fauna, and the Cypriniformes replace the Characiformes as the most numerous

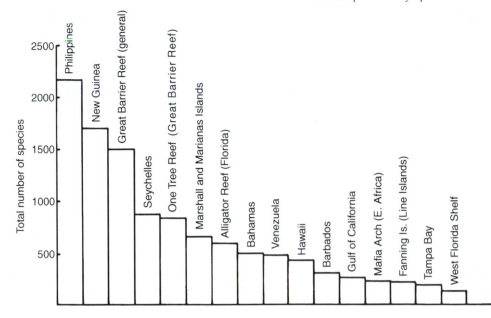

Fig. 12.1 Total species richness in reef fish faunas of each of 16 geographical regions. Redrawn from Sale (1980).

order. The Great Lakes of Africa are the scene of the spectacular adaptive radiation of the perciform cichlids (Chapter 3). In South East Asia, the Cypriniformes are also a major component of the rich freshwater fish fauna.

The decline in species richness with latitude in fresh waters is illustrated by a comparison of the African and North American Great Lakes. While the former, situated between the Equator and 15 °S each contain more than 250 species (Table 12.1), the total for all the North American Great Lakes, 42–49°N, is about 170 (Hocutt and Wiley, 1986; Payne, 1986). There are similar comparisons for riverine assemblages. Species richness at any latitude also tends to decrease with increasing altitude.

Species richness in relation to area

River basins, lakes and coral reefs all show a positive correlation between their area and their species richness. This relationship can

be described by a species–area curve, typically of the form:

$$S = cA^z \tag{12.4}$$

or:

$$\log S = \log c + z \log A \tag{12.5}$$

where S is the number of species, A is the area and c and z are parameters determined empirically. The exponent z is typically less than 1.0, indicating a decrease in the rate at which species richness increases with area. For a group of 45 rivers from South America, Africa, Asia and Europe, the relationship took the form:

$$S = 0.297A^{0.471} \tag{12.6}$$

and accounted for more than 80% of the variance in species number (Welcomme, 1979, 1985). Both the parameters c and z showed some geographical variation when the curves were calculated separately for the different continents, with the z value varying from

Table 12.1 Species richness of African Great Lakes, showing relative representation of family Cichlidae*

Lake	Cichlid species	Non-cichlid species	Total species	Non-cichlid families
Victoria[†]	>250	38	>288	11
Tanganyika	165	75	240	13
Malawi	>450	45	>495	8
Turkana	12	35	47	14
Edward + George	> 60	17	> 77	7
Albert	12	36	48	13

*Source: Greenwood (1991).
[†] Values for L. Victoria are the richness prior to the effect of Nile perch.

0.24 for European rivers ($N = 7$) to 0.55 ($N = 11$) for South American rivers. Habitat volume proved a better predictor of species richness than area in an analysis of the stream fishes from Minnesota, Illinois and Panama (Angermeier and Schlosser, 1989).

A survey of 70 lakes of the world yielded a z value of 0.15, but the value for a subsample of North American lakes of 0.16 was lower than the value of 0.35 for African lakes (Barbour and Brown, 1974).

At least three hypotheses may account for the positive correlation between species richness and area (Angermeier and Schlosser, 1989). Firstly, larger areas of a particular habitat will support more individuals, and so rare species have a better chance of being represented. Larger areas have a higher α diversity. Secondly, larger areas often contain a greater variety of habitats than small areas. If each environment has its own representative fauna, then there will be an increase in the β diversity with area. Thirdly, assemblages in small areas are likely to suffer high rates of extinction and have lower rates of immigration than in larger areas.

A striking effect of area on species richness, albeit in a restricted geographical locality, was found in a study of the fish fauna of Australian desert springs (Kodric-Brown and Brown, 1993). Five species occurred in the locality. As spring area increased, the number of species present increased. Each species had an approximate area threshold and was found only in those springs with an area exceeding the threshold. The species richness of springs increased by the addition of species, rather than replacement of species by other species. Each fish species probably had a minimum spring area that provided suitable habitats and could support a sustainable population.

The global patterns and the effect of area (or volume) on the numbers of fish species call for an explanation. In any region, the number of species present will be a consequence of historical and contemporary processes. Historical processes will determine the number of species present in a given region – its species pool. Contemporary processes will determine the composition of assemblages found at particular localities within that region, each assemblage forming a subset of the total pool. A useful conceptual model introduced the concept of filters (Fig. 12.2). Species found at a given locality are viewed as having passed through a hierarchical series of filters. It was initially applied to the origin and maintenance of fish assemblages in small forest lakes in Minnesota and Finland (Tonn et al., 1990). Regional processes, such as speciation and patterns of

REGIONAL PROCESSES

Pleistocene events
Dispersal barriers
Geomorphic/edaphic
   limits

Regional species
pool

e.g. northern Wisconsin
   lake district

LAKE-TYPE
CHARACTERISTICS

Abiotic conditions
Resource distribution
Habitat stability and
   complexity

Small lake species
pool

e.g. small lakes in
   northern Wisconsin

LOCAL PROCESSES

Area
Structural complexity
Isolation
Abiotic conditions
Biotic interactions

Local community
structure

e.g. Jude Lake

Fig. 12.2 Tonn *et al.*'s (1990) concept of filters (in capitals), generating species pools (in italics) through action of processes operating on different spatial and temporal scales, using northern Wisconsin lakes as example (simplified from Tonn *et al.*, 1990).

colonization, determine the regional species pool, for example the species pool of fresh waters in Finland. The characteristics of a habitat type, including its abiotic conditions, structural complexity and stability, then define a subset of the regional species pool found in that type of habitat, for example species typical of small lakes. Local processes, including area, structural complexity, abiotic conditions and biotic interactions, determine the local assemblage composition, for example the assemblage of a given lake.

12.3 HISTORICAL FACTORS IN SPECIES DIVERSITY

Cladogenesis, the process of speciation that generates two or more reproductively isolated gene pools from a previously single pool, generates species richness. Other things being equal, areas with higher rates of spe-

ciation will be richer in species. Older (in terms of geological time) environments may also be richer because there has been more time for speciation to occur. There is a continuing debate about the mechanisms of cladogenesis in fish (Echelle and Kornfield, 1984; Lowe-McConnell, 1987; Sinclair, 1988; Greenwood, 1991). This is a subject largely outside the scope of this text, but speciation will be favoured by physical or biotic conditions that prevent or restrict the gene flow between populations that could originally interbreed. Cladogenesis is seen at its most exuberant in the evolution of species flocks. Numerous closely related species have evolved from a common ancestor in a restricted geographical area. The cichlids of the Great Lakes of Africa are a good example (Chapter 3). Of approximately 240 species of fish in Lake Tanganyika, over 80% are endemic, having evolved within that lake, and nearly 70% are cichlids (Payne, 1986; Greenwood, 1991). Cichlids formed or form even higher proportions of the fish faunas of Lakes Victoria and Malawi (Table 12.1). Nor are species flocks confined to cichlids in large tropical lakes. There is a species flock of sculpins (Cottidae) in Lake Baikal in Asiatic Russia (Smith and Todd, 1984) and of cyprinodonts, *Orestias*, in Lake Titicaca, South America, at 3800 m above sea level (Parenti, 1984).

Several features of lakes may make them suitable sites for the evolution of species flocks. They are often more stable in terms of abiotic environmental factors than streams or rivers. Some lakes occupy ancient sites, although fluctuations in water levels cause variation in the area or even the number of lakes at a given site. Lacustrine faunas are essentially closed in terms of recruitment – there is little or no immigration from other lakes – so gene flow is minimized. This contrasts sharply with many marine species that have a pelagic larval stage, which allows progeny to become dispersed over wide geographical areas (page 219). Coral reef fishes, however, also have pelagic larvae yet are highly diverse (Fig. 12.1).

Speciation in lakes can be astonishingly rapid in terms of geological time. Lake Nabugabo, Uganda, became isolated from Lake Victoria some 4000 years ago. Five cichlids endemic to Lake Nabugabo have evolved in that period, each presumably from a cichlid species in Lake Victoria (Greenwood, 1965; Dominey, 1984).

The ease of colonization of an aquatic system may also be an important factor in determining its species richness. The fish fauna of USA is sharply divided by the Rocky Mountains (Smith, 1981; Hocutt and Wiley, 1986). To the east, there is a species-rich area centred on the Mississippi River basin, in which the rivers mostly run north–south. To the west, there is a species-poor region through which the rivers tend to run east–west and drain into the Pacific. A significant factor in generating this pattern is the ease with which species driven into the southern parts of the Mississippi basin with the advance of the Pleistocene glaciations have been able, as the ice retreated, to recolonize northern lakes and rivers along the north–south corridors.

In western and central Europe, the grain of the country also runs east–west, with the major rivers such as the Rhine and Danube running along the grain. The freshwater fish fauna of Europe is impoverished compared with that of eastern North America and again this is probably because fish could not easily retreat southwards to refuges, from the effects of the Pleistocene glaciations. Grand River, Ontario, and Nida River, southern Poland, are similar in size and are in similar climatic regions. A comparison of the species richness of the fish assemblages in the two rivers showed that the Canadian river is richer in species than the Polish river when the comparisons are made for the same distance from the source (Mahon, 1984). The increased richness of the Canadian fauna is largely a consequence of the higher number of small-sized species.

An analysis of the species richness of lakes in Ontario suggested that historical factors played a major role (Mandrak, 1995). Species richness tended to be lower in lakes situated at longer distances from the corridors that had allowed fish to disperse into the region as the ice of the last Pleistocene glaciation retreated. Species richness was higher in regions that had been ice-free and covered by glacial lakes for the longer periods. The main contemporary factor correlated with species richness of Ontario lakes was climate, particularly mean annual air temperature.

The probable effects of historical factors on the species richness in lakes or drainages are summarized by Smith (1981) (Fig. 12.3).

12.4 CONTEMPORARY DETERMINANTS OF SPECIES DIVERSITY

Role of abiotic and biotic factors

The relative role of abiotic factors, competition and predation as factors in community organization has been the subject of a sometimes impassioned debate (McIntosh, 1995). This debate has hinged largely on the question of the importance of interspecific competition (Strong *et al.*, 1984; Diamond and Case, 1986). At risk of caricaturing the positions of the protagonists, their basic models of community structure should be defined.

Many ecologists see competition between species that exploit similar and limited resources as the major, organizing interaction in communities (Giller, 1984). Species partition resources such as food or living space. There is a limit to how similar two species can be in their resource requirements and still coexist (Gause's Principle): if two species are too similar, interspecific competition will lead to the exclusion of one or the other (see also Chapter 9). Species can be packed into a community until the limiting similarities between the species are reached. Thereafter, new species can only invade by being com-

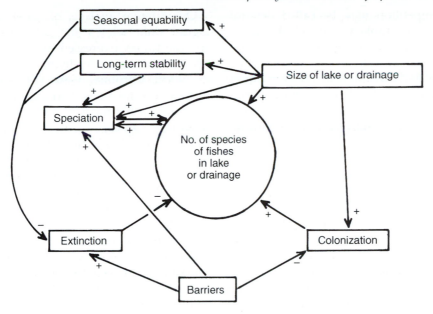

Fig. 12.3 Diagram of interaction of environmental stability and barriers in controlling extinction, colonization and speciation in freshwater fishes. Redrawn from Smith (1981).

petitively superior to resident species. For such a community, species richness is at an equilibrium, with the number of successful invasions balancing the number of extinctions. The species in the community show resource partitioning, which under normal circumstances minimizes interspecific competition sufficiently to allow the species to coexist. In principle, the species composition of the community is predictable because it is determined by the outcome of interspecific competition for limited resources.

An alternative view is that the effects of predators, natural disturbances, periods of unfavourable abiotic conditions or inadequate recruitment mean that populations rarely, if ever, reach densities at which interspecific competition becomes sufficiently strong to cause competitive exclusion. (This does not exclude the possibility that from time to time, interspecific competition becomes sufficiently intense to act as an important selective force

(Wiens, 1977). Such periods have been called 'competitive crunches'.) The species composition of the fish assemblage and the relative abundances of the species change as the age-specific rates of survival, growth and fecundity change in response to environmental factors. Some of these factors may change in unpredictable ways, for example the strength of a current in a stream prone to flooding, or the effect of tidal surge on the structure of a stand of coral. Community structure is essentially unpredictable. A community does not attain an equilibrium state in which the species composition and relative abundances are determined predominantly by interspecific competitive interactions. This view emphasizes the role that chance events play in determining the nature of the interactions between the members of an assemblage at any given time.

Communities in which the species diversity is determined by the patterns of inter-

specific competition may be called deterministic while assemblages in which chance
events play a major role may be called stochastic (Grossman *et al.*, 1982). An alternative
terminology is to distinguish between equilibrial and non-equilibrial communities. These
two models of community structure may
represent fundamentally different philoso-
phies about the nature of biological organization. The equilibrium (or deterministic)
model tends to emphasize the harmonious,
self-organizing properties of living systems.
The non-equilibrium (stochastic) model
emphasizes the opportunistic nature of
natural selection and the capricious qualities
of the non-living world. The crucial empirical
problem is to relate the number of species,
and their relative abundances in fish assemblages, to identifiable causal factors.

Resource partitioning and species diversity

Indirect and direct evidence show that the
species in an assemblage of fishes partition
the resources available to them. Resources in
this context can mean not only such obvious
factors as food and shelter but also temperature zones or time of day. Resource partitioning is usually linked to the concept that
interspecific competition is a major organizing interaction in communities, although the
two concepts are logically separate. Resources may be partitioned because of selective
segregation rather than interactive segregation caused by interspecific competition
(Chapter 9) (Nilsson, 1967).

The indirect evidence of resource partitioning is provided by differences in morphology
among species in an assemblage where the
differences are in traits that are known, or
suspected, to be related to the ways in which
species use resources. The most frequently
used traits are body and fin shape, which
can be related to modes of locomotion
(Chapter 2), and mouth size, shape and position which can be related to trophic ecology
(Chapter 3) (Winemiller, 1991).

Quantitative analysis of such ecomorphological data allows the differences among
species to be expressed numerically. From
his analysis of species living in three streams
in North Carolina, Gatz (1979) concluded
that the species present in a stream were not
a random assortment of species and did not
randomly divide their resources. Similar
studies of species in rain-forest streams in Sri
Lanka also revealed morphological specializations, especially in traits related to feeding
and position in the water column (Moyle and
Senanayake, 1984; Wikramanayake, 1990). A
quantitative analysis of morphological similarity between stream fishes in the Red River
basin in the USA found that morphologically
similar species occurred together significantly
less often than would be expected by chance
(Winston, 1995). A feature of this analysis
was that it showed that phylogenetic relationships between species had no significant
effect on co-occurrence, suggesting that historical factors such as speciation were not
responsible for the morphological pattern
detected. A plausible hypothesis is that the
pattern is generated by interspecific competition.

Ecomorphological studies assume that
traits that differ significantly between two
species indicate different patterns of resource
utilization. The definition of a species' niche
in morphological terms encounters two problems. Firstly, there is the possibility that
more than one suite of morphological characteristics may be suitable for the exploitation of particular resources. Secondly, a
specialized morphology does not necessarily
mean that the use of resources is specialized.
The morphological traits define how resources are exploited rather than what resources
are used (Chapter 3).

Direct evidence of resource partitioning
comes from observations on the diets, spatial
distributions, patterns of diel activity and use
of other relevant dimensions by fishes in an
assemblage. The resource axes (or niche
dimensions) that are important in many fish

assemblages are food, habitat and to a lesser extent, time and abiotic factors such as temperature (Ross, 1986). In 37 studies in which diet, habitat and temporal resource axes were studied, 57% showed the greatest separation of species along the diet axis, 32% along the habitat axis and in 11% temporal segregation was most important (Ross, 1986). Several examples of resource partitioning along these three axes have already been described either implicitly or explicitly (Chapters 3, 5, 8 and 9). The fish assemblage of Lake Opinicon (Fig. 3.1) and those of other temperate North American lakes provide well-described examples (Keast, 1970, 1978). These lakes also provide evidence of segregation along an abiotic axis.

Although fish cannot show exploitation competition for temperature, a competitively subordinate species might be excluded from a preferred temperature regime (Chapter 4) through interference competition by a dominant species (Magnuson *et al.*, 1979). This would have consequences for rates of food consumption, growth and reproduction (Chapters 3, 6 and 7) of the excluded species. For North American freshwater fishes, three groups can be recognized on the basis of their thermal preference behaviour (see also Chapter 4): Centrarchidae and lctaluridae are representatives of the warm-water group, Percidae and Esocidae of the cool-water group and Salmonidae of the cold-water group. Cyprinids have representatives in all three groups. Within a single lake, members of these different groups will be segregated by their thermal preferences. Such thermal partitioning was described for the fish assemblage of Lake Michigan (Brandt *et al.*, 1980). In this assemblage, not only do different species show different temperature preferences, but there are ontogenetic changes within species, with young alewives occupying warmer temperatures than adults. If thermal discharges from industrial complexes cause distortions of the thermal regime in lakes, estuaries and rivers, these distortions may cause changes in the pattern of interactions occurring within the fish assemblages (Coutant, 1987).

Even in systems that are richer in species than the lakes of North America, detailed observations usually reveal resource partitioning. In a small Panamanian stream, the fish showed spatial segregation and a segregation between diurnal and nocturnal species (Fig. 12.4(a)) (Zaret and Rand, 1971). Cyprinid species in a Borneo stream illustrated a segregation by depth (Fig. 12.4(b)) (Welcomme, 1985).

The assemblage in rain-forest streams of Sri Lanka demonstrated resource partitioning along several niche dimensions (Fig. 12.4(c)) (Moyle and Senanayake, 1984). Diet and position in the water column accounted for the majority of differences between the Sri Lankan species. In this assemblage, there was also evidence of associations between morphological characteristics and resource partitioning, as predicted by the ecomorphological hypothesis (Wikramanayake, 1990). Species with short guts, dorsally orientated or terminal mouths and lacking barbels fed predominantly on invertebrates. Species with long guts, ventral or subterminal mouths and barbels had a herbivorous diet (see also Chapter 3). Fish with deep, laterally compressed bodies, together with small fishes, occupied slower waters (Chapter 2).

Detailed observations on the members of feeding guilds in the species flocks of the African Great Lakes cichlids frequently suggest some dietary, habitat or temporal segregation between species that superficially seem to be unusually similar in their utilization of resources (Witte, 1984; Barel *et al.*, 1991).

Structurally complex habitats such as rocky or coral reefs offer more opportunities for resource partitioning and usually have a higher species diversity than more homogeneous habitats such as the pelagic zone or sandy, level bottoms (Fryer and Iles, 1972;

(a)

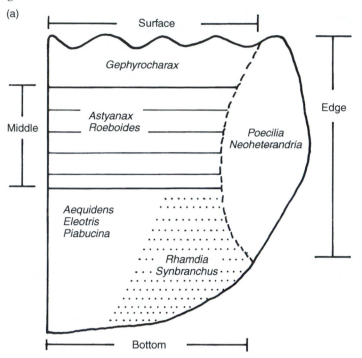

Fig. 12.4 Three examples of resource partitioning in tropical riverine communities. (a) Distribution of fish species in a Panamanian stream in a dry season (stippled area indicates nocturnal species). Redrawn from Zaret and Rand (1971). (b) Segregation by depth of cyprinid species in a Borneo stream. Redrawn from Welcomme (1985). (c) (Page 296) Resource partitioning by an assemblage of Sri Lankan rain-forest fishes showing partitioning by diet, depth and current (MWV, mean water velocity). Redrawn from Moyle and Senanayake (1984).

Lowe-McConnell, 1975, 1987; Sale, 1980; Ebeling and Hixon, 1991). In complex environments, there are more different ways of making a living, especially for species with a small body size that allows them to exploit interstices in the habitat (Miller, 1979a, 1996).

The effect of structural complexity can be seen when modifications are made that change this complexity. In many streams and rivers in Britain, sections have been straightened and dredged to improve land drainage. These changes make the waterways more homogeneous habitats and there is a decrease in the diversity and biomass of the fishes present (Swales and O'Hara, 1983; O'Hara, 1993). The introduction into such

sections of structures that increase the complexity of the habitat by increasing the amount of cover and changing the flow characteristics can, even over the short period of a year, increase the biomass and number of species present (Swales and O'Hara, 1983). A field experiment in Lake Malawi illustrated effects of both the structural complexity and size of habitat on species diversity (McKaye and Gray, 1984). Artificial reefs were built from cement blocks on a sandy substrate 1 km from the nearest rocky outcrop. Over a 5 year period, 75 different species of cichlid visited the reefs; about 40–50 species were present at any one time. Some of the species had never been recorded over a sandy sub-

(b)

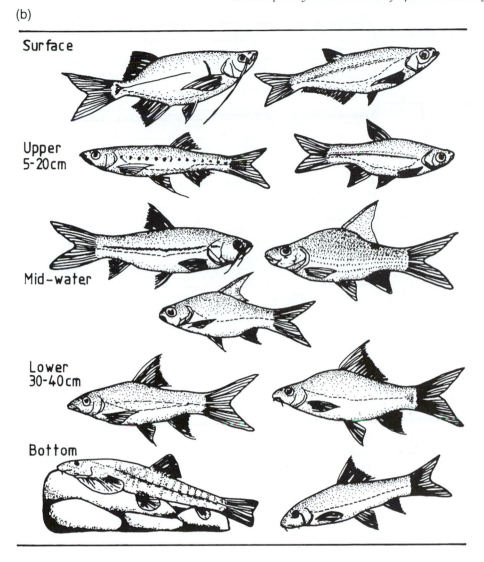

Surface

Upper
5-20cm

Mid-water

Lower
30-40cm

Bottom

strate. The number of species associated with an artificial reef was higher for larger reefs and higher for reefs that were more heterogeneous in their structure. Comparable studies show that coral reef fish rapidly colonize artificial reefs in previously unused areas. There is interest in constructing artificial reefs to increase fish stocks in coastal areas. The problem is to determine whether such structures lead to an absolute increase in fish abundance or merely concentrate fish already present.

Resource partitioning in fish assemblages is complicated both by ontogenetic changes in the pattern of resource utilization by a species (Chapters 3 and 9) and by the phenotypic plasticity shown by many species (Werner, 1984, 1986). A species may fill several ecological roles during its ontogeny or change roles if environmental conditions

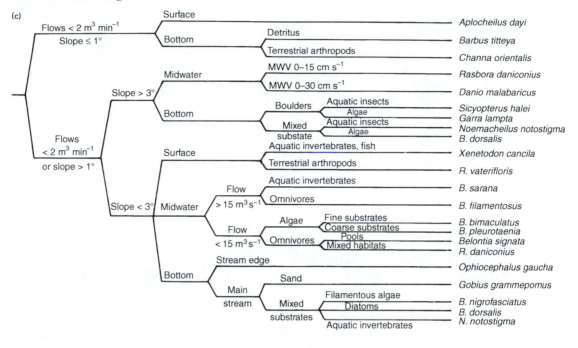

change. Such changes relate to the costs and benefits of the occupancy of different habitats, including their profitability in terms of foraging, the metabolic costs of occupancy and the danger of predation. In an analysis of the Lake Opinicon assemblage (Fig. 3.1), Keast (1978) suggested that 17 Linnaean species effectively represent 24–27 ecological species because of such ontogenetic and plastic changes in the utilization of resources.

Competition and species diversity

In itself, resource partitioning provides only circumstantial evidence that competition is a major, contemporary organizing interaction in an assemblage (Schoener, 1986; Werner, 1986). Interspecific competition may cause resource partitioning, but so may other processes, including predator avoidance (Chapters 8 and 9). Resource partitioning may even be a result of competition in the past, including competitive crunches, which has led to

the evolution of specialization such that segregation is now selective rather than interactive.

A crucial question is the extent to which interspecific competition determines the observed pattern of resource partitioning. Are some species prevented from joining the community by competitive exclusion because their resource requirements are too similar to those of species already present? If this is the case, species richness will be higher if the resource utilization curves (or niche breadths) are narrow. A possible explanation for the species richness observed in tropical lakes and streams or on coral reefs is that the species are usually specialists with narrow niche breadths. Fishes in rain-forest streams of Sri Lanka mostly do show narrow niche breadths although there are also a few generalist species present (Moyle and Senanayake, 1984). Specializations can evolve because of the relative predictability in such environments of both the abiotic conditions and the resource base. At higher latitudes, the strong seasonality, with its correlated changes in abiotic conditions and production, imposes on species the need to be more generalist in their resource requirements and so they should have wider resource utilization curves. Fewer species can be packed into a given habitat. For example, the die-back of most aquatic vegetation during the winter would exclude an obligate herbivore from high-latitude environments. Species from higher latitudes may be behaviourally, physiologically and morphologically more plastic than low-latitude species.

Winemiller (1991) used an ecomorphological approach to compare the fish assemblages of lowland stream and backwater habitats at five localities ranging from species-poor Alaska to species-rich Venezuela. He measured 30 morphological traits on individuals and used the multivariate technique of principal components analysis to define the morphological space occupied by the species measured. The entire morphological space occupied by species in the two habitats increased as species richness increased. There was a greater range of body forms in the species-rich assemblages, with the high-latitude assemblages consisting of small numbers of generalist species. The higher species richness was not associated with an increase in morphological similarity between species. The analysis suggested that the higher number of species at low latitudes is associated with a greater diversity of niches. This greater diversity may reflect an expansion in the range of food categories that are exploited or a finer subdivision of the available resources. Winemiller (1991) argued that the patterns observed have evolved because of the effects of interspecific competition and defence against predators. The higher species richness of the backwater compared with the stream habitat at all localities may reflect the greater structural complexity of backwaters.

In the extreme environment of the Antarctic waters that are permanently below the freezing point of pure water, a highly specialized, but species-poor, fauna has evolved (MacDonald *et al.*, 1987).

Observational and experimental evidence that interspecific competition can affect the distribution and relative abundances of species have already been described (Chapter 9). If competition is setting an upper limit to the number of species in an assemblage, then another species could only be accommodated by the loss of a species. This replacement is occasionally observed. In a few cases, the introduction of whitefish into Swedish lakes was followed by the extinction of the Arctic charr population, although more frequently the charr merely declined in abundance (Svardson, 1976). In Lake Texoma, Oklahoma, inland silversides, *Menidia beryllina*, replaced brook silversides, *Labidesthes sicculus* (McComas and Drenner, 1982). The inland silverside is a more effective predator on copepods. Other introductions of a new species have been followed by changes in the abundance, diet or habitat of the resident species, but not by extinctions (Chapter 9). The plasticity which many species show in their feeding and growth allows them to respond to the appearance of potential competitors by phenotypic changes that reduce the risk of extinction (Weatherley, 1972). The importance of competition as a determinant of the species richness of an assemblage is an empirical question, which can be answered only if introductions take place. There are strong ethical arguments against the deliberate introduction of exotic species into preexisting assemblages. Consequently information on the effects of accidental introductions or any natural extension in the range of a species will be valuable (Evans *et al.*, 1987).

Effect of abiotic harshness and/or unpredictability on species diversity

Decreases in diversity with increases in latitude, depth or altitude are correlated with increases in either the harshness or the unpredictability of the abiotic conditions. Harshness means that abiotic factors such as temperature, oxygen concentration, salinity or pressure are close to the limits at which life is possible. Unpredictability means that there is a poor correlation between conditions experienced at one time and conditions experienced at subsequent times. An indication of the effect of abiotic factors was given by a comparison of the number of species found in rivers from four geographical areas (Fig. 12.5) (Zalewski and Naiman, 1985). The number of species in the desert and boreal rivers was similar although the rivers are at different latitudes and have different temperature regimes. The harshness and unpredictability of a desert stream can be judged by data from the Mojave River, California (Castleberry and Cech, 1986). Water temperature varied from 0.0 to 36.0 °C, with a 10–15 °C monthly variation, there were periods of hypoxia and there was occasional torrential flash flooding.

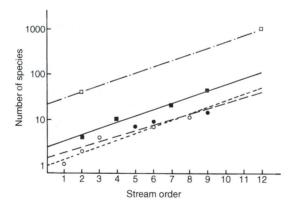

Fig. 12.5 Comparison of species richness (log scale) in rivers from different climatic regions according to stream order (see page 301). Symbols: ○, tropical rain forest – Amazon (South America); ■ temperate zone – Vistula River (central Europe); ○ (short dashes), desert streams – Nevada (south-western USA); ● (long dashes), boreal zone – Moise River (Labrador, Canada). Redrawn from Zalewski and Naiman (1985).

Several factors might cause low species diversity in harsh or unpredictable abiotic conditions, but it is difficult to assess their relative importance. Only a few phylogenetic lineages might evolve the adaptations necessary for life in such conditions, for example the need for defence against the freezing of body fluids in polar waters (DeVries, 1980) or the ability to tolerate wide and rapid fluctuations in water temperature in desert rivers. The importance of this factor is likely to be partly dependent on the age of the environment. In older environments, more time has been available for organisms to adapt to the conditions. In the Mojave River, the introduced arroyo chub, *Gila orcutti*, has displaced the native tui chub, *G. bicolor mohavensis*, perhaps because the former has had a longer evolutionary history in fluctuating, desert streams (Castleberry and Cech, 1986).

At high latitudes, the range of species may be determined by the length of the summer growing period. If this period is too short, it will not allow adequate time for growth and accumulation of reserves sufficient to sustain survival though the low temperatures and low resource abundance of winter (Mandrak, 1995).

Harsh or unpredictable environments are frequently long distances from centres of species richness or difficult to colonize. In unpredictable environments, the risk of extinction will be high, so populations will have difficulty in becoming established for enough time to become adapted to the variability. The risk of extinction might be increased if production in the environment is low, making the abundance of the fish populations necessarily low.

Effect of disturbance on species diversity

Although highly unpredictable environments have low species diversities, intermediate levels of disturbance might promote diversity (Connell, 1978; Chesson, 1986). Disturbance could prevent a competitively dominant species from excluding other species by creating patches of habitat in which competition is sufficiently relaxed to allow other species to increase in abundance. In the presence of moderate levels of disturbance, species that would be unable to coexist under stable conditions because of competitive interactions might all show positive population growth rates at low densities (Chesson, 1986). The importance of this for fish assemblages has not been critically assessed, although the species richness of some assemblages in lakes in Wisconsin may provide an example (Tonn and Magnuson, 1982). Some lakes have assemblages of the mudminnow and cyprinids. A higher species richness of these assemblages in summer is associated with lower levels of dissolved oxygen in the winter. The variation in oxygen content of the water was interpreted as a form of disturbance, which permitted a higher number of species to coexist than would be possible if competitive interactions were determining species richness.

Studies on regulated and unregulated rivers in northern California suggested that disturbance in the form of scouring winter floods influenced the trophic relationships, but these studies did not provide evidence for an effect on the species richness of the fish assemblage (Power *et al.*, 1996). In the summer following a flood, the most prominent trophic relationships formed a four-level chain. Steelhead trout and California roach, *Hesperoleucas symmetricus*, fed on small carnivores, including predatory invertebrates. The small carnivores consumed chironomids that fed on algae. When steelhead and roach were excluded by cages, the algae showed increased growth because the small carnivores reached higher densities and controlled the algae-feeding chironomids. After winters without scouring floods and in regulated channels, the dominant trophic link in the summer was from algae to grazing invertebrates, which were less vulnerable to predation.

Effects of predation on species diversity

Predation might also help to maintain the species richness of an assemblage by preventing a competitively dominant species from reaching levels of abundance that cause the exclusion of other species. Predation acts as a moderate disturbance. The alewife, a zooplanktivore, was first observed in Lake Michigan in 1949 and increased rapidly in abundance in the 1960s. Native fish, which at some stage in their ontogeny depend on zooplankton, declined or became locally extinct. Subsequently, piscivores were stocked in the lake, including the native lake trout and other exotic salmonids such as coho and chinook salmon from the Pacific North-west. As these piscivores became established, the alewife declined in abundance and other species which had declined during the alewife explosion recovered (Kitchell and Crowder, 1986). An alternative explanation of these changes was that the decline in alewife abundance was caused by a series of cooler summers that reduced the growth of the young-of-the-year with a consequent reduction in overwintering survival (Eck and Wells, 1987). This shows the difficulty of determining causation from observational data.

A factor that originally allowed the alewife to increase was probably the absence of native piscivores such as the lake trout. The recent history of the lake trout illustrates the other effect that predation can have – a decrease, sometimes catastrophic, in species richness. Such declines have occurred in three well-documented cases in which an exotic predator has been introduced, either intentionally or accidentally, into a fish assemblage.

The St Lawrence Seaway links the North American Great Lakes with the Atlantic Ocean. Using this corridor, the sea lamprey moved into the Great Lakes. Although this route was available in the 19th century, the lamprey was not recorded in the Great Lakes until the 1920s and 1930s. Its numbers then increased dramatically over the next 25 years (Scott and Crossman, 1973). This lamprey, an agnathan (jawless) predator of fish, attaches itself to a victim with its suctorial mouth. The effect of this predator on the lake trout, one of the two important, deep-water, native piscivores in the Great Lakes, is shown by the fishery statistics. In Lake Huron, the yield from the trout fishery declined from 2248 tonnes in 1938 to 76 t in 1954 and the fishery finally collapsed in 1959. The other deep-water piscivore, the burbot, *Lota lota*, and many other species were adversely affected. A programme of lamprey control initiated in the 1950s has allowed the restocking of the lake with both native and exotic salmonids. Smith and Tibbles (1980) provided a history of the effect of the lamprey and its subsequent control.

Gatún Lake forms part of the Panama Canal system. In 1967, a piscivorous cichlid, *Cichla ocellaris*, a native of the Amazon Basin, was introduced into the Chagres River. By 1969 it had entered Gatún Lake where it fed on the smaller, native fish species. Prior to this invasion, the lake contained 12 species. The effects of the predator on this assemblage were observed as it spread through the lake (Zaret and Paine, 1973; Zaret, 1979). At one site, a 99% reduction in the abundance of the native species occurred within six months of the arrival of *Cichla* and only one of the original, native species remained. The adults of this species, a cichlid, were too large for *Cichla* to swallow. Other species, which had potential refuges from predation in deeper water or in dense vegetation, were also expected to recolonize the area eventually. In contrast, the presence of *Cichla* in the Chagres River seems to have had minimal effect on the species composition or abundance of the native fish. Zaret (1979) suggested that this lack of a major effect in the riverine environment is caused by the limited number of suitable breeding sites for *Cichla* and because the turbidity of the river

water during the rainy season reduces the efficiency of *Cichla* as a predator (Chapter 3).

The introduction in about 1960 of the piscivorous *Lates niloticus* (Nile perch) to Lake Victoria has had a catastrophic consequences for the species flock of haplochromine cichlids (Barel *et al.*, 1985, 1991; Payne, 1987). This unique species assemblage is probably in danger of being lost, with the Nile perch now making up 80% of the fish in the lake. Ironically, the cichlid fauna of L. Victoria before the introduction of *Lates* was rich in piscivores (Witte, 1984). They represented about 40% of the species and each species probably had its own unique combination of feeding behaviour, food preferences and habitat. Predation has even been suggested as a factor promoting the evolution of species richness in the African Great Lakes cichlids (Fryer and Iles, 1972) by reducing interspecific competition and by restricting the movement of fish away from their natal habitats and so reducing gene flow.

Both in the North American Great Lakes and in Lake Victoria, the effects of the invasion by a new piscivore were complicated by the existence of fisheries for its prey (Bundy and Pitcher, 1995). Overfishing may have made the exploited populations more susceptible to the effects of another, exotic predator. Other changes were also taking place because of changes in land use and water quality, which further complicate the interpretation of the events. The effect of a given predator on species diversity can only be understood from a knowledge of the biology of that predator in relation to the biology of its potential prey, in the context of the environment in which the predator–prey interaction takes place. A distinction should also be made between the effects of the introduction of an alien predator into an assemblage of fishes and the effects of predatory species, which already form part of that assemblage and so are a characteristic of the environment to which the prey species have become adapted (Chapter 8).

Variation in recruitment

If recruitment is more than adequate to replace any losses in the postrecruitment phases of the life history in all the species present in an assemblage, then the species composition and the relative abundances will be independent of recruitment. But if recruitment is low, then both the composition and abundances in an assemblage will reflect the availability of recruits, but modified in some species by postrecruitment processes such as predation (Sale, 1980; Roughgarden, 1986; Doherty and Williams, 1988; Caley *et al.*, 1996). The composition of the assemblage would then be related to the factors that determine the supply and composition of the recruits. This mechanism has been evoked in explanations of the diversity of coral reef assemblages (Sale, 1991) (page 305). The wide variation in recruitment that is typical of many populations (Chapter 10) may also help to maintain the species richness of an assemblage if the fish are longlived and iteroparous (Warner, 1984; Chesson, 1986). A rare but long-lived species may be able to maintain its presence in an assemblage if it enjoys occasional high recruitment. The long life with extended iteroparity would tend to buffer the effects of periods of poor recruitment (Chapter 11).

Many species of fish show major, ontogenetic changes in their choice of habitats. Individuals are recruited from elsewhere into the habitat occupied by sexually mature fish (Chapter 10). For example, most species of coral reef fishes have a pelagic larval phase so the larvae from an area of reef may be dispersed over a wide geographic area (Sale, 1980; Booth and Brosnan, 1995; Caley *et al.*, 1996). This decouples the recruitment at a particular area from both the species composition and the relative abundances in that area (Chapter 10). A contrast is provided by some of the cichlid species of the African Great Lakes that are associated with rocky shores. These species show little tendency to stray and because the eggs and young are brooded in the mouth by the females (Fryer and Iles, 1972; Lowe-McConnell, 1987; Keenleyside, 1991), recruitment in an area will be related to the density of adults present, as there is no other source of recruits.

12.5 EXAMPLES OF FISH ASSEMBLAGES

There is no consensus on the factors that control the species diversity of fish assemblages, but those in streams and rivers and on coral reefs are used to illustrate the ways in which the processes discussed above may operate. A recurring theme in these examples is the unresolved question of the relative importance of predictable biotic interactions and unpredictable, chance events in determining the number of species and their relative abundances (Schoener, 1987). As these studies have continued, the importance of multifactorial explanations of species diversity has become appreciated.

Riverine assemblages

The dominant feature of river systems is the essentially unidirectional flow of water: passive transport is downstream (Matthews and Heins, 1987). A second typical feature is that the gradient reduces from the head waters to the estuary. This basic pattern is often modified by geological and geomorphological features specific to individual rivers. From source to estuary, a stretch of river can be classified in terms of its stream order. A first-order stream has no tributaries. When two first-order streams join, they form a second-order stream, a third-order stream is formed when two second-order streams join and so on. The order is not changed when a stream of lower order joins one of a higher order. In general, stream width is positively correlated with stream order, that is living space tends to increase.

An early classification of fish assemblages in a river system related, in a descriptive

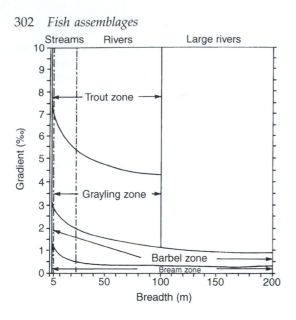

Fig. 12.6 Huet's classification of faunal zones in western European rivers showing relationship between gradient, breadth and faunal zone. Characteristic species: trout zone, *Salmo trutta*; grayling zone, *Thymallus thymallus*; barbel zone, *Barbus barbus*; bream zone, *Abramis brama*. Redrawn from Hawkes (1975).

way, assemblages to the physical factors of temperature and gradient (Huet, 1959). Zones characterized by typical species succeed each other from the head waters to the coastal plain (Fig. 12.6). This scheme was developed for continental European rivers, but the principle of zonation has been adapted for other regions, for example the rivers in the Pacific North-west of America (Li *et al.*, 1987).

An alternative model for the richness of riverine fish assemblages emerged from studies of the relationship of species richness to stream order (Lotrich, 1973; Horwitz, 1978). Species diversity increases with stream order. This increase in richness is primarily by the addition of species to the assemblage of species present in lower-order sections. In Clemon's Fork, the small Kentucky stream

studied by Lotrich (1973), the first-order stream contained one resident species, the second-order eight species and the third-order fifteen. All the changes were by additions to the assemblage (Table 12.2).

An increase in species richness with stream order is not always solely by the successive addition of species. In streams on the coastal plain of South Carolina, species richness increased with stream order. This increase reflected the addition of species to the stream assemblage, but also the replacement of species between first-order and fourth-order streams (Paller, 1994). The tendency was for small-bodied species of the head waters to be replaced by larger-bodied species. The species richness in the low order streams was also high compared with streams from regions of the USA having greater extremes of temperature and rainfall than South Carolina.

Where streams have been studied over long distances or large altitudinal gradients, both zonation and patterns of species addition were identified, for example along Horse Creek, Wyoming (Rahel and Hubert, 1991). Zones reflected the tolerance of species to abiotic conditions, particularly temperature. Cold-water species such as salmonids defined zones in the headwater regions of a watershed, whereas warm-water species defined zones further down the course of the stream. Within zones, an increase in species richness was usually, but not exclusively, by addition.

What causes an increase in species richness with stream order? Three factors seem to be crucial. There is an increase in the number of types of habitat downstream (Schlosser, 1987b). Two important additions are pools of deeper water and backwaters in which macrophytes can flourish because currents are slack. Such backwaters frequently act as important nursery areas for riverine fish. The increase in the size of the stream with an increase in stream order may, simply by providing greater living space, promote higher

Table 12.2 Distribution of species of fish in the first-, second- and third-order segments of Clemon's Fork, Kentucky*

Species[†]	Order		
	1st	2nd	3rd
Semotilus atromaculatus	+	+	+
Campostoma anomalum		+	+
Etheostoma sagitta		+	+
Etheostoma nigrum		+	+
Etheostoma flabellare		+	+
Etheostoma caeruleum		+	+
Hypentelium nigricans		+	+
Catostomus commersoni		+	+
Ericymba buccata			+
Notropis ardens			+
Notropis chrysocephalus			+
Pimephales notatus			+
Ambloplites rupestris			+
Lepomis megalotis			+
Micropterus dolomieui			+

*Source: Lotrich (1973).
[†] + denotes presence.

species richness (Rahel and Hubert, 1991). The energy base for fish assemblages also widens. In the head waters, the primary income is usually in the form of allochthonous material. Further downstream, with the development of backwaters and other habitats with slack water, autochthonous energy sources become more important. *In situ* primary production by macrophytes provides both living vegetation and detritus derived from dead plants. With the increase in volume of the river and longer residence times, a phytoplankton assemblage develops with associated zooplankton. A third factor is the stability of the physical environment, which tends to increase with the increase in stream order (Horwitz, 1978). Low-order streams frequently show wide variations in water discharge, and periods of high discharge cause disturbance of the substrate. Temperature fluctuations are usually greater in shallow, low-order streams.

The concept of the stream continuum argues that these changes in the physical aspects and energy basis along the length of a river system are predictable and result in consistent patterns of assemblage organization (Vannote *et al.*, 1980). There can be reversals in the typical upstream–downstream pattern. In the Amazon system, the head waters are at a higher altitude than the forest and many of the fish are algal grazers dependent on *in situ* primary production (Lowe-McConnell, 1987). Further downstream, the dense forest shade and the nutrient-poor waters inhibit primary production; the fish assemblages rely on allochthonous foods provided by the forest.

One interpretation of the pattern of species richness in relation to stream order was that at each stream order, there is an equilibrium fish community in which interspecific competition is the organizing interaction (Lotrich, 1973; Horwitz, 1978). The number of species reflects the range of habitats and potential food sources available at each order. A further factor reducing the species richness at low stream orders is that the relatively high variability that characterizes these environments obliges the fish present to be generalists with wide niches. Consequently, only a low number of generalist species can be accommodated.

An alternative interpretation was that the increase in species richness with stream order reflects the increase in stability of the environment. This increase in stability reduces the possibility of a species going extinct as a result of some chance event and so more species accumulate in the more stable, higher-order streams. A related factor is that the low-order streams are at a greater distance from the species-rich downstream areas and this reduces the rate of colonization (Horwitz, 1978). When a given stream order is compared, rivers with more variable discharges or temperature regimes have a lower species richness than more stable rivers (Horwitz, 1978; Zalewski and Naiman, 1985).

A compromise explanation of riverine species richness argued that the importance of biotic interactions, including predation and interspecific competition, increases with stream order as the abiotic conditions become more stable (Zawleski and Naiman, 1985; Schlosser, 1987b). This explanation recalls the suggestion that the relative importance of density-independent and density-dependent effects on fish populations depends on the harshness of the abiotic conditions (Chapter 10).

A study of a second-order stream, Jordan Creek in Illinois, related an increase in species richness to the diversity of habitats and food resources available (Schlosser, 1982, 1987b). Four regions of the stream of increasing habitat complexity were identified. This increase took the form of an increased definition and development of pools and riffles. The downstream areas were also more stable in that depth, substrate, current and water volume varied less over time than in the upstream habitats. Species diversity was positively correlated with habitat diversity, but this correlation was lower in the autumn when there was an increase in fish diversity because of the influx of young fish, but a decrease in habitat diversity because of reduced discharge. The downstream increase in species richness was primarily a result of an increase in the number of pool-dwelling species including large-bodied Centrarchidae and Catostomidae. Pools were inhabited by guilds feeding on insects and fish or on benthic insects, while the shallow, unstable habitats were inhabited predominantly by a guild of generalized insectivores, especially small-bodied cyprinids.

Surveys of a fish assemblage in a small area of Otter Creek, Indiana, found that the ten species that made up over 75% of the assemblage showed major changes in their relative abundances during a 12 year period. Over this time the stream showed no major changes in structure (Grossman *et al.*, 1982). There were also changes in the relative abundance of the different trophic guilds within the assemblage. The changes were interpreted as evidence that unpredictable changes in the abiotic conditions were responsible for the changes in the abundance of species and that interspecific interactions were of lesser importance. In contrast, a similar study of Martis Creek, a stream in California, found evidence of the persistence of the assemblage even though the sampling period included a time of extreme flooding (Moyle and Vondracek, 1985). For this stream it was concluded that the assemblage was structured by biotic interactions and showed a pattern of resource segregation that was compatible with this interpretation. The difference between the streams in Indiana and California may be related to the difference in the predictability of the periods of high discharge. In Indiana, floods are unpredictable in both severity and timing, whereas the Californian stream shows a clear seasonal pattern with high discharges in winter and spring and little rainfall in the summer (Grossman *et al.*, 1982; Moyle and Vondracek, 1985).

Studies that have compared the species diversity of upstream and downstream areas within a single stream, or have compared streams that differ in stability, suggest that changes in the relative abundances of species in an assemblage are more pronounced in more unstable habitats. But species richness can be persistent even in the face of major environmental disturbances (Schlosser, 1982; Ross *et al.*, 1985). A feature of some streams that suffer from periods when sections dry up completely is that recolonization is rapid and the previous species richness, if not the relative abundances, is re-established (Ross *et al.*, 1985; Pearsons *et al.*, 1992).

In river systems, the species composition of recruits is largely determined by the species that reproduce within the system. In the more unstable upstream areas, some of the changes in diversity may reflect annual changes in the success of the recruitment of the species present. Unpredictable flooding during the summer breeding season could have major effects because fish in the first month or so of life are probably most vulnerable to high discharge rates (Grossman *et al.*, 1982; Moyle and Vondracek, 1985; Schlosser, 1985). In high desert streams in Oregon, complex habitats acted as refuges during floods and may act as sources of fish that colonize less complex habitats after floods (Pearsons *et al.*, 1992).

Circumstantial evidence that interspecific competition is or has been a factor in determining the composition and relative abundances of stream assemblages is provided by the resource partitioning shown within such assemblages (Fig. 12.4). In Martis Creek, the species were segregated by habitat and diet (Moyle and Vondracek, 1985). In this creek, the introduction of exotic salmonids, rainbow and brown trout, led to the extinction of an endemic form, the Lahotan cutthroat trout, *Oncorhynchus clarki.* This extinction suggests a case of competitive exclusion. Furthermore, species from an upstream reservoir and a downstream river were occasionally found in the study section, but had not established themselves permanently.

Many studies of riverine assemblages have been observational (Moyle and Li, 1979), and the observations have been open to alternative interpretations (Grossman *et al.*, 1985). Matthews (1998) provides a comprehensive review of freshwater fish assemblages. Experimental studies will be helpful in distinguishing between rival interpretations. However, as studies on the fish assemblages of coral reefs show, even experimental studies may not easily resolve the problem of the relative importance of deterministic and stochastic processes for diversity of the assemblages.

Coral reef assemblages

Assemblages of coral reef fish are so rich in species (Fig. 12.1) that they pose a major challenge to any theory of species diversity (Ehrlich, 1975; Sale, 1980, 1991; Warner, 1984). There are more than 2000 reef fish species in the Philippines. At the southern extremity of the Great Barrier Reef, Australia, more than 800 species occur. A collection from less than 50 m^2 can contain as many as 150 species (Talbot *et al.*, 1978).

One model argued that the species diversity of coral reef assemblages is an example of fine partitioning of available resources through interspecific competition. This model assumes that populations are limited by resources, either food or more probably space, and are at equilibrium densities determined by the resource levels. The long, stable history of this productive, physically benign, spatially complex environment permits the evolution of species specialized in their utilization of resources. Examples of such specialists include the anemone fishes with their mutualistic relationship with coelenterates, and the obligate cleaner fishes (Chapter 9).

A study of the butterflyfishes (Chaetodontidae) along a 50 km transect of the Great Barrier Reef from the outer reef to a nearshore reef provided some evidence for resource partitioning within this group of fishes (Anderson *et al.*, 1981). Two patterns are relevant. Although more than 20 species of butterflyfishes were found in the region, at any one site there were typically six or seven in appreciable abundance. These locally sympatric species were divided among three or four trophic guilds. Along the transect, there was a tendency for geographic replacement of species within a given trophic guild. Species from the near-shore reef were deeper bodied than those from the outer barrier. This difference in body form may relate to adaptations to the differences in the turbulence of the water along the transect (Chapter 2). These patterns can be interpreted as showing that at a given site, several species of butterflyfish can coexist because they partition the resources including food. On a geographical scale, species that are better adapted to local conditions replace less-well-adapted species that have similar patterns of resource utilization.

A detailed study of the planktivorous pomacentrid *Acanthochromis polyacanthus* at a site on the Great Barrier Reef provided indirect evidence that interspecific competition for food may occur (Thresher, 1983, 1985). *Acanthochromis* is an unusual reef fish because it lacks a pelagic larval phase; the young remain close to their parents, which tend them. Several components of the fitness of *Acanthochromis* pairs, including adult size, spawning date and initial brood size, were inversely correlated with one or more of the number, diversity or density of other planktivorous species. Survival of adults was reduced in the presence of a piscivorous species, *Plectropomus leopardus*. Thresher suggested that at the site he studied, the local populations of *Acanthochromis* are resource limited, with predation as an important cause of adult mortality.

A contrasting model of coral reef assemblages was developed by Sale and his associates (Sale, 1977, 1978, 1980, 1988). This model emphasized the role that unpredictable events have in maintaining the species diversity of these assemblages. This model was developed from two types of observation made at Great Barrier Reef sites. Firstly, although acknowledging that reef species do show resource partitioning, Sale noted that many species have quite similar patterns of resource utilization (Sale, 1977). Secondly, a study of the distribution of three, territorial damselfish (Pomacentridae) species over a patch of rubble showed that their territories occupied most of the available space. Inter- and intraspecific aggression maintained the territorial boundaries (Chapter 5). The crucial observation was that when an individual of one species disappeared, the space was reoccupied, but not always by an individual of the same species. This suggested that, although space was limiting, the species were competitively equivalent. One species could not be competitively displaced by another. Occupancy of a site depended on which species settled in it when it became vacant rather than on competitive interactions between settlers. Most reef fishes have a planktonic larval stage, so the species composition of the larval fish at a locality does not reflect the composition of the assemblage at that immediate locality, but the composition over a much wider geographical area. When a vacant space on the reef becomes available for colonization because of the loss of the resident fish, it will be filled by an individual that happens to be ready to occupy that vacancy. In some cases, this will be a juvenile fish ready to settle. Vacancies arise at unpredictable times and in unpredictable places, and chance determines which species is ready to colonize each one. Once an individual has settled, its prior residency usually allows it to exclude other species from usurping the space. Sale (1978) has called this a lottery for pieces of resources in short supply – usually the resource is space.

Subsequent studies have modified Sale's model, while retaining the view that coral reef assemblages are non-equilibrial systems, not equilibium communities structured mainly by interspecific competition. Space is not always a limiting resource (Doherty, 1983; Sale, 1984; Doherty and Williams, 1988). Under these conditions, the species composition of an assemblage at a site will be determined by the species composition of any larvae ready to settle at that site, as modified by any post-settlement processes such as predation and age-specific or size-specific movement. For a given species, the number of larvae ready to settle can vary widely, both in time and spatially. In the bluehead wrasse, a Caribbean reef species, settlement takes place in short bursts, often around the time of the new moon (Victor, 1983, 1986). In one area, the number settling was not correlated with the local density of fish, but with the abundance of fish larvae in the water over the reef over a wide geographical range. This abundance was related to exposure to the onshore current. On the Great Barrier Reef, a study of nine species showed that the number of recruits to a reef varied both between reefs and between years at a given reef (Sale *et al.*, 1984). The mechanisms that lead to these temporal and spatial variations are poorly understood, because the ecology of the pelagic larvae of reef fishes is not well known (see also Chapter 10). However, the temporal patterning of settlement but not the abundance of settlers in the Caribbean damselfish, *Stegastes partitus*, was related to a lunar periodicity in spawning (Robertson *et al.*, 1988).

Settlement of larvae is not totally a random process (Sweatman, 1985; Forrester, 1990). Two processes may influence larval settlement. The first is habitat selection by the settling larvae, and the second is the effect of the resident fish in the area of settlement. In his study of four planktivorous species, Sweatman (1985) found a tendency for larvae to settle on corals already occupied by conspecifics.

There is also evidence that events taking place after settlement play a role in determining the abundance of the reproductively active adults (Shulman and Ogden, 1987; Jones, 1987, 1991). Problems of sampling mean that the process of settlement is rarely observed directly, rather the number of recruits to the reef environment is observed some time after their settlement. Booth and Brosnan (1995) argued that immediately after settlement, there is a period of high mortality that defines the number of recruits to the juvenile age classes on the reef. Predation on juveniles and adults may further modify the species richness or relative abundance of the assemblage on the reef (Hixon, 1991). On artificial reefs at St Thomas in the Virgin Isles in the Caribbean, the number of prey species was inversely related to the number of piscivores (Hixon, 1991). The survival of adults of the small goby, *Coryphopterus glaucofraenium*, on artificial reefs was inversely related to their density, although there was no evidence of density-dependent growth (Forrester, 1995). Recruitment of juveniles to the reef was also lower at high adult densities. In this situation, these two density-dependent processes had the potential to regulate adult densities in the face of fluctuations in recruitment (Chapter 10).

An assumption of the lottery model that species are competitively equivalent was challenged by a study of the territorial damselfish, *Stegastes*, on reefs off the Caribbean coast of Panama (Robertson, 1996). The abundances of *S. partitus* and *S. variabilis* increased when territorial *S. planifrons* were removed from patch reefs. The interaction was asymmetrical because removal of *S. partitus* was not followed by an increase in *S. planifrons*. The abundance of *S. partitus* increased within a year of the removal *S. planifrons*, but maximum abundance was reached after 4 years.

Evidence that chance (stochastic) events can play a role in assemblage composition came from studies of the changes in assemblages on small artificial or natural reefs. When small artificial reefs were built at some distance from natural reefs, they were rapidly colonized by reef-dwelling fishes, usually by juveniles. The species composition and the number of individuals per species on the artificial reefs showed both inter-reef and temporal variation (Talbot *et al.*, 1978). There was no tendency for the reefs to acquire the same species with the same relative abundances. Such a result might be expected if predictable, competitive, interspecific interactions are the major organizing processes on reefs. The species diversity on small patches of reef within a single lagoon can vary both between reefs and with time on a single reef (Sale and Douglas, 1984). Comparable changes in diversity have been observed on other reefs.

The debate about the relative importance of recruitment, predation, intra- and interspecific competition in determining the species diversity of coral reefs has led to an emphasis on multifactorial explanations of diversity. The concept of filters, originally developed in the context of lake diversity, is also relevant to reef diversity. An assemblage at a particular locality is the result of both regional and local processes. Regional processes will define what species are available, but local processes, including chance events, will determine what particular species are found at a locality at a given time. Roughgarden (1986) argued that coral reef fish assemblages share some of the characteristics of closed, resource-limited, terrestrial vertebrate populations and some of the characteristics of open, space-limited, marine invertebrate populations. He suggested that carnivorous species at the top of the food web such as groupers (Serranidae) may form food-limited subcommunities, which demonstrate local resource partitioning as a mechanism of coexistence. At other trophic levels, the subcommunities may be open and space-limited with disturbance acting as a mechanism of local coexistence.

12.6 CHARACTERISTICS OF FISH ASSEMBLAGES

The fish fauna of the Great Lakes of North America has undergone many changes over the last 100 years, both in its species richness and in the relative abundance of species. There have been invasions by exotic species such as the alewife, and changes in water quality caused by pollution. These changes have stimulated intensive research on the biology of the Great Lakes fishes. On the basis of this research Evans *et al.* (1987) proposed that the following key characteristics of the assemblages may be identified, and as the previous pages show, some of these are probably applicable to other fish assemblages.

1. Species richness is a function of area, but the species composition is also a function of habitat characteristics. The availability and distribution of suitable habitats may influence the intensity of species interactions. The habitats used early in life (nursery areas) are especially important, but these may be located remote from the habitat of adults.

2. The total production of fish assemblages is primarily determined by energy inputs, nutrients, edaphic factors and habitat variables. But the division of total production among species is strongly influenced by size-dependent metabolic processes and species interactions.

3. One species can affect the abundance of other species directly by predation or indirectly through competition. Large, indigenous piscivores may have a stabilizing effect on their prey. But invasion by exotics disrupts community structure, causing changes in relative abundances and even extinctions.

4. Coexistence of species may depend on complementarity of form and behaviour. Similarity in habitat use and/or trophic structure, perhaps coupled with some environmental change, can result in displacement of one species by another.
5. Ontogenetic niche shifts can be affected by changes in growth and survival rates and other demographic traits.
6. Food or food availability appears to be limited in most fish communities, and species respond to changes in food availability by changes in diet.
7. Human influences through fishing, pollution and habitat modification etc. can disrupt fish communities by disrupting their size- and niche-based organization.

12.7 COMMUNITY-WIDE PROPERTIES

Bioenergetic and trophic relationships

Fish form part of the network of consumers and consumed that makes up the food web within a habitat. The total, maximum biomass that can be supported in a habitat will be determined by the energy produced by the primary producers plus any energy that is imported. Examples of such imports include leaves, flowers and fruits falling into streams or lakes, terrestrial insects trapped by surface tension at the air–water interface, or prey that have migrated from another habitat.

Nutrient dynamics

Fish play a role in determining the general level of production in a system when, by their activities, they alter the availability of nutrients to other components of the system (see also Table 12.3, below). An example of this is the pulse of nutrients released into the head waters of rivers by the decaying carcasses of spawned-out Pacific salmon (Northcote, 1978). On coral reefs, which are highly productive but where the water is often poor in nutrients, some fishes shoal and rest over and in coral reefs, but feed elsewhere. A study of grunts, *Haemulon* spp., on a Caribbean reef showed that the fish excreted significant amounts of ammonium ions and egested significant amounts of particulate nitrogen and phosphorus, while resting over coral (Meyer and Schultz, 1985). The corals that sheltered shoals of grunts showed better growth and condition than corals that lacked resident grunts, perhaps because of the fertilizing effects of the waste products of the fish.

In temperate lakes, primary production by phytoplankton is frequently limited by phosphorus. Fish make phosphorus available directly through excretion and defecation. Benthic feeders, including fish feeding on detritus, may act as nutrient links by making phosphorus and other nutrients derived from their food available to plankton through excretion and soluble faeces (Vanni, 1996). Planktivorous fish may also make an indirect contribution (Vanni and Findlay, 1990; Kitchell and Carpenter, 1993). By feeding on large-bodied zooplankton (Chapter 3), the fish increase the proportion of smaller forms. These smaller zooplankters have higher rates of phosphorus excretion per unit body weight.

Fish can also act as nutrient sinks, tying up in their flesh significant quantities of nutrients such as phosphorus (Bartell and Kitchell, 1978) and reducing the availability to the primary producers. The phosphorus incorporated in the tissue or sedimented in the faeces of yellow perch in Lake Memphremagog (Canada) accounted for 15–40% of the phosphorus lost by the seston (suspended particulate matter smaller than large zooplankters) (Nakashima and Leggett, 1980). The phosphorus excreted by the perch and other fish accounted for less than 0.5% of the requirements of the seston, although phosphorus released from the decaying corpses of perch may have represented an important source.

Food webs

Food webs in aquatic systems differ primarily in the extent to which the main energy source is within the system, that is autochthonous, or is imported into the system, that is allochthonous. The pelagic zones of the sea or large lakes contain food webs based on the net primary production by phytoplankton. In contrast, the benthic zone tends to depend on organic material, including faeces and dead organisms, imported from pelagic communities. Streams, rivers and small lakes often have food webs heavily dependent on allochthonous material, particularly organic material washed into the water from terrestrial systems. The two types of food web can be illustrated by a simplified web for the North Sea and for the River Thames near Reading (UK) (Fig. 12.7).

Theoretical analyses have proposed that food webs are characterized by certain static properties that are independent of the size of the food web (see discussions in Polis and Winemiller, 1996). An example is the species scaling property. This suggests that in any given web, the proportion of species that do not feed (species at the bottom of the web) is about 0.19. The proportion of intermediate species, that is species that act as both prey and predators, is about 0.53, and the proportion of top predators about 0.29. If such properties could be confirmed, they would indicate that constraints operate on how species can be assembled into food webs.

A comprehensive study of food webs in freshwater fish assemblages in Venezuela and Costa Rica sought to identify food web properties (Winemiller, 1990). In each country, a swamp creek and a stream were sampled, with fish collected in dry and wet seasons. The described food webs were complex. The number of species placed in a web ranged from 58 for the Costa Rican stream to 104 for the Venezuelan swamp. The number of trophic links identified ranged from 208 to 1243. The study did not

find scale-invariant properties common to all the webs. The properties of the webs differed between sites, probably because of differences in species richness, rainfall patterns, physiography of the sites and gross primary production. The main inputs into the webs were aquatic primary production, allochthonous material of terrestrial origin and detritus largely derived from the breakdown of aquatic macrophytes. On an annual basis, the swamp creek communities used aquatic, terrestrial and detrital inputs approximately equally. The stream communities made greater use of terrestrial and detrital inputs. However, there were seasonal shifts. These were most marked for the Venezuelan swamp, with a shift from aquatic primary production in the wet season to detritus in the dry season as the major inputs to the web (Winemiller, 1996).

These food webs were used in an analysis of resource overlap and degree of organization (Winemiller and Pianka, 1990; Pianka, 1994). This analysis compared the observed dietary overlap with the pattern that would be expected if the trophic links were randomized. The analysis suggested that the Venezuelan and Puerto Rican webs deviated significantly from the overlap expected under the random hypothesis.

Trophic levels

Food webs are commonly simplified into links between trophic levels such as primary producers, herbivores, primary carnivores, secondary and tertiary carnivores. This scheme was applied to lakes in Scandinavia to demonstrate a relationship between the level of primary production and the number of trophic levels in the pelagic zone (Persson *et al.*, 1992). In lakes with low levels of production, only three trophic levels were present in the pelagic zone: phytoplankton, zooplankton and planktivorous fish, but no piscivorous fish. At higher levels of production, pelagic piscivores, represented mainly

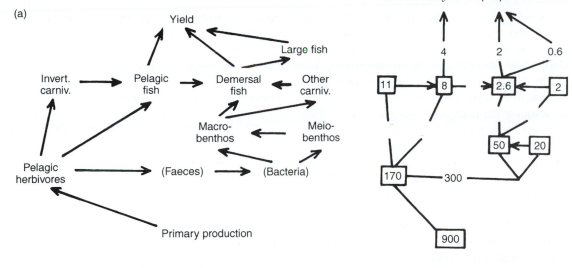

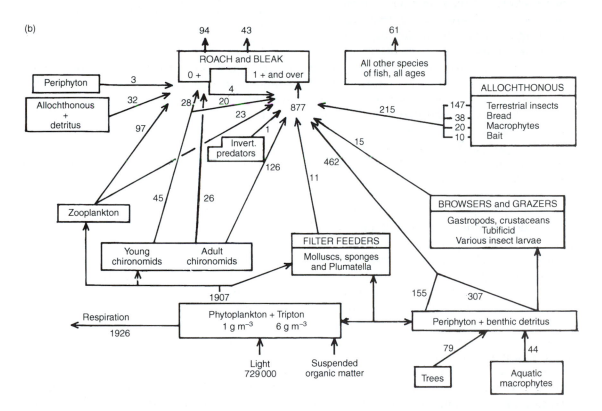

Fig. 12.7 Examples of a pelagic and a riverine food web. (a) North Sea food web based on main groups of organisms (left) and showing values of yearly production in kcal m⁻² year⁻¹ (right). Redrawn from Steele (1974). (b) River Thames (at Reading) food web, showing energy flows in kcal m⁻² year⁻¹ Redrawn from Mann *et al.* (1972).

by perch, occupied the fourth trophic level. But at the highest levels of production, piscivores were rare, possibly because of size-dependent interactions with cyprinids, like roach (Chapter 9). In the benthic zone, piscivorous fish were present even at low levels of primary production. This study and others showed that the effect of productivity on the number of trophic levels, and hence species richness, will be modified by other factors. These include habitat heterogeneity, the frequency of disturbance, the flexible foraging (Chapter 3) and antipredator behaviour (Chapter 8) of fish, and size-structured interactions within and between populations (Chapters 8 and 9) (Persson *et al.*, 1996).

Studies on food webs in tropical rivers and lakes and in estuaries show that nutrients and energy derived from detritus can form important inputs into a fish food web. For such systems, the trophic level model has to incorporate detritivory. In his study of food webs in Venezuela and Costa Rica, Winemiller (1990) also showed that omnivory was relatively common, with omnivorous species less abundant than herbivores but more common than secondary and tertiary carnivores.

A simple trophic model is often less appropriate in the study of fish assemblages, because of the difficulty of assigning given species to given trophic levels. Firstly, many species are flexible in their foraging behaviour, switching to foods as they become available and profitable (Chapter 3). Secondly, many species show ontogenetic changes in their diet and their preferred habitat (Chapters 3 and 5). A classic and much-reproduced example of ontogenetic changes in the diet of the herring is shown in Fig. 12.8. In future, it may be informative to map trophic interactions in aquatic systems in terms of the size spectrum of prey and predators rather than simply in terms of trophic level. In his virtuoso study of food webs in Venezuela and Costa Rica, Winemiller (1990) found: 'Prey size appeared to be of much greater impor-

tance in structuring aquatic food webs than was trophic level *per se.*'

Assemblage-level bioenergetics

The energy (and material) transfers mapped by food webs are subject to the laws of thermodynamics. At each energy transfer, some energy is dissipated from the system in the form of heat. Clearly, the biomass that can be supported by an energy base of a given size will depend on the number of transfers and on the efficiency with which energy is transferred through the food web. Ecological efficiency is defined as:

(energy content of prey consumed by a predator population) / (energy content of the food consumed by the prey population).

If ecological efficiencies were constrained within particular values, the potential yields available to predators including fisheries could be predicted from the size of the energy base for the system and the number of transfers.

Experimental studies of invertebrate systems led to the generalization that the ecological efficiency between populations was about 10% (Slobodkin, 1961). A preliminary study of the food web of the North Sea showed that such efficiencies were too low and that values around 20% might characterize some transfers (Fig. 12.7(a)) (Steele, 1974). Ecological efficiency is a quantity that is averaged over populations and is not itself subject to natural selection (Slobodkin, 1972). Its value depends on the characteristics of individual organisms, including the efficiency with which food is absorbed and assimilated and the efficiency with which the assimilated energy is used for growth (Chapters 4 and 6). The efficiency of the transfer of energy through an ecological system is dependent on the bioenergetics of individual organisms (Chapter 4), rather than a constrained global value.

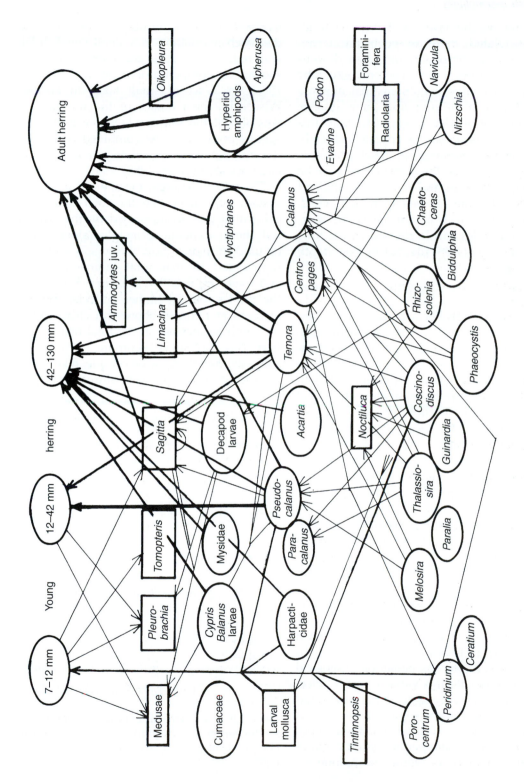

Fig. 12.8 Food web of North Sea herring, *Clupea harengus*, showing ontogenetic changes in diet associated with increase in body size. Arrows point in direction of predator. Redrawn from Cushing (1975).

As long as the laws of thermodynamics are not violated, there are no physical constraints on the complexity of the food webs of which fish form a part. The only energetic constraint is on the maximum total biomass of fish that can be supported in a system. Some indirect evidence suggests that this maximum biomass may be approached. Fisheries biologists seeking to develop simple ways of predicting the fishing yield to be expected from lakes and rivers have defined variables that can be used to predict yield. An example of such a variable is the morphoedaphic index (*MEI*), defined as the total dissolved solids (*TDS*) divided by lake depth (*D*):

$$MEI = TDS / D \qquad (12.7)$$

(Ryder *et al.*, 1974). For a given set of lakes, the yield is positively correlated with *MEI*, and the likely mechanism for this relationship is that *TDS* is related to the fertility of the lake. Better empirical predictors of both yield and total fish biomass are either total phosphorus concentration or the ratio of macrobenthos biomass to lake depth (Hanson and Leggett, 1982). In the seas, the highest production is usually associated with nutrient-rich waters such as those associated with areas of upwelling (Cushing, 1982, 1995).

The strong relationship between total fish biomass and macrobenthos biomass/depth was derived from lakes over a wide geographic range (0–56 °N). The relationship suggests a near-constant energy transfer from the benthos to fish, which is independent of the number and type of species present. This is a reflection of the similarities between many species in their absorption and growth efficiencies (Chapter 6), although ecological efficiency is not itself a variable under selection. Hanson and Leggett (1982) suggested that individual species expand their diet to include previously unexploited food types when interspecific competition is reduced. A constant fish biomass would be maintained irrespective of the number of species. When the roach population in a small Swedish lake was reduced by 70%, the biomass of perch showed a compensatory increase (Persson, 1986). In a set of small lakes in northern Wisconsin, the biomass of the mudminnow was almost the same, when it was the only species present, as the combined biomass of the mudminow and yellow perch when the two coexisted (Tonn, 1985).

The rate of energy income to an aquatic system may determine the total biomass of fishes that can be supported by that system. But thermodynamic principles alone cannot be used to deduce how many species will be represented in that total biomass. Streams and rivers are usually richer in fish species in the tropics than at higher latitudes, but this increase in richness shows no clear correlation with an increase in the total fish production (Fig. 12.9) (Welcomme, 1985). This suggests that there is no simple relationship between production and species richness.

Biomass size spectra and production

Most studies of the feeding relationships in communities that include fish continue to use

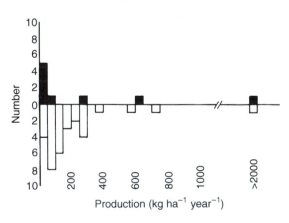

Fig. 12.9 Frequency histogram of total annual production in nine tropical rivers (black columns) and 31 temperate rivers (open columns). Data from Welcomme (1985).

the concepts of food webs and trophic levels. A complementary tradition has emphasized the size spectrum of communities in association with size-dependent processes such as consumption, respiration, growth and production. A central aim of such studies is to develop models that allow the prediction of fish biomass and production from studies of the biomass of smaller organisms in the community, while avoiding the need for a detailed taxonomic analysis of those organisms. This approach has most often been applied to the structurally simple pelagic systems of seas and lakes. Organisms in a community are classified on the basis of size, using a logarithmic scale for size categories. The biomass spectrum describes the distribution between these size categories of the total biomass in the community. Spectra for pelagic communities are typically rather flat, which implies that the biomass in each logarithmic size class is approximately equal (Boudreau and Dickie, 1992). Attempts to predict the biomass of fish communities using biomass spectra have ranged from poor to reasonably successful (Cyr and Peters, 1996). The circumstances under which this approach is likely to provide good predictions of fish biomass, production and yield have still to be defined.

Populations or groups of populations show statistically significant relationships between mean body weight and variables which include the ratio of production to biomass (P/B) and the ratio of assimilated food to biomass [$(R + P) / B$] (Dickie *et al.*, 1987). These relationships reflect the weight dependency of food consumption (Chapter 3), respiration (Chapter 4) and growth (Chapter 6) of individuals in populations and the similarity of these weight-dependent relationships for different fish species. The size distribution of species in an assemblage will be an important factor in determining the pattern of energy flow through the aquatic system to which that assemblage belongs. Conversely, changes in the rate of energy

income into the system will drive changes in the size distribution of species present and so possibly in the species composition of the assemblage. However, the physical properties of the environment, including the complexity of its physical structure, will also influence the size distribution of species.

The study of the interrelationships between species diversity, the size composition of a species assemblage and the pattern of energy flow is in its infancy. It remains to be seen whether such studies will generate insights into the diversity and composition of fish assemblages.

Trophic-cascade model

The links between species in a community mapped by food webs (Figs 12.7 and 12.8) suggest that changes in the abundance or species composition at one level can have consequences at other levels because of effects on the reproductive success of organisms at these other levels. The effect of planktivorous fish on the species composition and size structure of zooplankton sub-communities is described in Chapters 3 and 9. This idea has led to the suggestion that production and species composition in some pelagic communities are controlled by the abundance and species composition at the higher trophic levels (piscivores and planktivores). This is called top-down control. On the other hand, the correlations between overall production and the amount of primary production suggest that abiotic factors, especially temperature, nutrients and light, will be the most important controlling factors through their effects on food availability (bottom-up control) (McQueen *et al.*, 1986).

The trophic-cascade model incorporates the concepts of top-down and bottom-up control, but recognizes that such a dichotomy is simplistic (Carpenter and Kitchell, 1993). The variance in primary production observed in lakes is generated partly by the variations in abiotic factors, including the availability of nutrients, and partly by food web interactions. When piscivores are present, feeding on planktivorous fishes, the abundance of large-bodied zooplankton such as *Daphnia* can increase. These have high grazing rates and so reduce the biomass of the phytoplankton. When piscivores are absent, the planktivorous fishes, by their size-selective feeding, eliminate the large zooplankton species, and the biomass of the phytoplankton can increase. The fish will also affect the phytoplankton more indirectly by their effects on nutrient dynamics. In the autumn of 1987, there was a massive die-off of the planktivorous cisco, *Coregonus artidii*, in Lake Mendota (Wisconsin). After the die-off, the large *Daphnia pulicaria*, previously rare, became abundant. In the following spring, the biovolume of the phytoplankton was reduced greatly compared with the previous year (Vanni *et al.*, 1990).

A comparative study suggested that for lakes in the Great Lakes region of North America, the size distribution of the zooplankton is a crucial factor in determining the abundance of phytoplankton (Kitchell and Carpenter, 1993). In lakes where the zooplankton mean length exceeds about 0.25 mm, the concentration of chlorophyll in the pelagic zone is constrained by grazing of phytoplankton to about 8 μg l^{-1}. In lakes where the zooplankton mean length is less than about 0.25 mm, the concentration of chlorophyll is variable, reflecting factors other than grazing.

Experimental studies using large enclosures in lakes or whole lakes showed that food-web effects account for some of the variability in total production over and above that caused by variations in abiotic factors (McQueen *et al.*, 1986; Carpenter *et al.*, 1987). However, the effects observed depended on the nutrient status of the lake and on details of the behaviour of the piscivorous and planktivorous species. When piscivores were introduced into a lake, the rate of pre-dation on pelagic zooplankton by planktivores reduced more quickly than could be accounted for by the rate of piscivory. The small planktivorous fishes showed predator-avoidance responses (Chapter 8), which reduced the impact of the planktivores on the zooplankton (Carpenter and Kitchell, 1993).

The trophic-cascade model also makes a link with the concept of a stock–recruitment relationship in fish populations. The variability in recruitment discussed in Chapter 10 will lead to variations at lower trophic levels and in water quality through trophic-cascade effects. Some of the variability in recruitment will be driven by variations in abiotic conditions such as temperature, wind stress and nutrient levels, which will also have effects on lower trophic levels (Kitchell and Carpenter, 1993).

Other mechanisms by which fishes can affect the water quality and the species composition in an aquatic community are given in Table 12.3 (Northcote, 1988) and have been described either implicitly or explicitly earlier.

Dynamic characteristics of communities

The studies of fish assemblages suggest that no single interaction or process is responsible for their species diversity. Both the species composition and the relative abundances in the assemblages change over time, partly as a consequence of unpredictable events. Ecologists have paid much attention to the likely responses of species assemblages to environmental changes, particularly those caused by human activities. If it is assumed that most communities are, in terms of species composition and relative abundances, at an equilibrium maintained by competition, the dynamic properties of that equilibrial state can be explored (Pimm and Hyman, 1987; Begon *et al.*, 1996). These properties include the ability of the assemblage to resist change (resistance), its ability to return rapidly to its

Table 12.3 Summary of potential effects of fish on inland waters and their consequences*

Process	Limnological parameter affected	Mechanism and consequences
Direct feeding	Water transparency	1. Food searching stirs up bottom sediments and lowers transparency 2. Intense phytophagous feeding may increase transparency: reverse effects may also be demonstrated depending on algal sizes grazed and extent of fertilization effects
	Nutrient release, cycling	1. Benthic food searching increases mud–water nutrient exchange 2. Littoral grazing increases nutrient cycling
	Phytoplankton	1. As in water transparency, item (2) 2. Heavy cropping commonly increases production
	Periphyton	1. Strong cropping effect on biomass in lakes and streams
	Macrophytes	1. As for periphyton
	Zooplankton	1. Strong cropping effect on abundance, especially of larger forms 2. Some evidence of increase in production
	Zoobenthos	1. Strong cropping effect on abundance is common, but not invariable, in lakes and streams 2. Distribution and size of feeding fish cause marked seasonality in effects 3. Production is often increased in lakes but not so in streams
Selective predation (size, motility, visibility)	Phytoplankton	1. Shifts in relative abundance of algal size and species composition
	Zooplankton	1. Shifts in relative abundance of species reduce algal grazing efficiency and water transparency 2. Changes in clutch size and maturation timing
	Zoobenthos	1. Heaviest predation on forms with large body size affects their cover selection, activity patterns and reproductive behaviour (in both lakes and streams)
	Nutrient release	1. Shift to smaller body size (zooplankton) increases nutrient release
Excretion	Nutrient release	1. Liquid release provides quick, patchy availability 2. Faeces release provides slower, patchy availability after remineralization 3. Epidermal mucus release increases iron availability to algae through chelation
Decomposition	Nutrient release	1. Carcass remineralization provides slow, patchy release
Migration with excretion/ decomposition	Nutrient release	1. Transport of excreta or body decomposition products from high-nutrient to low-nutrient regions (sea to inland waters, lower stream reaches to upper, lake layers)

*Source: Northcote (1988).

equilibrium state after a change (resilience), and the stability of the equilibrium in the face of large or small disturbances (global and local stability). The possible relationships between these dynamic characteristics of communities and the species diversity, together with the pattern of interactions that define the communities, are the subject of much, as yet unresolved debate (May, 1973; Krebs, 1994; Begon *et al.*, 1996). The principle of natural selection makes it possible to speak of the goal of an individual as the maximization of its lifetime production of offspring. The function of the traits of individuals can then be studied in the context of that goal. There is no process equivalent to natural selection for communities and no community goal (Łomnicki, 1988). The dynamics of communities in response to environmental changes or species invasions depend on the responses of the individual organisms that make up the community. The dynamics are an effect of the individual responses, not a function of those responses.

If, as much of the evidence for fish assemblages suggests, the species diversity of an assemblage partly reflects the effects of unpredictable events, an equilibrium composition may never be reached. The consequences of an environmental change can then only be predicted by a painstaking analysis of the likely effects of the change on the age- and size-specific growth, survival and fecundity of the species in the assemblage. These effects will have consequences for the interactions in that assemblage (Schoener, 1987). The effect of an environmental change will depend on the composition of the assemblage at the time the disturbance occurs.

12.8 TOWARDS A SYNTHESIS

The debate about the importance of interspecific competition in determining the species diversity of fish assemblages is being superseded as a result of the theoretical and empirical studies stimulated by that debate. Three conceptual models provide a framework for further studies, although a synthesis of the three has still to be realized.

The concept of filters was originally developed by Tonn *et al.* (1990) in relation to the species richness of small lakes. This model argues that the species composition at a locality is a consequence of processes operating on a series of spatial and temporal scales ranging from the scale required for speciation to the local abiotic and biotic conditions. The model emphasizes that several factors, including the abiotic conditions, the presence of potential competitors and predators, together with the availability of resources, will interact to determine what species are present in a locality. The pattern of use of resources and the predator–prey relationships define the food web at a locality. Ecomorphological analyses and studies of the partitioning of food resources provide clues on the degree of organization of the food web (Winemiller and Pianka, 1990; Winemiller, 1991). The food web provides the basis for the trophic cascade and related models. These seek to predict the consequences of size-selective predation on overall production in a habitat in relation to nutrient levels and other abiotic factors.

These three models emphasize the need to understand the mechanistic basis of species interactions and processes occurring at a locality. The central theme of this book is that an understanding of the dynamics of a population or a group of populations will require a knowledge of how individual fish are likely to respond to the abiotic and biotic environmental factors they encounter during their lifetime.

12.9 SUMMARY AND CONCLUSIONS

1. All the fish species in an area constitute an assemblage. Those species that inter-

act form a subcommunity embedded in the community formed by all those organisms, irrespective of taxonomic identity, that interact either directly or indirectly. Species richness, species diversity and equitability provide partial quantitative descriptions of assemblages (or communities).

2. In both fresh waters and the seas, the species richness of fish assemblages tends to decline with an increase in latitude and depth (or altitude). There is a positive correlation between species richness and area.

3. The processes of speciation and colonization provide the species pool of a given area. At a particular locality, the assemblage may be a subset of the total species pool of the area because of contemporary factors.

4. The relative importance of interspecific competition and other factors in determining the species richness of assemblages has stimulated much debate. A distinction between deterministic (where interspecific competition is the dominant interaction) and stochastic communities has been proposed.

5. Resource partitioning, circumstantial evidence for interspecific competition, is a common characteristic of fish assemblages. Food and habitat are the two resources that most commonly seem to be partitioned by coexisting species. Time and temperature (or other abiotic factors) may also be relevant niche axes. Structurally complex habitats usually have a higher species richness than more homogeneous habitats because the former seem to provide fishes with more different ways of making a living.

6. There is only limited direct evidence that interspecific competition sets the limits to the number of species that can coexist stably in fish assemblages.

7. Abiotic factors, disturbance, predation and variable recruitment may all play roles in preventing an assemblage from reaching an equilibrial species richness determined by interspecific competition.

8. Studies of riverine and coral reef communities yield equivocal evidence for the role of deterministic and stochastic processes in controlling species richness. However, studies of the fish assemblages of the North American Great Lakes have suggested a number of potentially useful generalizations about fish assemblages.

9. Fish assemblages have an influence on the total production of a community, modifying the dependence of total production on the energy and nutrient inputs into the community.

10. Areas in which primary production is high usually have a high biomass of fishes, but species richness is not strongly correlated with production *per se*. There are statistical regularities in the relationships between the adult body sizes of fishes in assemblages and bioenergetic variables. The relevance, if any, of these size-dependent relationships to species richness is not understood.

11. The prediction of the effects of abiotic and biotic disturbances on the species richness and diversity of an assemblage will depend on an analysis of the effects on the size-specific (or age-specific) schedules of survival, growth and fecundity of the constituent species. These effects will depend, in turn, on the changes in the time and resource (energy) allocations of individual fish. These effects will also determine the dynamic characteristics of the assemblage.

12. Three multifactorial models now provide the basis for further understanding of the processes that determine the structure and dynamics of communities that include fishes. The concept of filters emphasizes the importance of regional and local processes in determining species diversity at a locality. Food web analyses, including ecomorphological

and resource partitioning studies, are defining patterns of energy and nutrient flow through communities. The trophic-cascade model targets the consequences of size-selective predation by fish on the size spectrum and biomass of herbivores, and hence on the dynamics of production and of nutrient flow.

REFERENCES

Adams, P.B. (1980) Life history patterns in marine fishes and their consequences for fisheries management. *Fish. Bull.*, **78**, 1–12.

Adelman, I.R. (1987) Uptake of radioactive amino acids as indices of current growth rate of fish: a review, in *The Age and Growth of Fish* (eds R.C. Summerfelt and G. Hall), Iowa State University Press, Ames, Iowa, pp. 65–79.

Alexander, G.R. (1979) Predators of fish in cold-water streams, in *Predator–Prey Systems in Fisheries Management* (ed. H. Clepper), Sport Fishing Institute, Washington, DC, pp. 153–170.

Alexander, R.McN. (1967) The functions and the mechanisms of the protrusible upper jaws of some actinopterygian fish. *J. Zool., Lond.*, **151**, 43–64.

Alexander, R.McN. (1974) *Functional Design in Fishes*, Hutchinson, London.

Allan, J. (1995) *Stream Ecology: Structure and Function of Running Waters*, Chapman & Hall, London.

Allen, G.R. (1972) *The Anemone Fishes: Their Classification and Biology*, TFH Publications, Neptune City, NJ.

Allen, J.R.M. and Wootton, R.J. (1982a) The effect of ration and temperature on the growth of the three-spined stickleback, *Gasterosteus aculeatus* L. *J. Fish Biol.*, **20**, 409–422.

Allen, J.R.M. and Wootton, R.J. (1982b) Age, growth and rate of food consumption in an upland population of the three-spined stickleback, *Gasterosteus aculeatus* L. *J. Fish Biol.*, **21**, 95–105.

Allen, J.R.M. and Wootton, R.J. (1982c) Effect of food on the growth of carcase, liver and ovary in female *Gasterosteus aculeatus* L. *J. Fish Biol.*, **21**, 537–547.

Allen, J.R.M. and Wootton, R.J. (1983) Rate of food consumption in a population of three-spined sticklebacks, *Gasterosteus aculeatus*, estimated from the faecal production. *Env. Biol. Fishes*, **8**, 157–162.

Allen, J.R.M. and Wootton, R.J. (1984) Temporal patterns in diet and rate of food consumption of the three-spined stickleback (*Gasterosteus aculeatus*) in Llyn Frongoch, an upland Welsh lake. *Freshwat. Biol.*, **14**, 335–346.

Alm, G. (1959) Connection between maturity, size and age in fishes. *Rep. Inst. Freshwat. Res. Drottningholm*, **40**, 5–145.

Al-Shamma, A.A.M.-H. (1986) The feeding ecology of fish in Llyn Frongoch, PhD thesis, University of Wales, 182 pp.

Anderson, G.R.V., Ehrlich, A.H., Ehrlich, P.R., Roughgarden, J.D., Russell, B.C. and Talbot, F.H. (1981) The community structure of coral reef fishes. *Am. Nat.*, **117**, 476–495.

Anderson, R.M. (1981) Population ecology of infectious agents, in *Theoretical Ecology* (ed. R.M. May), Blackwell, Oxford, pp. 318–355.

Andersson, M. (1994) *Sexual Selection*, Princeton University Press, Princeton, NJ.

Andrewartha, H.G. and Birch. L.C. (1954) *The Distribution and Abundance of Animals*, University of Chicago Press, Chicago.

Andrewartha, H.G. and Birch, L..C. (1984) *The Ecological Web*, University of Chicago Press, Chicago.

Andrusak, H. and Northcote, T.G. (1971) Segregation between adult cutthroat trout (*Salmo clarki*) and dolly varden (*Salvelinus malma*) in small coastal British Columbian lakes. *J. Fish. Res. Bd Can.*, **28**, 1259–1268.

Angermeier, P.L. and Schlosser, I.J. (1989) Species–area relationships for stream fishes. *Ecology* **70**, 1450–1462.

Armstrong, J.D. (1997) Self-thinning in juvenile sea trout and other salmonid fishes revisited. *J. Anim. Ecol.*, **66**, 519–526.

Armstrong, J.D., Braithwaite, V.A. and Huntingford, F.A. (1997) Spatial strategies of wild Atlantic salmon parr: exploration and settlement. *J. Anim. Ecol.*, **66**, 203–211.

Arnold, G.P. and Cook, P.H. (1984) Fish migration

by selective tidal stream transport: first results with a computer simulation model for the European continental shelf, in *Mechanisms of Migration in Fishes* (eds J.D. McCleave, G.P. Arnold, J.J. Dodson and W.H. Neill), Plenum, New York, pp. 227–261.

Aronson, L.R. (1951) Orientation and jumping behavior in the gobiid fish *Bathygobius saporator, American Museum Novitates*, No. 1486, 1–21.

Backiel, T. and Le Cren, E.D. (1978) Some density relationships for fish population parameters in *Ecology of Freshwater Fish Production* (ed. S.D. Gerking), Blackwell, Oxford, pp. 279–302.

Baerends, G.P. (1986) On causation and function of pre-spawning behaviour of cichlid fish. *J. Fish Biol.*, **28** (Suppl. A), 107–121.

Bagenal, T.B. (1969) The relationship between food supply and fecundity in brown trout *Salmo trutta* L. *J. Fish Biol.*, **1**, 169–182.

Bagenal, T.B. (1971) The interrelation of the size of fish eggs, the date of spawning and the production cycle. *J. Fish Biol.*, **3**, 207–219.

Bagenal, T.B. (1974) *The Ageing of Fish*, Unwin Brothers, London.

Bagenal, T.B. (1978) Aspects of fish fecundity, in *Ecology of Freshwater Fish Production* (ed. S.D. Gerking), Blackwell, Oxford, pp. 75–101.

Bagenal, T.B. and Braum, E. (1971) Eggs and early life history, in *Fish Production in Fresh Waters* (ed. W.E. Ricker), Blackwell, Oxford, pp. 166–198.

Baggerman, B. (1980) Photoperiodic and endogenous control of the annual reproductive cycle, in teleost fishes in *Environmental Physiology of Fishes* (ed. M.A. Ali), Plenum, New York, pp. 533–567.

Baggerman, B. (1990) Sticklebacks, in *Reproductive Seasonality in Teleosts : Environmental Influences* (eds A.D. Munro, A.P. Scott and T.J. Lam), CRC Press, Boca Raton, FL, pp. 79–107.

Bailey, K.M. (1994) Predation on juvenile flatfish and recruitment variability. *Neth. J. Sea Res.*, **32**, 175–189.

Bailey, K.M. and Houde, E.D. (1989) Predation on eggs and larvae of marine fishes and the recruitment problem. *Adv. Mar. Biol.*, **25**, 1–83.

Baker, J.A. (1994) Life history variation in female threespine stickleback, in *The Evolutionary Biology of the Threespine Stickleback* (eds M.A. Bell and S.A. Foster), Oxford University Press, Oxford, pp. 144–187.

Baker, J.R. and Schofield, C.L. (1985) Acidification impacts on fish populations: a review, in *Acid Deposition, Environmental, Economic and Policy Issues* (eds D.D. Adams and W.P. Page), Plenum, New York, pp. 183–221.

Balon, E.K. (1975) Reproductive guilds of fishes: a proposal and definition. *J. Fish. Res. Bd Can.*, **32**, 821–864.

Balon, E.K. (1981) Additions and amendments to the classification of reproductive styles in fishes. *Env. Biol. Fishes*, **6**, 377–389.

Balvay, G. (1983) L'alimentation naturelle des alevins de brochet (*Esox lucius* L.) durant leur premier mois de vie, in *Le Brochet* (ed. R. Billard), INRA, Paris, pp. 179–198.

Bannister, R.C.A., Harding, D. and Lockwood, S.J. (1974) Larval mortality and subsequent year-class strength in the plaice (*Pleuronectes platessa* L.), in *The Early Life History of Fish* (ed. J.H.S. Blaxter), Springer-Verlag, Berlin, pp. 21–37.

Barber, W.E. and Walker, R.J. (1988) Circuli spacing and annulus formation: is there more than meets the eye? The case of sockeye salmon, *Oncorhynchus nerka*. *J. Fish Biol.*, **22**, 237–245.

Barbour. C.D. and Brown, J.H. (1974) Fish diversity in lakes. *Am. Nat.*, **108**, 473–489.

Barel, C.D.N., Dorit, R., Greenwood, P.H., Fryer, G., Hughes, N., Jackson, P.B.N., Kawanabe, H., Lowe-McConnell, R.H., Nagoshi, N., Ribbink, A.J., Trewavas, E., Witte, F. and Yamaoka, K. (1985) Destruction of fisheries in Africa's lakes. *Nature*, **315**, 19–20.

Barel, C.D.N., Ligtvoet, W., Goldschmidt, T., Witte, F. and Goudswaard, P.C. (1991) The haplochromine cichlids of Lake Victoria: an assessment of biological and fisheries interests, in *Cichlid Fishes: Behaviour, Ecology and Evolution* (ed. M.H.A. Keenleyside), Chapman and Hall, London, pp. 258–279.

Bartell, S.M. and Kitchell, J.F. (1978) Seasonal impact of planktivory on phosphorus release by Lake Wingra zooplankton. *Verh. Int. Verein. Theor. angew. Limnol.*, **20**, 466–474.

Bartell, S.M., Breck, J.E., Gardner, R.H. and Brenkert, A.L. (1986) Individual parameter perturbation and error analysis of fish bioenergetics models. *Can. J. Fish. Aquat. Sci.*, **43**, 160–168.

Baylis, J.R. (1981) The evolution of parental care in fishes, with reference to Darwin's rule of male sexual selection. *Env. Biol. Fish*, **6**, 223–257.

Beamish, F.W.H. (1974) Apparent specific dynamic action of largemouth bass, *Micropterus salmoides*. *J. Fish. Res. Bd Can.*, **31**, 1763–1769.

Beamish, F.W.H. (1978) Swimming capacity, in

Fish Physiology, Vol. VIII (eds W.S. Hoar and D.J. Randall), Academic Press, London, pp. 101–187.

Beamish, R.J. and McFarlane, G.A. (1987) Current trends in age determination methodology, in *The Age and Growth of Fish* (eds R.C. Summerfelt and G.E. Hall), Iowa State University Press, Ames, IA, pp. 15–41.

Beckman, D.W. and Wilson, C.A. (1995) Seasonal timing of opaque zone formation in fish otoliths, in *Recent Developments in Fish Otolith Research* (eds D.H. Secor, J.M. Dean and S.E. Campana), University of South Carolina Press, Columbia, SC, pp. 27–43.

Beddow, T.A., Leeuwen, J.L. van and Johnston, I.A. (1995) Swimming kinetics of fast starts are altered by temperature acclimation in the marine fish *Myoxocepahlus scorpius*. *J. Exp. Biol.*, **198**, 203–208.

Begon, M. (1979) *Investigating Animal Abundance*, Edward Arnold, London.

Begon, M., Harper, J.L. and Townsend, C.R. (1996) *Ecology, Individuals, Populations and Communities*, Blackwell, Oxford.

Beitinger, T.L. and FitzPatrick, L.C. (1979) Physiological and ecological correlates of preferred temperature in fish. *Am. Zool.*, **19**, 319–329.

Bell, G. (1984) Measuring the cost of reproduction. The correlation structure of the life tables of five freshwater invertebrates. *Evolution*, **38**, 314–326.

Bell, G. and Koufopanou, V. (1986) The cost of reproduction, in *Oxford Surveys of Evolutionary Biology*, Vol. 3 (ed. R. Dawkins), Oxford University Press, Oxford, pp. 83–131.

Bergman, E. and Greenberg, L.A. (1994) Competition between a planktivore, a benthivore and a species with ontogenetic niche shifts. *Ecology*, **75**, 1233–1245.

Bernardo, J. (1996) The particular maternal effect of propagule size, especially egg size: patterns, models, quality of evidence and interpretations. *Am. Zool.*, **36**, 216–236.

Bertalanffy, L. von (1957) Quantitative laws in metabolism and growth. *Q. Rev. Biol.*, **32**, 217–231.

Beukema, J.J. (1968) Predation by the three-spined stickleback (*Gasterosteus aculeatus*): the influence of hunger and experience. *Behaviour*, **31**, 1–126.

Bevelhimer, M.S., Stein, R.A. and Carline, R.F. (1985) Assessing significance of physiological differences among three esocids with a bioenergetics model. *Can. J. Fish. Aquat. Sci.*, **42**, 57–69.

Beverton, R.J.H. (1963) Maturation, growth and mortality of clupeid and engraulid stocks in relation to fishing. *Rapp. P.-v. Réun. Cons. Int. Explor. Mer*, **154**, 44–67.

Beverton, R.J.H. (1984) Dynamics of single species, in *Exploitation of Marine Communities* (ed. R.M. May), Springer-Verlag, Berlin, pp. 13–58.

Beverton, R.J.H. (1987) Longevity in fish: some ecological and evolutionary considerations, in *Evolution of Longevity in Animals* (eds. A.D. Woodhead and K.H. Thompson), Plenum Press, New York, pp. 161–186.

Beverton, R.J.H. (1990) Small marine pelagic fish and the threat of fishing; are they endangered. *J. Fish Biol.*, **37** (Suppl. A), 5–16.

Beverton, R.J.H. (1992) Patterns of reproductive strategy parameters in some marine teleost fishes. *J. Fish Biol.*, 41 (Suppl. B), 137–160.

Beverton, R.J.H. and Holt, S.J. (1957) On the dynamics of exploited fish populations. *Fishery Invest., Lond.*, **19**, 533 pp.

Beverton, R.J.H. and Iles, T.C. (1992a) Mortality rates of 0-group plaice (*Pleuronectes platessa* L.), dab (*Limanda limanda* L.) and turbot (*Scophthalmus maximus* L.) in European waters. II. Comparison of mortality rates and construction of life table for 0-group plaice. *Neth. J. Sea Res.*, **29**, 49–59.

Beverton, R.J.H. and Iles, T.C. (1992b) Mortality rates of 0-group plaice (*Pleuronectes platessa* L.), dab (*Limanda limanda* L.) and turbot (*Scophthalmus maximus* L.) in European waters. III. Density dependence of mortality rates of 0-group plaice and some demographic implications. *Neth. J. Sea Res.*, **29**, 61–79.

Beyer, J.E. (1989) Recruitment stability and survival – simple size-specific theory with examples from the early life dynamics of marine fish. *Dana*, **7**, 45–147.

Biette, R.M. and Geen, G.H. (1980) Growth of underyearling sockeye salmon (*Oncorhynchus nerka*) under constant and cyclic temperature in relation to live zooplankton ration size. *Can. J. Fish. Aquat. Sci.*, **37**, 203–210.

Billard, R., Bry, C. and Gillet, C. (1981) Stress, environment and reproduction in teleost fish, in *Stress and Fish* (ed. A.D. Pickering), Academic Press, London, pp. 185–208.

Birkeland, K. (1996) Consequences of premature return by sea trout (*Salmo trutta*) infested with salmon louse (*Lepeophtheirus salmonis* Kroyer): migration, growth and mortality. *Can. J. Fish. Aquat. Sci.*, 53, 2808–2813.

Blacker, R.W. (1974) Recent advances in otolith

studies, in *Sea Fisheries Research* (ed. F.R. Harden Jones), Elek, London, pp. 67–90.

Blaxter, J.H.S. (1969) Development: eggs and larvae, in *Fish Physiology*, Vol. III (eds W.S. Hoar and D.J. Randall), Academic Press, London, pp. 177–252.

Blaxter, J.H.S. and Hempel, G. (1963) The influence of egg size on herring larvae (*Clupea harengus* L.). *J. Cons. Perm. Int. Explor. Mer*, 28, 211–240.

Bleckmann, H. (1993) Role of the lateral line in fish behaviour, in *The Behaviour of Teleost Fishes*, 2nd edn (ed. T.J. Pitcher), Chapman and Hall, London, pp. 201–246.

Blumer, L.S. (1979) Male parental care in the bony fishes. *Q. Rev. Biol.*, 54, 149–161.

Boer, P.J. den (1987) Density dependence and the stabilization of animal numbers 2. The pine looper. *Neth. J. Zool.*, 37, 220–237.

Boisclair, D. and Leggett, W.C. (1988) An in situ experimental evaluation of the Elliott and Persson and the Eggers models for estimating fish daily ration. *Can. J. Fish. Aquat. Sci.*, 45, 138–145.

Boisclair, D. and Leggett, W.C. (1989a) The importance of activity in bioenergetics models applied to actively foraging fishes. *Can. J. Fish. Aquat. Sci.*, 46, 1859–1867.

Boisclair, D. and Leggett, W.C. (1989b) Among-population variability of fish growth: I. Influence of the quantity of food consumed. *Can. J. Fish. Aquat. Sci.*, 46, 457–467.

Boisclair, D. and Leggett, W.C. (1989c) Among-population variability of fish growth: III. Influence of fish community. *Can. J. Fish. Aquat. Sci.*, 46, 1539–1550.

Boisclair, D. and Marchand, F. (1993) The guts to estimate daily ration. *Can. J. Fish. Aquat. Sci.*, 50, 1969–1975.

Bond, C.E. (1979) *Biology of Fishes*, Saunders, Philadelphia.

Bone, Q., Marshall, N.B. and Blaxter, J.H.S. (1995) *Biology of Fishes*, 2nd edn, Blackie A&P, Glasgow.

Booth, D.J. and Brosnan, D.M. (1995) The role of recruitment dynamics in rocky shore and coral reef communities. *Adv. Ecol. Res.*, 26, 309–385.

Borowsky, R.L. (1973) Social control of adult size in males of *Xiphophorus variatus*. *Nature*, 245, 332–335.

Boudreau, P.R. and Dickie, L.M. (1992) Biomass spectra of aquatic ecosystems in relation to fisheries yield. *Can. J. Fish. Aquat. Sci.*, 49, 1528–1538.

Bowen, S.H. (1979) A nutritional constraint in detritivory by fishes: the stunted population of *Sarotherodon mossambicus* in Lake Sibaya, South Africa. *Ecol. Monogr.*, 49, 17–31.

Brabrand, A., Bakke, T.A. and Faafeng, B.A. (1994) The ectoparasite *Ichthyophthirius multifiliis* and the abundance of roach (*Rutilus rutilus*): larval fish epidemics in relation to host behaviour. *Fish. Res.*, 20, 49–61.

Bradbury, J.W. and Andersson, M.B. (1987) *Sexual Selection: Testing the Alternatives*, J. Wiley, Chichester.

Brafield, A.E. (1985) Laboratory studies of energy budgets, in *Fish Energetics: New Perspectives* (eds P. Tytler and P. Calow), Croom Helm, London, pp. 257–281.

Brandt, S.B. and Kirsch, J. (1993) Spatially explicit models of striped bass growth potential in Chesapeake Bay. *Trans. Am. Fish. Soc.*, 122, 845–869.

Brandt, S.B., Magnuson, J.J. and Crowder, L.B. (1980) Thermal habitat partitioning by fishes in Lake Michigan. *Can. J. Fish. Aquat. Sci.*, 37, 1557–1564.

Breder, C.M., Jr and Rosen, D.E. (1966) *Modes of Reproduction in Fishes*, Natural History Press, New York.

Breitburg, D.L. (1992) Episodic hypoxia in Chesapeake Bay: interacting effects of recruitment, behavior, and physical disturbance. *Ecol. Monogr.*, 62, 525–546.

Brett, J.R. (1964) The respiratory metabolism and swimming performance of young sockeye salmon. *J. Fish. Res. Bd Can.*, 21, 1183–1226.

Brett, J.R. (1971a) Satiation time, appetite and maximum food intake of sockeye salmon (*Oncorhynchus nerka*). *J. Fish. Res. Bd Can.*, 28, 409–415.

Brett, J.R. (1971b) Energetic responses of salmon to temperature. A study of some thermal relations in the physiology and freshwater ecology of the sockeye salmon (*Oncorhynchus nerka*). *Am. Zool.*, 11, 99–113.

Brett, J.R. (1979) Environmental factors and growth, in *Fish Physiology*, Vol. VIII (eds W.S. Hoar, D.J. Randall and J.R. Brett), Academic Press, London, pp. 599–675.

Brett, J.R. (1983) Life energetics of sockeye salmon, *Oncorhynchus nerka*, in *Behavioural Energetics: The Cost of Survival in Vertebrates* (eds W.P. Aspey and S.I. Lustick), Ohio State University Press, Columbus, OH, pp. 29–63.

Brett, J.R. (1986) Production energetics of population of sockeye salmon, *Oncorhynchus nerka*. *Can. J. Zool.*, **64**, 555–564.

Brett, J.R. and Blackburn, J.M. (1981) Oxygen requirements for growth of young coho (*Oncorhynchus kisutch*) and sockeye (*O. nerka*) salmon at 15 °C. *Can. J. Fish. Aquat. Sci.*, **38**, 399–404.

Brett, J.R. and Groves, T.D.D. (1979) Physiological energetics, in *Fish Physiology*, Vol. VIII (eds W.S. Hoar, D.J. Randall and J.R. Brett), Academic Press, London, pp. 279–352.

Brett, J.R. and Higgs, D.A. (1970) Effect of temperature on the rate of gastric digestion in fingerling sockeye salmon, *Oncorhynchus nerka*. *J. Fish. Res. Bd Can.*, **27**, 1767–1779.

Brett, J.R. and Zala, C.A. (1975) Daily pattern of nitrogen excretion and oxygen consumption of sockeye salmon (*Oncorhynchus nerka*) under controlled conditions. *J. Fish. Res. Bd Can.*, **32**, 2479–2486.

Brett, J.R., Shelbourn, J.E. and Shoop, C.T. (1969) Growth rate and body composition of fingerling sockeye salmon. *Oncorhynchus nerka*, in relation to temperature and ration size. *J. Fish. Res. Bd Can.*, **26**, 2363–2394.

Broekhuizen, N., Gurney, W.S.C., Jones, A. and Bryant, A.D. (1994) Modelling compensatory growth. *Funct. Ecol.*, **8**, 770–782.

Bromage, N.R. (1995) Broodstock management and seed quality – general considerations, in *Broodstock Management and Egg and Larval Quality* (eds N.R. Bromage and R.J. Roberts), Blackwell Science, Oxford, pp. 1–24.

Bromage, N.R. and Roberts, R.J. (1995) *Broodstock Management and Egg and Larval Quality*, Blackwell Science, Oxford.

Bromley, P.J. (1994) The role of gastric evacuation experiments in quantifying feeding rates of predatory fish. *Rev. Fish Biol. Fisheries*, **4**, 36–66.

Bronmark, C.R. and Miner, J.G. (1992) Predator-induced phenotypical change in body morphology in crucian carp. *Science*, **258**, 1348–1350.

Brown, J.H., Marquet, P.A. and Taper, M.L. (1993) Evolution of body size: consequences of an energetic definition of fitness. *Am. Nat.*, **142**, 573–584.

Brown, M.E. (1946) The growth of brown trout (*Salmo trutta* Linn.) II. The growth of two year old trout at a constant temperature of 11.5 °C. *J. Exp. Biol.*, **22**, 130–144.

Brown, M.E. (1957) Experimental studies on growth, in *The Physiology of Fishes*, Vol. I; (ed. M.E. Brown), Academic Press, London, pp. 361–400.

Bryan, J.E. (1973) Feeding history, parental stock and food selection in rainbow trout. *Behaviour*, **45**, 123–153.

Bryan, J.E. and Larkin, P.A. (1972) Food specialization by individual trout. *J. Fish. Res. Bd Can.*, **20**, 1615–1624.

Bulow, F.J. (1987) RNA-DNA ratios as indicators of growth in fish: a review, in *The Age and Growth of Fish* (eds. R.C. Summerfelt and G.E. Hall), Iowa State University Press, Ames, IA, pp. 45–64.

Bundy, A. and Pitcher, T.J. (1995) An analysis of species changes in Lake Victoria: did the Nile perch act alone? in *The Impact of Species Change in African Lakes* (eds T.J. Pitcher and P.J.B. Hart), Chapman & Hall, London, pp. 11–135.

Burns, J.R. (1975) Seasonal changes in the respiration of pumpkinseed, *Lepomis gibbosus*, correlated with temperature, day length, and stage of reproductive development. *Physiol. Zool.*, **48**, 142–149.

Burns, J.R. (1985) The effect of low-latitude photoperiods on the reproduction of female and male *Poeciliopsis gracilis* and *Poecilia sphenops*. *Copeia*, 1985, 961–965.

Burrough, R.J. and Kennedy, C.R. (1979) The occurrence and natural alleviation of stunting in a population of roach, *Rutilus rutilus* (L.). *J. Fish Biol.*, **15**, 93–109.

Burrough, R.J., Bregazzi, P.R. and Kennedy, C.R. (1979) Interspecific dominance amongst three species of coarse fish in Slapton Ley, Devon. *J. Fish Biol.*, **15**, 535–544.

Bye, V.J. (1984) The role of environmental factors in the timing of reproductive cycles, in *Fish Reproduction: Strategies and Tactics* (eds G.W. Potts and R.J. Wootton), Academic Press, London, pp. 187–205.

Caley, M.J., Carr, M.H., Hixon, M.A., Hughes, T.P., Jones, G.P. and Menge, B.A. (1996) Recruitment and the local dynamics of open marine populations. *Ann. Rev. Ecol. Syst.*, **27**, 477–500.

Calow, P. (1976) *Biological Machines A Cybernetic Approach to Life*, Edward Arnold, London.

Campana, S.E. and Neilson, J.D. (1985) Microstructure of fish otoliths. *Can. J. Fish. Aquat. Sci.*, **42**, 1014–1032.

Campbell, R.N. (1971) The growth of brown trout *Salmo trutta* L. in northern Scottish lochs with special reference to the improvement of fisheries. *J. Fish Biol.*, **3**, 1–28.

Cao Wen-xuan, Chan Yi-yu, Wu Yun-fei and Zhu

Song-quan (1981) Origin and evolution of schizothoracine fishes in relation to the upheaval of Qinghai-Xizang Plateau in *Geological and Ecological Studies of Qinghai-Xizang Plateau*, Gordon and Beach, New York, pp. 1053–1060.

Cargnelli, L.M. and Gross, M.R. (1996) The temporal dimension in fish recruitment: birth date, body size, and size-dependent survival in a sunfish (bluegill: *Lepomis macrochirus*). *Can. J. Fish. Aquat. Sci.*, **53**, 360–367.

Carpenter, R.C. (1988) Mass mortality of a Caribbean sea urchin: immediate effects on community metabolism and other herbivores. *Proc. Natn Acad. Sci. USA*, **85**, 511–514.

Carpenter, S.R. and Kitchell, J.F. (1993) *The Trophic Cascade in Lakes*, Cambridge University Press, Cambridge.

Carpenter, S.R., Kitchell, J.F., Hodgson, J.R., Cochrane, P.A., Elser, J.J., Elser, M.M., Lodge, D.M., Kretchmer, D., He, X. and Ende, C.N. von (1987) Regulation of lake primary productivity by food web structure. *Ecology*, **68**, 1863–1876.

Carr, M.H. and Reed, D.C. (1993) Conceptual issues relevant to marine harvest refuges: examples from temperate reef fishes. *Can. J. Fish. Aquat. Sci.*, **50**, 2019–2028.

Carvalho, G.R. (1993) Evolutionary aspects of fish distribution: genetic variability and adaptation. *J. Fish Biol.*, **43** (Suppl. A), 53–73.

Carvalho, G.R. and Hauser, L. (1994) Molecular genetics and the stock concept in fisheries. *Rev. Fish Biol. Fisheries*, **4**, 326–350.

Cassie, R.M. (1954) Some uses of probability papers in analysis of size frequency distribution. *Aust. J. Mar. Freshwat. Res.*, **5**, 513–522.

Castleberry, D.T. and Cech, J.J., Jr (1986) Physiological responses of a native and an introduced desert fish to environmental stressors. *Ecology*, **67**, 912–918.

Caswell, H. (1983) Phenotypic plasticity in life-history traits: demographic effects and evolutionary consequences. *Am. Zool.*, **23**, 35–46.

Caswell, H. (1989) *Matrix Population Models*, Sinauer, Sunderland, MA.

Caulton, M.S. (1978) The importance of habitat temperatures for growth in the tropical cichlid *Tilapia rendalli* Boulenger. *J. Fish Biol.*, **13**, 99–112.

Caulton, M.S. (1981) Feeding, metabolism and growth of tilapias: some quantitative considerations, in *The Biology and Culture of Tilapias* (eds R.S.V. Pullin and R.H. Lowe-McConnell), ICLARM, Manila, pp. 157–175.

Causton, D.R., Elias, C.O. and Hadley, P. (1978) Biometrical studies of plant growth 1. The Richards function and its application in analysing the effects of temperature on leaf growth. *Plant Cell Environ.*, **1**, 163–184.

Chambers, R.C. and Leggett, W.C. (1996) Maternal influences on variation in egg sizes in temperate marine fishes. *Am. Zool.*, **36**, 180–196.

Chambers, R.C., Leggett, W.C. and Brown, J.A. (1989) Egg size, female effects, and the correlations between early life history traits of capelin, *Mallotus villosus* : an appraisal at the individual level. *Fish. Bull., U.S.*, **87**, 515–523.

Chapman, D.W. (1978) Production in fish populations, in *Ecology of Freshwater Fish Production* (ed. S.D. Gerking), Blackwell, Oxford, pp. 5–25.

Charlesworth, B. (1994) *Evolution in Age-Structured Populations*, 2nd edn, Cambridge University Press, Cambridge.

Charnov, E.L. (1976) Optimal foraging: the marginal value theorem. *Theor. Pop. Biol.*, **9**, 129–136.

Charnov, E.L. (1993) *Life History Invariants*, Oxford University Press, Oxford.

Charnov, E.L. (1997) Trade-off invariant rules for evolutionary stable life histories. *Nature*, **387**, 393–394.

Charnov, E.L. and Berrigan, D. (1991) Dimensionless numbers and assembly rules for life histories. *Phil. Trans. R. Soc. Lond. B*, **332**, 41–48.

Chesson, J. (1983) The estimation and analysis of preference and its relationship to foraging models. *Ecology*, **64**, 1297–1304.

Chesson, P.L. (1986) Environmental variation and the coexistence of species, in *Community Ecology* (eds J. Diamond and T.J. Case), Harper and Row, New York, pp. 240–256.

Choat, J.H. (1991) The biology of herbivorous fishes on coral reefs, in *The Ecology of Fishes on Coral Reefs* (ed. P.F. Sale), Academic Press, London, pp. 120–155.

Christie, W.J. (1974) Changes in the fish species composition of the Great Lakes. *J. Fish. Res. Bd Can.*, **31**, 827–854.

Clark, C.W. and Levy, D.A. (1988) Diel vertical migrations by juvenile sockeye salmon and the antipredator window. *Am. Nat.*, **131**, 271–290.

Clarke, A. (1993) Seasonal acclimatization and latitudinal compensation in metabolism: do they exist? *Funct. Ecol.*, **7**, 139–149.

Clutton-Brock, T.H. (1991) *The Evolution of Parental Care*, Princeton University Press, Princeton, NJ.

Colgan, P.W. (1973) Motivational analysis of fish feeding. *Behaviour*, **45**, 38–66.

Cone, R.S. (1989) The need to reconsider the use of condition indices in fishery science. *Trans. Am. Fish. Soc.*, **118**, 510–514.

Connell, J.H. (1978) Diversity in tropical rainforests and coral reefs. *Science*, **199**, 1302–1310.

Conover, D.O. (1990) The relation between capacity for growth and length of growing season: evidence for and implications of countergradient variation. *Trans. Am. Fish. Soc.*, **119**, 416–430.

Conover, D.O. (1992) Seasonality and scheduling of life history at different latitudes. *J. Fish Biol.*, **41** (Suppl. B), 161–178.

Conover, D.O. and Heins, S.W. (1987) Adaptive variation in environmental and genetic sex determination in a fish. *Nature*, **326**, 496–498.

Conover, D.O. and Kynard, B.E. (1981) Environmental sex determination: interaction of temperature and genotype in a fish. *Science*, **213**, 577–589.

Conover, D.O. and Present, T.M.C. (1990) Countergradient variation in growth rate: compensation for length of growing season among Atlantic silversides from different latitudes. *Oecologia*, **83**, 316–324.

Conover, D.O., Voorhees, D.A. van and Ehtisham, A. (1992) Sex ratio selection and evolution of environmental sex determination in laboratory populations of *Menidia menidia*. *Evolution*, **46**, 1722–1730.

Constantz, G.D. (1980) Energetics of viviparity in the Gila topminnow (Pisces: Poeciliidae). *Copeia*, 1980, 676–678.

Cornell, H., Hurd, L.E. and Lotrich, V.A. (1976) A measure of response to perturbation used to assess structural change in some polluted and unpolluted stream fish communities. *Oecologia*, **23**, 335–342.

Costello, M.J. (1990) Predator feeding strategy and prey importance: a new graphical analysis. *J. Fish Biol.*, **36**, 261–263.

Coutant, C.C. (1987) Thermal preference: when does an asset become a liability. *Env. Biol. Fishes*, **18**, 161–172.

Cowey, C.B., Mackie, A.M. and Bell, J.G. (1985) *Nutrition and Feeding in Fish*, Academic Press, London.

Cowx, I.G. (1983) Review of methods for estimating fish population size from survey removal methods. *Fish. Manage.*, **14**, 67–82.

Cowx, I.G. and LaMarque, P. (1990) *Fishing with Electricity*, Fishing News Books, Oxford.

Cox, D.K. and Coutant, C.C. (1981) Growth dynamics of juvenile striped bass as functions of temperature and ration. *Trans. Am. Fish. Soc.*, **110**, 226–238.

Craig, J.F. (1974) Population dynamics of perch, *Perca fluviatilis* L., in Slapton Ley. 1. Trapping behaviour, reproduction, migration, population estimates, mortality and food. *Freshwat. Biol.*, **4**, 417–431.

Craig, J.F. (1978) A study of the food and feeding of perch, *Perca fluviatilis* L., in Windermere. *Freshwat. Biol.*, **8**, 59–68.

Craig, J.F. (1980) Growth and production of the 1955–1972 cohorts of perch, *Perca fluviatilis* L., in Windermere. *J. Anim. Ecol.*, **49**, 291–315.

Craig, J.F. (1985) Ageing in fish. *Can. J. Zool.*, **63**, 1–8.

Craig, J.F. (1987) *The Biology of the Perch and Related Fish*, Croom Helm, London.

Craig, J.F. (1996) *Pike*, Chapman and Hall, London.

Craig, J.F., Kenley, M.J. and Talling, J.F. (1978) Comparative estimations of the energy content of fish tissue from bomb calorimetry, wet oxidation and proximate analysis. *Freshwat. Biol.*, **8**, 585–590.

Craig, P.C., Griffiths, W.B., Haldorson, L. and McElderry, H. (1982) Ecological studies of Arctic cod (*Boreogadus saida*) in Beaufort Sea coastal waters, Alaska. *Can. J. Fish. Aquat. Sci.*, **39**, 395–406.

Crow, M.E. (1982) Some statistical techniques for analysing the stomach contents of fish, in *Gutshop '81 Fish Food Habits Studies* (eds G.M. Cailliett and C.A. Simenstad), Washington Sea Grant Program, Seattle, pp. 8–15.

Crowden, A.E. and Broom, D.M. (1980) Effects of the eyefluke, *Diplostomum spathaceum*, on the behaviour of dace (*Leuciscus leuciscus*). *Anim. Behav.*, **28**, 287–294.

Crowder, L.B. and Magnuson, J.J. (1983) Cost–benefit analysis of temperature and food resource use: a synthesis with examples from the fishes, in *Behavioural Energetics: The Cost of Survival in Vertebrates* (eds W.P. Asprey and S.I. Lustick), Ohio State University Press, Columbus, OH, pp. 189–221.

Croy, M.I. and Hughes, R.N. (1991) The role of learning and memory in the feeding behaviour of the fifteen-spined stickleback (*Spinachia spinachia* L.). *Anim. Behav.* **41**, 161–170.

Cui, Y. (1987) Bioenergetics and growth of a teleost, *Phoxinus phoxinus* (Cyprinidae), PhD thesis, University of Wales, 240 pp.

Cui, Y. and Wootton, R.J. (1988) Effects of ration, temperature and body size on the body composition, energy content and condition of the minnow, *Phoxinus phoxinus* (L.). *J. Fish Biol.*, **32**, 749–764.

Cui, Y. and Wootton, R.J. (1989) Bioenergetics of growth of a cyprinid, *Phoxinus phoxinus* (L.): development and testing of a growth model. *J. Fish Biol.*, **34**, 46–64.

Cui, Y., Chen, S., Wang, S. and Liu, X. (1993) Laboratory observations on the circadian feeding patterns in the grass carp (*Ctenopharyngodon idella* Val.) fed three different diets. *Aquaculture*, **113**, 57–64.

Cui, Y., Chen, S. and Wang, S. (1994) Effect of ration size on growth and energy budget of the grass carp, *Ctenopharyngodon idella* Val. *Aquaculture*, **123**, 95–107.

Curio, E. (1976) *The Ethology of Predation*, Springer-Verlag, Berlin.

Cushing, D.H. (1969) The regularity of the spawning season in some fishes. *J. Cons. Perm. Int. Explor. Mer*, **33**, 81–87.

Cushing, D.H. (1975) *Marine Ecology and Fisheries*, Cambridge University Press, Cambridge.

Cushing, D.H. (1982) *Climate and Fisheries*, Academic Press, London.

Cushing, D.H. (1990) Plankton production and year class strength in fish populations: an update of the match–mismatch hypothesis. *Adv. Mar. Biol.*, **26**, 249–293.

Cushing, D.H. (1995) *Population Production and Regulation in the Sea: A Fisheries Perspective*, Cambridge University Press, Cambridge.

Cyr, H. and Peters, R.H. (1996) Biomass-size spectra and the prediction of fish biomass in lakes. *Can. J. Fish. Aquat. Sci.*, **53**, 994–1006.

Daan, N. (1987) Multispecies versus single-species assessment of North Sea fish stocks. *Can. J. Fish. Aquat. Sci.*, **44** (Suppl. 2), 360–370.

Daan, N. and Sissenwine, M.P. (1991) Multispecies models relevant to management of living resources. *ICES Mar. Sci. Symp.*, **193**, 358 pp.

Dabrowski, K.R. (1982a) The influence of light intensity on feeding of fish larvae and fry. I. *Coregonus pollan* (Thompson) and *Esox lucius* (L.). *Zoologische Jarbucher, Abteilung fur Physiologie*, **86**, 341–351.

Dabrowski, K.R. (1982b) The influence of light intensity on feeding of fish larvae and fry. II. *Rutilus rutilus* L. and *Perca fluviatilis* L. *Zoologische Jarbucher Abteilung fur Physiologie*, **86**, 353–360.

Davies, N.B. and Houston, A.I. (1984) Territory economics, in *Behavioural Ecology. An Evolutionary Approach* (eds J.R. Krebs and N.B. Davies), Blackwell, Oxford, pp. 148–169.

DeAngelis, D.L. and Gross, L.J. (1992) *Individual-based Models and Approaches in Ecology*, Chapman and Hall, London.

Dempster, J.P. (1975) *Animal Population Ecology*, Academic Press, London.

Dempster, P., Baird, D.J. and Beveridge, M.C.M. (1995) Can fish survive by filter-feeding on microparticles? Energy balance in tilapia grazing on algal suspensions. *J. Fish. Biol.*, **47**, 7–17.

Denny, M.W. (1993) *Air and Water*, Princeton University Press, Princeton, NJ.

Des Clers, S. (1993) Modelling the impact of disease-induced mortality on the population size of wild salmonids. *Fish. Res.*, **17**, 237–248.

DeVries, A.L. (1980) Biological antifreezes and survival in freezing environments, in *Animals and Environmental Fitness* (ed. R. Gilles), Pergamon Press, New York, pp. 583–607.

Diamond, J. and Case, T.J. (eds) (1986) *Community Ecology*, Harper and Row, New York.

Diana, J.S. (1980) Diel activity pattern and swimming speeds of northern pike (*Esox lucius*) in Lac Ste. Anne, Alberta. *Can. J. Fish. Aquat. Sci.*, **37**, 1454–1458.

Diana, J.S. (1983a) An energy budget for northern pike (*Esox lucius*). *Can. J. Zool.*, **61**, 1968–1975.

Diana, J.S. (1983b) Growth, maturation and production of northern pike in three Michigan lakes. *Trans. Am. Fish. Soc.*, **112**, 38–46.

Diana, J.S. and MacKay, W.C. (1979) Timing and magnitude of energy deposition and loss in the body, liver and gonads of northern pike (*Esox lucius*). *J. Fish. Res. Bd Can.*, **36**, 481–487.

Diana, J.S., MacKay, W.C. and Ehrman, M. (1977) Movements and habitat preference of northern pike (*Esox lucius*) in Lac Ste. Anne, Alberta. *Trans. Am. Fish. Soc.*, **106**, 560–565.

Dickie, L.M., Kerr, S.R. and Boudreau, P.R. (1987) Size-dependent processes underlying regularities in ecosystem structure. *Ecol. Monogr.*, **57**, 233–250.

Dill, L.M. (1983) Adaptive flexibility in the foraging behaviour of fishes. *Can. J. Fish. Aquat. Sci.*, **40**, 398–408.

Dill, L.M., Ydenberg, R.C. and Fraser, A.H.G. (1981) Food abundance and territory size in juvenile coho salmon (*Oncorhynchus kisutch*). *Can. J. Zool.*, 59, 1801–1809.

Dobson, A.P. and May, R.M. (1987) The effects of parasites on fish populations – theoretical aspects. *Int. J. Parasitol.*, **17**, 363–370.

Doherty, P.J. (1983) Tropical territorial damsel fishes: is density limited by aggression or recruitment? *Ecology*, **64**, 176–190.

Doherty, P.J. and Williams, D.McB. (1988) The replenishment of coral reef populations. *Oceanogr. Mar. Biol. Ann. Rev.*, **26**, 487–551.

Dominey, W.J. (1980) Female mimicry in male bluegill sunfish – a genetic polymorphism? *Nature*, **284**, 546–548.

Dominey, W.J. (1981) Anti-predator function of bluegill nesting colonies. *Nature*, **290**, 586–588.

Dominey, W.J. (1984) Effects of sexual selection and life history on speciation: species flocks in African cichlids and Hawaiian *Drosophila*, in *Evolution of Species Flocks* (eds A.A. Echelle and I. Kornfield), University of Maine Press, Orono, ME, pp. 231–249.

Donald, D.B., Anderson, R.S. and Mayhood, D.W. (1980) Correlations between brook trout growth and environmental variables for mountain lakes in Alberta. *Trans. Am. Fish. Soc.*, **109**, 603–610.

Dower, J.F., Miller, T.J. and Leggett, W.C. (1997) The role of microscale turbulence in the feeding ecology of larval fish. *Adv. Mar. Biol.*, **31**, 169–220.

Downhower, J.F. and Brown, L. (1981) The timing of reproduction and its behavioural consequences for mottled sculpins, in *Natural Selection and Social Behaviour* (eds R.D. Alexander and D.W. Tinkle), Chiron Press, Newton, pp. 78–95.

Downhower, J.F., Brown, L., Pederson, R. and Staples, G. (1983) Sexual selection and sexual dimorphism in mottled sculpins. *Evolution*, **37**, 96–103.

Duarte, C.M. and Alcaraz, M. (1989) To produce many small or few large eggs: a size-independent reproductive tactic of fish. *Oecologia*, **80**, 401–404.

Duman, J.G. and DeVries, A.L. (1974) The effects of temperature and photoperiod on antifreeze production in cold-water fishes. *J. Exp. Zool.*, **190**, 89–98.

Dunbrack, R.L. and Dill, L.M. (1983) A model of size dependent feeding in a stream dwelling salmonid. *Env. Biol. Fishes*, **8**, 203–216.

Dutil, J.D. (1986) Energetic constraints and spawning interval in the anadromous Arctic charr (*Salvelinus alpinus*). *Copeia*, 1986, 954–955.

East, P. and Magnan, P. (1987) The effect of locomotor activity on the growth of the brook charr, *Salvelinus fontinalis* Mitchill. *Can. J. Zool.*, **65**, 843–846.

Ebeling, A.W. and Hixon, M.A. (1991) Tropical and temperate reef fishes: comparison of community structures, in *The Ecology of Fishes on Coral Reefs* (ed. P.F. Sale), Academic Press, London, pp. 509–563.

Ebersole, J.P. (1977) The adaptive significance of interspecific territoriality in the reef fish *Eupomacentrus leucosticus*. *Ecology*, **58**, 914–920.

Echelle, A.A. and Kornfield, I.A. (eds) (1984) *Evolution of Fish Species Flocks*, University of Maine Press, Orono, ME.

Eck, G.W. and Wells, L. (1987) Recent changes in Lake Michigan's fish community and their probable causes, with emphasis on the role of the alewife (*Alosa pseudoharengus*). *Can. J. Fish. Aquat. Sci.*, **44** (Suppl. 2), 53–60.

Edmunds, M. (1974) *Defence in Animals*, Longman, London.

Edwards, R.W., Densem, J.W. and Russell, P.A. (1979) An assessment of the importance of temperature as a factor controlling the growth rate of brown trout in streams. *J. Anim. Ecol.*, **48**, 501–507.

Eggers, D.M. (1977) The nature of prey selection by planktivorous fish. *Ecology*, **58**, 46–59.

Eggers, D.M. (1979) Comments on some recent methods for estimating food consumption by fish. *J. Fish. Res. Bd Can.*, **36**, 1018–1019.

Eggers, D.M. (1982) Planktivore preference by prey size. *Ecology*, **63**, 381–390.

Ehlinger, T.J. (1990) Habitat choice and phenotype limited feeding efficiency in bluegill: individual differences and trophic polymorphism. *Ecology*, **71**, 886–896.

Ehlinger, T.J. and Wilson, D.S. (1988) Complex foraging polymorphism in bluegill sunfish. *Proc. Natn Acad. Sci. USA*, **85**, 1878–1882.

Ehrlich, P.R. (1975) The population biology of coral reef fishes. *Ann. Rev. Ecol. Syst.*, **6**, 211–247.

Eklov, P. and Persson, L. (1996) The response of prey to the risk of predation: proximate cues for refuging juvenile fish. *Anim. Behav.*, **51**, 105–115.

Elder, J.F. Jr and Schlosser, I.J. (1995) Extreme clonal uniformity of *Phoxinus eos/neogaeus* gynogens (Pisces: Cyprinidae) among variable habitats in northern Minnesota beaver ponds. *Proc. Natn Acad. Sci. USA*, **92**, 5001–5005.

Elgar, M.A. (1989) Evolutionary compromise between few large and many small eggs: comparative evidence in teleost fish. *Oikos*, **59**, 283–287.

Elliott, J.K. and Mariscal, R.N. (1996) Ontogenetic and interspecific variation in the protection of anemone fishes from sea anemones. *J. Exp. Mar. Biol. Ecol.*, **208**, 57–72.

Elliott, J.M. (1972) Rates of gastric evacuation in brown trout, *Salmo trutta* L. *Freshwat. Biol.*, **2**, 1–18.

Elliott, J.M. (1975a) Weight of food and time required to satiate brown trout, *Salmo trutta* L. *Freshwat. Biol.*, **5**, 51–64.

Elliott, J.M. (1975b) Number of meals in a day, maximum weight of food consumed in a day and maximum rate of feeding for brown trout *Salmo trutta* L. *Freshwat. Biol.*, **5**, 287–303.

Elliott, J.M. (1975c) The growth rate of brown trout (*Salmo trutta* L.) fed on reduced rations. *J. Anim. Ecol.*, **44**, 823–842.

Elliott, J.M. (1975d) The growth rate of brown trout, *Salmo trutta* L., fed on maximum rations. *J. Anim. Ecol.*, **44**, 805–821.

Elliott, J.M. (1976a) Energy losses in waste products of brown trout (*Salmo trutta* L.). *J. Anim. Ecol.*, **45**, 561–580.

Elliott, J.M. (1976b) The energetics of feeding, metabolism and growth of brown trout (*Salmo trutta* L.) in relation to body weight, water temperature and ration size. *J. Anim. Ecol.*, **45**, 923–948.

Elliott, J.M. (1979) Energetics of freshwater teleosts. *Symp. Zool. Soc. Lond.*, **44**, 29–61.

Elliott, J.M. (1981) Some aspects of thermal stress on freshwater teleosts, in *Stress and Fish* (ed. A.D. Pickering), Academic Press, London, pp. 209–245.

Elliott, J.M. (1984) Numerical changes and population regulation in young migratory trout *Salmo trutta* in a Lake District stream 1966–1983. *J. Anim. Ecol.*, **53**, 327–350.

Elliott, J.M. (1985a) Population dynamics of migratory trout, *Salmo trutta*, and their implications for fisheries management. *J. Fish Biol.*, **27** (Suppl. A), 35–43.

Elliott, J.M. (1985b) The choice of a stock–recruitment model for migratory trout, *Salmo trutta*, in an English Lake District stream. *Arch. Hydrobiol.*, **104**, 145–168.

Elliott, J.M. (1985c) Population regulation for dif-

ferent life-stages of migratory trout, *Salmo trutta*, in a Lake District stream, 1966–1983. *J. Anim. Ecol.*, **54**, 617–638.

Elliott, J.M. (1985d) Growth, size, biomass and production for different life-stages of migratory trout, *Salmo trutta*, in a Lake District stream, 1966–1983. *J. Anim. Ecol.*, **54**, 985–1002.

Elliott, J.M. (1986) Spatial distribution and behavioural movements of migratory trout, *Salmo trutta*, in a Lake District stream. *J. Anim. Ecol.*, **55**, 907–922.

Elliott, J.M. (1987) Population regulation in contrasting populations of trout, *Salmo trutta*, in two Lake District streams. *J. Anim. Ecol.*, **56**, 83–98.

Elliott, J.M. (1989a) Mechanisms responsible for population regulation in young migratory trout, *Salmo trutta*. I. The critical time for survival. *J. Anim. Ecol.*, **58**, 987–1001.

Elliott, J.M. (1989b) The critical-period concept for juvenile survival and its relevance for population regulation in young sea trout, *Salmo trutta*. *J. Fish Biol.*, **35** (Suppl. A), 91–98.

Elliott, J.M. (1990a) Mechanisms responsible for population regulation in young migratory trout, *Salmo trutta*. II. Fish growth and size variation. *J. Anim. Ecol.*, **59**, 171–185.

Elliott, J.M. (1990b) Mechanisms responsible for population regulation in young migratory trout, *Salmo trutta*. III. The role of territorial behaviour. *J. Anim. Ecol.*, **59**, 803–818.

Elliott, J.M. (1993) The self-thinning rule applied to juvenile sea-trout, *Salmo trutta*. *J. Anim. Ecol.*, **62**, 371–379.

Elliott, J.M. (1994) *Quantitative Ecology and the Brown Trout*, Oxford University Press, Oxford.

Elliott, J.M. and Elliott, J.A. (1995) The critical thermal limits for the bullhead, *Cottus gobio*, from three populations in north-west England. *Freshwat. Ecol.*, **33**, 411–418.

Elliott, J.M. and Persson, L. (1978) The estimation of daily rates of food consumption for fish. *J. Anim. Ecol.*, **47**, 977–991.

Elliott, J.M., Humpesch, U.H. and Hurley, M.A. (1987) A comparative study of eight mathematical models for the relationship between water temperature and hatching time of eggs of freshwater fish. *Archiv fur Hydrobiologie*, **109**, 257–277.

Elliott, J.M., Hurley, M.A. and Fryer, R.J. (1995) A new improved model for brown trout, *Salmo trutta*. *Funct. Ecol.*, **9**, 290–298.

Elson, P.F. (1962) Predator–prey relationships

between fish-eating birds and Atlantic salmon (with a supplement on the fundamentals of merganser control). *Bull. Fish. Res. Bd Can.*, no. **133**, 1–87.

Endler, J.A. (1980) Natural selection on colour patterns in *Poecilia reticulata*. *Evolution*, **34**, 76–91.

Endler, J.A. (1983) Natural and sexual selection on colour patterns in poeciliid fishes. *Env. Biol. Fishes*, **9**, 173–190.

Ensign, W.E., Angermeier, P.L. and Dolloff, C.A. (1995) Use of line transect methods to estimate abundance of benthic stream fishes. *Can. J. Fish. Aquat. Sci.*, **52**, 213–222.

Evans, D.O. (1984) Temperature independence of the annual cycle of standard metabolism in the pumpkinseed. *Trans. Am. Fish. Soc.*, **113**, 494–512.

Evans, D.O., Henderson, B.A., Bax, N.J., Marshall, T.R., Oglesby, R.T. and Christie, W.J. (1987) Concepts and methods of community ecology applied to freshwater fisheries management. *Can. J. Fish. Aquat. Sci.*, **44** (Suppl. 2), 448–470.

Everest, F.H. and Chapman, D.W. (1972) Habitat selection and spatial interaction by juvenile chinook salmon and steelhead trout in two Idaho streams. *J. Fish. Res. Bd Can.*, **29**, 91–100.

Fahy, E. (1979) Prey selection by young trout fry (*Salmo trutta*). *J. Zool., Lond.*, **190**, 27–37.

Fahy, E. (1985) *Child of the Tides*, Glendale Press, Dun Laoghaire, Eire.

Falconer, D.S. (1989) *Introduction to Quantitative Genetics*, 3rd edn, Longman, London.

Farbridge, K.J. and Leatherland, J.F. (1987) Lunar cycles of coho salmon, *Oncorhynchus kisutch* 1. Growth and feeding. *J. Exp. Biol.*, **129**, 165–178.

Faris, A.A. (1986) Some effects of acid water on the biology of *Gasterosteus aculeatus* (Pisces), PhD Thesis, University of Wales, 143 pp.

Farmer, G.J. and Beamish, F.W H. (1969) Oxygen consumption of *Tilapia nilotica* in relation to swimming speed and salinity. *J. Fish. Res. Bd Can.*, **26**, 2807–2821.

Fausch, K.D. (1984) Profitable stream positions for salmonids relating specific growth rate to net energy gain. *Can. J. Zool.*, **62**, 441–451.

Fausch, K.D. and White, R.J. (1981) Competition between brook trout (*Salvelinus fontinalis*) and brown trout (*Salmo trutta*) for positions in a Michigan stream. *Can. J. Fish. Aquat. Sci.*, **38**, 1220–1227.

Feder, H.M. (1966) Cleaning symbiosis in the marine environment, in *Symbiosis*, Vol. I (ed. S.M. Henry), Academic Press, London, pp. 327–380.

Feinsinger, P., Spears, E.E. and Poole, R.W. (1981) A simple measure of niche breadth. *Ecology*, **62**, 27–32.

Feldmeth, C.R. (1983) Costs of aggression in trout and pupfish, in *Behavioural Energetics: The Cost of Survival in Vertebrates* (eds W.P. Asprey and S.I. Lustick), Ohio State University Press, Columbus, OH, pp. 117–138.

Ferguson, A. (1980) *Biochemical Systematics and Evolution*, Blackie, Glasgow.

Ferguson, A. and Mason, F.M. (1981) Allozyme evidence for reproductively isolated sympatric populations of brown trout *Salmo trutta* L. in Lough Melvin, Ireland. *J. Fish Biol.*, **18**, 629–642.

Ferguson, A., Taggart, J.B., Prodohl, P.A., McMeel, O., Thompson, C., Stone, C., McGinnity, P. and Hynes, R.A. (1995) The application of molecular markers to the study and conservation of fish populations, with special reference to *Salmo*. *J. Fish Biol.*, **47** (Suppl. A), 103–126.

Ferraro, S.P. (1980) Daily time of spawning of 12 fishes in the Peconic Bays, New York. *Fish. Bull. Fish. Wildl. Serv. US*, **78**, 455–464.

Flath, L.E. and Diana, J.S. (1985) Seasonal energy dynamics of the alewife in southeastern Lake Michigan. *Trans. Am. Fish. Soc.*, **114**, 328–337.

Fleming, I.A. (1996) Reproductive strategies of Atlantic salmon: ecology and evolution. *Rev. Fish Biol. Fisheries*, **6**, 379–416.

Fleming, I.A. and Gross, M.R. (1990) Latitudinal clines: a trade-off between egg number and size in Pacific salmon. *Ecology*, **71**, 1–11.

Fletcher, D.A. and Wootton, R.J. (1995) A hierarchical response to differences in ration size in the reproductive performance of female three-spined sticklebacks. *J. Fish Biol.*, **46**, 657–668.

Fletcher, G.L. and King, M.J.(1978) Seasonal dynamics of Cu^{2+}, Zn^{2+}, Ca^{2+}, and Mg^{2+} in gonads and liver of winter flounder (*Pseudopleuronectes americanus*): evidence for summer storage of Zn^{2+}, for winter gonad development in females. *Can. J. Zool.*, **56**, 284–290.

Foerster, R.E. (1968) The sockeye salmon, *Oncorhynchus nerka*. *Bull. Fish. Res. Bd Can.*, no. **162**, 1–422.

Forrester, G.E. (1990) Factors influencing the juvenile demography of a coral reef fish population. *Ecology*, **71**, 1666–1681.

Forrester, G.E. (1995) Strong density dependent survival and recruitment regulate the abundance of a coral reef fish. *Oecologia*, **103**, 275–282.

Forseth, T. and Jonsson, B. (1994) The growth and food ration of piscivorous brown trout (*Salmo trutta*). *Funct. Ecol.*, **8**, 171–177.

Forseth, T., Jonsson, B., Naeumann, R. and Ugedal, O. (1992) Radioisotope method of estimating food consumption by brown trout (*Salmo trutta*). *Can. J. Fish. Aquat. Sci.*, **49**, 1328–1335.

Fortier, L. and Leggett, W.C. (1985) A drift study of larval survival. *Mar. Ecol. Progr. Ser.*, **25**, 245–257.

Foster, S.A. (1985a) Size-dependent territory defense by a damselfish. *Oecologia*, **67**, 499–505.

Foster, S.A. (1985b) Group foraging by a coral reef fish: a mechanism for gaining access to defended resources. *Anim. Behav.*, **33**, 782–792.

Foster, S.A. (1985c) Wound healing: a possible role of cleaning stations. *Copeia*, 1985, 875–880.

Foster, S.A. (1987) Acquisition of a defended resource: a benefit of group foraging for the neotropical wrasse, *Thalassoma lucasanum*. *Env. Biol. Fishes*, **19**, 215–222.

Francis, R.I.C.C. (1990) Back-calculation of fish length: a critical review. *J. Fish Biol.*, **36**, 883–902.

Francis, R.I.C.C. (1995) The analysis of otolith data – a mathematician's perspective (what, precisely *is* your model?), in *Recent Developments in Fish Otolith Research* (eds D.H. Secor, J.M. Dean and S.E. Campana), University of South Carolina Press, Columbia, SC, pp. 81–95.

Frank, K.T. and Leggett, W.C. (1986) Effect of prey abundance and size on the growth and survival of larval fish: an experimental study employing large volume enclosures. *Mar. Ecol. Progr. Ser.*, **34**, 11–22.

Fraser, D.F. and Gilliam, J.F. (1992) Nonlethal impacts of predator invasion: facultative supression of growth and reproduction. *Ecology*, **73**, 959–970.

Fraser, N.H.C., Metcalfe, N.B. and Thorpe, J.E. (1993) Temperature-dependent switch between diurnal and nocturnal foraging in salmon. *Proc. R. Soc. Lond. B*, **252**, 135–139.

Fraser, N.H.C. and Metcalfe, N.B. (1997) The costs of becoming nocturnal: feeding efficiency in relation to light intensity in juvenile salmon. *Funct. Ecol.*, **11**, 385–391.

Fretwell, S.D. and Lucas, J.R. (1970) On territorial behaviour and other factors influencing habitat distribution in birds. 1. Theoretical developments. *Acta Biotheor.*, **19**, 16–36.

Friedland, K.D. (1985) Functional morphology of the branchial basket structure associated with feeding in the Atlantic menhaden, *Brevoortia tyrannus* (Pisces: Clupeidae). *Copeia*, 1985, 1018–1027.

From, J. and Rasmussen, G. (1984) A growth model, gastric evacuation, and body composition in rainbow trout, *Salmo gairdneri* Richardson, 1836. *Dana*, **3**, 61–139.

Frost, W.E. and Brown, M.E. (1967) *The Trout*, Collins, London.

Fry, F.E.J. (1971) The effect of environmental factors on the physiology of fish, in *Fish Physiology*, Vol. VI (eds W.S. Hoar and D.J. Randall), Academic Press, London, pp. 1–98.

Fryer, G. and Iles, T.D. (1972) *The Cichlid Fishes of the Great Lakes of Africa*, Oliver and Boyd, Edinburgh.

Fuiman, L.A. and Magurran, A. (1994) Development of predator defences in fishes. *Rev. Fish Biol. Fisheries*, **4**, 145–183.

Fuiman, L.A. and Webb, P.A. (1988) Ontogeny of routine swimming activity and performance in zebra danios (Teleostei: Cyprinidae). *Anim. Behav.*, **36**, 250–261.

Galbraith, M.G. (1967) Size-selective predation on *Daphnia* by rainbow trout and yellow perch. *Trans. Am. Fish. Soc.*, **96**, 1–10.

Garrod, D.J. (1982) Stock and recruitment – again. *Fish. Tech. Rep.*, **68**, 22.

Garrod, D.J. and Knights, B.J. (1979) Fish stocks: their life-history characteristics and response to exploitation, *Symp. Zool. Soc. Lond.*, **44**, 361–382.

Gatz, A.J., Jr (1979) Community organization in fishes as indicated by morphological features. *Ecology*, **60**, 711–718.

Gatz, A.J., Jr and Loar, J.M. (1988) Petersen and removal population size estimates: combining methods to adjust and interpret results when assumptions are violated. *Env. Biol. Fishes*, **21**, 293–307.

Gerking, S.D. (1955) Influence of rate of feeding on body composition and protein metabolism of bluegill sunfish. *Physiol. Zool.*, **28**, 267–282.

Gerking, S.D. (1959) The restricted movements of fish populations. *Biol. Rev.*, **34**, 221–242.

Gerking, S.D. (1962) Production and food utilization in a population of bluegill sunfish. *Ecol. Monogr.*, **32**, 31–78.

Gerking, S.D. (1972) Revised food consumption estimate of a bluegill sunfish population in Wyland Lake, Indiana, U.S.A. *J. Fish Biol.*, **4**, 301–308.

Gerking, S.D. (1980) Fish reproduction and stress, in *Environmental Physiology of Fishes* (ed. M.A. Ali), Plenum, New York, pp. 569–587.

Gerking, S.D. (1984) Assimilation and maintenance ration of an herbivorous fish, *Sarpa salpa*, feeding on green alga. *Trans. Am. Fish. Soc.*, **113**, 378–387.

Gerking, S.D. (1994) *Feeding Ecology of Fish*, Academic Press, London.

Gibson, R.M. (1980) Optimal prey-size selection by three-spined sticklebacks (*Gasterosteus aculeatus*): a test of the apparent size hypothesis. *Z. Tierpsychol.*, **52**, 291–307.

Gibson, R.N. (1969) The biology and behaviour of littoral fish. *Oceanogr. Mar. Biol. Ann. Rev.*, **7**, 367–410.

Gibson, R.N. (1993) Intertidal teleosts: life in a fluctuating environment, in *The Behaviour of Teleost Fishes*, 2nd edn (ed. T.J. Pitcher), Chapman and Hall, London, pp. 512–536.

Giles, N. (1983) Behavioural effects of the parasite *Schistocephalus solidus* (Cestoda) on an intermediate host, the three-spined stickleback, *Gasterosteus aculeatus*. *Anim. Behav.*, **31**, 1192–1194.

Giles, N. (1987) A comparison of the behavioural responses of parasitized and nonparasitized three-spine sticklebacks, *Gasterosteus aculeatus* L., to progressive hypoxia. *J. Fish Biol.*, **30**, 631–638.

Giles, N., Wright, R.M., and Nord, M.E. (1986) Cannibalism in pike fry, *Esox lucius* L.: some experiments with fry densities. *J. Fish Biol.*, **29**, 107–113.

Giller, P.S. (1984) *Community Structure and the Niche*, Chapman and Hall, London.

Giller, P.S. and Gee, J.H.R. (1987) The analysis of community organization: the influence of equilibrium, scale and terminology, in *Organization of Communities Past and Present* (eds J.H.R. Gee and P.S. Giller), Blackwell, Oxford, pp. 519–542.

Gilliam, J.F. and Fraser, D.F. (1987) Habitat selection under predation hazard: test of a model with foraging minnows. *Ecology*, **68**, 1856–1862.

Gjerde, B. (1986) Growth and reproduction in fish and shellfish. *Aquaculture*, **57**, 37–55.

Glass, N.R. (1969) Discussion of calculation of power function with special reference to respiratory metabolism in fish. *J. Fish. Res. Bd Can.*, **26**, 2643–2650.

Glebe, B.D. and Leggett, W.C. (1981a) Temporal intra-population differences in energy allocation and use by American shad (*Alosa sapidissima*) during the spawning migration. *Can. J. Fish. Aquat. Sci.*, **38**, 795–805.

Glebe, B.D. and Leggett. W.C. (1981b) Latitudinal differences in energy allocation and use during the freshwater migrations of American shad (*Alosa sapidissima*) and their life history consequences. *Can. J. Fish. Aquat. Sci.*, **38**, 806–820.

Goodey, W. and Liley, N.R. (1986) The influence of early experience on escape behaviour in the guppy (*Poecilia reticulata*). *Can. J. Zool.*, **64**, 885–888.

Goolish, E.M. (1991) Aerobic and anaerobic scaling in fish. *Biol. Rev.*, **66**, 33–56.

Goolish, E.M. and Adelman, I.R. (1984) Effect of ration size and temperature on growth of juvenile common carp (*Cyprinus carpio* L.). *Aquaculture*, **36**, 27–35.

Gorlick, D.L.. Atkins, P.D. and Losey, G.S. Jr (1978) Cleaning stations as water holes, garbage dumps, and sites for the evolution of reciprocal altruism. *Am. Nat.*, **112**, 341–353.

Gorman, G.C. and Nielson, L.A. (1982) Piscivory by stocked brown trout (*Salmo trutta*) and its impact on the nongame fish community of Bottom Creek, Virginia. *Can. J. Fish. Aquat. Sci.*, **39**, 862–869.

Goulding. M. (1980) *The Fishes and The Forest*, University of California Press, Berkeley.

Graham, J.B. (1983) Heat transfer, in *Fish Biomechanics* (eds P.W. Webb and D. Weihs), Praeger, New York, pp. 248–279.

Grant, J.W.A. (1993) Whether or not to defend? The influence of resource distribution. *Mar. Behav. Physiol.*, **23**, 137–153.

Grant, J.W.A. and Kramer, D.L. (1990) Territory size as a predictor of the upper limit to population density of juvenile salmonids in streams. *Can. J. Fish. Aquat. Sci.*, **47**, 1724–1737.

Graves, J.E. and Somero, G.N. (1982) Electrophoretic and functional enzymic evolution in four species of Eastern Pacific barracudas from different thermal environments. *Evolution*, **36**, 97–106.

Greenwood, P.H. (1965) The cichlid fishes of Lake Nabugabo, Uganda. *Bull. Br. Mus. Nat. Hist. (D. Zoology)*, **12**, 315–357.

Greenwood, P.H. (1984) African cichlids and evolutionary theories, in *Evolution of Fish Species Flocks* (eds A.A. Echelle and I. Kornfield), University of Maine Press, Orono, ME, pp. 141–154.

Greenwood, P.H. (1991) Speciation, in *Cichlid Fishes: Behaviour, Ecology and Evolution* (ed. M.H.A. Keenleyside), Chapman and Hall, London, pp. 86–102.

Griffith, J.S. (1972) Comparative behaviour and

habitat utilization of brook trout (*Salvelinus fontinalis*) and cutthroat trout (*Salmo clarki*) in small streams in northern Idaho. *J. Fish. Res. Bd Can.*, **29**, 265–273.

Grimm, M.P. and Klinge, M. (1996) Pike and some aspects of its dependence on vegetation, in *Pike Biology and Exploitation* (ed. J.F. Craig), Chapman and Hall, London, pp. 125–156.

Groot, C. (1965) On the orientation of young sockeye salmon (*Oncorhynchus nerka*) during their seaward migration out of lakes. *Behaviour (Supp.)*, **14**, 1–198.

Gross, M.L. and Kapuscinski, A.R. (1997) Reproductive success of smallmouth bass estimated from family-specific DNA fingerprints. *Ecology*, **78**, 1424–1430.

Gross, M.R. (1982) Sneakers, satellites and parentals: polymorphic mating strategies in North American sunfishes. *Z. Tierpsychol.*, **60**, 1–26.

Gross, M.R. (1984) Sunfish, salmon and the evolution of alternative reproductive strategies and tactics in fishes, in *Fish Reproduction: Strategies and Tactics* (eds G.W. Potts and R.J. Wootton), Academic Press, London, pp. 55–75.

Gross, M.R. (1985) Disruptive selection for alternative life histories in salmon. *Nature*, **313**, 47–48.

Gross, M.R. (1987) Evolution of diadromy in fishes. *Am. Fish. Soc. Symp.*, **1**, 14–25.

Gross, M.R. (1991) Evolution of alternative reproductive strategies: frequency-dependent sexual selection in male bluegill sunfish. *Phil. Trans. R. Soc. Lond. B*, **332**, 59–66.

Gross, M.R., Coleman, R.M. and McDowall, R.M. (1988) Aquatic productivity and the evolution of diadromous fish migration. *Science*, **239**, 1291–1293.

Grossman, G.D., Moyle, P.B. and Whitaker, J.O., Jr (1982) Stochasticity in structural and functional characteristics of an Indiana stream fish assemblage: a test of community theory. *Am. Nat.*, **120**, 423–554.

Grossman, G.D., Freeman, M.C., Moyle, P.B. and Whitaker, J.O., Jr (1985) Stochasticity and assemblage organization in an Indiana stream fish assemblage. *Am. Nat.*, **126**, 275–285.

Gulland, J.A. (1977) *Fish Population Dynamics*, J. Wiley, London.

Guthrie, D.M. and Muntz, W.R.A. (1993) Role of vision in fish behaviour, in *The Behaviour of Teleost Fishes*. 2nd edn (ed. T.J. Pitcher), Chapman and Hall, London, pp. 89–128.

Hackney, P.A. (1979) Influence of piscivorous fish on fish community structure of ponds, in *Predator–prey Systems in Fisheries Management* (ed. H. Clepper), Sport Fishing Institute, Washington, DC, pp. 111–121.

Hagen, D.W. (1967) Isolating mechanisms in threespine sticklebacks (*Gasterosteus*). *J. Fish. Res. Bd Can.*, **24**, 1637–1692.

Haines, T.A. and Baker, J.P. (1986) Evidence of fish population responses to acidification in eastern United States. *Water Air Soil Pollut.*, **31**, 605–629.

Hairston, N.G. Jr, Li, K.T. and Easter, S.S. Jr (1982) Fish vision and the detection of planktonic prey. *Science*, **218**, 1240–1242.

Hamrin, S.F. and Persson, L. (1986) Asymmetrical competition between age classes as a factor causing population oscillations in an obligate planktivorous fish species. *Oikos*, **47**, 223–232.

Hansen, M.J., Boisclair, D., Brandt, S.B., Hewett, S.W., Kitchell, J.F., Lucas, M.C. and Ney, J.J. (1993) Applications of bioenergetics models to fish ecology and management: where do we go from here? *Trans. Am. Fish. Soc.*, **122**, 1019–1030.

Hanski, I. and Gilpin, M. (1991) Metapopulation dynamics: brief history and conceptual domain. *Biol. J. Linn. Soc.*, **42**, 3–16.

Hanson, A.J. and Smith, H.D. (1967) Mate selection in a population of sockeye salmon (*Oncorhynchus nerka*) of mixed age-groups. *J. Fish. Res. Bd Can.*, **24**, 1955–1977.

Hanson, J.M. and Leggett, W.C. (1982) Empirical prediction of fish biomass and yield. *Can. J. Fish. Aquat. Sci.*, **39**, 257–263.

Hanson, J.M. and Leggett, W.C. (1985) Experimental and field evidence for inter- and intraspecific competition in two freshwater fishes. *Can. J. Fish. Aquat. Sci.*, **42**, 280–286.

Hanson, J.M. and Leggett, W.C. (1986) Effect of competition between two freshwater fishes on prey consumption and abundance. *Can. J. Fish. Aquat. Sci.*, **43**, 1363–1372.

Hansson, S. (1984) Competition as a factor regulating the geographical distribution of fish species in a Baltic Archipelago: a neutral model. *J. Biogeogr.*, **11**, 367–381.

Hara, T.J. (1993) Role of olfaction in fish behaviour, in *Behaviour of Teleost Fishes*, 2nd edn (ed. T.J. Pitcher), Chapman and Hall, London, pp. 171–199.

Harden Jones, F.R. (1968) *Fish Migration*, Edward Arnold, London.

Hart, P.J.B. (1993) Foraging in teleost fishes, in *Behaviour of Teleost Fishes*, 2nd edn (ed. T.J.

Pitcher), Chapman and Hall, London, pp. 253–284.

Hart, P.J.B. and Gill, A.B. (1993) Choosing prey size: a comparison of static and dynamic foraging models for predicting prey choice by fish. *Mar. Behav. Physiol.*, **23**, 91–104.

Hart, P.J.B. and Pitcher, T.J. (1969) Field trials of fish marking using a jet inoculator. *J. Fish Biol.*, **1**, 383–385.

Hartman, G. (1965) The role of behaviour in the ecology and interaction of underyearling coho salmon (*Oncorhynchus kisutch*) and steelhead trout (*Salmo gairdneri*). *J. Fish. Res. Bd Can.*, **22**, 1035–1081.

Hartmann, J. (1983) Two feeding strategies of young fishes. *Arch. Hydrobiol.*, **96**, 496–509.

Harvey, P.H. and Pagel, M.D. (1991) *The Comparative Method in Evolutionary Biology*, Oxford University Press, Oxford.

Hasler, A.D. and Scholz, A.T. (1983) *Olfactory Imprinting and Homing in Salmon*, Springer-Verlag, Berlin.

Hassell, M.P. (1976) *The Dynamics of Competition and Predation*, Edward Arnold, London.

Hauser, L., Carvalho, G.R. and Pitcher, T.J. (1995) Morphological and genetic differentiation of the African clupeid, *Limnothrissa miodon*, 34 years after its introduction to Lake Kivu. *J. Fish Biol.*, **47** (Suppl. A), 127–144.

Hawkes, H.A. (1975) River zonation and classification, in *River Ecology* (ed. B.A. Whitton), Blackwell, Oxford, pp. 312–374.

Hawkins, A.D. (1993) Underwater sound and fish behaviour, in *Behaviour of Teleost Fishes*, 2nd edn (ed. T.J. Pitcher), Chapman and Hall, London, pp. 129–169.

Hay, D.E. and Brett, J.R. (1988) Maturation and fecundity of Pacific herring (*Clupea harengus pallasi*): an experimental study with comparisons to natural populations. *Can. J. Fish. Aquat. Sci.*, **45**, 399–406.

Hay, D.E., Brett, J.R., Bilinski, E., Smith, D.T., Donaldson, E.M., Hunter, G.A. and Solmie, A.V. (1988) Experimental impoundments of pre-spawning Pacific herring (*Clupea harengus pallasi*): effects of feeding and density on maturation, growth, and proximate analysis. *Can. J. Fish. Aquat. Sci.*, **45**, 388–398.

Hearn, W.E. (1987) Interspecific competition and habitat segregation among stream dwelling trout and salmon: a review. *Fisheries*, **12**, 24–31.

Hearn, W.E. and Kynard, B.E. (1986) Habitat utilization and behavioural interaction of juvenile Atlantic salmon (*Salmo salar*) and rainbow trout (*S. gairdneri*) in tributaries of the White River of Vermont. *Can. J. Fish. Aquat. Sci.*, **43**, 1988–1998.

Heath, M.R. (1992) Field investigations of the early life stages of marine fish. *Adv. Mar. Biol.*, **28**, 1–174.

Heidinger, R.C. and Crawford, S.D. (1977) Effect of temperature and feeding rate on the liver-somatic index of the largemouth bass, *Micropterus salmoides*. *J. Fish. Res. Bd Can.*, **34**, 633–638.

Heisler, N. (1984) Acid–base regulation in fishes, in *Fish Physiology*. Vol. X, Part A (eds W.S. Hoar and D.J. Randall), Academic Press, London, pp. 315–401.

Helfman, G.S. (1978) Patterns of community structure in fishes: summary and overview. *Env. Biol. Fishes*, **3**, 129–148.

Helfman. G.S. (1981) Twilight activities and temporal structure in a freshwater fish community. *Can. J. Fish. Aquat. Sci.*, **40**, 888–894.

Helfman, G.S. (1993) Fish behaviour by day, night and twilight, in *Behaviour of Teleost Fishes*, 2nd edn (ed. T.J. Pitcher), Chapman and Hall, London, pp. 479–512.

Helfman, G.S. and Schultz, E.T. (1984) Social transmission of behavioural traditions in a coral reef fish. *Anim. Behav.*, **32**, 379–384.

Heller, R. and Milinski, M. (1979) Optimal foraging of sticklebacks on swarming prey. *Anim. Behav.*, **27**, 1127–1141.

Henderson, B.A., Wong, J.L. and Nepszy, S.J. (1996) Reproduction of walleye in Lake Erie: allocation of energy. *Can. J. Fish. Aquat. Sci.*, **53**, 127–133.

Hester, F.J. (1964) Effects of food supply on fecundity in the female guppy, *Lebistes reticulatus* (Peters). *J. Fish. Res. Bd Can.*, **21**, 757–764.

Hew, C.L. and Fletcher, G.L. (1979) The role of the pituitary in regulating antifreeze protein synthesis in the winter flounder. *FEBS Lett.*, **99**, 337–339.

Hewett, S.W. and Johnson, B.L. (1989) A general bioenergetics model for fishes. *Am. Fish. Soc. Symp.*, **6**, 206–208.

Higgins, P.J. (1985) An interactive computer program for population estimation using the Zippin method. *Aquacult. Fish. Manage.*, **1**, 287–295.

Higgins, P.J. and Talbot, C. (1985) Growth and feeding in juvenile Atlantic salmon (*Salmo salar*), in *Nutrition and Feeding in Fish* (eds C.B. Cowey, A.M. Mackie and J.G. Bell), Academic Press, London, pp. 243–263.

Hilborn, R. and Walters, C.J. (1992) *Quantitative Fisheries Stock Assessment*, Chapman and Hall, London.

Hill, J. and Grossman, G.D. (1993) An energetic model of microhabitat use for rainbow trout and rosyside dace. *Ecology*, **74**, 685–698.

Hill, M.O. (1973) Diversity and evenness: a unifying notation and its consequences. *Ecology*, **54**, 427–432.

Hindar, K., Jonsson, B., Andrew, J.H. and Northcote, T.G. (1988) Resource utilization of sympatric and experimentally allopatric cutthroat trout and dolly varden. *Oecologia*, **74**, 481–491.

Hirshfield, M.F. (1980) An experimental analysis of reproductive effort and cost in the Japanese medaka, *Oryzias latipes*. *Ecology*, **61**, 282–292.

Hislop, J.R.G. (1984) A comparison of the reproductive tactics and strategies of cod, haddock, whiting and Norway pout in the North Sea, in *Fish Reproduction: Strategies and Tactics* (eds G.W. Potts and R.J. Wootton), Academic Press, London, pp. 311–329.

Hixon, M.A. (1980) Food production and competitor density as the determinants of feeding territory size. *Am. Nat.*, **115**, 510–530.

Hixon, M.A. (1991) Predation as a process structuring coral reef fish communities, in *The Ecology of Fishes on Coral Reefs* (ed. P.F. Sale), Academic Press, London, pp. 475–508.

Hixon, M.A. and Brostoff, W.N. (1996). Succession and herbivory: effects of differential fish grazing on Hawaiian coral-reef algae. *Ecol. Monogr.*, **66**, 67–90.

Hobson, E.S. (1972) Activity of Hawaiian reef fishes during evening and morning transitions between daylight and darkness. *Fish. Bull. Fish Wildl. Serv. US*, **70**, 715–740.

Hobson, E.S. (1973) Diel feeding migrations in tropical reef fishes. *Helgolander wiss. Meeresunters.*, **24**, 361–370.

Hobson, E.S., Chess, J.R. and McFarland, W.N. (1981) Crepuscular and nocturnal activities of Californian nearshore fishes with consideration of their scotopic visual pigments and their photic environment. *Fish. Bull. Fish Wildl. Serv. US*, **79**, 1–30.

Hochachka, P.W. (1980) *Living Without Oxygen*, Harvard University Press, Cambridge, MA.

Hochachka, P.W. and Somero, G.N. (1984) *Biochemical Adaptation*, Princeton University Press, Princeton, NJ.

Hocutt, C.H. and Wiley, E.O. (1986) *The Zoogeography of North American Freshwater Fishes*, J. Wiley, New York.

Hodgkiss, I.J. and Mann, H.S.H. (1978) Reproductive biology of *Sarotherodon mossambicus* (Cichlidae) in Plover Creek Reservoir, Hong Kong. *Env. Biol. Fishes*, **3**, 287–292.

Hofer, R., Krewedl, G. and Koch, F. (1985) An energy budget for an omnivorous cyprinid: *Rutilus rutilus* (L.). *Hydrobiologia*, **122**, 53–59.

Hogendoorn, H. (1983) Growth and production of the African catfish, *Clarias lazera* (C. & V.) III. Bioenergetics relations of body weight and feeding level. *Aquaculture*, **35**, 1–17.

Hogendoorn, H., Jansen, J.A.J., Koops, W.J., Machiels, M.A.M., Ewijk, P.H. van and Hess, J.P. van (1983) Growth and production of the African catfish, *Clarias lazera* (C. & V.) II. Effects of body weight, temperature and feeding level in intensive tank culture. *Aquaculture*, **34**, 265–285.

Hogman, W.J. (1968) Annulus formation on scales of four species of coregonids reared under artificial conditions. *J. Fish. Res. Bd Can.*, **25**, 2111–2112.

Hokanson, K.E.F., Kleiner, C.F. and Thorslund, T.W. (1977) Effects of constant temperatures and diel temperature fluctuations on specific growth and mortality rates and yield of juvenile rainbow trout, *Salmo gairdneri*. *J. Fish. Res. Bd Can.*, **34**, 639–648.

Holbrook, S.J. and Schmitt, R.J. (1984) Experimental analyses of patch selection by foraging black surfperch (*Embiotoca jacksoni* Agazzi). *J. Exp. Mar. Biol. Ecol.*, **79**, 39–64.

Holbrook, S.J. and Schmitt, R.J. (1989) Resource overlap, prey dynamics and the strength of competition. *Ecology*, **70**, 1943–1953.

Holeton, G.F. (1980) Oxygen as an environmental factor of fishes, in *Environmental Physiology of Fishes* (ed. M.A. Ali), Plenum Press, New York, pp. 7–32.

Holling, C.S. (1959) Some characteristics of simple types of predation and parasitism. *Can. Ent.*, **91**, 385–398.

Holopainen, I.J., Aho, J., Vornanen, M. and Fujioka, R.S. (1997) Phenotypic plasticity and predator effects on morphology and physiology of crucian carp in nature and in the laboratory. *J. Fish Biol.*, **50**, 781–798.

Hoogland, R.D., Morris, D. and Tinbergen, N. (1957) The spines of sticklebacks (*Gasterosteus* and *Pygosteus*) as a means of defence against predators (*Perca* and *Esox*). *Behaviour*, **10**, 205–237.

Hopkins, C.A. (1959) Seasonal variation in the incidence and development of the cestode *Proteocephalus filicollis* (Rud, 1810) in *Gasterosteus aculeatus* (L., 1776). *Parasitology*, **49**, 529–542.

Horn, M.H. (1989) Biology of herbivorous fishes. *Oceanogr. Mar. Biol. Ann. Rev.*, **27**, 167–272.

Horn, M.H., and Neighbor, M.A. (1984) Protein and nitrogen assimilation as a factor in predicting the seasonal macroalgal diet of the monkeyface prickleback. *Trans. Am. Fish. Soc.*, **113**, 388–396.

Horwitz, R.J. (1978) Temporal variability patterns and the distributional patterns of stream fishes. *Ecol. Monogr.*, **48**, 307–321.

Horwood, J.W., Bannister, R.C.A. and Howlett, G.J. (1986) Comparative fecundity of North Sea plaice (*Pleuronectes platessa* L). *Proc. R. Soc. B*, **228**, 401–431.

Houde, E.D. (1978) Critical food concentrations for larvae of three species of subtropical marine fishes. *Bull. Mar. Sci.*, **28**, 395–411.

Houston, A.I., McNamara, J.M. and Hutchinson, J.M.C. (1993) General results concerning the trade-off between gaining energy and avoiding predation. *Phil. Trans. R. Soc. Lond. B*, **341**, 375–397.

Howells. G. (1983) Acid waters – the effect of low pH and acid associated factors on fisheries. *Appl. Biol.*, **9**, 143–255.

Howells, G. and Dalziel, T.R.K. (1992) *Restoring Acid Waters: Loch Fleet 1984–1990*, Elsevier, London.

Hubbell. S.P. (1971) Of sowbugs and systems: the ecological bioenergetics of a terrestrial isopod, in *Systems Analysis and Simulation in Ecology*, Vol. 1 (ed. B.C. Patten), Academic Press, London, pp. 269–324.

Huet, M. (1959) Profiles and biology of Western European streams as related to fish management. *Trans. Am. Fish. Soc.*, **88**, 153–163.

Hughes, G.M. (1974) *Comparative Physiology of Vertebrate Respiration*, 2nd edn, Heinemann, London.

Hughes, G.M. (1984) General anatomy of the gills, in *Fish Physiology*, Vol. X, Part A (eds W.S. Hoar and D.J. Randall), Academic Press, London, pp. 1–72.

Hughes, G.M. and Morgan, M. (1973) The structure of fish gills in relation to their respiratory function. *Biol. Rev.*, **48**, 419–475.

Hughes, N.F. (1992) Ranking of feeding positions by drift-feeding Arctic grayling (*Thymallus arcticus*) in dominance hierarchies. *Can. J. Fish. Aquat. Sci.*, **49**, 1994–1998.

Hughes, N.F. and Dill, L.M. (1990) Position choice by drift-feeding salmonids: model and test for Arctic grayling (*Thymallus arcticus*) in subarctic mountain streams, Interior Alaska. *Can. J. Fish. Aquat. Sci.*, **47**, 2039–2048.

Hume, J.M.B. and Northcote, T.G. (1985) Initial changes in the use of space and food by experimentally segregated populations of dolly varden (*Salvelinus malma*) and cutthroat trout (*Salmo clarki*). *Can. J. Fish. Aquat. Sci.*, **42**, 101–109.

Humphreys, W.F. (1979) Production and respiration in animal populations. *J. Anim. Ecol.*, **48**, 427–454.

Hunter, J.R. (1972) Swimming and feeding behaviour of larval anchovy, *Engraulis mordax*. *Fish. Bull. Fish Wildl. Serv. US*, **70**, 821–838.

Hunter, J.R. (1981) Feeding ecology and predation of marine fish larvae, in *Marine Fish Larvae* (ed. R. Lasker), Washington Sea Grant Program, Seattle, WA, pp. 33–77.

Hunter, J.R. and Leong, R. (1981) The spawning energetics of female northern anchovy, *Engraulis mordax*. *Fish. Bull. Fish Wildl. Serv. US*, **79**, 215–230.

Hunter, J.R. and Macewicz, B.J. (1985) Measurement of spawning frequency in multiple spawning fishes. *NOAA Tech. Rep. NMFS SSRF*, **36**, 79–94.

Hunter, J.R., Macewicz, B.J., Lo, N.C.-H. and Kimbrell, C.A. (1992) Fecundity, spawning, and maturity of female Dover sole *Microstomus pacificus*, with an evaluation of assumptions and precision. *Fish. Bull. US*, **90**, 101–128.

Huntingford, F.A. (1993) Can cost–benefit analysis explain fish distribution patterns? *J. Fish Biol.*, **43** (Suppl. A), 289–308.

Huntingford, F.A., Wright, P.J. and Tierney, J.F. (1994) Adaptive variation in antipredator behaviour in threespine stickleback, in *The Evolutionary Biology of the Threespine Stickleback* (eds M.A. Bell and S.A. Foster), Oxford University Press, Oxford, pp. 277–296.

Hurlbert, S.H. (1978) The measurement of niche overlap and some relatives. *Ecology*, **59**, 67–77.

Hutchings, J.A. (1991) Fitness consequences of variation in egg size and food abundance in brook trout, *Salvelinus fontinalis*. *Evolution*, **45**, 1162–1168.

Hutchings, J.A. (1993a) Behavioural implications of intraspecific life history variation. *Mar. Behav. Physiol.*, **23**, 187–203.

Hutchings, J.A. (1993b) Adaptive life histories as

effected by age-specific survival and growth rate. *Ecology*, **74**, 673–684.

Hutchings, J.A. (1994) Age- and size-specific costs of reproduction within populations of brook trout, *Salvelinus fontinalis*. *Oikos*, **70**, 12–20.

Hutchings, J.A. and Myers, R.A. (1994) The evolution of alternative mating strategies in variable environments. *Evol. Ecol.*, **8**, 256–269.

Hutchinson, G.E. (1958) Concluding remarks. *Cold Spring Harb. Symp. Quant. Biol.*, **22**, 415–427.

Hynes, H.B.N. (1950) The food of freshwater sticklebacks (*Gasterosteus aculeatus* and *Pygosteus pungitius*) with a review of methods used in studies of the food of fishes. *J. Anim. Ecol.*, **19**, 36–58.

Hynes, H.B.N. (1970) *The Ecology of Running Waters*, Liverpool University Press, Liverpool.

Hyslop, E.J. (1980) Stomach contents analysis – a review of methods and their application. *J. Fish Biol.*, **17**, 411–429.

Ibrahim, A.A. and Huntingford, F.A. (1989) The role of visual cues in prey selection in three-spined sticklebacks (*Gasterosteus aculeatus*). *Ethology*, **81**, 265–272.

Iersel, J.J.A. van (1953) An analysis of the parental behaviour of the male three-spined stickleback (*Gasterosteus aculeatus* L.). *Behaviour Suppl.*, **3**, 1–159.

Iles, T.C. (1994) A review of stock–recruitment relationships with reference to flatfish populations. *Neth. J. Sea Res.*, **32**, 399–420.

Iles, T.C. and Beverton, R.J.H. (1991) Mortality rates of 0-group plaice (*Pleuronectes platessa* L.), dab (*Limanda limanda* L.) and turbot (*Scophthalmus maximus* L.) in European waters. I. Statistical estimation of the data and estimation of parameters. *Neth. J. Sea Res.*, **27**, 217–235.

Iles, T.D. (1974) The tactics and strategy of growth in fishes, in *Sea Fisheries Research* (ed. F.R. Harden Jones), Elek, London, pp. 331–345.

Iles, T.D. (1984) Allocation of resources to gonad and soma in Atlantic herring *Clupea harengus* L, in *Fish Reproduction: Strategies and Tactics* (eds G.W. Potts and R.J. Wootton), Academic Press, London, pp. 331–347.

Ivlev, V.W. (1961) *Experimental Ecology of the Feeding of Fishes*, Yale University Press, New Haven, CT.

Jenkins, T.M. (1969) Social structure, position choice and microdistribution of two trout species (*Salmo trutta* and *Salmo gairdneri*) resident in mountain streams. *Anim. Behav. Monogr.*, **2**, 57–123.

Jobling, M. (1980) Gastric evacuation in plaice,

Pleuronectes platessa L.: effects of dietary energy level and food composition. *J. Fish Biol.*, **17**, 187–196.

Jobling, M. (1981) The influence of feeding on the metabolic rate of fishes: a short review. *J. Fish Biol.*, **18**, 385–400.

Jobling, M. (1983a) Towards an explanation of specific dynamic action (SDA). *J. Fish Biol.*, **23**, 549–555.

Jobling, M. (1983b) Growth studies with fish – overcoming the problems of size variation. *J. Fish Biol.*, **22**, 153–157.

Jobling, M. (1985a) Growth, in *Fish Energetics: New Perspectives* (eds P. Tytler and P. Calow), Croom Helm, London, pp. 213–230.

Jobling, M. (1985b) Physiological and social constraints on growth of fish with special reference to Arctic charr, *Salvelinus alpinus* L. *Aquaculture*, **44**, 83–90.

Jobling, M. (1986) Mythical models of gastric emptying and implications for food consumption studies. *Env. Biol. Fishes*, **16**, 35–50.

Jobling, M. (1994) *Fish Bioenergetics*, Chapman and Hall, London.

Jobling, M. (1995) *Environmental Biology of Fishes*, Chapman and Hall, London.

Jobling, M. and Baardvik, B.M. (1994) The influence of environmental manipulations on inter- and intra-individual variation in food acquisition and growth performance of Arctic charr, *Salvelinus alpinus*. *J. Fish Biol*, **44**, 1069–1087.

Jobling, M. and Koskela, J. (1996) Interindividual variations in feeding and growth in rainbow trout during restricted feeding and in a subsequent period of compensatory growth. *J. Fish Biol.*, **49**, 658–667.

Jobling, M. and Spencer Davies, P. (1980) Effects of feeding on metabolic rate and specific dynamic action in plaice, *Pleuronectes platessa* L. *J. Fish Biol.*, **16**, 629–638.

Johnsen, B.O. and Jensen, A.J. (1991) The *Gyrodactylus* story in Norway. *Aquaculture*, **97**, 289–302.

Johnson, T.P. and Bennett, A.F. (1995) The thermal acclimation of burst speed performance: an integrated study of molecular and cellular physiology and organismal performance. *J. Exp. Biol.*, **198**, 2165–2175.

Johnston, I.A. (1993) Phenotypic plasticity of fish muscle to temperature change, in *Fish Ecophysiology* (eds J.C. Rankin and F.B. Jensen), Chapman and Hall, London, pp. 322–340.

Jones, G.P. (1987) Competitive interactions among

adults and juveniles in coral reef fish. *Ecology*, **68**, 1534–1547.

Jones, G.P. (1991) Postrecruitment processes in the ecology of coral reef fish populations: a multifactorial perspective, in *The Ecology of Fishes on Coral Reefs* (ed. P.F. Sale), Academic Press, London, pp. 294–328.

Jones, J.W. (1959) *The Salmon*, Collins, London.

Jones, R. and Hall, W.B. (1974) Some observations on the population dynamics of the larval stage in the common gadoids, in *The Early Life History of Fish* (ed. J.H.S. Blaxter), Springer-Verlag, Berlin, pp. 3–19.

Jonsson, B. and L'Abee-Lund, J.H. (1993) Latitudinal clines in life-history variables of anadromous brown trout in Europe. *J. Fish. Biol.*, **43** (Suppl. A), 1–16.

Jonsson, B. and Jonsson, N. (1993) Partial migration: niche shift versus sexual maturation in fishes. *Rev. Fish Biol. Fisheries*, **3**, 348–365.

Jonsson, N., Hansen, L.P. and Jonsson, B. (1991) Variation in age, size and repeat spawning of adult Atlantic salmon in relation to river discharge. *J. Anim. Ecol.*, **60**, 937–947.

Jonsson, N., Jonsson, B. and Hansen, L.P. (1997) Changes in proximate composition and estimates of energetic costs during upstream migration and spawning in Atlantic salmon *Salmo salar*. *J. Anim. Ecol.*, **66**, 425–436.

Kamler, E. (1992) *Early Life History of Fish*, Chapman and Hall, London.

Kapoor, B.G., Smit, H. and Verighina, I.A. (1975) The alimentary canal and digestion in teleosts. *Adv. Mar. Biol.*, **13**, 109–239.

Keast, A.J. (1970) Food specializations and bioenergetic inter-relations in the fish fauna of some small Ontario waterways, in *Marine Food Chains* (ed. J.J. Steele), Oliver and Boyd, Edinburgh, pp. 377–411.

Keast, A. (1978) Trophic and spatial interrelationships in the fish species of an Ontario temperate lake. *Env. Biol. Fishes*, **3**, 7–31.

Keast, A. (1980) Food and feeding relationships of young fish in the first weeks after the beginning of exogenous feeding in Lake Opinicon, Ontario. *Env. Biol. Fishes*, **5**, 305–314.

Keast, A. (1985) Development of dietary specializations in a summer community of juvenile fishes. *Env. Biol. Fishes*, **13**, 211–224.

Keast, A. and Eadie, J. (1984) Growth in the first summer of life: comparison of nine co-occurring fish species. *Can. J. Zool.*, **62**, 1242–1250.

Keast, A. and Webb, D. (1966) Mouth and body form relative to feeding ecology in the fish fauna of a small lake, Lake Opinicon, Ontario. *J. Fish. Res. Bd Can.*, **23**, 1845–1874.

Keast, A. and Welsh, L. (1968) Daily feeding periodicities, food uptake rates and dietary changes with hour of day in some lake fishes. *J. Fish. Res. Bd Can.*, **25**, 1133–1144.

Kedney, G.I., Boule, V. and FitzGerald, G.J. (1987) The reproductive ecology of threespine sticklebacks breeding in fresh and brackish water. *Am. Fish. Soc. Symp.*, **1**, 151–161.

Keeley, E.R. and Grant, J.W.A. (1995) Allometric and environmental correlates of territory size in juvenile Atlantic salmon (*Salmo salar*). *Can. J. Fish. Aquat. Sci.*, **52**, 186–196.

Keenleyside, M.H.A. (1979) *Diversity and Adaptation in Fish Behaviour*, Springer Verlag, Berlin.

Keenleyside, M.H.A. (1991) Parental care, in *Cichlid Fishes: Behaviour, Ecology and Evolution* (ed. M.H.A. Keenleyside), Chapman and Hall, London, pp. 191–208.

Kerr, S.R. (1980) Niche theory in fisheries ecology. *Trans. Am. Fish. Soc.*, **109**, 254–260.

Kerr, T. (1948) The pituitary in normal and parasitized roach (*Leuciscus rutilus* Flem.). *Q. J. Microsc. Sci.*, **89**, 129–137.

Ketola, H.G. (1978) Nutritional requirements and feeding of selected coolwater fishes: a review. *Progve Fish-cult.*, **40**, 127–132.

Kinne, O. (1960) Growth, food intake and food conversion in a euryplastic fish exposed to different temperatures and salinities. *Physiol. Zool.*, **33**, 288–317.

Kislalioglu, M. and Gibson, R.N. (1976a) Some factors governing prey selection by the 15-spined stickleback (*Spinachia spinachia* L.). *J. Exp. Mar. Biol. Ecol.*, **25**, 159–169.

Kislalioglu, M. and Gibson, R.N. (1976b) Prey 'handling time' and its importance in food selection by the 15-spined stickleback, *Spinachia spinachia* L. *J. Exp. Mar. Biol. Ecol.*, **25**, 115–158.

Kitchell, J.F. (1983) Energetics, in *Fish Biomechanics* (eds P.W. Webb and D. Weihs), Praeger, New York, pp. 312–338.

Kitchell, J.F. and Breck, J.E. (1980) Bioenergetics model and foraging hypothesis for sea lamprey (*Petromyzon marinus*). *Can. J. Fish. Aquat. Sci.*, **37**, 2159–268.

Kitchell, J.F. and Carpenter, S.R. (1993) Synthesis and new directions, in *The Trophic Cascade in Lakes* (eds S.R. Carpenter and J.F. Kitchell), Cambridge University Press, Cambridge, pp. 332–350.

Kitchell, J.F. and Crowder, L.B. (1986) Predator–prey interactions in Lake Michigan: model predictions and recent dynamics. *Env. Biol. Fishes*, **16**, 205–211.

Kitchell, J.F., Stewart, D.J. and Weininger, D. (1977) Applications of a bioenergetics model to yellow perch (*Perca flavescens*) and walleye (*Stizostedion vitreum vitreum*). *J. Fish. Res. Bd Can.*, **34**, 1922–1935.

Kjesbu, O.S. and Holm, J.C. (1994) Oocyte recruitment in first-time spawning Atlantic cod (*Gadus morhua*) in relation to feeding regime. *Can. J. Fish. Aquat. Sci.*, **51**, 1893–1898.

Kjesbu, O.S., Klungsoyr, J., Kryvi, H., Witthames, P.R. and Greer Walker, M. (1991) Fecundity, atresia, and egg size of captive Atlantic cod (*Gadus morhua*) in relation to proximate body composition. *Can. J. Fish. Aquat. Sci.*, **48**, 2333–2343.

Kjorsvik, E., Mangor-Jensen, A. and Holmefjord, I. (1990) Egg quality in fishes. *Adv. Mar. Biol.*, **26**, 71–113.

Klinger, S.A., Magnuson, J.J. and Gallepp, G.W. (1982) Survival mechanisms of the central mudminnow (*Umbra limi*), fathead minnow (*Pimephales promelas*) and brook stickleback (*Culea inconstans*) for low oxygen in winter. *Env. Biol. Fishes*, **7**, 113–120.

Koch, F. and Wieser, W. (1983) Partitioning of energy in fish: can reduction of swimming activity compensate for the cost of production. *J. Exp. Biol.*, **107**, 141–146.

Kodric-Brown, A. and Brown, J.H. (1993) Highly structured fish communities in Australian desert springs. *Ecology*, **74**, 1847–1855.

Kohler, C.C. and Ney, J.J. (1982) A comparison of methods for quantitative analysis of feeding selection of fishes. *Env. Biol. Fishes*, **7**, 363–368.

Kooijman, S.A.L.M. (1993) *Dynamic Energy Budgets in Biological Systems*, Cambridge University Press, Cambridge.

Koslow, J.A. (1992) Fecundity and the stock–recruitment relationship. *Can. J. Fish. Aquat. Sci.*, **49**, 210–217.

Kotrschal, K., Adam, H., Brandstätter, R., Junger, H., Zaunreiter, M. and Goldschmid, A. (1990) Larval size constraints determine directional ontogenetic shifts in the visual system of teleosts. *Z. zool. Syst. Evolut.-forsch.*, **28**, 166–182.

Kozlowski, J. (1991) Optimal energy allocation models – an alternative to concepts of reproductive effort and cost of reproduction. *Acta Oecol.*, **12**, 11–33.

Kraak, S.B.M. and Videler, J.J. (1991) Mate choice in *Aidablennius sphynx* (Teleostei, Blennidae); females prefer nests containing more eggs. *Behaviour*, **119**, 243–266.

Kramer, D.L. (1978) Reproductive seasonality in the fishes of a tropical stream. *Ecology*, **59**, 976–985.

Kramer, D.L. (1983) The evolutionary ecology of respiratory mode in fishes: an analysis based on the cost of breathing. *Env. Biol. Fishes*, **9**, 145–158.

Kramer, D.L.. (1987) Dissolved oxygen and fish behaviour. *Env. Biol. Fishes.*, **18**, 81–92.

Kramer, D.L. and Bryant, M.J. (1995a). Intestine length in the fishes of a tropical stream: 1. Ontogenetic allometry. *Env. Biol. Fishes*, **42**, 115–127.

Kramer, D.L. and Bryant, M.J. (1995b) Intestine length in the fishes of a tropical stream: 2. Relationships to diet – the long and short of a convoluted issue. *Env. Biol. Fishes*, **42**, 129–141.

Kramer, D.L., Lindsey, C.C., Moodie, G.E.E. and Stevens, E.D. (1978) The fishes and the aquatic environment of the central Amazon Basin, with particular reference to respiratory patterns. *Can. J. Zool.*, **56**, 717–729.

Kramer, D.L., Rangeley, R.W. and Chapman, L.J. (1991) Habitat selection: patterns of spatial distribution from behavioural decisions, in *Behavioural Ecology of Fishes* (ed. J.-G.J. Godin) Oxford University Press, Oxford, pp. 37–80..

Krause, J. (1993) Positioning in fish shoals: a cost–benefit analysis. *J. Fish Biol*, **43** (Suppl. A), 309–314.

Krause, J. (1994) Differential fitness returns in relation to spatial position in groups. *Biol. Rev.*, **69**, 187–206.

Krebs, C.J. (1994) *Ecology: The Experimental Analysis of Distribution and Abundance*, 4th edn, Harper and Row, New York.

Krebs, J.R. and Davies, N.B. (1993) *An Introduction to Behavioural Ecology*, Blackwell, Oxford.

Krebs, J.R. and McCleery, R.H. (1984) Optimization in behavioural ecology, in *Behavioural Ecology. An Evolutionary Approach* (eds J.R. Krebs and N.B. Davies), Blackwell, Oxford, pp. 91–121.

Krohn, M.M. and Boisclair, D. (1994) Use of a stereo-video system to estimate the energy expenditure of free-swimming fish. *Can. J. Fish. Aquat. Sci.*, **51**, 1119–1127.

Lafferty, K.D. and Morris, A.K. (1996) Altered behaviour of parasitized killifish increases susceptibility to predation. *Ecology*, **77**, 1390–1397.

Lagler, K.F. (1971) Capture, sampling and examination of fishes, in *Methods of Assessment of Fish Production in Fresh Waters* (ed. W.E. Ricker), Blackwell, Oxford, pp. 7–44.

Lagomarsino, I.V. and Conover, D.O. (1993) Variation in environmental and genotypic sex-determining mechanisms across a latitudinal gradient in the fish, *Menidia menidia. Evolution,* **47**, 487–494.

Lam, T.J. (1983) Environmental influences on gonadal activity in fish, in *Fish Physiology*, Vol. IX, Part B (eds W.S. Hoar, D.J. Randall and E.M. Donaldson), Academic Press, London, pp. 65–116.

Langeland, A. and Nost, T. (1995). Gill raker structure and selective predation on zooplankton by particulate feeding fish. *J. Fish Biol.,* **47**, 719–732.

Lasker, R. (1975) Field criteria for survival of anchovy larvae: the relation between inshore chlorophyll maximum layers and successful first feeding. *Fish. Bull. Fish Wildl. Serv. US,* **73**, 453–462.

Lasker, R. (ed.) (1985) An egg production method for estimating spawning biomass of pelagic fish: application to the northern anchovy, *Engraulis mordax. NOAA Tech. Rep. NWFS* No. **36**, 1–99.

Lauder, G.V. (1983) Food capture, in *Fish Biomechanics* (eds P.W. Webb and D. Weihs), Praeger, New York, pp. 280–311.

Laurence, G.C. (1981) Modelling – an esoteric or potentially utilitarian approach to understanding larval fish dynamics. *Rapp. P.-v. Réun. Cons. Perm. Int. Explor. Mer,* **178**, 3–6.

Lavin, P.A. and McPhail, J.D. (1985) The evolution of freshwater diversity in the threespine stickleback (*Gasterosteus aculeatus*): site-specific differentiation of trophic morphology. *Can. J. Zool.,* **63**, 2632–2638.

Lavin, P.A. and McPhail, J.D. (1986) Adaptive divergence of trophic phenotype among freshwater populations of the threespine stickleback (*Gasterosteus aculeatus*). *Can. J. Fish. Aquat. Sci.,* **43**, 2455–2463.

Law, R. and Grey, D.R. (1989) Evolution of yields from populations with age-specific cropping. *Evol. Ecol.,* **3**, 343–359.

Lazzaro, X. (1987) A review of planktivorous fishes: their evolution, feeding behaviours, selectivities, and impacts. *Hydrobiologia,* **146**, 97–167.

Leatherland, J.F. (1994) Reflections on the thyroidology of fishes: from molecules to humankind. *Guelph Ichthyol. Rev.,* **2**, 1–67.

Lechowicz, M.J. (1982) The sampling characteristics of electivity indices. *Oecologia,* **52**, 22–30.

LeCren, E.D. (1947) The determination of the age and growth of the perch (*Perca fluviatilis*) from the opercular bone. *J. Anim. Ecol.,* **16**, 188–204.

LeCren, E.D. (1951) The length–weight relationship and seasonal cycle in gonad weight and condition in the perch (*Perca fluviatilis*). *J. Anim. Ecol.,* **20**, 201–219.

LeCren, E.D. (1987) Perch (*Perca fluviatilis*) and pike (*Esox lucius*) in Windermere from 1940 to 1985: studies in population dynamics. *Can. J. Fish. Aquat. Sci.,* **44** (Suppl. 2), 216–228.

Leggett, W.C. (1977) The ecology of fish migrations. *Ann. Rev. Ecol. Syst.,* 8, 285–308.

Leggett, W.C. (1986) The dependence of fish larval survival on food and predator densities, in *The Role of Freshwater Outflow in Coastal Marine Ecosystems* (ed. S. Skreslet), Springer-Verlag, Berlin, pp. 117–137.

Leggett, W.C. and Carscadden, J.E. (1978) Latitudinal variation in reproductive characteristics of American shad (*Alosa sapidissima*): evidence for population specific life history strategies in fish. *J. Fish. Res. Bd Can.,* **35**, 1469–1478.

Leggett, W.C. and DeBlois, E. (1994) Recruitment in marine fishes: is it regulated by starvation and predation in the egg and larval stages? *Neth. J. Sea Res.,* **32**, 119–134.

Lester, R.J.G. (1971) The influence of *Schistocephalus* plerocercoids on the respiration of *Gasterosteus* and a possible resulting effect on the behaviour of the fish. *Can. J. Zool.,* **49**, 361–366.

Lester, R.J.G. (1984) A review of methods for estimating mortality due to parasites in wild fish populations. *Helgolander Wiss. Meeresunters.,* **37**, 53–64.

Levine, J.S., Lobel, P.S., and MacNichol, E.F., Jr (1980) Visual communication in fishes, in *Environmental Physiology of Fishes* (ed. M.A. Ali), Plenum, New York, pp. 447–475.

Levins, R. (1968) *Evolution in Changing Environments*, Princeton University Press, Princeton, NJ.

Lewontin, R.C. (1974) *The Genetic Basis of Evolutionary Change*, Columbia University Press, New York.

Li, H.W. and Brocksen, R.W. (1977) Approaches to the analysis of energetic costs of intraspecific competition for space by rainbow trout (*Salmo gairdneri*). *J. Fish Biol.,* **11**, 329–341.

Li, H.W., Schreck, C.B., Bond, C.E. and Rexstad, E. (1987) Factors influencing changes in fish assemblages of Pacific northwest streams, in

Community and Evolutionary Ecology of North American Stream Fishes (eds. W.J. Matthews and D.C. Heins), University of Oklahoma Press, Normal, OK, pp. 193–202.

Li, K.T., Wetterer, J.K. and Hairston, N.G. Jr (1985) Fish size, visual resolution, and prey selectivity. *Ecology*, **66**, 1729–1735.

Li, S. and Mathias, J.A. (1982) Causes of high mortality among cultured larval walleyes. *Trans. Am. Fish. Soc.*, **111**, 710–721.

Liem, K.F. (1973) Evolutionary strategies and morphological innovations: cichlid pharyngeal jaws. *Syst. Zool.*, **22**, 425–441.

Liem, K.F. (1980) Acquisition of energy by teleosts: adaptive mechanisms and evolutionary patterns, in *Environmental Physiology of Fishes* (ed. M.A. Ali), Plenum, New York, pp. 299–334.

Liem, K.F. (1991) Functional morphology, in *Cichlid Fishes: Behaviour, Ecology and Evolution* (ed. M.H.A. Keenleyside), Chapman and Hall, London, pp. 129–150.

Liley, R.N. and Seghers, B.H. (1975) Factors affecting the morphology and behaviour of guppies in Trinidad, in *Evolution and Behaviour* (eds G.P. Baerends, C. Beer and A. Manning), Oxford University Press, Oxford, pp. 92–118.

Lima, S.L. and Dill, L.M. (1990) Behavioural decisions made under risk of predation: a review and prospectus. *Can. J. Zool.*, **68**, 619–640.

Limbaugh, C. (1961) Cleaning symbiosis. *Scient. Am.*, **205**, 42–49.

Lindsey, C.C. (1978) Form, function and locomotory habits in fish, in *Fish Physiology*, Vol. VII (eds W.S. Hoar and D.J. Randall), Academic Press, London, pp. 1–100.

Linfield, R.S.J. (1985) An alternative concept to home range theory with respect to populations of cyprinids in major river systems. *J. Fish Biol.*, **27** (Suppl. A), 187–196.

Lister, D.B. and Genoe, H.S. (1970) Stream habitat utilization by cohabiting underyearlings of chinook (*Oncorhynchus tshawytscha*) and coho (*O. kisutch*) salmon in the Big Qualicum River, British Columbia. *J. Fish. Res. Bd Can.*, **27**, 1215–1224.

Lo, N.C.H. (1986) Modeling life-stage-specific instantaneous mortality rates, an application to northern anchovy, *Engraulis mordax*, eggs and larvae. *Fish. Bull. Fish Wildl. Serv. US*, **84**, 395–407.

Loew, E.R. and McFarland, W.N. (1990) The underwater visual environment, in *The Visual System of Fish* (eds R.H. Douglas and M.B.A. Djamgoz), Chapman and Hall, London, pp. 1–43.

Loiselle, P.V. and Barlow, G.W. (1978) Do fishes lek like birds?, in *Contrasts in Behaviour* (eds E.S. Reese and F.J. Lighter), John Wiley, New York, pp. 31–73.

Łomnicki, A. (1988) *Population Ecology of Individuals*, Princeton University Press, Princeton, NJ.

Longhurst, A.R. and Pauly, D. (1987) *Ecology of Tropical Oceans*, Academic Press, London.

Losey, G.S. Jr (1972) The ecological importance of cleaning symbiosis. *Copeia*, **1972**, 820–833.

Lotrich, V.A. (1973) Growth, production and community composition of fishes inhabiting a first, second and third order stream of eastern Kentucky. *Ecol. Monogr.*, **43**, 377–397.

Love, R.M. (1980) *The Chemical Biology of Fishes*, Vol. 2, Academic Press, London.

Low, R.M. (1971) Interspecific territoriality in a pomacentrid reef fish, *Pomacentrus flavicauda* Whitely. *Ecology*, **52**, 648–654.

Lowe-McConnell, R.M. (1975) *Fish Communities in Tropical Freshwaters*, Longman, London.

Lowe-McConnell, R.M. (1979) Ecological aspects of seasonality in fishes of tropical waters. *Symp. Zool. Soc. Lond.*, **44**, 219–241.

Lowe-McConnell, R.M. (1987) *Ecological Studies in Tropical Fish Communities*, Cambridge University Press, Cambridge.

Lucas, A. (1996) *Bioenergetics of Aquatic Animals*, Taylor and Francis, London.

Lucas, M.C., Johnstone, A.D. and Priede, I.G. (1993) Use of physiological telemetry as a method of estimating metabolism of fish in the natural environment. *Trans. Am. Fish. Soc.*, **122**, 822–833.

Lythgoe, J.N. (1979) *The Ecology of Vision*, Oxford University Press, Oxford.

MacArthur, R.H. and Wilson, E.O. (1967) *The Theory of Island Biogeography*, Princeton University Press, Princeton, NJ.

MacCall, A.D. (1990) *Dynamic Geography of Marine Fish Populations*, University of Washington Press, Seattle.

McCarthy, I.D., Carter, C.G. and Houlihan, D.F. (1992) The effect of feeding hierarchy on individual variability in daily feeding in rainbow trout, *Oncorlynchus mykiss* (Walbaum). *J. Fish Biol.*, **41**, 257–263.

McCleave, J.D., Arnold, G.P., Dodson, J.J. and Neill, W.H. (eds) (1984) *Mechanisms of Migration in Fishes*, Plenum, New York.

McComas, S.R. and Drenner, R.W. (1982) Species

replacement in a reservoir fish community: silverside feeding mechanisms and competition. *Can. J. Fish. Aquat. Sci.*, **39**, 815–821.

MacDonald, J.A., Montgomery, J.C. and Wells, R.M.G. (1987) Comparative physiology of Antarctic fishes. *Adv. Mar. Biol*, **24**, 321–388.

MacDonald, J.F., Bekkers, J., MacIsaac, S.M. and Blouw, D.M. (1995) Intertidal breeding and aerial development of embryos of a stickleback fish (*Gasterosteus*). *Behaviour*, **132**, 1183–1206.

MacDonald, J.S. and Green, R.H. (1986) Food resource utilization by five species of benthic feeding fish in Passamaquoddy Bay, New Brunswick. *Can. J. Fish. Aquat. Sci.*, **43**, 1534–1546.

MacDonald, P.D.M. and Pitcher, T.J. (1979) Age-groups from size-frequency data: a versatile and efficient method of analysing distribution mixtures. *J. Fish. Res. Bd Can.*, **36**, 987–1001.

McDowall, R.M. (1987) The occurrence and distribution of diadromy among fishes. *Am. Fish. Soc. Symp.*, **1**, 1–13.

McFadden, J.T., Alexander, G.R. and Shetter, D.S. (1967) Numerical changes and population regulation in brook trout *Salvelinus fontinalis*. *J. Fish. Res. Bd Can.*, **24**, 1425–1459.

McFarland, W.N. (1991) The visual world of coral reef fishes, in *The Ecology of Fishes of Coral Reefs* (ed. P.F. Sale), Academic Press, London, pp. 16–38.

McFarland, W.N. and Munz, F.W. (1976) The visible spectrum during twilight and its implications to vision, in *Light as an Ecological Factor: II* (eds G.C. Evans, R. Bainbridge and O. Rackham), Blackwell, Oxford, pp. 249–270.

McGurk, M.D. (1986) Natural mortality of marine pelagic eggs and larvae: role of spatial patchiness. *Mar. Ecol. Progr. Ser.*, **34**, 227–242.

Machiels, M.A.M. and Henken, A.M. (1986) A dynamic simulation model of growth of the African catfish, *Clarias gariepinus* (Burchell) I. Effect of feeding level on growth and energy metabolism. *Aquaculture*, **56**, 29–52.

McIntosh, R.P. (1995) H.A. Gleason's 'individualistic concept' and theory of animal communities: a continuing controversy. *Biol. Rev.*, **70**, 317–357.

McKay, L.R. and Gjerde, B. (1985) The effect of salinity on growth of rainbow trout. *Aquaculture*, **49**, 325–331.

McKaye, K.R. (1977) Competition for breeding sites between the cichlid fishes of Lake Jiloa, Nicaragua. *Ecology*, **58**, 291–302.

McKaye, K.R. (1984) Behavioural aspects of cichlid reproductive strategies: patterns of territoriality and brood defence in Central American substratum spawners and African mouth brooders, in *Fish Reproduction: Strategies and Tactics* (eds G.W. Potts and R.J. Wootton), Academic Press, London, pp. 245–273.

McKaye, K.R. and Gray, W.N. (1984) Extrinsic barriers to gene flow in rock-dwelling cichlids of Lake Malawi: macrohabitat heterogeneity and reef colonisation, in *Evolution of Fish Species Flocks* (eds A.A. Echelle and I. Kornfield), University of Maine Press, Orono, ME, pp. 169–183.

MacKenzie, K. (1987) Relationships between the herring, *Clupea harengus* L., and its parasites. *Adv. Mar. Biol.*, **24**, 263–319.

McKenzie, W.D., Crews, D., Kallman, K.D., Policansky, D. and Sohn, J.J. (1983) Age, weight and the genetics of sexual maturation in the platyfish, *Xiphophorus maculatus*. *Copeia*, 1983, 770–774.

McKeown, B.A. (1984) *Fish Migration*, Croom Helm, London.

MacKinnon, J.C. (1973) Analysis of energy flow and production in an unexploited marine flatfish population. *J. Fish. Res. Bd Can.*, **30**, 1717–1728.

McLaughlin, R.L., Grant, J.W.A. and Kramer, D.L. (1992) Individual variation and alternative patterns of foraging movements in recently-emerged brook charr (*Salvelinus fontinalis*). *Behaviour*, **120**, 286–301.

MacLean, J.A. and Evans, D.O. (1981) The stock concept, discreteness of fish stocks, and fisheries management. *Can. J. Fish. Aquat. Sci.*, **38**, 1889–1898.

MacLennan, D.N. and Simmonds, E.J. (1992). *Fisheries Acoustics*, Chapman and Hall, London.

McNamara, J.M. and Houston, A.I. (1996) State-dependent life histories. *Nature*, **380**, 215–221.

McPhail, J.D. (1994) Speciation and the evolution of reproductive isolation in the sticklebacks (*Gasterosteus*) of south-western British Columbia, in *The Evolutionary Biology of the Threespine Stickleback* (eds M.A. Bell and S.A. Foster), Oxford University Press, Oxford, pp. 399–437.

McQueen, D.J., Post, J.R. and Mills, E.L. (1986) Trophic relationships in freshwater pelagic ecosystems. *Can. J. Fish. Aquat. Sci.*, **43**, 1571–1581.

McQuinn, I.H. (1997) Metapopulations and the Atlantic herring. *Rev. Fish Biol. Fisheries*, **7**, 297–329.

Madenjian, C.P., Carpenter, S.R., Eck, G.W. and

Miller, M.A. (1993) Accumulation of PCBs by lake trout (*Salvelinus namaycush*): an individual-based model approach. *Can. J. Fish. Aquat. Sci.*, **50**, 97–109.

Magnuson, J.J. (1962) An analysis of aggressive behaviour, growth, and competition for food and space in medaka (*Oryzias latipes* (Pisces, Cyprinodontidae)). *Can. J. Zool.*, **40**, 313–363.

Magnuson, J.J., Crowder, L.B. and Medvick, P.A. (1979) Temperature as an ecological resource. *Am. Zool.*, **19**, 331–343.

Magurran, A. E. (1988) *Ecological Diversity and its Measurement*, Croom Helm, London.

Magurran, A.E. and Pitcher, T.J. (1983) Foraging, timidity and shoal size in minnows and goldfish. *Behav. Ecol. Sociobiol.*, **12**, 142–152.

Magurran, A.E., Oulton, W. and Pitcher, T.J. (1985) Vigilant behaviour and shoal size in minnows. *Z. Tierpsychol.*, 67, 167–178.

Magurran, A.E., Seghers, B., Shaw, P. and Carvalho, G. (1995) Evolutionary basis of the behavioural diversity of the guppy, *Poecilia reticulata*, populations in Trinidad. *Adv. Study Behav.*, **24**, 155–202.

Mahon, R. (1984) Divergent structure in fish taxocenes of north temperate streams. *Can. J. Fish. Aquat. Sci.*, **41**, 330–350.

Majkowski, J. and Waiwood, K.G. (1981) A procedure for evaluating the food biomass consumed by a fish population. *Can. J. Fish Aquat. Sci.*, **38**, 1199–1208.

Major, P.F. (1978) Predator–prey interactions in two schooling fishes, *Caranx ignobilis* and *Stolephorus purpureus*. *Anim. Behav.*, **26**, 760–777.

Mandrak, N.E. (1995) Biogeographic patterns of fish species richness in Ontario lakes in relation to historical and environmental factors. *Can. J. Fish. Aquat. Sci.*, **52**, 1462–1474.

Manly, B.F.J. (1994) *Multivariate Statistical Methods*, 2nd edn. Chapman and Hall, London.

Mann, K.H. (1965) Energy transformations by a population of fish in the River Thames. *J. Anim. Ecol.*, **34**, 253–275.

Mann, K.H., Britton, R.H., Kowakzewski, A., Lack, T.J., Mathews, C.P. and McDonald, I. (1972) Productivity and energy flow at all trophic levels in the River Thames, England, in *Productivity Problems of Freshwaters* (eds Z. Kajak and A. Hillbricht-Ilkowska), PWN, Warsaw, pp. 579–596.

Mann, R.H.K. (1971) The populations, growth and production of fish in four small streams in southern England. *J. Anim. Ecol.*, **40**, 155–190.

Mann, R.H.K. (1976a) Observations on the age, growth, reproduction and food of the pike *Esox lucius* (L.) in two rivers in southern England. *J. Fish Biol.*, **8**. 179–197.

Mann, R.H.K. (1976b) Observations on the age, growth, reproduction and food of the chub, *Squalius cephalus* (L.) in the River Stour, Dorset. *J. Fish Biol.*, **8**, 265–288.

Mann, R.H.K. and Mills, C.A. (1979) Demographic aspects of fish fecundity. *Symp. Zool. Soc. Lond.*, **44**, 161–177.

Mann, R.H.K. and Mills, C.A. (1985) Variations in the sizes of gonads, eggs and larvae of the dace, *Leuciscus leuciscus*. *Env. Biol. Fishes*, **13**, 277–287.

Mann, R.H.K. and Penczak, T. (1986) Fish production in rivers: a review. *Polskie Archm Hydrobiol.*, **33**, 233–247.

Mann, R.H.K., Mills, C.A. and Crisp, D.T. (1984) Geographical variation in the life history tactics of some species of freshwater fish, in *Fish Reproduction: Strategies and Tactics* (eds G.W. Potts and R.J. Wootton), Academic Press, London, pp. 171–186.

Marsh, E. (1986) Effects of egg size on offspring fitness and maternal fecundity in the orangethroat darter, *Etheostoma spectabile* (Pisces: Percidae). *Copeia*, 1986, 18–30.

Mathur, G.B. (1967) Anaerobic respiration in a cyprinoid fish *Rasbora daniconius* (Ham). *Nature*, **214**, 318–319.

Matthews, W.J. (1998) *Patterns in Freshwater Fish Ecology*. Chapman & Hall, New York.

Matthews, W.J. and Heins, D.C. (eds) (1987) *Community and Evolutionary Ecology of North American Stream Fishes*, University of Oklahoma Press, Normal and London.

Matty, A.J. (1985) *Fish Endocrinology*, Croom Helm, London.

Matty, A.J. and Lone, K.P. (1985) The hormonal control of metabolism and feeding, in *Fish Energetics New Perspectives* (eds P. Tytler and P. Calow), Croom Helm, London, pp. 185–209.

May, R.C. (1974) Larval mortality in marine fishes and the critical period concept, in *The Early Life History of Fish* (ed. J.H.S. Blaxter), Springer-Verlag, Berlin, pp. 3–19.

May, R.M. (1973) *Stability and Complexity in Model Ecosystems*, Princeton University Press, Princeton, NJ.

Maynard Smith, J. (1976) Group selection. *Q. Rev. Biol.*, **51**, 277–283.

Maynard Smith, J. (1982) *Evolution and The Theory of Games*, Cambridge University Press, Cambridge.

Maynard Smith, J. (1991) The evolution of reproductive strategies: a commentary. *Phil. Trans. Roy. Soc. B*, **332**, 103–104.

Meakins, R.H. (1974) The bioenergetics of the *Gasterosteus/Schistocephalus* host–parasite system. *Polskie Archm Hydrobiol.*, **21**, 455–466.

Meakins, R.H. and Walkey, M. (1975) The effects of plerocercoid of *Schistocephalus solidus* Muller 1776 (Pseudophyllidea) on the respiration of the three-spined stickleback *Gasterosteus aculeatus* L. *J. Fish Biol.*, **7**, 817–824.

Medland, T.E. and Beamish, F.W.H. (1985) The influence of diet and fish density on apparent heat increment in rainbow trout, *Salmo gairdneri*. *Aquaculture*, **47**, 1–10.

Meien, V.A. (1939) On the annual cycle of ovarian changes in teleosts. *Izv. ANSSSR*, 3.

Metcalfe, N.B. and Thorpe, J.E. (1992) Anorexia and defended energy levels in over-wintering juvenile salmon. *J. Anim. Ecol.*, **61**, 175–181.

Metcalfe, N.B., Huntingford, F.A. and Thorpe, J.E. (1986) Seasonal changes in feeding motivation of juvenile Atlantic salmon (*Salmo salar*). *Can. J. Zool.*, 64, 2439–2446.

Metcalfe, N.B., Huntingford, F.A. and Thorpe, J.E. (1987) The influence of predation risk on the feeding motivation and foraging strategy of juvenile Atlantic salmon. *Anim. Behav.*, **35**, 901–911.

Metcalfe, N.B., Huntingford, F.A. and Thorpe, J.E. (1988) Feeding intensity, growth rates, and the establishment of life-history patterns in juvenile Atlantic salmon *Salmo salar*. *J. Anim. Ecol.*, **57**, 463–474.

Metcalfe, N.B., Huntingford, F.A., Graham, W.D. and Thorpe, J.E. (1989) Early social status and the development of life-history strategies in Atlantic salmon. *Proc. R. Soc. Lond. B*, **236**, 7–19.

Metcalfe, N.B., Taylor, A.C. and Thorpe, J.E. (1995) Metabolic rate, social status and life-history strategies in Atlantic salmon. *Anim. Behav.*, **49**, 431–436.

Meyer, A. (1987) Phenotypic plasticity and heterochrony in *Cichlasoma managuense*. *Evolution*, **41**, 1357–1369.

Meyer, J.L. and Schultz, E.T. (1985) Migrating haemulid fishes as a source of nutrients and organic matter on coral reefs. *Limnol. Oceanogr.*, **30**, 146–156.

Middaugh, D.P. (1981) Reproductive ecology and spawning periodicity of the Atlantic silverside, *Menidia menidia* (Pisces: Atherinidae). *Copeia*, 1981, 766–776.

Miglavs, I. and Jobling, M. (1989) Effects of feeding regime on food consumption, growth rates and tissue nucleic acids in juvenile charr, *Salvelinus alpinus*, with particular respect to compensatory growth. *J. Fish Biol.*, **34**, 947–957.

Milinski, M. (1977a) Experiments on the selection by predators against spatial oddity of their prey. *Z. Tierpsychol.*, **43**, 311–325.

Milinski, M. (1977b) Do all members of a swarm suffer the same predation? *Z. Tierpsychol.*, **45**, 373–388.

Milinski, M. (1979) An evolutionary stable feeding strategy in sticklebacks. *Z. Tierpsychol.*, **51**, 36–40.

Milinski, M. (1982) Optimal foraging: the influence of intraspecific competition on diet selection. *Behav. Ecol. Sociobiol.*, **11**, 109–115.

Milinski, M. (1984a) A predator's cost of overcoming the confusion effect of swarming prey. *Anim. Behav.*, **32**, 1157–1162.

Milinski, M. (1984b) Competitive resource sharing: an experimental test of a learning rule for ESSs. *Anim. Behav.*, **32**, 233–242.

Milinski, M. (1985) Risks of predation of parasitized sticklebacks (*Gasterosteus aculeatus* L.) under competition for food. *Behaviour*, **93**, 203–116.

Milinski, M. (1986a) Constraints placed by predators on feeding behaviour, in *Behaviour of Teleost Fishes* (ed. T.J. Pitcher), Croom Helm, London, pp. 236–252.

Milinski, M. (1986b) A review of competitive resource sharing under constraints in sticklebacks. *J. Fish Biol.*, **29** (Suppl. A), 1–14.

Milinski, M. (1993) Predation risk and feeding behaviour, in *Behaviour of Teleost Fishes*, 2nd edn (ed. T.J. Pitcher), Chapman and Hall, London, pp. 285–305.

Milinski, M. (1996) By-product mutualism, Tit-for-tat reciprocity and co-operative behaviour: a reply to Connor. *Anim. Behav.*, **51**, 458–461.

Milinski, M. and Heller, R. (1978) Influences of a predator on optimal foraging behaviour of sticklebacks (*Gasterosteus aculeatus* L). *Nature*, **275**, 642–644.

Miller, P.J. (1979a) Adaptiveness and implications of small size in teleosts. *Symp. Zool. Soc. Lond.*, **44**, 263–306.

Miller, P.J. (1979b) A concept of fish phenology. *Symp. Zool. Soc. Lond.*, **44**, 1–28.

Miller, P.J. (1984) The tokology of gobioid fishes, in *Fish Reproduction: Strategies and Tactics* (eds G.W. Potts and R.J. Wootton), Academic Press, London, pp. 119–153.

Miller, P.J. (1996) The functional ecology of small fish: some opportunities and consequences. *Symp. Zool. Soc. Lond.*, **69**, 175–199.

Miller, T.J., Crowder, L.B., Rice, J.A. and Marschall, E.A. (1988) Larval size and recruitment mechanisms in fishes: towards a conceptual framework. *Can. J. Fish. Aquat. Sci.*, **52**, 1083–1093.

Milliken, M.R. (1982) Qualitative and quantitative nutrient requirements of fishes: a review. *Fish. Bull. Fish Wildl. Serv. US*, **80**, 655–686.

Mills, C.A. (1987) The life history of the minnow *Phoxinus phoxinus* in a productive stream. *Freshwat. Biol.*, **17**, 53–67.

Mills, C.A. and Eloranta, A. (1985) Reproductive strategies in the stone loach *Noemacheilus barbalulus*. *Oikos*, **44**, 341–349.

Mills, C.A. and Mann, R.H.K. (1985) Environmentally-induced fluctuations in yearclass strength and their implications for management. *J. Fish Biol.*, **27** (Suppl. A), 209–226.

Milner, N.J., Wyatt, R.J. and Scott, M.D. (1993) Variability in the distribution and abundance of stream salmonids, and the associated use of habitat models. *J. Fish Biol.*, **43** (Suppl. A), 103–119.

Mins, C.K. (1995) Allometry of home range size in lake and river fishes. *Can. J. Fish. Aquat. Sci.*, **52**, 1499–1508.

Minton, J.W. and McLean, R.B. (1982) Measurements of growth and consumption of sauger (*Stizostedion canadense*): implications for fish energetics studies. *Can. J. Fish. Aquat. Sci.*, **39**, 1396–1403.

Misra, R.K. and Carscadden, J.E. (1987) A multivariate analysis of morphometrics to detect differences in populations of capelin (*Mallotus villosus*). *J. Cons. Perm. Int. Explor. Mer*, **43**, 99–106.

Misund, O.A. (1997) Underwater acoustics in marine fisheries and fisheries research. *Rev. Fish Biol. Fisheries*, **7**, 1–34.

Mittelbach, G.G. (1981a) Foraging efficiency and body size: a study of optimal diet and habitat use by bluegills. *Ecology*, **62**, 1370–1386.

Mittelbach, G.G. (1981b) Patterns of invertebrate size and abundance in aquatic habitats. *Can. J. Fish. Aquat. Sci.*, **38**, 896–904.

Mittelbach, G.G. (1983) Optimal foraging and growth in bluegills. *Oecologia*, **59**, 157–162.

Mittelbach, G.G. (1984) Predation and resource partitioning in two sunfishes (Centrarchidae). *Ecology*, **65**, 499–513.

Mittelbach, G.G. (1986) Predator-mediated habitat use: some consequences for species interactions. *Env. Biol. Fishes*, **16**, 159–169.

Mittelbach, G.G. (1988) Competition among refuging sunfishes and effects of fish density on littoral zone invertebrates. *Ecology*, **69**, 614–623.

Mittelbach, G.C., Osenberg, C.W. and Wainwright, P.C. (1992) Variation in resource abundance affects diet and feeding morphology in the pumpkinseed sunfish (*Lepomis gibbosus*). *Oecologia*, **90**, 8–13.

Moffat, N.M. and Thomson, D.A. (1978) Tidal influence on the evolution of egg size in the grunions (*Leuresthes* : Atherinidae). *Env. Biol. Fishes*, **3**, 267–273.

Moksness, E., Rukan, K., Ystanes, L., Folkvord, A. and Johannessen, A. (1995) Comparison of somatic and otolith growth in North Sea herring (*Clupea harengus* L.) larvae: evaluation of growth dynamics in mesocosms, in *Recent Developments in Fish Otolith Research* (eds D.H. Secor, J.M. Dean and S.E. Campana), University of South Carolina Press, Columbia, SC, pp. 119–134.

Moller, P. (1995) *Electric Fishes*, Chapman and Hall, London.

Moodie, G.E.E. and Salfert, I.G. (1982) Evaluation of fluorescent pigment for marking a small scaleless fish, the brook stickleback. *Progve Fish-Cult.*, **44**, 192–195.

Moore, J.W. and Moore, I.A. (1976) The basis of food selection in some estuarine fishes. Eels, *Anguilla anguilla* (L.), whiting, *Merlangius merlangus* (L.), sprat, *Sprattus sprattus* (L.) and stickleback, *Gasterosteus aculeatus* L. *J. Fish Biol.*, **9**, 375–390.

Moore, W.S. (1984) Evolutionary ecology of unisexual fishes, in *Evolutionary Genetics of Fishes* (ed. B.J. Turner), Plenum, New York, pp. 329–398.

Moriarty, D.J.W. and Moriarty, C.M. (1973) The assimilation of carbon from phytoplankton by two herbivorous fishes: *Tilapia nilotica* and *Haplochromis nigripinnus*. *J. Zool., Lond.*, **171**, 41–45.

Morris, D. (1958) The reproductive behaviour of the ten-spined stickleback (*Pygosteus pungitius* L). *Behaviour* (Suppl.), **6**, 1–154.

Morse, W.W. (1980) Spawning and fecundity of Atlantic mackerel, *Scomber scombrus*, in the Middle Atlantic Bight. *Fish. Bull. Fish Wildl. Serv. US*, **78**, 103–108.

Motta, P.J. (1984) Mechanics and functions of jaw protrusion in teleost fishes: a review. *Copeia*, **1984**, 1–18.

Moyle, P.B. and Cech, J.J. Jr (1996) *Fishes: An Introduction to Ichthyology*, 3rd edn, Prentice Hall, Englewood Cliffs, NJ.

Moyle, P.B. and Li, H.W. (1979) Community ecology and predator–prey relations in warmwater streams, in *Predator–prey Systems in Fisheries Management* (ed. H. Clepper), Sport Fishing Institute, Washington, DC, pp. 171–180.

Moyle, P.B. and Senanayake, R. (1984) Resource partitioning amongst the fishes of the rainforest streams of Sri Lanka. *J. Zool., Lond.*, **202**, 195–223.

Moyle, P.B. and Vondracek, B. (1985) Persistence and structure of the fish assemblage in a small Californian stream. *Ecology*, **66**, 1–13.

Muller, K. (1978) The flexibility of the circadian system of fish at different latitudes, in *Rhythmic Activity of Fishes* (ed. J.E. Thorpe), Academic Press, London, pp. 91–104.

Müller, U.K. and Videler, J.J. (1996) Inertia as a 'safe harbour': do fish larvae increase length growth to escape viscous drag? *Rev. Fish Biol. Fisheries*, **6**, 353–360.

Munro, A.D. (1990) Tropical freshwater fish, in *Reproductive Seasonality in Teleosts: Environmental Influences* (eds A.D. Munro, A.P. Scott and T.J. Lam), CRC Press, Boca Raton, FL, pp. 145–239.

Munro, A.D., Scott, A.P. and Lam, T.J. (eds) (1990) *Reproductive Seasonality in Teleosts: Environmental Influences*, CRC Press, Boca Raton, FL.

Murdoch, W.W. and Oaten, A. (1975) Predation and population stability. *Adv. Ecol. Res.*, **9**, 1–131.

Murphy, G.I. (1967) Vital statistics of the Pacific sardine (*Sardinops caerulea*) and the population consequences. *Ecology*, **48**, 731–736.

Murphy, G.I. (1968) Pattern in life history and the environment. *Am. Nat.*, **102**, 391–403.

Murphy, G.I. (1977) Clupeoids, in *Fish Population Dynamics* (ed. J.A. Gulland), J. Wiley, New York, pp. 283–308.

Murray, R.W. (1971) Temperature receptors, in *Fish Physiology*, Vol. V (eds W.S. Hoar and D.J. Randall), Academic Press, London, pp. 121–133.

Myers, R.A. and Barrowman, N.J. (1996) Is fish recruitment related to abundance? *Fish. Bull.*, **94**, 707–724.

Myers, R.A. and Cadigan, N.G. (1993) Densitydependent juvenile mortality in marine demersal fish. *Can. J. Fish. Aquat. Sci.*, **50**, 1576–1590.

Myers, R.A., Hutchings, J.A. and Barrowman, N.J. (1997) Why do fish stocks collapse? The example of cod in Atlantic Canada. *Ecol. Appl.*, **7**, 91–106.

Myrberg, A.A., Jr and Thresher, R.E. (1974) Interspecific aggression and its relevance to the concept of territoriality in fishes. *Am. Zool.*, **14**, 81–96.

Nakashima, B.S. and Leggett, W.C. (1980) The role of fishes in the regulation of phosphorus availability in lakes. *Can. J. Fish. Aquat. Sci.*, **37**, 679–686.

Nakatsuru, K. and Kramer, D.L. (1982) Is sperm cheap? Limited male fertility and female choice in the lemon tetra (Pisces: Characidae). *Science*, **216**, 753–755.

Neill, S.R.St J. and Cullen, J.M. (1974) Experiments on whether schooling by their prey affects the hunting behaviour of cephalopods and fish predators. *J. Zool., Lond.*, **172**, 549–569.

Neilson, J.D., Geen, G.H. and Bottom, D. (1985) Estuarine growth of juvenile chinook salmon (*Oncorhynchus tshawytscha*) as inferred from otolith microstructure. *Can. J. Fish. Aquat. Sci.*, **42**, 899–908.

Nelson, J.S. (1994) *Fishes of The World*, 3rd edn, J. Wiley, New York

Nilsson, N.-A. (1963) Interaction between trout and char in Scandinavia. *Trans. Am. Fish. Soc.*, **92**, 276–285.

Nilsson, N.-A. (1965) Food segregation between salmonoid species in North Sweden. *Rep. Inst. Freshwat. Res. Drottningholm*, **46**, 58–78.

Nilsson, N.-A. (1967) Interactive segregation between fish species, in *The Biological Basis of Freshwater Fish Production* (ed. S.D. Gerking), Blackwell, Oxford, pp. 295–313.

Nilsson, N.-A. (1978) The role of size-biased predation in competition and interactive segregation in fish, in *Ecology of Freshwater Fish Production* (ed. S.D. Gerking), Blackwell, Oxford, pp. 303–325.

Nilsson, N.-A. and Northcote, T.G. (1981) Rainbow trout (*Salmo gairdneri*) and cutthroat trout (*S. clarki*) interactions in coastal British Columbia lakes. *Can. J. Fish. Aquat. Sci.*, **38**, 1228–1246.

Nordeng, H. (1977) A pheromone hypothesis for homeward migration in anadromous salmonids. *Oikos*, **28**, 155–159.

Northcote, T.G. (1978) Migratory strategies and production in freshwater fishes, in *Ecology of Freshwater Fish Production* (ed. S.D. Gerking), Blackwell, Oxford, pp. 326–359.

Northcote, T.G. (1984) Mechanisms of fish migration in rivers, in *Mechanisms of Migration in Fishes* (eds J.D. McCleave, G.P. Arnold, J.J. Dodson and W.H. Neill), Plenum, New York, pp. 317–355.

Northcote, T.G. (1988) Fish in the structure and function of freshwater ecosystems: a 'top-down' view. *Can. J. Fish. Aquat. Sci.,* **45**, 361–379.

Nursall, J.R. (1977) Territoriality in redlip blennies (*Ophioblennius atlanticus*) – Pisces: Bleniidae. *J. Zool., Lond.,* **182**, 205–23.

Nursall, J.R. (1981) The activity budget and use of territoriality by a tropical blenniid fish. *Zool. J. Linn. Soc.,* **72**, 69–92.

O'Brien, W.J. (1987) Planktivory by freshwater fish: thrust and parry in pelagia, in *Predation. Direct and Indirect Effects on Aquatic Communities* (eds W.C. Kerfoot and A. Sih), University Press of New England, Hanover, NH, pp. 3–16.

O'Brien, W.J., Slade, N.A. and Vinyard, G.L. (1976) Apparent size as the determinant of prey selection by bluegill sunfish (*Lepomis macrochirus*). *Ecology,* **57**, 1304–1310.

O'Brien, W.J., Evans, B. and Luecke, C. (1985) Apparent size choice of zooplankton by planktivorous sunfish: exceptions to the rule. *Env. Biol. Fishes,* **13**, 225–233.

O'Connell, C.P. (1980) Percentage of starving northern anchovy, *Engraulis mordax*, larvae in the sea as estimated by histological methods. *Fish. Bull. Fish Wildl. Serv. US,* **78**, 475–489.

O'Connell, C.P. (1981) Estimation by histological methods of the percent of starving larvae of the northern anchovy (*Engraulis mordax*) in the sea. *Rapp. P.-v. Réun. Cons. Perm. Int. Explor. Mer,* **178**, 357–360.

Odum, W.E. (1970) Utilization of the direct grazing and plant detritus food chains by the striped mullet, *Mugil cephalus,* in *Marine Food Chains* (ed. J.H. Steele), Oliver and Boyd, Edinburgh, pp. 222–240.

O'Hara, K. (1993) Fish behaviour and the management of freshwater fisheries, in *Behaviour of Teleost Fishes* 2nd edn (ed. T.J. Pitcher), Chapman and Hall, London, pp. 645–670.

Ohguchi, O. (1981) Prey density and selection against oddity by three-spined sticklebacks. *Adv. Ethol.,* **23**, 1–79.

Olson, M.H., Mittelbach, G.G. and Osenberg, C.W. (1995) Competition between predator and prey: resource-based mechanisms and implications for stage-structured dynamics. *Ecology,* **76**, 1758–1771.

Osenberg, C.W., Werner, E.E., Mittelbach, G.G.

and Hall, D.J. (1988) Growth patterns in bluegill (*Lepomis macrochirus*) and pumpkinseed (*L. gibbosus*) sunfish: environmental variation and the importance of ontogenetic niche shifts. *Can. J. Fish. Aquat. Sci.,* **45**, 17–26.

Osenberg, C.W., Mittelbach, G.G. and Wainwright, P.C. (1992) Two-stage life histories in fish: the interaction between juvenile competition and adult performance. *Ecology,* **73**, 255–267.

Osse, J. W.M. (1985) Jaw protrusion, an optimization of the feeding apparatus of teleosts? *Acta Biotheor.,* **34**, 219–232.

Ottaway, E.M. and Simkiss, K. (1977) 'Instantaneous' growth rates of fish scales and their use in studies of fish populations. *J. Zool., Lond.,* **181**, 407–419.

Otto, R.G. (1971) Effects of salinity on the survival and growth of pre-smolt coho salmon (*Oncorhynchus kisutch*). *J. Fish. Res. Bd Can.,* **28**, 343–349.

Otto, R.G. and McInerney, J.E. (1970) Development of salinity preference in pre-smolt coho salmon, *Oncorhynchus kisutch. J. Fish. Res. Bd Can.,* **27**, 793–800.

Paller, M.H. (1994) Relationship between fish assemblage structure and stream order in South Carolina coastal plain streams. *Trans. Am. Fish. Soc.,* **123**, 150–161.

Paloheimo, J.E. and Dickie, L.M. (1965) Food and growth of fishes. 1. A growth curve derived from experimental data. *J. Fish. Res. Bd Can.,* **22**, 521–542.

Pandian, T.J. and Vivekanandan, E. (1985) Energetics of feeding and digestion, in *Fish Energetics New Perspectives* (eds P. Tytler and P. Calow), Croom Helm, London, pp. 99–124.

Pannella, G. (1971) Fish otoliths: daily growth layers and periodical patterns. *Science,* **173**, 1124–1127.

Paradis, A.R., Pepin, P. and Brown, J.A. (1996) Vulnerability of fish eggs and larvae to predation: review of the influence of the relative size of prey and predator. *Can. J. Fish. Aquat. Sci.,* **53**, 1226–1235.

Parenti, L.R. (1984) Biogeography of the Andean killifish genus *Orestias* with comment on the species flock concept, in *Evolution of Fish Species Flocks* (eds A.A. Echelle and I. Kornfield), University of Maine Press, Orono, ME, pp. 85–92.

Parin, N.V. (1984) Oceanic ichthyologeography: an attempt to review the distribution and origin of pelagic and bottom fishes outside continental shelves and neritic zones. *Arch. FischWiss.*, **35**, 5–41.

Parker, G.A. and Begon, M. (1986) Optimal egg size and clutch size: effects of environmental and maternal phenotype. *Am. Nat.*, **128**, 573–592.

Parker, N.C.A., Giorgi, A.E., Heidinger, R.C., Jester, D.B. Jr, Prince, E.D. and Winans, G.A. (eds) (1990). *Fish-marking Techniques* (Am. Fish. Soc. Symp., 7) American Fisheries Society, Bethesda, Ma.

Parsons, T.R. and LeBrasseur, R.J. (1970) The availability of food to different trophic levels in the marine food chain, in *Marine Food Chains* (ed. J. H. Steele), Oliver and Boyd, Edinburgh, pp. 325–343.

Pascoe, D. and Cram, P. (1977) The effect of parasitism on the toxicity of cadmium to the three-spined stickleback, *Gasterosteus aculeatus* L. *J. Fish Biol.*, **10**, 467–472.

Paszkowski, C.A. (1986) Foraging site use and interspecific competition between bluegills and golden shiners. *Env. Biol. Fishes*, **17**, 227–233.

Patterson, K.R. (1996) Modelling the impact of disease-induced mortality in an exploited population: the outbreak of the fungal parasite *Ichthyophonus hoferi* in the North Sea herring. *Can. J. Fish. Aquat. Sci.*, **53**, 2870–2887.

Pauly, D. (1980) On the interrelationships between natural mortality, growth parameters and mean environmental temperature in 175 fish stocks. *J. Cons. Perm. Int. Explor. Mer*, **39**, 175–192.

Pauly, D. (1981) The relationship between gill surface area and growth performance in fish: a generalization of von Bertalanffy's theory of growth. *Meeresforschung*, **28**, 251–282.

Pauly, D. (1987) A review of the ELEFAN system for analysis of length-frequency data in fish and aquatic invertebrates, in *Length-Based Methods in Fisheries Research* (eds D. Pauly and G.R. Morgan), ICLARM, Manila, Philippines and Kuwait Institute Scientific Research, Safat, Kuwait, pp. 7–34.

Pauly, D. (1994) *On the Sex of Fish and the Gender of Scientists*, Chapman and Hall, London.

Payne, A.I. (1979) Physiological and ecological factors in the development of fish culture. *Symp. Zool. Soc. Lond.*, **44**, 383–415.

Payne, A.I. (1986) *The Ecology of Tropical Lakes and Rivers*, J. Wiley, New York.

Payne, A.I. (1987) A lake perched on piscine peril. *New Scient.*, **115**, 50–54.

Pearsons, T.N., Li, H.W. and Lamberti, G.A.

(1992) Influence of habitat complexity on resistance to flooding and resilience of stream fish assemblages. *Trans. Am. Fish. Soc.*, **121**, 427–436.

Pennycuick, L. (1971) Frequency distributions of parasites in a population of the threespined stickleback, *Gasterosteus aculealus* L., with special reference to the negative binomial distribution. *Parasitology*, **63**, 389–406.

Pepin, P. (1991) Effect of temperature and size on development, mortality, and survival rates of the pelagic early life-history of marine fish. *Can. J. Fish. Aquat. Sci.*, **48**, 503–518.

Pepin, P., Helbig, J.A., Laprise, R., Colbourne, E. and Shears, T.H. (1995) Variations in the contribution of transport to changes in planktonic animal abundance: a study of the flux of fish larvae in Conception Bay, Newfoundland. *Can. J. Fish. Aquat. Sci.*, **52**, 1475–1486.

Persat, H., Olivier, J.-M. and Pont, D. (1994) Theoretical habitat templets, species traits, and species richness: fish in the Upper Rhone River and its floodplain. *Freshwat. Biol.*, **31**, 439–454.

Persson, L. (1983a) Food consumption and competition between age classes in a perch *Perca fluviatilis* population in a shallow eutrophic lake. *Oikos*, **40**, 197–207.

Persson, L. (1983b) Food consumption and the significance of detritus and algae to intraspecific competition in roach *Rutilus rutilus* in a shallow eutrophic lake. *Oikos*, **41**, 118–125.

Persson, L. (1984) Food evacuation and models for multiple meals in fishes. *Env. Biol. Fishes*, **10**, 305–309.

Persson, L. (1985) Optimal foraging: the difficulty of exploiting different feeding strategies simultaneously. *Oecologia*, **67**, 338–341.

Persson, L. (1986) Effects of reduced interspecific competition on resource utilization in perch (*Perca fluviatilis*). *Ecology*, **67**, 355–364.

Persson, L. (1987) Effects of habitat and season on competitive interactions between roach (*Rutilus rutilus*) and perch (*Perca fluviatilis*). *Oecologia*, **73**, 170–177.

Persson, L. and Greenberg, L.A. (1990a) Juvenile competitive bottlenecks: the perch (*Perca fluviatilis*) – roach (*Rutilus rutilus*) interaction. *Ecology*, **71**, 44–56.

Persson, L. and Greenberg, L.A. (1990b) Interspecific and intraspecific size class competition affecting resource use and growth of perch, *Perca fluviatilis*. *Oikos*, **59**, 97–106.

Persson, L., Diehl, S., Johansson, L., Andersson, G. and Hamrin, S.F. (1992) Trophic interactions in temperate lake ecosystems: a test of food chain theory. *Am. Nat.*, **140**, 59–84.

Persson, L., Andersson, J., Wahlstrom, E. and Eklov, P. (1996) Size-specific interactions in lake systems: predator gape limitation and prey growth and mortality. *Ecology*, **77**, 900–911.

Peter, R.E. (1979) The brain and feeding behavior, in *Fish Physiology*, Vol. VIII (eds W.S. Hoar, D.J. Randall and J.R. Brett), Academic Press, London, pp. 121–159.

Peterman, R.M., Bradford, M.J., Lo, N.C.H. and Methot, R.D. (1988) Contribution of early life history stages to interannual variability in recruitment of northern anchovy (*Engraulis mordax*). *Can. J. Fish. Aquat. Sci.*, **45**, 8–16.

Peters, D.S. and Boyd, M.T. (1972) The effects of temperature and salinity and availability of food on the feeding and growth of the hog-choker, *Trinectes maculatus* (Bloch and Schneider). *J. Exp. Mar. Biol. Ecol.*, **7**, 201–207.

Peterson, I. and Wroblewski, J.S. (1984) Mortality rate of fishes in the pelagic ecosystem. *Can. J. Fish. Aquat. Sci.*, **41**, 1117–1120.

Pfister, C.A. (1996) The role and importance of recruitment variability to a guild of tide pool fishes. *Ecology*, **77**, 1928–1941.

Pianka, E.R. (1976) Natural selection of optimal reproductive tactics. *Am. Zool.*, **16**, 775–784.

Pianka, E.R. (1981) Competition and niche theory, in *Theoretical Ecology* (ed. R.M. May), Blackwell, Oxford, pp. 167–196.

Pianka, E.R. (1994) *Evolutionary Ecology*, 5th edn, Harper Collins, New York.

Pimm, S.L. and Hyman, J.B. (1987) Ecological stability in the context of multispecies fisheries. *Can. J. Fish. Aquat. Sci.*, **44** (Suppl. 2), 84–94.

Pitcher, T.J. (1980) Some ecological consequences of fish school volumes. *Freshwat. Biol.*, **10**, 539–544.

Pitcher, T.J. (1986) Functions of shoaling behaviour in teleosts, in *The Behaviour of Teleost Fishes*, 1st edn. (ed. T.J. Pitcher) Croom Helm, London. pp. 294–338.

Pitcher, T.J. (1992) Who dares wins: the function and evolution of predator inspection behaviour in shoaling fish. *Neth. J. Zool.*, **42**, 371–391.

Pitcher, T.J. and Hart, P.J.B. (1982) *Fisheries Ecology*, Croom Helm, London.

Pitcher, T.J. and Hart, P.J.B. (1995) *The Impact of Species Changes in African Lakes*, Chapman and Hall, London.

Pitcher, T.J. and Parrish, J.K. (1993) Functions of shoaling behaviour in teleosts, in *Behaviour of*

Teleost Fishes, 2nd edn (ed. T.J. Pitcher), Chapman and Hall, London, pp. 363–439.

Pitcher, T.J. and Turner, J.R. (1986) Danger at dawn: experimental support for the twilight hypothesis in shoaling minnows. *J. Fish Biol.*, **29** (Suppl. A), 59–70.

Pitcher, T.J., Magurran, A.E. and Winfield, I. (1982) Fish in larger shoals find food faster. *Behav. Ecol. Sociobiol.*, **10**, 149–157.

Pitt, T.K. (1966) Sexual maturity and spawning of the American plaice, *Hippoglossoides platessoides* (Fabricus), from Newfoundland and the Grand Banks areas. *J. Fish. Res. Bd Can.*, **23**, 651–672.

Pitt, T.K. (1975) Changes in abundance and certain biological characters of Grand Bank American plaice, *Hippoglossoides platessoides*. *J. Fish. Res. Bd Can.*, **32**, 1383–1398.

Place, A.R. and Powers, D.A. (1979) Genetic variation and relative catalytic efficiencies: lactate dehydrogenase B allozymes of *Fundulus heteroclitus*. *Proc. Natn Acad. Sci. USA*, **76**, 2354–2358.

Policansky, D. (1983) Size, age and demography of metamorphosis and sexual maturation in fishes. *Am. Zool.*, **23**, 57–63.

Polis, G.A. (1981) The evolution and dynamics of intraspecific predation. *Ann. Rev. Ecol. Syst.*, **12**, 225–251.

Polis, G.A. and Winemiller, K.O. (eds) (1996) *Food Webs Integration of Patterns and Dynamics*, Chapman and Hall, London.

Pollard, D.A. (1973) The biology of a landlocked form of the normally catadromous salmoniform fish *Galaxias maculatus* (Jenyns). V. Composition of the diet. *Aust. J. Mar. Freshwat. Res.*, **24**, 281–295.

Pollock, K.H. and Mann, R.H.K. (1983) Use of an age-dependent mark–recapture model in fisheries research. *Can. J. Fish. Aquat. Sci.*, **40**, 1449–1455.

Polunin, N.V.C. (1988) Efficient uptake of algal production by a single resident herbivorous fish on the reef. *J. Exp. Mar. Biol. Ecol.*, **123**, 61–76.

Popova, O.A. (1978) The predator–prey relationship among fish, in *The Biological Basis of Freshwater Fish Production* (ed. S.D. Gerking), Blackwell, Oxford, pp. 359–376.

Popper, A.N. and Coombs, S. (1980) Acoustic detection by fishes, in *Environmental Physiology of Fishes* (ed. M.A. Ali), Plenum, New York, pp. 403–430.

Post, J.R. and Lee, J.A. (1996) Metabolic ontogeny of teleost fishes. *Can. J. Fish. Aquat. Sci.*, **53**, 910–923.

Post, J.R., Vandenbos, R. and McQueen, D.J. (1996) Uptake rates of food-chain and water-borne mercury by fish: field measurements, a mechanistic model, and an assessment of uncertainties. *Can. J. Fish. Aquat. Sci.*, **53**, 395–407.

Potts, G.W. (1984) Parental behaviour in temperate marine teleosts with specific reference to the development of nest structures, in *Fish Reproduction: Strategies and Tactics* (eds G.W. Potts and R.J. Wootton), Academic Press, London, pp. 223–244.

Power, M. (1984) Habitat quality and the distribution of algae-grazing catfish in a Panamanian stream. *J. Anim. Ecol.*, **53**, 357–374.

Power, M. (1987) Predator avoidance by grazing fishes in temperate and tropical streams: importance of stream depth and prey size, in *Predation. Direct and Indirect Effects on Aquatic Communities* (eds W.C. Kerfoot and A. Sih), University Press of New England, Hanover, NH, pp. 333–351.

Power, M.E., Parker, M.S. and Wootton, J.T. (1996) Disturbance and food chain length in rivers, in *Food Webs Integration of Patterns and Dynamics* (eds G.A. Polis and K.O. Winemiller), Chapman and Hall, London, pp. 286–297.

Powers, D.A., DiMichele, L. and Place, A.R. (1983) The use of enzyme kinetics to predict differences in cellular metabolism, developmental rate and swimming performance between LDH-B genotypes of the fish *Fundulus heteroclitus. Isozymes. Current Topics in Biological and Medical Research. Vol. 10: Genetics and Evolution*, pp. 147–170.

Powers, D.A., Ropson, I., Brown, D.C., Beneden, R. van, Gonzalez-Villasenor, L.I. and DiMichele, I.A. (1986) Genetic variation in *Fundulus heteroclitus*: geographic distribution. *Am. Zool.*, **26**, 131–144.

Powers, D.A., Smith, M., Gonzalez-Villasenor, I., DiMichele, L., Crawford, D.L., Bernardi, G. and Lauerman, T.A. (1993) Multidisciplinary approach to the selectionist/neutralist controversy using the model teleost *Fundulus heteroclitus*, in *Oxford Surveys of Evolutionary Biology*, Vol. 9 (eds D. Futuyma and J. Antonovics), Oxford University Press, Oxford, pp. 43–107.

Pressley, P.H. (1981) Parental effort and the evolution of nest guarding tactics in the threespine stickleback *Gasterosteus aculealus. Evolution*, **35**, 282–295.

Preston, J.L. (1978) Communication systems and social interactions in a goby–shrimp symbiosis. *Anim. Behav.*, **26**, 791–802.

Price, L.W. (1981) *Mountains and Man: A Study of Process and Environment*, University of California Press, Berkeley, CA.

Priede, I.G. (1985) Metabolic scope in fishes, in *Fish Energetics New Perspectives* (eds P. Tytler and P. Calow), Croom Helm, London, pp. 33–64.

Priede, I.G. and Merrett, N.R. (1996) Estimation of abundance of abyssal demersal fishes; a comparison of data from trawls and baited cameras. *J. Fish Biol.*, **49** (Suppl. A), 207–216.

Priede, I.G. and Smith, K.L., Jr (1986) Behaviour of the abyssal grenadier, *Coryphaenoides yaquinae*, monitored using ingestible acoustic transmitters in the Pacific Ocean. *J. Fish Biol.*, **29** (Suppl. A), 199–206.

Pullin, R.S.V. and Lowe-McConnell, R.H. (1981) *The Biology and Culture of Tilapias*, ICLARM, Manila.

Purdom, C.E. (1974) Variation in fish, in *Sea Fisheries Research* (ed. F.R. Harden Jones), Elek, London, pp. 347–355.

Purdom, C.E. (1979) Genetics of growth and reproduction in teleosts. *Symp. Zool. Soc. Lond.*, **44**, 207–217.

Purdom, C.E. (1993) *Genetics and Fish Breeding*, Chapman and Hall, London.

Rabeni, C.F. and Jacobson, R.B. (1993) Geomorphic and hydraulic influences on the abundance and distribution of stream centrarchids in Ozark USA streams. *Pol. Arch. Hydrobiol.*, **40**, 87–99.

Rahel, F.J. and Hubert, W.A. (1991) Fish assemblages and habitat gradients in a Rocky Mountain–Great Plains stream: biotic zonation and additive patterns of community change. *Trans. Am. Fish. Soc.*, **120**, 319–332.

Ralston, S. and Miyamoto, G.T. (1983) Analyzing the width of daily otolith increments to age the Hawaiian snapper, *Pristipomoides filamentosus. Fish. Bull. Fish Wildl. Serv. US*, **81**, 523–535.

Randall, J.E. (1965) Grazing effects of sea grasses by herbivorous reef fishes in the West Indies. *Ecology*, **46**, 255–260.

Rao, G.M.M. (1968) Oxygen consumption of rainbow trout (*Salmo gairdneri*) in relation to activity and salinity. *Can. J. Zool.*, **46**, 781–786.

Ratkowsky, D.A. (1986) Statistical properties of alternative parameterizations of the von Bertalanffy growth curve. *Can. J. Fish. Aquat. Sci.*, **43**, 742–747.

Reebs, S.G., Whoriskey, F.G. and FitzGerald, G.J. (1984) Diel patterns of fanning activity, egg respiration and the nocturnal behaviour of male three-spined sticklebacks, *Gasterosteus aculeatus* L. (f. *trachurus*). *Can. J. Zool.*, **62**, 329–334.

Reimchen, T.E. (1980) Spine deficiency and polymorphism in a population of *Gasterosteus aculeatus*, an adaptation to predators? *Can. J. Zool.*, **58**, 1232–1244.

Reimchen, T.E. (1994) Predators and morphological evolution in threespine stickleback, in *The Evolutionary Biology of the Threespine Stickleback* (eds M.A. Bell and S.A. Foster), Oxford University Press, Oxford, pp. 240–276.

Reynolds, W.W. and Casterlin, M.E. (1979) Behavioural thermoregulation and the 'final preferendum' paradigm. *Am. Zool.*, **19**, 211–224.

Reynolds, W.W. and Casterlin, M.E. (1980) The role of temperature in the environmental physiology of fishes, in *Environmental Physiology of Fishes* (ed. M.A. Ali), Plenum, New York, pp. 497–518.

Reznick, D.N. (1983) The structure of guppy life histories: the tradeoff between growth and reproduction. *Ecology*, **64**, 862–873.

Reznick, D.N. (1985) Cost of reproduction: an evaluation of the empirical evidence. *Oikos*, **44**, 257–267.

Reznick, D.N. and Braun, B. (1987) Fat cycling in the mosquitofish (*Gambusia affinis*): fat storage as a reproductive adaptation. *Oecologia*, **73**, 401–413.

Reznick, D.N. and Brygla, H. (1996) Life history evolution in guppies (*Poecilia reticulata* : Poeciliidae). V. Genetic basis of parallelism in life histories. *Amer. Nat.*, **147**, 339–359.

Reznick, D.N. and Endler, J.A. (1982) The impact of predation on life history evolution in Trinidadian guppies (*Poecilia reticulata*). *Evolution*, **36**, 160–177.

Reznick, D.N. and Yang, A.P. (1993) The influence of fluctuating resources on life history: patterns of allocation and plasticity in female guppies. *Ecology*, **74**, 2011–2019.

Reznick, D.N., Bryga, H. and Endler, J.A. (1990) Experimentally induced life-history evolution in a natural population. *Nature*, **346**, 357–359.

Reznick, D.N., Callahan, H. and Llauredo, R. (1996a) Maternal effects on offspring quality in Poeciliid fishes. *Am. Zool.*, **36**, 147–156.

Reznick, D.N., Butler, M.J. IV, Rodd, H. and Ross, P. (1996b) Life history evolution in guppies (*Poecilia reticulata*). 6. Differential mortality as a mechanism for natural selection. *Evolution*, **50**, 1651–1660.

Reznick, D.N., Rodd, H. and Cardenas, M. (1996c) Life history evolution in guppies (*Poecilia reticulata* : Poeciliidae). IV. Parallelism in life history phenotype. *Am. Nat.*, **147**, 319–338.

Reznick, D.N., Shaw, F.H., Rodd, H. and Shaw, R.G. (1997) Evaluation of the rate of evolution in natural populations of guppies (*Poecilia reticulata*). *Science*, **275**, 1934–1937.

Ribble, D.O. and Smith, M.H. (1983) Relative intestine length and feeding ecology of freshwater fishes. *Growth*, **47**, 292–300.

Rice, J.A. and Cochran, P.A. (1984) Independent evaluation of a bioenergetics model for largemouth bass. *Ecology*, **65**, 732–739.

Rice, J.A., Breck, J.E., Bartell, S.M. and Kitchell, J.F. (1983) Evaluating the constraints of temperature, activity and consumption on growth of largemouth bass. *Env. Biol. Fishes*, **9**, 263–275.

Ricker, W.E. (1962) Regulation of the abundance of pink salmon stocks, in *Symposium on Pink Salmon* (ed. N.I. Wilimovsky), University of British Columbia, Vancouver, pp. 155–201.

Ricker, W.E. (1975) Computation and interpretation of biological statistics of fish populations. *Bull. Fish. Res. Bd Can.*, no. 191, 1–382.

Ricker, W.E. (1979) Growth rates and models, in *Fish Physiology*, Vol. VIII (eds W.S. Hoar, D.J. Randall and J.R. Brett), Academic Press, London, pp. 677–743.

Ridell, B.E. and Leggett. W.C. (1981) Evidence for an adaptive basis for geographic variation in body morphology and time of downstream migration of juvenile Atlantic salmon (*Salmo salar*). *Can. J. Fish. Aquat. Sci.*, **38**, 308–320.

Ridell, B.E., Leggett, W.C. and Saunders, R.L. (1981) Evidence for adaptive polygenic variation between two populations of Atlantic salmon (*Salmo salar*) native to tributaries of the S. W. Miramichi River, N. B. *Can. J. Fish. Aquat. Sci.*, **38**, 321–333.

Rijnsdorp, A.D. (1994) Population-regulating processes during the adult phase in flatfish. *Neth. J. Sea Res.*, **32**, 207–223.

Ringler, N.H. (1979) Selective predation by drift-feeding brown trout (*Salmo trutta*). *J. Fish. Res. Bd. Can.*, **36**, 392–403.

Ringler, N.H. (1983) Variation in foraging tactics of fishes, in *Predators and Prey in Fishes* (eds D.L.G. Noakes, D.G. Undquist, G.S. Helfman and J.A. Ward), Junk, The Hague, pp. 159–171.

Ringler, N.H. (1985) Individual and temporal variation in prey switching by brown trout, *Salmo trutta. Copeia*, 1985, 918–926.

Ringler, N.H. and Brodowski, D.F. (1983) Functional response of brown trout (*Salmo trutta* L.) to invertebrate drift. *J. Freshwat. Ecol.*, **2**, 45–57.

Robertson, D.R. (1996) Interspecific competition controls abundance and habitat use of territorial Caribbean damselfishes. *Ecology*, **77**, 885–899.

Robertson, D.R. and Warner, R.R. (1978) Sexual patterns in labroid fishes of the western Caribbean II: the parrotfishes (Scaridae). *Smithsonian Contr. Zool.*, **255**, 1–26.

Robertson, D.R., Sweatman, H.P.A., Fletcher, E.A. and Cleland, M.G. (1976) Schooling as a mechanism for circumventing the territoriality of competitors. *Ecology*, **57**, 1208–1220.

Robertson, D.R., Green, D.G. and Victor, B.C. (1988) Temporal coupling of production and recruitment of larvae of a Caribbean reef fish. *Ecology*, **69**, 370–381.

Robinson, B.W. and Wilson, D.S. (1994) Character release and displacement in fishes: a neglected literature. *Am. Nat.*, **144**, 596–627.

Robinson, B.W., Wilson, D.S., Margosian, A.S. and Loyito, P.T. (1993) Ecological and morphological differentiation of pumpkinseed sunfish in lakes without bluegill sunfish. *Evol. Ecol.*, **7**, 451–464.

Robinson, B.W., Wilson, D.S. and Shea, G.O. (1996) Trade-offs of ecological specialization: an intraspecific comparison of pumpkinseed sunfish phenotypes. *Ecology*, **77**, 170–178.

Rodriguez, M.A. and Lewis, W.M. Jr (1997) Structure in fish assemblages along environmental gradients in floodplain lakes of the Orinoco River. *Ecology*, **67**, 109–128.

Roff, D.A. (1981) Reproductive uncertainty and the evolution of iteroparity: why don't flatfish put all their eggs in one basket? *Can. J. Fish. Aquat. Sci.*, **38**, 968–977.

Roff, D.A. (1982) Reproductive strategies in flatfish: a first synthesis. *Can. J. Fish. Aquat. Sci.*, **39**, 1686–1698.

Roff, D.A. (1983) An allocation model of growth and reproduction in fish. *Can. J. Fish. Aquat. Sci.*, **40**, 1395–1404.

Roff, D.A. (1984) The evolution of life history parameters. *Can. J. Fish. Aquat. Sci.*, **41**, 989–1000.

Roff, D.A. (1988) The evolution of migration and some life history parameters in marine fishes. *Env. Biol. Fishes*, **22**, 133–146.

Roff, D.A. (1992) *The Evolution of Life Histories*, Chapman and Hall, London.

Rombough, P.J. (1994) Energy partitioning during fish development: additive or compensatory allocation of energy to support growth? *Funct. Ecol.*, **8**, 178–186.

Root, R.B. (1967) The niche exploitation pattern of the blue–gray gnatcatcher. *Ecol. Monogr.*, **37**, 317–350.

Rose, G.A. and Leggett, W.C. (1989) Interactive effects of geophysically-forced sea temperatures and prey abundance on mesoscale coastal distributions of a marine predator, Atlantic cod (*Gadus morhua*). *Can. J. Fish. Aquat. Sci.*, **46**, 1904–1913.

Rose, G.A. and Leggett, W.C. (1990) The importance of scale to predator–prey spatial correlations: an example of Atlantic fishes. *Ecology*, **71**, 33–43.

Rose, K.A., Tyler, J.A., Chambers, R.C., Klein-MacPhee, G. and Danila, D.J. (1996) Simulating winter flounder population dynamics using coupled individual-based young-of-the-year and age-structured adult models. *Can. J. Fish. Aquat. Sci.*, **53**, 1071–1091.

Ross, S.T. (1986) Resource partitioning in fish assemblages: a review of field studies. *Copeia*, **1986**, 352–388.

Ross, S.T., Matthews, W.J. and Echelle, A.A. (1985) Persistence of stream fish assemblages: effects of environmental change. *Am. Nat.*, **126**, 24–40.

Rothschild, B.I. (1986) *Dynamics of Marine Fish Populations*, Harvard University Press, Cambridge, MA.

Rothschild, B.I. and Osborn, T.R. (1988) Small-scale turbulence and plankton contact rates. *J. Plankt. Res.*, **10**, 465–474.

Roughgarden, J. (1979) *Theory of Population Genetics and Evolutionary Ecology: An Introduction*, MacMillan, London.

Roughgarden, J. (1986) A comparison of food-limited and space-limited animal competition communities, in *Community Ecology* (eds J. Diamond and T.J. Case), Harper-Row, New York, pp. 492–516.

Rubenstein, D.I. (1981) Individual variation and competition in the Everglades pygmy sunfish. *J. Anim. Ecol.*, **50**, 337–350.

Russell, F.S. (1976) *The Eggs and Planktonic Stages of British Marine Fishes*, Academic Press, London.

Russell, N.R. and Wootton, R.J. (1992) Appetite and growth compensation in the European minnow, *Phoxinus phoxinus*, following short periods of food restriction. *Env. Biol. Fishes*, **34**, 277–285.

Ruzzante, D.E., Taggart, C.T. and Cook, D. (1996) Spatial and temporal variation in the genetic composition of a larval cod (*Gadus morhua*) aggregation: cohort contribution and genetic stability. *Can. J. Fish. Aquat. Sci.*, **53**, 2695–2705.

Ryder, R.A., Kerr, S.R., Loftus, K.H. and Regier, H.A. (1974) The morphoedaphic index, a fish yield estimator – review and evaluation. *J. Fish. Res. Bd Can.*, **31**, 663–688.

Sadler, K. (1983) A model relating the results of low pH bioassay experiments to the fishery status of Norwegian lakes. *Freshwat. Biol.*, **13**, 453–463.

Sadler, K. and Lynam, S. (1986) Some effects of low ph and calcium on the growth and mineral content of yearling brown trout, *Salmo trutta*. *J. Fish Biol.*, **29**, 313–324.

Saether, B.-S., Johnsen, H.K. and Jobling, M. (1996) Seasonal changes in food consumption and growth of Arctic charr exposed to either simulated natural or 12:12 LD photoperiod at constant water temperature. *J. Fish Biol.*, **48**, 1113–1122.

Sagnes, P., Gaudin, P. and Statzner, B. (1997) Shifts in morphometrics and their relation to hydrodynamic potential and habitat use during grayling ontogenesis. *J. Fish Biol.*, **50**, 846–858.

Saint-Paul, U. and Soares, G. (1987) Diurnal distribution and behavioural responses of fishes to extreme hypoxia in an Amazon floodplain lake. *Env. Biol. Fishes*, **20**, 91–104.

Saldana, J. and Venables, B. (1983) Energy compartmentalization in a migratory fish, *Prochilodus mariae* (Prochilodontidae), of the Orinoco River. *Copeia*, **1983**, 617–625.

Sale, P.F. (1969) A suggested mechanism for habitat selection by juvenile manini *Acanthurus tristegus sandvicensis* Streets. *Behaviour*, **35**, 27–44.

Sale, P.F. (1977) Maintenance of high diversity in coral reef fish communities. *Am. Nat.*, **111**, 337–359.

Sale, P.F. (1978) Coexistence of coral reef fishes – a lottery for living space. *Env. Biol. Fishes*, **3**, 85–102.

Sale, P.F. (1980) The ecology of fishes on coral reefs. *Oceanogr. Mar. Biol. Ann. Rev.*, **18**, 367–421.

Sale, P.F. (1984) The structure of communities of fish on coral reefs and the merit of a hypothesis testing, manipulative approach to ecology, in *Ecological Communities: Conceptual Issues and the Evidence* (eds D.R. Strong, D. Simberloff, L.G. Abele and A.B. Thistle), Princeton University Press, Princeton, NJ, pp. 478–490.

Sale, P.F. (1988) Perception, pattern. chance and structure of reef fish communities. *Env. Biol. Fishes*, **21**, 3–15.

Sale, P.F. (1991) Reef fish communities: open nonequilibrial systems, in *The Ecology of Fishes on Coral Reefs* (ed. P.F. Sale), Academic Press, London, pp. 564–598.

Sale, P.F. and Douglas, W.A. (1984) Temporal variability in the community structure of fish on coral patch reefs and the relation of community structure to reef structure. *Ecology*, **65**, 409–422.

Sale, P.F., Doherty, P.J., Eckert, G.J., Douglas, W.A. and Ferrell, D.J. (1984) Large scale spatial and temporal variation in recruitment to fish populations on coral reefs. *Oecologia*, **64**, 191–198.

Sargent, R.C. and Gross, M.R. (1986) William's principle: an explanation of parental care in teleost fishes, in *The Behaviour of Teleost Fishes* (ed. T.J. Pitcher), Croom Helm, London, pp. 275–293.

Sargent, R.C., Taylor, P.D. and Gross, M.R. (1987) Parental care and evolution of egg size in fishes. *Am. Nat.*, **129**, 32–46.

Savino, J.F. and Stein, R.A. (1982) Predator–prey interactions between largemouth bass and bluegills as influenced by simulated submersed vegetation. *Trans. Am. Fish. Soc.*, **111**, 255–266.

Savino, J.F. and Stein, R.A. (1989) Behavioural interactions between fish predators and their prey: effects of plant density. *Anim. Behav.*, **37**, 311–321.

Schaaf, W.E., Peters, D.S., Vaughan, D.S., Coston-Clements, L. and Krouse, C.W. (1987) Fish population responses to chronic and acute pollution: the influence of life history strategies. *Estuaries*, **10**, 267–275.

Schaffer, W.M. (1974a) Selection for optimal life histories: the effects of age structure. *Ecology*, **55**, 291–303.

Schaffer, W.M. (1974b) Optimal reproductive effort in fluctuating environments. *Am. Nat.*, **108**, 783–790.

Schaffer, W.M. (1979) The theory of life-history evolution and its application to Atlantic salmon. *Symp. Zool. Soc. Lond.*, **44**, 307–326.

Schaffer, W.M. and Rosenzweig, M.K. (1977) Selection for optimal life histories, II. Multiple equilibria and the evolution of alternative reproductive strategies. *Ecology*, **58**, 60–62.

Scheidegger, K.J. and Bain, M.B. (1995) Larval fish distribution and microhabitat use in free-flowing and regulated rivers. *Copeia*, **1995**, 125–135.

Schlosser, I.J. (1982) Fish community structure along two habitat gradients in a headwater stream. *Ecol. Monogr.*, **52**, 395–414.

Schlosser, I.J. (1985) Flow regime, juvenile abundance, and the assemblage structure of stream fishes. *Ecology*, **66**, 1484–1490.

Schlosser, I.J. (1987a) The role of predation in age- and size-related habitat use by stream fishes. *Ecology*, **68**, 651–659.

Schlosser, I.J. (1987b) A conceptual framework for fish communities in small warmwater streams, in *Community Evolutionary Ecology of North American Stream Fishes* (eds W.J. Matthews and D.C. Heins), University of Oklahoma Press, Normal, OK, pp. 17–24.

Schluter, D. (1993) Adaptive radiation in sticklebacks: size, shape and habitat use efficiency. *Ecology*, **74**, 699–709.

Schluter, D. (1994) Experimental evidence that competition promotes divergence in adaptive radiation. *Science*, **266**, 798–801.

Schluter, D. (1995) Adaptive radiation in sticklebacks: trade-offs in feeding performance and growth. *Ecology*, **76**, 82–90.

Schmitt, R.J. and Holbrook, S.J. (1984) Ontogeny of prey selection by black surf perch *Embiotoca jacksoni* (Pisces: Embiotocidae): the roles of fish morphology, foraging behaviour, and patch selection. *Mar. Ecol. Progr. Ser.*, **18**, 225–239.

Schmitt, R.J. and Holbrook, S.J. (1990) Population responses of surfperch released from competition. *Ecology*, **71**, 1653–1665.

Schnute, J. (1981) A versatile growth model with statistically stable parameters. *Can. J. Fish. Aquat. Sci.*, **38**, 1128–1140.

Schoener, T.W. (1986) Resource partitioning, in *Community Ecology: Pattern and Process* (eds J. Kikkawa and D.J. Anderson), Blackwell, Oxford, pp. 91–126.

Schoener, T.W. (1987) Axes of controversy in community ecology, in *Community and Evolutionary Ecology of North American Stream Fishes* (eds W.J. Matthews and D.C. Heins), University of Oklahoma Press, Normal, OK, pp. 8–16.

Schultz, E.T. and Warner, R.R. (1989) Phenotypic plasticity in life-history traits of female *Thalassoma bifasciatum* (Pisces: Labridae). I. Manipulations of social structure in tests for adaptive shifts of life-history allocations. *Evolution*, **43**, 1497–1506.

Schultz, E.T. and Warner, R.R. (1991) Phenotypic plasticity in life-history traits of female *Thalassoma bifasciatum* (Pisces: Labridae). II. Correlation of fecundity and growth rate in comparative studies. *Env. Biol. Fishes*, **30**, 333–344.

Schultz, E.T., Reynolds, K.E. and Conover, D.O. (1996) Countergradient variation in growth among newly hatched *Fundulus heteroclitus* : geographic differences revealed by common-environment experiments. *Funct. Ecol.*, **10**, 366–374.

Schutz, D.C. and Northcote, T.G. (1972) An experimental study of feeding behaviour and interaction of coastal cutthroat trout (*Salmo clarki clarki*) and dolly varden (*Salvelinus malma*). *J. Fish. Res. Bd Can.*, **29**, 555–565.

Schwassman, H.O. (1974) Refractive state, accommodation and resolving power of the fish eye, in *Vision in Fishes* (ed. M.A. Ali), Plenum, New York, pp. 279–288.

Schwassmann, H.O. (1978) Times of annual spawning and reproductive strategies in Amazonian fishes, in *Rhythmic Activity of Fishes* (ed. J.E. Thorpe), Academic Press, London, pp. 187–200.

Scott, D.B.C. (1979) Environmental timing and the control of reproduction in teleost fish. *Symp. Zool. Soc. Lond.*, **44**, 105–132.

Scott, W.B. and Crossman, E.J. (1973) Freshwater Fishes of Canada. *Bull. Fish. Res. Bd Can.* no. 184, 1–966.

Seber, G.A.F. (1973) *The Estimation of Animal Abundance and Related Parameters*, Griffin, London.

Seber, G.A.F. and LeCren, E.D. (1967) Estimating population parameters from catches large relative to the population. *J. Anim. Ecol.*, **36**, 631–643.

Secor, D.H., Dean, J.M. and Campana, S.E. (eds) (1995) *Recent Developments in Fish Otolith Research*, University of South Carolina Press, Columbus, SC.

Seghers, B.H. (1974a) Role of gill rakers in size-selective predation by lake whitefish, *Coregonus clupeaformis* (Mitchill). *Verh. Int. Verein. Theor. angew. Limnol.*, **19**, 2401–2405.

Seghers, B.H. (1974b) Geographical variation in the responses of guppies (*Poecilia reticulata*) to aerial predators. *Oecologia*, **14**, 93–98.

Seghers, B.H. (1974c) Schooling behaviour in the guppy (*Poecilia reticulata*): an evolutionary response to predation. *Evolution*, **28**, 486–489.

Semler, D.E. (1971) Some aspects of adaptation of a polymorphism for breeding colours in the threespine stickleback (*Gasterosteus aculeatus*). *J. Zool. Lond.*, **165**, 291–302.

Sempeski, P. and Gaudin, P. (1995). Habitat selection by grayling – II. Preliminary results on larval and juvenile daytime habitats. *J. Fish Biol.*, **47**, 345–349.

Shapiro, D.Y. (1984) Sex reversal and sociodemographic processes in coral reef fishes, in *Fish Reproduction: Strategies and Tactics* (eds G.W. Potts and R.J. Wootton), Academic Press, London, pp. 103–118.

Shepherd, J.G. (1982) A versatile new stock–recruitment relationship for fisheries and the construction of sustainable yield curves. *J. Cons. Perm. Int. Explor. Mer*, **40**, 67–75.

Shepherd, J.G. and Cushing, D.H. (1980) A mechanism for density dependent survival of larval fish as the basis of a stock–recruitment relationship. *J. Cons. Perm. Int. Explor. Mer*, **39**, 160–167.

Shepherd, J.G., Pope, J.G. and Cousens, R.D. (1984) Variations in fish stocks and hypotheses concerning their link with climate. *Rapp. P.-v. Réun. Cons. Perm. Int. Explor. Mer*, **185**, 255–267.

Shrode, J.B. and Gerking, S.D. (1977) Effects of constant and fluctuating temperatures on reproductive performance of a desert pupfish *Cyprinodon n. nevadensis*. *Physiol. Zool.*, **50**, 1–10.

Shul'man, G.E. (1974) *Life Cycles of Fish*, John Wiley, New York.

Shulman, M.J. and Ogden, J.C. (1987) What controls tropical reef populations: recruitment or benthic mortality? An example in the Caribbean reef fish *Haemulon flavolineatum*. *Mar. Ecol. Progr. Ser.*, **39**, 233–242.

Shuter, B.J. and Post, J.R. (1990) Climate, population variability, and the zoogeography of temperate fishes. *Trans. Am. Fish. Soc.*, **119**, 314–336.

Sibly, R.M. (1981) Strategies of digestion and defecation, in *Physiological Ecology* (eds C.R. Townsend and P. Calow), Blackwell, Oxford, pp. 109–139.

Sibly, R.M. and Calow, P. (1983) An integrated approach to life-cycle evolution using selective landscapes. *J. Theor. Biol.*, **102**, 527–547.

Sibly, R.M. and Calow, P. (1986) *Physiological Ecology of Animals*, Blackwell, Oxford.

Sih, A. (1994) Predation risk and the evolutionary ecology of reproductive behaviour. *J. Fish Biol.*, **45** (Suppl. A), 111–130.

Sillah, A.B.S. (1981) The feeding ecology of cyprinids with particular reference to the dace, *Leuciscus leuciscus* (L.), PhD thesis, University of Liverpool, 165 pp.

Simpson, B.R.C. (1979) The phenology of annual killifishes. *Symp. Zool. Soc. Lond.*, 44, 243–261.

Sinclair, M. (1988) *Marine Populations*, University of Washington Press, Seattle.

Sindermann, C.J. (1966) Diseases of marine fishes. *Adv. Mar. Biol.*, **4**, 1–89.

Sindermann, C.J. (1987) Effects of parasites on fish populations: practical considerations. *Int. J. Parasitol.*, **17**, 371–382.

Slobodkin, L.B. (1961) *Growth and Regulation of Animal Populations*, Holt Rinehart and Winston, New York.

Slobodkin, L.B. (1972) On the inconstancy of ecological efficiency and the form of ecological theories. *Trans. Conn. Acad. Arts Sci.*, **44**, 291–305.

Slobodkin, L.B. and Rapoport, A. (1974) An optimal strategy of evolution. *Q. Rev. Biol.*, **49**, 181–200.

Smagula, C.M. and Adelman, I.R. (1982) Day-to-day variation in food consumption by largemouth bass. *Trans. Am. Fish. Soc.*, **111**, 543–548.

Smale, M.A. and Rabeni, C.F. (1995) Influences of hypoxia and hyperthermia on fish species composition in headwater streams. *Trans. Am. Fish. Soc.*, **124**, 711–725.

Smart, G.R. (1981) Aspects of water quality producing stress in intensive fish culture, in *Stress and Fish* (ed. A.D. Pickering), Academic Press, London, pp. 277–293.

Smit, H. (1980) Some aspects of environmental (phenotypic) adaptations in fishes. *Neth. J. Zool.*, **30**, 179–207.

Smith, B.R. and Tibbles, J.J. (1980) Sea lamprey (*Petromyzon marinus*) in Lakes Huron, Michigan and Superior: history of invasion and control, 1936–78. *Can. J. Fish. Aquat. Sci.*, **37**, 1780–1801.

Smith, C. and Reay, P. (1991) Cannibalism in teleost fish. *Rev. Fish Biol. Fisheries*, **1**, 41–64.

Smith, C. and Wootton, R.J. (1995) The costs of parental care in teleost fishes. *Rev. Fish Biol. Fisheries*, **5**, 7–22.

Smith, G.R. (1981) Late Cenozoic freshwater fishes of North America. *Ann. Rev. Ecol. Syst.*, **12**, 163–193.

Smith, G.R. and Todd, T.N. (1984) Evolution of species flocks of fishes in north temperate lakes, in *Evolution of Fish Species Flocks* (eds A.A. Echelle and I. Kornfield), University of Maine Press, Orono, ME, pp. 45–68.

Smith, P.E., Flerx, W. and Hewitt, R.P. (1985) The CalCOFI vertical egg tow (CalVET) net. *NOAA Tech. Rep. NMFS*, **36**, 27–32.

Smith, R.J.F. (1974) Effects of 17a-methyltestosterone on the dorsal pad and tubercules of fathead minnows (*Pimephales promelas*), *Can. J. Zool.*, **52**, 1031–1038.

Smith, R.J.F. (1985) *The Control of Fish Migration*, Springer-Verlag, Berlin.

Smith, R.F.J. (1992) Alarm signals in fishes. *Rev. Fish Biol. Fisheries*, **2**, 33–63.

Smith, T.B. and Skulason, S. (1996) Evolutionary significance of resource polymorphisms in fishes, amphibians, and birds. *Ann. Rev. Ecol. Syst.*, **27**, 111–133.

Snorrason, S.S., Skulason, S., Jonsson, B., Malmquist, H.J., Jonasson, P.M., Sandlund, O.T. and Lindem, T. (1994) Trophic specialization in Arctic charr, *Salvelinus alpinus* (Pisces; Salmonidae): morphological divergence and ontogenetic niche shifts. *Biol. J. Linn. Soc.*, **52**, 1–18.

Sohn, J.J. (1977) Socially induced inhibition of genetically determined maturation in the platyfish, *Xiphophorus maculatus*. *Science*, **195**, 199–201.

Solomon, D.J. and Brafield, A.E. (1972) The energetics of feeding, metabolism and growth of perch (*Perca fluviatilis*). *J. Anim. Ecol.*, **41**, 699–718.

Soofiani, N.M. and Hawkins, A.D. (1985) Field studies of energy budgets, in *Fish Energetics New Perspectives* (eds P. Tytler and P. Calow), Croom Helm, London, pp. 283–307.

Soutar, A. and Isaacs, J.D. (1974) Abundance of pelagic fish during the 19th and 20th centuries as recorded in anaerobic sediments off California. *Fish. Bull. Fish. Wildl. Serv. US*, **72**, 257–275.

Southwood, T.R.E. (1977) Habitat, the templet for ecological strategies? *J. Anim. Ecol.*, **46**, 337–365.

Southwood, T.R.E. (1988) Tactics, strategies and templets. *Oikos*, **52**, 3–18.

Springate, J.R.C. and Bromage, N.R. (1985) Effects of egg size on early growth and survival in rainbow trout (*Salmo gairdneri* Richardson). *Aquaculture*, **47**, 163–172.

Springate, J.R.C., Bromage, N.R. and Cumaranatunga, P.R.T. (1985) The effects of different rations on fecundity and egg quality in the rainbow trout (*Salmo gairdneri*), in *Nutrition and Feeding in Fish* (eds C.B. Cowey, A.M. MacKie and J.G. Bell), Academic Press, London, pp. 371–393.

Stacey, N.E. (1984) Control of the timing of ovulation by exogenous and endogenous factors, in *Fish Reproduction: Strategies and Tactics* (eds G.W. Potts and R.J. Wootton), Academic Press, London, pp. 207–222.

Stanley, B.V. and Wootton, R.J. (1986) Effects of ration and male density on the territoriality and nest-building of male three-spined sticklebacks (*Gasterosteus aculeatus* L.). *Anim. Behav.*, **34**, 527–535.

Statzner, B., Resh, V.H. and Doledec, S. (1994) Ecology of the Upper Rhone River: a test of habitat templet theories. *Freshwat. Biol.*, **31**, 253–556.

Stearns, S.C. (1983a) A natural experiment in life-history evolution: field data on the introduction of mosquito fish (*Gambusia affinis*) to Hawaii. *Evolution*, **37**, 601–617.

Stearns, S.C. (1983b) The genetic basis of differences in life-history traits among six populations of mosquito fish (*Gambusia affinis*) that shared ancestors in 1905. *Evolution*, **37**, 618–627.

Stearns, S.C. (1983c) The evolution of life-history traits in mosquito fish since their introduction to Hawaii in 1905: rates of evolution, heritabilities, and developmental plasticity. *Am. Zool.*, **23**, 65–75.

Stearns, S.C. (1992) *The Evolution of Life Histories*. Oxford University Press, Oxford.

Stearns, S.C. and Crandall, R.E. (1984) Plasticity for age and size at sexual maturity: a life-history response to unavoidable stress, in *Fish Reproduction: Strategies and Tactics* (eds G.W. Potts and R.J. Wootton), Academic Press, London, pp. 13–33.

Stearns, S.C. and Koella, J. (1986) The evolution of phenotypic plasticity in life-history traits: predictions for norms of reaction for age- and size-at-maturity. *Evolution*, **40**, 893–913.

Steele, J.H. (1974) *The Structure of Marine Ecosystems*, Blackwell, Oxford.

Steele, M.A. (1997) The relative importance of processes affecting recruitment of two temperate reef fishes. *Ecology*, **78**, 129–145.

Stephens, D.W. and Krebs, J.R. (1986) *Foraging Theory*, Princeton University Press, Princeton, NJ.

Stevens, E.D. and Neill, W.H. (1978) Body temperature relations of tunas, especially skipjack, in *Fish Physiology*, Vol. VII (eds W.S. Hoar and D.J. Randall), Academic Press, London, pp. 316–359.

Stewart, D.J., Weininger, D., Rottiers, D.V. and Edsall, T.A. (1983) An energetics model for lake trout, *Salvelinus namaycush* : application to Lake Michigan population. *Can. J. Fish. Aquat. Sci.*, **40**, 681–698.

Stirling, H.P. (1977) Growth, food utilization and

effect of social interaction in European bass *Dicentrarchus labrax*. *Mar. Biol.*, **40**, 173–184.

Strauss, R.E. (1979) Reliability estimates for lvlev's electivity index, the forage ratio and a proposed linear index to food selection. *Trans. Am. Fish. Soc.*, **108**, 344–352.

Strong, D.R., Simbertoff, D., Abele, L.G. and Thistle, A.B. (1984) *Ecological Communities: Conceptual Issues and the Evidence*, Princeton University Press, Princeton, NJ.

Sutherland, W.J. (1996). *From Individual Behaviour to Population Ecology*, Oxford University Press, Oxford.

Svardson, G. (1976) Interspecific population dominance in fish communities of Scandinavian lakes. *Rep. Inst. Freshwat. Res. Drottningholm*, **56**, 144–171.

Swain, D.P. and Kramer, D.L. (1995) Annual variation in temperature selection by Atlantic cod *Gadus morhua* in the southern Gulf of St. Lawrence, Canada, in relation to population size. *Mar. Ecol Progr. Ser.*, **116**, 11–23.

Swain, D.P. and Morin, R. (1996) Relationships between geographic distribution and abundance of American plaice (*Hippoglossoides platessoides*) in the southern Gulf of St. Lawrence. *Can. J. Fish. Aquat. Sci.*, **53**, 106–119.

Swales, S. (1986) Population dynamics, production and angling catch of brown trout, *Salmo trutta*, in a mature upland reservoir in Mid-Wales. *Env. Biol. Fishes*, **16**, 279–293.

Swales, S. and O'Hara, K. (1983) A short-term study of the effects of a habitat improvement programme on the distribution and abundance of fish stocks in a small lowland stream in Shropshire. *Fish. Manage.*, **14**, 135–144.

Sweatman, H.P.A. (1985) The influence of adults of some coral reef fishes on larval recruitment. *Ecol. Monogr.*, **55**, 469–485.

Tacon, G. and Cowey C.B. (1985) Protein and amino acid requirements, in *Fish Energetics New Perspectives* (eds P. Tytler and P. Calow), Croom Helm, London, pp. 155–183.

Talbot, C. (1985) Laboratory methods in fish feeding and nutritional studies, in *Fish Energetics New Perspectives* (eds P. Tytler and P. Calow), Croom Helm, London, pp. 125–154.

Talbot, F.H., Russell, B.C. and Anderson, G.R.V. (1978) Coral reef fish communities: unstable high diversity systems *Ecol. Monogr.*, **48**, 425–440.

Tavolga, W.N. (1971) Sound production and detection, in *Fish Physiology*, Vol. V (eds W.S.

Hoar and D.J. Randall), Academic Press, London, pp. 135–202.

Taylor, R.J. (1984) *Predation*, Chapman and Hall, London.

Tesch, F.W. (1971) Age and growth, in *Fish Production in Fresh Waters* (ed. W.E. Ricker), Blackwell, Oxford, pp. 98–130.

Thomas, G. (1974) The influence of encountering a food object on subsequent searching behaviour in *Gasterosteus aculeatus* L. *Anim. Behav.*, **22**, 941–952.

Thomas, G. (1977) The influence of eating and rejecting prey items upon feeding behaviour and food searching behaviour in *Gasterosteus aculeatus* L. *Anim. Behav.*, **25**, 52–66.

Thorpe, J.E. (1977) Bimodal distribution of length of juvenile Atlantic salmon (*Salmo salar* L.) under artificial rearing conditions. *J. Fish Biol.*, **11**, 175–184.

Thorpe, J.E. (1978) *Rhythmic Activity of Fishes*, Academic Press, London.

Thorpe, J.E. (1980) *Salmon Ranching*, Academic Press, London.

Thorpe, J.E., Miles, M.S. and Keay, D.S. (1984) Developmental rate, fecundity and egg size in Atlantic salmon, *Salmo salar* L. *Aquaculture*, **43**, 289–305.

Thresher, R.E. (1983) Habitat effects on reproductive success in the coral reef fish, *Acanthochromis polyacanthus* (Pomatocentridae). *Ecology*, **64**, 1184–1199.

Thresher, R.E. (1984) *Reproduction in Reef Fishes*. TFH Publications, Neptune City, NJ.

Thresher, R.E. (1985) Distribution, abundance and reproductive success in the coral reef fish *Acanthochromis polyacanthus*. *Ecology*, **66**, 1139–1150.

Thresher, R.E. (1988) Latitudinal variations in egg size of tropical and sub-tropical North Atlantic shore fishes. *Env. Biol. Fishes*, **21**, 17–25.

Thresher, R.E. (1991) Geographical variability in the ecology of coral reef fishes: evidence, evolution, and possible implications, in *The Ecology of Fishes on Coral Reefs* (ed. P.F. Sale), Academic Press, London, pp. 401–436.

Tierney, J.F., Huntingford, F.A. and Crompton, D.W.T. (1996) Body condition and reproductive status in sticklebacks exposed to a single wave of *Schistocephalus solidus* infection. *J. Fish Biol.*, **49**, 483–493.

Tonn, W.M. (1985) Density compensation in *Umbra–Perca* fish assemblages of northern Wisconsin lakes. *Ecology*, **66**, 415–429.

Tonn, W.M. and Magnuson, J.J. (1982) Patterns in species composition and richness of fish assemblages in northern Wisconsin lakes. *Ecology*, **63**, 1149–1166.

Tonn, W.M., Paszkowski, C.A. and Moermond, T.C. (1986) Competition in *Umbra–Perca* fish assemblages: experimental and field evidence. *Oecologia*, **69**, 126–133.

Tonn, W.M., Magnuson, J.J., Rask, M. and Toivonen, J. (1990) Intercontinental comparison of small lake fish assemblages: the balance between local and regional processes. *Am. Nat.*, **136**, 345–375.

Tonn, W.M., Paszkowski, C.A. and Holopainen, I.J. (1992) Piscivory and recruitment: mechanisms structuring populations in small lakes. *Ecology*, **73**, 951–958.

Townsend, C.R. (1989) Population cycles in freshwater fish. *J. Fish Biol.*, **35** (Suppl. A), 125–131.

Townshend, T.J. (1984) Effects of food availability on reproduction in Central American cichlid fishes, PhD thesis, University of Wales, 216 pp.

Townshend, T.J. and Wootton, R.J. (1985a) Effects of food supply on the reproduction of the convict cichlid, *Cichlasoma nigrofasciatum*. *J. Fish Biol.*, **24**, 91–104.

Townshend, T.J. and Wootton, R.J. (1985b) Adjusting parental investment to changing environmental conditions: the effect of food ration on parental behaviour of the convict cichlid, *Cichlasoma nigrofasciatum*. *Anim. Behav.*, **33**, 494–501.

Townshend, T.J. and Wootton, R.J. (1986) Variation in the mating system of a biparental cichlid fish, *Cichlasoma panamense*. *Behaviour*, **95**, 181–197.

Tuomi, J., Hakala, T. and Haukioja, E. (1983) Alternative concepts of reproductive effort, costs of reproduction, and selection in life-history evolution. *Am. Zool.*, **23**, 25–34.

Turner, G.F. (1986) Territory dynamics and cost of reproduction in a captive population of the colonial nesting mouthbrooder *Oreochromis mossambicus* (Peters). *J. Fish Biol.*, **29**, 573–587.

Turner, G.F (1993) Teleost mating behaviour, in *Behaviour of Teleost Fishes*, 2nd edn (ed. T.J. Pitcher), Chapman and Hall, London, pp. 253–274.

Tyler, A.V. (1966) Some lethal temperature relations of two minnows of the genus *Chromus*. *Can. J. Zool.*, **44**, 349–364.

Tyler, A.V. and Dunn, R.S. (1976) Ration, growth and measures of somatic and organ condition in relation to meal frequency in winter flounder, *Pseudopleuronectes americanus*, with hypotheses regarding population homeostasis *J. Fish. Res. Bd Can.*, **33**, 63–75.

Tyler, C.A. and Sumpter, J.P. (1996) Oocyte growth and development in teleosts. *Rev. Fish Biol. Fisheries*, **6**, 287–318.

Tyler, J.A. and Gilliam, J.F. (1995) Ideal free distribution of stream fish: a model and test with minnows, *Rhinichthys atratulus*. *Ecology*, **76**, 580–592.

Tyler, J.A. and Rose, K.A. (1994). Individual variability and spatial heterogeneity in fish population models. *Rev. Fish Biol. Fisheries*, **4**, 91–123.

Tyler, J.A. and Rose, K.A. (1997) Effects of individual habitat selection in a heterogeneous environment on fish cohort survival: a modelling analysis. *J. Anim. Ecol.*, **66**, 122–136.

Ursin, E. (1979) Principles of growth in fishes. *Symp. Zool. Soc. Lond.*, **44**, 63–87.

Van Oosten, J. (1957) The skin and scales, in *The Physiology of Fishes*, Vol. I (ed. M.E. Brown), Academic Press, London, pp. 207–244.

Van Winkle, W., Rose, K.A.. and Chambers, R.C. (1993) Individual-based approach to fish population dynamics. *Trans. Am. Fish. Soc.*, **122**, 397–403.

Vanni, M.J. (1996) Nutrient transport and recycling by consumers in lake food webs: implications for algal communities, in *Food Webs: Integration of Patterns and Dynamics* (eds G.A. Polis and K.O. Winemiller), Chapman and Hall, London, pp. 81–95.

Vanni, M.J. and Findlay, D.L. (1990) Trophic cascades and phytoplankton community structure. *Ecology*, **71**, 921–937.

Vanni, M.J., Luecke, C., Kitchell, J.F., Allen, Y., Temte, J. and Magnuson, J.J. (1990) Effects on lower trophic levels of massive fish mortality. *Nature*, **344**, 333–335.

Vannote, R.L., Minshall, G.W., Cummins, K.W., Sedell, J.R. and Cushing, C.E. (1980) The river continuum concept. *Can. J. Fish. Aquat. Sci.*, **37**, 130–137.

Varley, G.C., Gradwell, G.R. and Hassell, M.P. (1973) *Insect Population Ecology: An Analytical Approach*, Blackwell, Oxford.

Victor, B.C. (1983) Recruitment and population dynamics of a coral reef fish. *Science*, **219**, 419–420.

Victor, B.C. (1986) Larval settlement and juvenile mortality in a recruitment-limited coral reef fish population. *Ecol. Monogr.*, **56**, 145–160.

Victor, R., Chan, G.L. and Fernando, C.H. (1979) Notes on the recovery of live ostracods from the gut of the white sucker (*Catostomus commersoni* Lacepède 1808) (Pisces: Catostomidae). *Can. J. Zool.*, **51**, 1745–1747.

Videler, J.J. (1993) *Fish Swimming*, Chapman and Hall, London.

Vollestad, L.A., L'Abee-Lund, J.H. and Saegrov, H. (1993) Dimensionless numbers and life history variation in brown trout: evaluation of a theory. *Evol. Ecol.*, **7**, 207–218.

Vrijenhoek, R.C. (1984) The evolution of clonal diversity in *Poeciliopsis*, in *Evolutionary Genetics of Fishes* (ed. B.J. Turner), Plenum, New York, pp. 399–429.

Vrijenhoek, R.C. (1989) Genotypic diversity and coexistence among sexual and clonal lineages of *Poeciliopsis*, in *Speciation and its Consequences* (eds D. Otte and J.A. Endler), Sinauer, Sunderland, MA, pp. 386–400.

Wagner, G.F. and McKeown, B.A. (1985) Cyclical growth in juvenile rainbow trout, *Salmo gairdneri*. *Can. J. Zool.*, **63**, 2473–2474.

Wainwright, P.C. and Richard, B.A. (1995) Predicting patterns of prey use from morphology of fishes. *Env. Biol. Fishes*, **44**, 97–113.

Walkey, M. and Meakins, R.H. (1970) An attempt to balance the energy budget of a host–parasite system. *J. Fish Biol.*, **2**, 361–372.

Walton, W.E., Hairston, N.G. Jr and Wetterer, J.K. (1992) Growth-related constraints on diet selection by sunfish. *Ecology*, **73**, 429–437.

Wankowski, J.W.J. (1979) Morphological limitations, prey size selectivity and growth response of juvenile Atlantic salmon, *Salmo salar*. *J. Fish. Biol.*, **14**, 89–100.

Wankowski, J.W.J. and Thorpe, J.E. (1979) The role of food particle size in the growth of juvenile Atlantic salmon (*Salmo salar* L.). *J. Fish Biol.*, **14**, 351–370.

Wardle, C.S. (1980) Effect of temperature on maximum swimming speed of fishes, in *Environmental Physiology of Fishes* (ed. M.A. Ali), Plenum, London, pp. 519–531.

Wardle, C.S. (1993) Fish behaviour and fishing gear, in *The Behaviour of Teleost Fishes*, 2nd edn (ed. T.J. Pitcher), Chapman and Hall, London, pp. 609–643.

Ware, D.M. (1971) Predation by rainbow trout (*Salmo gairdneri*): the effect of experience. *J. Fish. Res. Bd Can.*, **28**, 1847–1852.

Ware, D.M. (1972) Predation by rainbow trout (*Salmo gairdneri*): the influence of hunger, prey density and prey size. *J. Fish. Res. Bd Can.*, **29**, 1193–1201.

Ware, D.M. (1973) Risk of epibenthic prey to predation by rainbow trout (*Salmo gairdneri*). *J. Fish. Res. Bd Can.*, **30**, 787–797.

Ware, D.M. (1975a) Growth, metabolism, and optimal swimming speed of a pelagic fish. *J. Fish. Res. Bd Can.*, **32**, 33–41.

Ware, D.M. (1975b) Relation between egg size, growth, and natural mortality of larval fish. *J. Fish. Res. Bd Can.*, **32**, 2503–2512.

Ware, D.M. (1977) Spawning time and egg size of Atlantic mackerel, *Scomber scombrus*, in relation to the plankton. *J. Fish. Res. Bd Can.*, **34**, 2308–2315.

Ware, D.M. (1978) Bioenergetics of pelagic fish: theoretical change in swimming speed and ration with body size. *J. Fish. Res. Bd Can.*, **35**, 220–228.

Ware, D.M. (1980) Bioenergetics of stock and recruitment. *Can. J. Fish. Aquat. Sci.*, **37**, 1012–1024.

Ware, D.M. (1982) Power and evolutionary fitness of teleosts. *Can. J. Fish. Aquat. Sci.*, **39**, 3–13.

Ware, D.M. (1984) Fitness of different reproductive strategies in teleost fishes, in *Fish Reproduction: Strategies and Tactics* (eds G.W. Potts and R.J. Wootton), Academic Press, London, pp. 349–366.

Warner, R.R. (1978) The evolution of hermaphroditism and unisexuality in aquatic and terrestrial vertebrates, in *Contrasts in Behaviour* (eds E.S. Reese and F.J. Lighter), J. Wiley, New York, pp. 77–101.

Warner, R.R. (1984) Recent developments in the ecology of tropical reef fishes. *Arch. FischWiss.*, **35**, 43–53.

Warner, R.R. and Downs, I.F. (1977) Comparative life histories: growth vs reproduction in normal males and sex-changing hermaphrodites of the striped parrotfish, *Scarus croicensis*. *Proceedings of the Third International Coral Reef Symposium*, 275–281.

Warner, R.R. and Hoffman, S.G. (1980) Population density and the economics of territorial defence in a coral reef fish. *Ecology*, **61**, 772–780.

Warner, R.R. and Robertson, D.R. (1978) Sexual patterns in the labroid fishes of the western Caribbean, 1: the wrasses (Labridae). *Smithsonian Contr. Zool.*, **254**, 1–27.

Warren, C.E. and Davis, G.E. (1967) Laboratory studies of the feeding, bioenergetics and growth of fish, in *The Biological Basis of Freshwater Fish*

Production (ed. S.D. Gerking), Blackwell, Oxford, pp. 175–214.

Watanabe, T., Ohhashi, S., Itoh, A., Kitajima, C. and Fujita, S. (1984) Effect of nutritional composition of diets on chemical components of red bream broodstock and eggs produced. *Bull. Jap. Soc. Scient. Fish.*, **50**, 503–515.

Weatherley, A.H. (1972) *Growth and Ecology of Fish Populations*, Academic Press, London.

Weatherley, A.H. and Gill, H.S. (1987) *The Biology of Fish Growth*, Academic Press, London.

Webb, P.W. (1975) Hydrodynamics and energetics of fish propulsion. *Bull. Fish. Res. Bd Can.*, no. 190, 1–159.

Webb, P.W. (1978a) Hydrodynamics: nonscombrid fish, in *Fish Physiology*, Vol. VII (eds W.S. Hoar and D.J. Randall), Academic Press, London, pp. 190–237.

Webb, P.W. (1978b) Temperature effects on acceleration of rainbow trout, *Salmo gairdneri*. *J. Fish. Res. Bd Can.*, **35**, 1417–1422.

Webb, P.W. (1984a) Body form, locomotion and foraging in aquatic vertebrates. *Am. Zool.*, **24**, 107–120.

Webb, P.W. (1984b) Form and function in fish swimming. *Scient. Am.*, **251**, 58–68.

Webb, P.W. (1988) Simple physical principles and vertebrate aquatic locomotion. *Am. Zool.*, **28**, 709–725.

Webb, P.W. and Weihs, D. (1986) Functional locomotor morphology of early life history stages of fishes. *Trans. Am. Fish. Soc.*, **115**, 115–127.

Weber, J.-M. and Kramer, D.L. (1983) Effects of hypoxia and surface access on growth, mortality, and behaviour of juvenile guppies, *Poecilia reticulata*. *Can. J. Fish. Aquat. Sci.*, **40**, 1583–1588.

Weeks, S.C. (1995) Comparisons of life-history traits between clonal and sexual fish (*Poeciliopsis*, Poeciliidae) raised in monoculture and mixed treatments. *Evol. Ecol.*, **9**, 258–274.

Welcomme, R.L. (1967) The relationship between fecundity and fertility in the mouthbrooding cichlid fish, *Tilapia leucosticta*. *J. Zool., Lond.*, **151**, 453–468.

Welcomme, R.L. (1979) *Fisheries Ecology of Floodplain Rivers*, Longman, London.

Welcomme, R.L. (1985) River fisheries. *F.A.O. Fish. Biol. Tech. Pap.*, no. 262, 1–330.

Werner, E.E. (1974) The fish size, prey size, handling time relation in several sunfishes and some implications. *J. Fish. Res. Bd Can.*, **31**, 1531–1536.

Werner, E.E. (1977) Species packing and niche complementarity in three sunfishes. *Am. Nat.*, **111**, 553–578.

Werner, E.E. (1980) Niche theory in fisheries ecology. *Trans. Am. Fish. Soc.*, **109**, 254–260.

Werner, E.E. (1984) The mechanisms of species interactions and community organization in fish, in *Ecological Communities: Conceptual Issues and the Evidence* (eds D.R. Strong, D. Simberloff, L.G. Abele and A.B. Thistle), Princeton University Press, Princeton, NJ, pp. 360–382.

Werner, E.E. (1986) Species interactions in freshwater fish communities, in *Community Ecology* (eds J. Diamond and T.J. Case), Princeton University Press, Princeton, NJ, pp. 344–357.

Werner, E.E. and Gilliam, J.F. (1984) The ontogenetic niche and species interactions in size-structured populations. *Ann. Rev. Ecol. Syst.*, **15**, 393–425.

Werner, E.E. and Hall, D.J. (1974) Optimal foraging and the size selection of prey by the bluegill sunfish (*Lepomis macrochirus*). *Ecology*, **55**, 1042–1052.

Werner, E.E. and Hall, D.J. (1976) Niche shifts in sunfishes: experimental evidence and significance. *Science*, **191**, 404–406.

Werner, E.E. and Hall, D.J. (1977) Competition and habitat shift in two sunfishes (Centrarchidae). *Ecology*, **58**, 869–876.

Werner, E.E. and Hall, D.J. (1979) Foraging efficiency and habitat switching in competing sunfish. *Ecology*, **60**, 256–264.

Werner, E.E. and Mittelbach, G.G. (1981) Optimal foraging: field tests of diet choice and habitat switching. *Am. Zool*, **21**, 813–829.

Werner, E.E., Mittelbach, G.G. and Hall, D.J. (1981) The role of foraging profitability and experience in habitat use by the bluegill sunfish. *Ecology*, **62**, 116–125.

Werner, E.E., Gilliam, J.F., Hall, D.J. and Mittelbach, G.G. (1983a) An experimental test of the effects of predation risk on habitat use in fish. *Ecology*, **64**, 1540–1548.

Werner, E.E., Mittelbach, G.G., Hall, D.J. and Gilliam, J.F. (1983b) Experimental tests of optimal habitat use in fish: the role of relative habitat profitability. *Ecology*, **64**, 1525–1539.

Westneat, M.W. (1995) Phylogenetic systematics and biomechanics in ecomorphology. *Env. Biol. Fishes*, **44**, 263–283.

Wetzel, R.G. (1983) *Limnology*, 2nd edn, Saunders, Philadelphia.

Whatley, R.C. (1983) An ostracod to catch a trout. *Brit. Micropalaeontol. Soc. Newsl. (mimeo)*, **20**, 2.

Whittaker, R.H. (1960) Vegetation of the Siskiyou Mountains, Oregon and California. *Ecol. Monogr.*, **30**, 279–338.

Whoriskey, F.G. and FitzGerald, G.J. (1985) Sex, cannibalism and sticklebacks. *Behav. Ecol. Sociobiol.*, **18**, 15–18.

Whoriskey, F.G. and FitzGerald, G.J. (1994) Ecology of the threespine stickleback on the breeding grounds, in *The Evolutionary Biology of the Threespine Stickleback* (eds M.A. Bell and S.A. Foster), Oxford University Press, Oxfrod, pp. 188–206.

Wiens, J.A. (1977) On competition and variable environments. *Am. Scient.*, **65**, 590–597.

Wieser, W. (1989) Energy allocation by addition and by compensation: an old principle revisited, in *Energy Transformation in Cells and Animals* (eds W. Wieser and E. Gnaiger), Thieme Verlag, Stuttgart, pp. 98–105.

Wieser, W. and Medgyesy, N. (1990) Aerobic maximum for growth in the larvae and juveniles of a cyprinid fish, *Rutilus rutilus* (L.): implications for energy budgeting in small poikilotherms. *Funct. Ecol.*, **4**, 233–242.

Wikramanayake, E.D. (1990) Ecomorphology and biogeography of a tropical stream fish assemblage: evolution of assemblage structure. *Ecology*, **71**, 1756–1764.

Wildhaber, M.L. and Crowder, L.B. (1990) Testing a bioenergetic-based habitat choice model: bluegill (*Lepomis macrochirus*) response to food availability and temperature. *Can. J. Fish. Aquat. Sci.*, **47**, 1664–1671.

Williams, G.C. (1966) *Adaptation and Natural Selection*, Princeton University Press, Princeton, NJ.

Williams, G.C. (1996) *Plan and Purpose in Nature*, Weidenfeld & Nicholson, London.

Williams, S.F. and Caldwell, R.S. (1978) Growth, food conversion and survival of 0 group English sole (*Parophrys ventulus* Girard) at five temperatures and five rations. *Aquaculture*, **15**, 129–139.

Winberg, G.G. (1956) Rate of metabolism and food requirements of fish. *Fish. Res. Bd Can. Transl.*, no. 194, 1–202.

Windell, J.T. (1971) Food analysis and rate of digestion, in *Fish Production in Freshwaters* (ed. W.E. Ricker), Blackwell, Oxford, pp. 215–226.

Winemiller, K.O. (1990) Spatial and temporal variation in tropical fish trophic networks. *Ecol. Monogr.*, **60**, 331–367.

Winemiller, K.O. (1991) Ecomorphological diversification in lowland freshwater fish assemblages from five biotic regions. *Ecol. Monogr.*, **61**, 343–365.

Winemiller, K.O. (1993) Seasonality of reproduction by livebearing fishes in tropical rainforest streams. *Oecologia* **95**, 266–276.

Winemiller, K.O. (1996) Factors driving temporal and spatial variation in aquatic floodplain food webs, in *Food Webs Integration of Patterns and Dynamics* (eds G.A. Polis and K.O. Winemiller), Chapman and Hall, London, pp. 298–312.

Winemiller, K.O. and Pianka, E.R. (1990) Organization in natural assemblages of desert lizards and tropical fishes. *Ecol. Monogr.*, **60**, 27–55.

Winemiller, K.O. and Rose, K.A. (1992) Patterns of life-history diversification in North American fishes: implications for population regulation. *Can. J. Fish Aquat. Sci.*, **49**, 2196–2218.

Winemiller, K.O. and Rose, K.A. (1993) Why do most fish produce so many tiny offspring? *Am. Nat.*, **142**, 585–603.

Winston, M.R. (1995) Co-occurrence of morphologically similar species of stream fishes. *Am. Nat.*, 145, 527–545.

Wisenden, B.D. (1994) Factors affecting mate desertion by males in free-ranging convict cichlids (*Cichlasoma nigrofasciatum*). *Behav. Ecol.*, **5**, 439–447.

Wisenden, B.D. (1995) Reproductive behaviour of free-ranging convict cichlids, *Cichlasoma nigrofasciatum*. *Env. Biol. Fishes*, **43**, 121–134.

Witte, F. (1984) Ecological differentiation in Lake Victoria haplochromines: comparison of cichlid species flocks in African lakes, in *Evolution of Fish Species Flocks* (eds A.A. Echelle and I. Kornfield), University of Maine Press, Orono, ME, pp. 155–167.

Wohlschlag, D.E. (1964) Respiratory metabolism and ecological characteristics of some fishes in McMurdo Sound, Antarctica, in *Biology of Antarctic Seas* (*Antarctic Research Series* No. 1) (ed. M.O. Lee), American Geophysics Union, Washington, DC, pp. 33–62.

Wood, C.C. (1985) Aggregative responses of common mergansers (*Mergus merganser*): predicting flock size and abundance on Vancouver Island streams. *Can. J. Fish. Aquat. Sci.*, **42**, 1259–1271.

Wood, C.C. (1987a) Predation of juvenile Pacific salmon by the common merganser (*Mergus merganser*) on eastern Vancouver Island. I: Predation during the seaward migration. *Can. J. Fish. Aquat. Sci.*, **44**, 941–949.

Wood, C.C. (1987b) Predation of juvenile Pacific

salmon by the common merganser (*Mergus merganser*) on eastern Vancouver Island. II: Predation of stream-resident juvenile salmon by merganser broods. *Can. J. Fish. Aquat. Sci.*, **44**, 950–959.

Wood, C.C. and Hand, C.M. (1985) Food-searching behaviour of the common merganser (*Mergus merganser*). I: Functional responses to prey and predator density. *Can. J. Zool.*, **63**, 1260–1270.

Wootton, R.J. (1973) Effect of size of food ration on egg production in the female threespined stickleback, *Gasterosteus aculeatus* L. *J. Fish Biol.*, **5**, 683–688.

Wootton, R.J. (1976) *The Biology of the Sticklebacks*, Academic Press, London.

Wootton, R.J. (1977) Effect of food ration during the breeding season on the size, body components and egg production of female sticklebacks (*Gasterosteus aculeatus* L.). *J. Anim. Ecol.*, **46**, 823–834.

Wootton, R.J. (1979) Energy costs of egg production and environmental determinants of fecundity in teleost fishes. *Symp. Zool. Soc. Lond.*, **44**, 133–159.

Wootton, R.J. (1982) Environmental factors in fish reproduction, in *Reproductive Physiology of Fish* (eds C.J.J. Richter and H.J.Th. Goos), Pudoc, Wageningen, pp. 210–219.

Wootton, R.J. (1984a) *A Functional Biology of Sticklebacks*, Croom Helm, London.

Wootton, R.J. (1984b) Introduction: tactics and strategies in fish reproduction, in *Fish Reproduction: Strategies and Tactics* (eds G.W. Potts and R.J. Wootton), Academic Press, London, pp. 1–12.

Wootton, R.J. (1985) Energetics of reproduction, in *Fish Energetics New Perspectives* (eds P. Tytler and P. Calow), Croom Helm, London, pp. 231–254.

Wootton, R.J. (1986) Problems in the estimation of food consumption and fecundity in fish production studies. *Polskie Arch. Hydrobiol.*, **33**, 263–276.

Wootton, R.J. (1992a) *Fish Ecology*, Blackie, Glasgow.

Wootton, R.J. (1992b) Constraints in the evolution of fish life histories. *Neth. J. Zool.*, **42**, 291–303.

Wootton, R.J. (1994a) Energy allocation in the threespine stickleback, in *The Evolutionary Biology of the Threespine Stickleback* (eds M.A. Bell and S.A. Foster), Oxford University Press, Oxford, pp. 114–143.

Wootton, R.J. (1994b) Life histories as sampling devices: optimum egg size in pelagic fishes. *J. Fish Biol.*, **45**, 1067–1077.

Wootton, R.J. and Evans, G.W. (1976) Cost of egg production in the three-spined stickleback (*Gasterosteus aculeatus*). *J. Fish Biol.*, **8**, 385–395.

Wootton, R.J. and Mills, L. (1979) Annual cycle in female minnows *Phoxinus phoxinus* (L.) from an upland Welsh lake. *J. Fish Biol.*, **14**, 607–618.

Wootton, R.J., Evans, G.W. and Mills, L. (1978) Annual cycle in female three-spined sticklebacks (*Gasterosteus aculeatus* L.) from an upland and lowland population. *J. Fish Biol.*, **12**, 331–343.

Wootton, R.J., Allen, J.R.M. and Cole, S.J. (1980) Energetics of the annual reproductive cycle in female sticklebacks, *Gasterosteus aculeatus* L. *J. Fish Biol.*, **17**, 387–394.

Worobec, M.N. (1984) Field estimates of the daily ration of winter flounder *Pseudopleuronectes americanus* (Walbaum) in a southern New England salt marsh. *J. Exp. Mar. Biol. Ecol.*, **77**, 183–196.

Wourms, J.P. (1972) The developmental biology of annual fishes Ill. Pre-embryonic and embryonic diapause of variable duration in the eggs of annual fishes. *J. Exp. Zool.*, **182**, 389–414.

Wourms, J.P. (1981) Viviparity: the maternal–fetal relationship in fishes. *Am. Zool.*, **21**, 473–515.

Wright, P.J., Metcalfe, N.B. and Thorpe, J.E. (1990) Otolith and somatic growth rates in Atlantic salmon parr, *Salmo salar* L.: evidence against coupling. *J. Fish Biol.*, **36**, 241–249.

Wright, R.M. and Giles, N. (1987) The survival, growth and diet of pike fry, *Esox lucius* L., stocked at different densities in experimental ponds. *J. Fish Biol.*, **30**, 617–629.

Wydoski, R.S. and Cooper, E.L. (1966) Maturation and fecundity of brook trout from infertile streams. *J. Fish. Res. Bd Can.*, **23**, 623–649.

Yamagishi, H. (1969) Post embryonal growth and its variability of the three marine fishes with special reference to the mechanism of growth variation in fishes. *Res. Popul. Ecol. Kyoto Univ.*, **11**, 14–33.

Yamaoka, K. (1991) Feeding relationships, in *Cichlid Fishes: Behaviour, Ecology and Evolution* (ed. M.H.A. Keenleyside), Chapman and Hall, London, pp. 151–172.

Yoshida, H. and Sakuri, Y. (1984) Relationship between food consumption and growth of adult walleye pollock *Theragra chalcogramma* in captivity. *Bull. Jap. Soc. Scient. Fish.*, **50**, 763–769.

Zalewski, M. and Naiman, R.J. (1985) The regulation of riverine fish communities by a continuum of abiotic–biotic factors, in *Habitat Modification and Freshwater Fisheries* (ed. J.S. Alabaster), Butterworths, London, pp. 3–9.

Zaret, T.M. (1972) Predator–prey interactions in a tropical lacustrine ecosystem. *Ecology*, **53**, 248–257.

Zaret, T.M. (1979) Predation in freshwater fish communities, in *Predator–prey Systems in Fisheries Management* (ed. H. Clepper), Sport Fishing Institute, Washington, DC, pp. 135–143.

Zaret, T.M. (1980) *Predation and Freshwater Communities*, Yale University Press, New Haven, CT.

Zaret, T.M. and Paine, R.T. (1973) Species introduction in a tropical lake. *Science*, **182**, 449–455.

Zaret, T.M. and Rand, A.S. (1971) Competition in tropical stream fishes: support for the competitive exclusion principle. *Ecology*, **52**, 336–342.

Zippin, C. (1956) An evaluation of the removal method of estimating animal populations. *Biometrics*, **12**, 163–169.

Zippin, C. (1958) The removal method of population estimation. *J. Wildl. Manage.*, **22**, 82–90.

AUTHOR INDEX

SYSTEMATIC INDEX

Non-fish taxa

SUBJECT INDEX

Fish and Fisheries Series

Fish and Fisheries Series

24. R.J. Wootton: *Ecology of Teleost Fishes*. Second Edition. 1999
     PB; ISBN 0-412-64200-X; HB; ISBN 0-412-84590-3

25. M.C.M. Beveridge and B.J. McAndrew (eds.): *Tilapias: Biology and Exploitation*.
     2000   PB; ISBN 0-7923-6391-4 HB; ISBN 0-412-80090-X

KLUWER ACADEMIC PUBLISHERS – BOSTON / DORDRECHT / LONDON